Energy Efficient
Computing & Electronics

Devices, Circuits, and Systems

Series Editor
Krzysztof Iniewski
Emerging Technologies CMOS Inc.
Vancouver, British Columbia, Canada

PUBLISHED TITLES:

PUBLISHED TITLES:

Energy Efficient
Computing & Electronics
Devices to Systems

Krzysztof Iniewski
Managing Editor

Edited by
Santosh K. Kurinec
Sumeet Walia

CRC Press
Taylor & Francis Group
Boca Raton London New York

CRC Press is an imprint of the
Taylor & Francis Group, an **informa** business

CRC Press
Taylor & Francis Group
6000 Broken Sound Parkway NW, Suite 300
Boca Raton, FL 33487-2742

Library of Congress Cataloging-in-Publication Data

Names: Kurinec, Santosh K., editor. | Walia, Sumeet, editor.
Title: Energy efficient computing & electronics : devices to systems / edited
by Santosh K. Kurinec and Sumeet Walia.
Other titles: Energy efficient computing and electronics
Description: Boca Raton : CRC/Taylor & Francis, [2019] | Series: Devices,
circuits, & systems | Includes bibliographical references and index.
Identifiers: LCCN 2018042978| ISBN 9781138710368 (hardback : alk. paper) |
ISBN 9781315200705 (ebook)
Subjects: LCSH: Electronic apparatus and appliances--Power supply. | Computer
systems--Energy conservation. | Low voltage systems. | Wireless
communication systems--Energy conservation.
Classification: LCC TK7868.P6 E54 2019 | DDC 621.381028/6--dc23
LC record available at https://lccn.loc.gov/2018042978

Visit the Taylor & Francis Web site at
http://www.taylorandfrancis.com

and the CRC Press Web site at
http://www.crcpress.com

Contents

Section I   Emerging Low Power Devices

Section II   Sensors, Interconnects, and Rectifiers

Section III   Systems Design and Applications

Preface

Performance of electronic systems are limited by energy inefficiencies that result in over-heating and thermal management problems. Energy efficiency is vital to improving performance at all levels. This includes transistors to devices and to large Internet Technology and electronic systems, as well from small sensors for the Internet-of-Things (IoT) to large data centers in cloud and supercomputing systems. The electronic circuits in computer chips still operate far from any fundamental limits to energy efficiency. A report issued by the Semiconductor Industry Association and Semiconductor Research Corporation bases its conclusions on system-level energy per bit operation, which are a combination of many components such as logic circuits, memory arrays, interfaces, and I/Os. Each of these contributes to the total energy budget. For the benchmark energy per bit, as shown in Figure 1, computing will not be sustainable by 2040. This is when the energy required for computing is estimated to exceed the world's estimated energy production. The "benchmark" curve shows the growing energy demand for the system level energy per bit values of mainstream systems. The target system curve uses the practical lower limit system level energy per bit value, set by factors such as materials. The Landauer limit curve uses the minimal device energy per bit value provided by the Landauer's Principle that relates to the Second Law of Thermodynamics to computation. As such, significant improvement in the energy efficiency of computing is needed.

There is a consensus across the many technologies touched by our ubiquitous computing infrastructure that future performance improvements across the board are now severely limited by the amount of energy it takes to manipulate, store, and critically transport data. Revolutionary device concepts, sensors, and associated circuits and architectures that will greatly extend the practical engineering limits of energy-efficient computation are being investigated. Disruptive new device architectures, semiconductor processes, and emerging

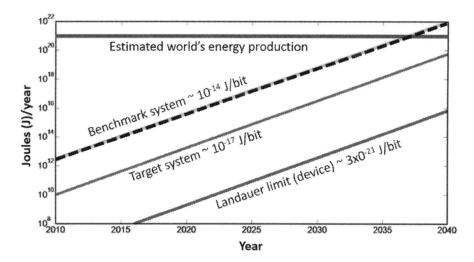

FIGURE 1
Estimated total energy expenditure for computing, directly related to the number of raw bit transitions. Source: SIA/SRC (*Rebooting the IT Revolution: A Call to Action,* Semiconductor Industry Association, September 2015).

new materials aimed at achieving the highest level of computational energy efficiency for general purpose computing systems need to be developed. This book will provide chapters dedicated to some of such efforts from devices to systems.

The book is divided into three sections each consisting of five chapters.

Section I is dedicated to device level research in developing energy efficient device structures.

Non-planar finFETs dominate highly scaled processes, such as sub-20 nm, CMOS processes, due to their ability to provide lower leakage and enable continued power supply scaling (VDD). Chapter 1 gives a comprehensive overview of the finFET-based predictive process design kit (PDK) that supports investigation into both the circuit as well as physical design, encompassing all aspects of digital design. Prevention of various degradation mechanisms in transistors is the key to increase the reliability and efficiency of electronic systems. Chapter 2 discusses the understanding of molecular phenomena at the MOSFET channel/dielectric interface, focussing on ZrO2 system aimed at minimizing the degrading mechanisms in transistors. Chapter 3 provides an important insight into the nanotube Tunneling field effect transistors (TFETs) device, which promise to exhibit steep slope faster than the Boltzmann *limit* of *60 mV/dec*. TFETs address two major challenges faced by aggressively scaled conventional CMOS technology; scaling the supply voltage (VDD) and minimizing the leakage currents. Chapter 4 gives an in-depth introduction and potential of spin based devices. It gives an overview of spintronic devices, circuits, and architecture levels that include thermally assisted (TA)-MRAM, STT-MRAM, domain wall (DW)-MRAM, spin-orbit torque (SOT)-MRAM, spin-transfer torque and spin Hall oscillators, logic-in-memory, all-spin logic, buffered magnetic logic gate grid, ternary content addressable memory (TCAM), and random number generators. A large bottleneck for energy efficiency has long been the information storage units, which typically rely on charge-storage for programming and erasing. Chapter 5 describes the promise of ferroelectric tunnel junctions (FTJs) as memristors. Memristor-based logic systems for XOR, XNOR, full-adder, DAC, and ADC outperform CMOS with as low as 50% the delay and 0.1% the power consumption.

Section II deals with sensors, interconnects, and rectifiers aimed at consuming lower power.

Chapter 6 provides insight into X-ray sensors based on chromium compensated GaAs for the development of modern X-ray imaging systems. Chapters 7 and 8 deal with optical interconnects, which are actively being pursued to reduce power requirements and increase speed. Chapter 7 provides an overview of the challenges and developments of various types of Vertical-Cavity Surface-Emitting Laser (VCSEL)-based interconnects. Optoelectronic interactions at interconnective hetero-interfaces between nanoparticles and two-dimensional semiconductors, motivated by their enhancement of electronic and photonic properties are described in Chapter 8. Chapter 9 investigates AlGaN/GaN heterostructures based Schottky diodes with low turn-on voltage of $V_f < 0.4V$, and high breakdown voltage of $V_{br} > 400V$ for applications in energy efficient 230V AC-DC rectifiers. In Chapter 10, the authors have discussed the stoichiometry-controlled crystal growth technique and its application to compound semiconductor oscillation devices for extending to THz region The THz wave generators can be used in applications of non-destructive evaluation, safe for human tissues.

Section III has five chapters aimed at low power system designs.

The imminent concern of inherent physical noise with low power biosensing mixed signal CMOS technology is addressed in Chapter 11. The information rate and bit energy have been incorporated into a design methodology for detecting weak signals in the presence of fixed system noise. Chapter 12 provides an overview of the high-level processor architecture design methodologies using Architecture Description Languages (ADLs). Chapter 13 focuses on challenging problems related to energy-efficient Cloud Data Centers (CDC). It presents how to minimize the total cost of a CDC provider in a market, and how to migrate to green cloud data centers (GCDCs). The authors propose a Temporal Request Scheduling algorithm (TRS) that can achieve higher throughput and lower grid energy cost for a GCDC. Chapter 14 presents an innovative brain inspired approach of implementing neural network in hardware implementation for ultra-low-voltage implementation of the perceptron and the inertial neuron. The final chapter (Chapter 15) addresses a large system-level problem of malware attacks to mobile devices. It presents a multi-pattern matching based dynamic malware detection mechanism in smart phones as an alternative to machine learning based methods. The proposed mechanism is more efficient and uses fewer resources.

Thus, this book brings together a wealth of information that will serve as a valuable resource for researchers, scientists, and engineers engaged in energy efficient designs of electronic devices, circuits, and systems. The editors express their sincere appreciation to the authors who have contributed their knowledge and expertise to this book. Special thanks to Nora Konopka and Erin Harris of Taylor & Francis Group/CRC Press for their publishing efforts and coordinating with the authors. The authors also express their sincere appreciation for Joanne Hakim, project manager of Lumina Datamatics, for coordinating the production of this book.

Editors

Santosh K. Kurinec is a professor of electrical and microelectronic engineering at Rochester Institute of Technology (RIT), Rochester, New York. She received a PhD in physics from University of Delhi, India and worked as a scientist at the National Physical Laboratory, New Delhi, India. She worked as a postdoctoral research associate in the Department of Materials Science and Engineering at the University of Florida, Gainesville, Florida where she researched thin metal film composites. Prior to joining RIT, she was assistant professor of electrical engineering at Florida State University/Florida A&M University College of Engineering in Tallahassee, Florida. She is a Fellow of Institute of Electrical and Electronics Engineers (IEEE), Member APS, NY State Academy of Sciences, and an IEEE EDS Distinguished Lecturer. She received the 2012 IEEE Technical Field Award. In 2016, she received the Medal of Honor from the International Association of Advanced Materials (IAAM). She was inducted into the International Women in Technology (WiTi) Hall of Fame in 2018. She has worked on a range of materials covering magnetic, ferroelectrics, semiconductors, photonic luminescent thin films, and metal composites for device applications. Her current research activities include nonvolatile memory, advanced integrated circuit materials and processes, and photovoltaics. She has over 100 publications in research journals and conference proceedings. She can be reached at Santosh.kurinec@rit.edu.

Sumeet Walia is a senior lecturer and a Vice Chancellor's Fellow at RMIT University, Melbourne, Australia. He is an expert in materials engineering for nanoelectronics, sensing, and wearable devices. He has been recognized as one of the Top 10 Innovators under 35 in Asia-Pacific by the MIT Technology Review in 2017, was awarded the Victorian Young Achiever Award for Research Impact in 2017, and named among the most innovative Engineers in Australia in 2018 by Engineers Australia. He can be reached at waliasumeet@gmail.com or sumeet.walia@rmit.edu.au.

Contributors

Jing Bi
Faculty of Information Technology
Beijing University of Technology
Beijing, China

Anupam Chattopadhyay
School of Computer Science and
   Engineering
Nanyang Technological University
Singapore

Lawrence T. Clark
School of Electrical, Computer, and Energy
   Engineering
Arizona State University
Tempe, Arizona, USA

Mohammad Rafiq Dar
Department of Electronics and
   Instrumentation Technology
University of Kashmir
Srinagar, India

Nishant Darvekar
Electrical and Microelectronic Engineering
Rochester Institute of Technology
Rochester, New York, USA

V. S. Devi
Department of Computer Science
Indian Institute of Information Technology
   and Management – Kerala (IIITM-K)
Thiruvananthapuram, India

Mattias Ekström
Department of Electronics
KTH Royal Institute of Technology
Kista, Sweden

Cristian Grecu
Department of Electrical and Computer
   Engineering
University of British Columbia
Vancouver, British Columbia, Canada

Amir N. Hanna
Department of Electrical Engineering
King Abdullah University of Science
   and Technology (KAUST)
Thuwal, Saudi Arabia

Werner H. E. Hofmann
Institute of Solid State Physics and
   Center of Nanophotonics
Technical University of Berlin
Berlin, Germany

Muhammad Mustafa Hussain
Department of Electrical Engineering
King Abdullah University of Science
   and Technology (KAUST)
Thuwal, Saudi Arabia

Andre Ivanov
Department of Electrical and
   Computer Engineering
University of British Columbia
Vancouver, British Columbia, Canada

Nasir Ali Kant
Department of Electronics and
   Instrumentation Technology
University of Kashmir
Srinagar, India

Farooq Ahmad Khanday
Department of Electronics and
   Instrumentation Technology
University of Kashmir
Srinagar, India

Santosh K. Kurinec
Electrical and Microelectronic
   Engineering
Rochester Institute of Technology
Rochester, New York, USA

Hiwa Mahmoudi
Institute of Electrodynamics, Microwave
    and Circuit Engineering
TU Wien
Vienna, Austria

Alexander Makarov
Institute for Microelectronics
TU Wien
Vienna, Austria

B. Gunnar Malm
Department of Electronics
KTH Royal Institute of Technology
Kista, Sweden

Hegoi Manzano
Department of Condensed Matter Physics
University of the Basque Country UPV/
    EHU
Barrio Sarriena s/n, 48330
Leioa, Bizkaia, Spain

Nicole McFarlane
Min H. Kao Department of Electrical
    Engineering & Computer Science
Tickle College of Engineering
The University of Tennessee
Knoxville, Tennessee, USA

Takeo Ohno
Department of Innovative Engineering
Oita University
Oita, Japan

Mikael Östling
Department of Electronics
KTH Royal Institute of Technology
Kista, Sweden

Yutaka Oyama
Department of Materials Science and
    Engineering
Tohoku University
Sendai, Japan

Spencer Allen Pringle
Electrical & Microelectronic
    Engineering
Rochester Institute of Technology
Rochester, New York, USA

S. Roopak
Department of Computer Science
Indian Institute of Information
    Technology and Management – Kerala
    (IIITM-K)
Thiruvananthapuram, India

D. Keith Roper
Ralph E. Martin Department of Chemical
    Engineering
and
Microelectronics-Photonics Graduate
    Program
and
Institute for Nanoscience and
    Engineering
University of Arkansas
Fayetteville, Arkansas, USA

Siegfried Selberherr
Institute for Microelectronics
TU Wien
Vienna, Austria

S. Arash Sheikholeslam
Department of Electrical and Computer
    Engineering
University of British Columbia
Vancouver, British Columbia, Canada

Tadao Tanabe
Department of Materials Science and
    Engineering
Tohoku University
Sendai, Japan

Tony Thomas
Department of Computer Science
Indian Institute of Information
    Technology and Management – Kerala
    (IIITM-K)
Thiruvananthapuram, India

Oleg Tolbanov
Functional Electronics Laboratory
Tomsk State University
Tomsk, Russia

Anton Tyazhev
Functional Electronics Laboratory
Tomsk State University
Tomsk, Russia

Md. Meraj Uddin
Department of Computer Science
Indian Institute of Information
    Technology and Management – Kerala
    (IIITM-K)
Thiruvananthapuram, India

Vinay Vashishtha
School of Electrical, Computer, and Energy
    Engineering
Arizona State University
Tempe, Arizona, USA

Zheng Wang
Shenzhen Institute of Advanced
    Technology
Chinese Academy of Sciences
Beijing, China

Thomas Windbacher
Institute for Microelectronics
TU Wien
Vienna, Austria

Arata Yasuda
Department of Creative Engineering
National Institute of Technology
Tsuruoka College
Tsuruoka, Japan

Haitao Yuan
School of Software Engineering
Beijing Jiaotong University
Beijing, China

MengChu Zhou
Helen and John C. Hartmann Department
    of Electrical and Computer Engineering
New Jersey Institute of Technology
Newark, New Jersey, USA

Section I

Emerging Low Power Devices

1

A FinFET-Based Framework for VLSI Design at the 7 nm Node

Vinay Vashishtha and Lawrence T. Clark

CONTENTS

1.1 Introduction

Recent years have seen finFETs dominate highly scaled (e.g., sub-20 nm) complementary metal-oxide-semiconductor (CMOS) processes (Wu et al. 2013; Lin et al. 2014) due to their ability to alleviate short channel effects, provide lower leakage, and enable some continued V_{DD} scaling. However, availability of a realistic finFET-based predictive process design kit (PDK) for academic use that supports investigation into both circuit, as well as physical design, encompassing all aspects of digital design, has been lacking. While the finFET-based FreePDK15 was supplemented with a standard cell library, it lacked full physical verification, layout vs. schematic check (LVS) and parasitic extraction (Bhanushali et al. 2015; Martins et al. 2015) at the time of development of the PDK described in this chapter. Consequently, the only available sub-45 nm educational PDKs are the planar CMOS-based Synopsys 32/28 nm and FreePDK45 (45 nm PDK) (Stine et al. 2007; Goldman et al. 2013). The cell libraries available for those processes are not very realistic since they use very large cell heights, in contrast to recent industry trends. Additionally, the static random access memory (SRAM) rules and cells provided by these PDKs are not realistic. Because finFETs have a 3-D structure and result in significant density impact, using planar libraries scaled to sub-22 nm dimensions for research is likely to give poor accuracy.

Commercial libraries and PDKs, especially for advanced nodes, are often difficult to obtain for academic use and access to the actual physical layouts is even more restricted. Furthermore, the necessary non-disclosure agreements (NDAs) are unmanageable for large university classes, and the plethora of design rules can distract from the key points. NDAs also make it difficult for the publication of physical design as these may disclose proprietary design rules and structures.

This chapter focuses on the development of a realistic PDK for academic use that overcomes these limitations. The PDK, developed for the N7 node even before 7 nm processes were available in the industry, is thus *predictive*. The predictions have been based on publications of the continually improving lithography, as well as our estimates of what would be available at N7. The original assumptions are described in Clark et al. (2016). For the most part, these assumptions have been accurate, except for the expectation that extreme ultraviolet (EUV) lithography would be widely available, which has turned out to be optimistic. The background and impact on design technology co-optimization (DTCO) for standard cells and SRAM comprises this chapter. The treatment here includes learning from using the cells originally derived in Clark et al. (2016) in realistic designs of SRAM arrays and large digital designs using automated place and route tools.

1.1.1 Chapter Outline

The chapter first outlines the important lithography considerations in Section 1.3. Metrics for overlay, mask errors and other effects that limit are described first. Then, modern liquid immersion optical lithography and its use in multiple patterning (MP) techniques that extend it beyond the standard 80 nm feature limit are discussed. This sets the stage for a discussion of EUV lithography, which can expose features down to about 16 nm in a single exposure, but at a high capital and throughput cost. This section ends with a brief overview of DTCO. DTCO has been required on recent processes to ensure that the very limited possible structures that can be practically fabricated are usable to build real designs. Thus, a key part of a process development is not just to determine transistor and interconnect structures that are lithographically possible, but also ensuring that successful designs can be built with those structures. This discussion is carried out by separating the front end of line (FEOL), middle of line (MOL), and back end of line (BEOL) portions of the process, which fabricate the transistors, contacts and local interconnect, and global interconnect metallization, respectively. The cell library architecture and automated placement and routing (APR) aspects comprise the next section, which with the SRAM results, comprise most of the discussion. The penultimate section describes the SRAM DTCO and array development and performance in the ASAP7 predictive PDK. The final chapter section summarizes.

1.2 ASAP7 Electrical Performance

The PDK uses BSIM-CMG SPICE models and the value used are derived from publicly available sources with appropriate assumptions (Paydavosi et al. 2013). Drive current increase from 14 to 7 nm node is assumed to be 15%, which corresponds to the diminished I_{dsat} improvement over time. In accordance with modern devices, saturation current was assumed to be 4.5× larger than that in the linear region (Clark et al. 2016). A relaxed 54 nm contacted poly pitch (CPP) allows a longer channel length and helps with the assumption of a near ideal subthreshold slope (SS) of 60 mV/decade at room temperature, along with a drain-induced barrier lowering (DIBL) of approximately 30 mV/V. P-type metal-oxide-semiconductor (PMOS) strain seems to be easier to obtain according to the 16 and 14 nm foundry data and larger Idsat values for PMOS than those for n-type metal-oxide-semiconductor (NMOS) have been reported (Wu et al. 2013; Lin et al. 2014). Following this trend, we assume a PMOS to NMOS drive ratio of 0.9:1. This value provides good slew rates at a fan-out of six (FO6), instead of the traditional four.

Despite the same drawn gate length, the PDK and library timing abstract views support four threshold voltage flavors, viz. super low voltage threshold (SLVT), low voltage threshold (LVT), regular threshold voltage (RVT), and SRAM, to allow investigation into both high performance and low-power designs. The threshold voltage is assumed to be changed through work function engineering. For SRAM devices the very low leakage using both a work function change and lightly doped drain (LDD) implant removal. The latter results in an effective channel length (L_{eff}) increase, GIDL reduction and, the overlap capacitance reduction. The drive strength reduces from SLVT to SRAM. The SRAM V_{th} transistors are convenient option for use in retention latches and designs that prioritize low-standby power. In addition to typical-typical (TT) models, fast-fast (FF) and slow-slow (SS) models are also provided for multi-corner APR optimization. Tables 1.1 and 1.2 show the electrical parameters for single fin NMOS and PMOS, respectively, for the TT corner at 25°C (Clark et al. 2016). The nominal operating voltage is $V_{DD} = 700$ mV.

TABLE 1.1

NMOS Typical Corner Parameters (per fin) at 25°C

Parameter	SRAM	RVT	LVT	SLVT
I_{dsat} (µA)	28.57	37.85	45.19	50.79
I_{eff} (µA)	13.07	18.13	23.56	28.67
I_{off} (nA)	0.001	0.019	0.242	2.444
V_{tsat} (V)	0.25	0.17	0.10	0.04
V_{tlin} (V)	0.27	0.19	0.12	0.06
SS (mV/decade)	62.44	63.03	62.90	63.33
DIBL (mV/V)	19.23	21.31	22.32	22.55

TABLE 1.2

PMOS Typical Corner Parameters (per fin) at 25°C

Parameter	SRAM	RVT	LVT	SLVT
$|I_{dsat}|$ (µA)	26.90	32.88	39.88	45.60
$|I_{eff}|$ (µA)	11.37	14.08	18.18	22.64
$|I_{off}|$ (nA)	0.004	0.023	0.230	2.410
V_{tsat} (V)	−0.20	−0.16	−0.10	−0.04
V_{tlin} (V)	−0.22	−0.19	−0.13	−0.07
SS (mV/decade)	64.34	64.48	64.44	64.94
DIBL (mV/V)	24.10	30.36	31.06	31.76

1.3 Lithography Considerations

Photolithography, hereinafter referred to simply as lithography, in a semiconductor industry context, refers to a process whereby a desired pattern is transferred to a target layer on the wafer through use of light. Interconnect metal, via, source-drain regions, and gate layers in a CMOS process stack are a few examples of the patterns defined, or "printed," using lithography.

A simplified pattern transfer flow is as follows. From among the pattern information that is stored in an electronic database file (GDSII) corresponding to all the layers of a given integrated circuit (IC) design, the enlarged pattern, or its photographic negative, corresponding to a single layer is inscribed onto a photomask or reticle. The shapes on the photomask, hereinafter referred to as mask, define the regions that are either opaque or transparent to light. Light from a suitable source is shone on the mask through an illuminator, which modifies the effective manner of illumination, and passes through the transparent mask regions. Thereafter, light passes through a projection lens, which shrinks the enlarged pattern geometries on the mask to their intended size, and exposes the photoresist that has been coated on the wafer atop the layer to be patterned. The photoresist is developed to either discard or retain its exposed regions corresponding to the pattern. This is followed by an etch that removes portions of the target layer not covered by the photoresist, which is then removed, leaving behind the intended pattern on the layer. Both lines and spaces can be patterned through this approach with some variations in the process steps.

Lithography plays a leading role in the scaling process, which is the industry's primary growth driver, as it determines the extent to which feature geometries can be shrunk in successive technology nodes. Lithography is one of the most expensive and complex procedures in semiconductor manufacturing, with mask manufacturing being the most expensive processing steps within lithography (Ma et al. 2010). Both complexity and the number of masks used for manufacturing at a node affect the cost, and an increase in either of these can increase the cost to the point of becoming the limiting factor in the overall cost of the product.

As in any other manufacturing process, the various lithography steps also suffer from variability. The lithographic resolution determines the minimum feature dimension, called critical dimension (CD), for a given layer and is based on the lithography technique employed at a particular technology node. Design rules (DRs) constitute design guidelines to minimize the effects from mask manufacturability issues, the impact of variability and layer misalignment, and ensure printed pattern fidelity to guarantee circuit operation at good yield. Thus, ascertaining these DRs requires consideration of the following lithography-related metrics that can cause final printed pattern on a layer to deviate from the intent and/or result in reliability issues.

1.3.1 Lithography Metrics and Other Considerations for Design Rule Determination

1.3.1.1 Critical Dimension Uniformity (CDU)

Critical dimension uniformity (CDU) relates to the consistency in the dimensions of a feature printed in resist. CD variations arise due to a number of factors—wafer temperature and photoresist thickness, to name a few. It is defined by

$$CDU = \frac{\sqrt{CDU_E^2 + CDU_F^2 + CDU_M^2}}{2},$$

(1.1)

where the CDU_E, CDU_F, and CDU_M are the CD variations due to dose, focus, and mask variations (Chiou et al. 2013). The required CDU is typically calculated as 7% of the target CD requirement, but the modern scanner systems continue to push the envelope beyond that requirement. The 3σ CDU for 40 nm isolated and dense lines can be as small as 0.58 and 0.55 nm (de Graaf et al. 2016), respectively, for ASML's TWINSCAN NXT:1980Di optical immersion lithography scanner released in 2016. For the ASAP7 PDK, we assumed a CDU of 2 nm for optical immersion lithography, which is in line with (Vandeweyer et al. 2010). For extreme ultra violet UV lithography (EUVL) patterned layers, we assumed the CDU to be 1 nm, which is close to the 1.2 nm CDU estimated by van Setten et al. (2014) and the later CDU specification of 1.1 nm for ASML's TWINSCAN NXE:3400B EUV scanner (ASML 2017).

1.3.1.2 Overlay

Overlay refers to the positional inaccuracy resulting from the misalignment between two subsequent mask steps and denotes the worst-case spacing between two non-self-aligned mask layers (Servin et al. 2009). Single machine overlay (SMO) refers to the overlay arising from both layers being printed on the same machine (scanner), which results in better alignment accuracy and, thus, smaller overlay. However, using the same machine for two layers or masks is slower from a processing perspective, and consequently, more expensive. Matched-machine overlay (MMO) refers to that arising from two successive layers being printed on different machines, resulting in a larger value than SMO. As two separate scanners are employed, the overall processing rate in an assembly line setting is faster.

Lin predicted 3σ SMO and MMO values at N7 to be 1.5 and 2 nm, respectively (Lin et al. 2015). ASML's TWINSCAN NXT:1980Di optical immersion lithography scanner released in 2016 has a 3σ SMO and MMO of 1.6 and 2.5 nm, respectively, while its Twinscan NXE:3400B EUV scanner released in 2017 has 3σ SMO and MMO of 1.4 and 2 nm, respectively (ASML 2016, 2017). For the PDK we assumed a 3σ MMO of 3.5 nm for optical immersion lithography, based on (ASML 2015). We assumed a 3σ MMO of 1.7 nm for the EUVL, based on the estimates by van Setten et al. (2014).

1.3.1.3 Mask Error Enhancement Factor (MEEF) and Edge Placement Error (EPE)

Mask error enhancement factor (MEEF) refers to the ratio of wafer or resist CD error to the mask CD error and is given as,

$$MEEF = \frac{\Delta CD_{wafer}}{\Delta CD_{mask}}. \tag{1.2}$$

Thus, it denotes the amount by which errors on the mask are magnified when they are transferred to the wafer and it depends on mask, optics, and the process. Its effects are more pronounced near the resolution limit for a specific patterning technique (Yeh and Loong 2006). Features such as the metal line-ends or tips are typically more adversely affected in optical immersion lithography (193i) systems. Van Setten et al. found the MEEF for 193i patterned layers to range from five to seven, but found it to be nearly one for EUVL patterned layers (van Setten et al. 2014).

The edge placement error (EPE) gives the deviation in edge placement of one layer relative to another, while accounting for both CDU and overlay contributions. For two layers, each patterned through single exposure steps, the EPE is given as,

$$EPE = \sqrt{\left(\frac{3\sigma CDU_{layer1}}{2}\right)^2 + \left(\frac{3\sigma CDU_{layer2}}{2}\right)^2 + \left(3\sigma Overlay_{layer1-2}\right)^2}.$$ (1.3)

1.3.1.4 Time-Dependent Dielectric Breakdown (TDDB)

The primary design rule limiter for metal layers is time dependent dielectric breakdown (TDDB). At the very small fabrication dimensions, very high electric fields are generated not just in the gate dielectric, but in all isolating dielectrics between metals. A key issue in the DTCO process is determining the worst-case spacing between any two metal structures with misalignment, so that the resulting process is reliable against TDDB. TDDB occurs due to the presence of a large (although not as high as in gate dielectrics) electric field between two conductors over a long duration. Its severity is more readily pronounced in conductor layers with large overlay issues, for instance between a via and metal at disparate voltage, or in the MOL layers (Standiford and Bürgel 2013). Obviously, sharp edges exacerbate the fields and, thus, are an important issue.

Although layer self-alignment can alleviate the TDDB to some extent, it does not guarantee complete mitigation and necessitates other measures. One such case is the self-aligned raised source-drain contact to gate separation, that must be increased through the addition of extra spacer thickness and gate cap (Demuynck et al. 2014). This is partially in anticipation of some erosion of the self-aligning spacer material. The final separation between two layers must therefore be not based on just overlay and CDU (i.e., EPE) but also on the TDDB requirement; that is, the expected potential differences between the structures. For the PDK, we assumed a 9 nm spacing requirement for TDDB prevention, a value similar to that assumed by Standiford and Bürgel (2013). Given that operating voltages are well below 1 V V_{DD}, this is conservative, which hopefully covers for any other small errors in the analysis.

1.3.2 Single Exposure Optical Immersion Lithography

The conventional lithography resolution limit, which determines the CD, is given by the Rayleigh equation as follows Ito and Okazaki (2000) by

$$CD = k_1 \frac{\lambda}{NA}$$ (1.4)

where λ is the illumination source wavelength, NA is the projection lens numerical aperture. NA is given as

$$NA = n_1 \sin\theta,$$ (1.5)

where θ is the maximum angle of the light diffracted from transparent mask regions, which can be captured by the lens, and n_1 is the refractive index of the material between the projection lens and wafer. The value of processing factor k_1 in Equation (1.4) depends on the illumination method and resist process.

The term optical lithography has become nearly synonymous with the use of argon fluoride (ArF) light sources in the industry, employed since the 90 nm technology node (Liebmann et al. 2014a). The use of water to boost the NA leads to the technique being termed as optical immersion lithography. NA for the present 193i toolsets is 1.35 and the k_1 value is 0.28. The present set of values for these terms are a result of enhancements over the years, arising from resist process improvements and resolution enhancement techniques such as optical proximity correction (OPC), off-axis illumination (OAI), and source mask optimization among others, each enabling a smaller CD at successive technology nodes. Ultimate limits for NA and k_1 are 1.35 and 0.25, respectively, but operating at these limits is challenging (Lin et al. 2015). Thus, as it stands, CD or half pitch for the layers patterned using a single exposure (SE) in 193i is about 40 nm. Although, the technical specifications for ASML's TWINSCAN NXT:1980Di optical immersion lithography scanner suggests that it can attain a resolution of about 38 nm (ASML 2016).

1.3.3 Multi-Patterning Approaches

The use of 193i single exposure, to achieve CD targets for all the patterned layers ended at the 22 nm technology node. This marked a severe restriction to continuing with the scaling trends for the subsequent nodes. Overcoming this limitation requires the use of multi-patterning (MP) techniques.

1.3.3.1 Litho-Etch^x (LE^x)

One of the most straightforward approaches to MP involves using multiple independent lithography and etch steps, where one litho-etch (LE) step refers to patterning shapes on a given layer through single exposure and etch step. The technique is termed LE^x, where x represents the number of litho-etch steps. It applies to any light source and is not specific to just 193i. LE^2, or LELE, technique is of more immediate use since it is used to pattern the target shapes just below the SE patterning limit. To prepare a design for the LELE process, the design layout (Figure 1.1a)—containing target shapes at a pitch that is smaller than the single exposure limit—is decomposed into two separate layers; for example, A and B, with different "colors" (Figure 1.1b). This decomposition, or "coloring," must produce shapes assigned to a specific color layer at a pitch that can be patterned through a single exposure, thereby "splitting" the pitch. Consequently, the two color layers, resulting from the LELE decomposition step, correspond to two masks that are used in consecutive LE steps to pattern all the shapes on a single layer. This approach results in the LELE steps

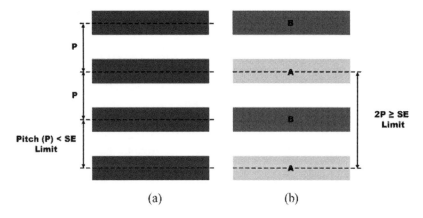

(a)   (b)

FIGURE 1.1
Litho-etch-litho-etch (LELE) multi-patterning (MP) approach. (a) Target layout shapes with inter-shape pitch below the single exposure (SE) patterning limit. (b) Target layout after decomposition of shapes into separate colors. Same-color shapes are at a pitch above the 193i SE patterning limit.

defining the line CD. Yet another approach involves specifying the space CD through LELE steps, instead of the line CD. While the process steps are simpler for the latter, the layout decomposition is more complex compared to the former approach.

The same basic principles used for LELE can be extended to LE^x process with an x value that is larger than two, where x denotes the number of distinct colors and corresponding masks. As multiple masks are used to pattern a single layer, the cost associated with LE^x is higher than the single exposure lithography and increases with the number of masks. Unsurprisingly, the greater the number of LE steps to pattern a single layer, the higher the complexity and overlay concerns. Moreover, misalignment between steps must be considered, and jagged edges can result. LE^x is prone to odd cycle conflicts, whereby decomposition may result in a coloring conflict when more than two shapes geometries exist so as to preclude topology patterning through single exposure. Such odd-cycle conflicts will be discussed in the later sections. Stitching can alleviate these odd-cycle conflicts to some extent by patterning a contiguous shape through different exposures. It requires that the disparate fragments of the same shape have a reasonable overlap to counter edge placement error. Stitching does not completely mitigate odd-cycle conflicts in all topologies. 1-D patterns are specially challenging from this perspective.

1.3.3.2 Self-Aligned Multiple Patterning (SAMP)

Self-aligned multiple patterning (SAMP) represents another MP approach that seeks to limit the mask-defined line or space CD, thereby reducing the overlay error as compared to LE^x. The technique derives its name from spacers that are deposited along the sidewalls of a mask-defined one-dimensional (1-D) or bi-directional (2-D) line; thus, self-aligned to it. These spacers subsequently define the layer as the actual mask. Self-aligned double patterning (SADP) and self-aligned quadruple patterning (SAQP) are two of the more common forms of SAMP technique and denote whether the pitch is split by a factor of two or four, respectively. In the latter case, a first spacer is used to produce two second spacers (i.e., pitch splitting) that is used to pattern the actual lines.

SAMP can be broadly categorized into "spacer positive tone" and "spacer negative tone" process flows (Ma et al. 2010). In the former, the spacers define the dielectric isolation or space between the lines, therefore, the process is also called "spacer-is-dielectric" (SID). It allows for multiple line and space CDs. In the latter process flow, also called "spacer-is-metal" (SIM), the spacers define the line. However, the latter only allows for two line widths and allows more variability in the intra-layer line spacing, which is a reliability risk for TDDB.

Figure 1.2 shows a generic SADP SID flow. Similar to LE^x, decomposition is also necessary for the SADP process (Figure 1.2b). In decomposition, one of these two masks (e.g., mask A) is selected as a candidate for a derivative photomask, which is then utilized to pattern the shape on the resist by using a single exposure (Figure 1.2c). The resist is trimmed in the event the target line CD is under the single exposure resolution limit, followed by an etch and resist strip for mandrel formation (Figure 1.2d) (Oyama et al. 2015). The mandrel is a sacrificial feature, around which the spacers are deposited as shown in Figure 1.2e. Note that the result includes loops. A second photomask, called the block mask, is then used in conjunction with the spacers to "block" the regions where the feature should not be present. In the case of metal lines, the remaining regions define the trenches in a damascene process and subsequently define the line widths. SAQP process involves two spacer deposition steps, where the first set of spacers serve as mandrels for the second set of spacers. This is evident in fin examples below.

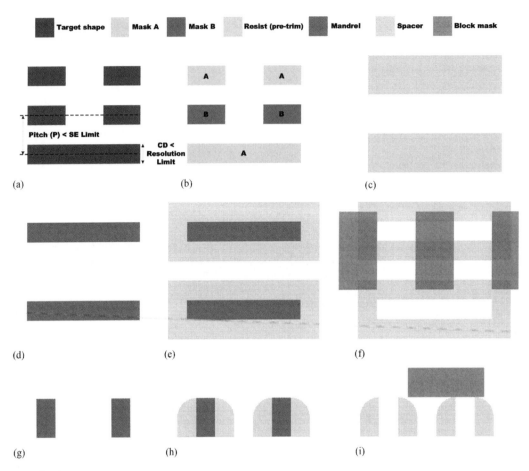

FIGURE 1.2
A "spacer-is-dielectric" (SID) or "spacer positive tone" self-aligned double patterning (SADP) process. (a) Target layout shapes with inter-shape pitch and line CD below the single exposure (SE) patterning limit. (b) Target layout after decomposition of shapes into separate colors. (c) Resist patterning (pre-trim) with CD obtainable through single exposure resolution limit. (d) Mandrel formation (post resist trim and etch). (e) Sidewall spacer deposition. (f) Spacer and block mask defined trench formation for line patterning. (g) Mandrel cross section. (h) Mandrel and spacer cross section after spacer deposition. (i) Spacer cross section after mandrel strip. Block mask is overlaid on top of the spacer to denote the regions where metal will not be deposited, but is not indicative of the process.

Note that the decomposition criteria for SAMP is different than LE^x, since the mandrel is continuous and any discontinuities in it can be marked using the block mask, unlike LE^x, where the lines with such discontinuities must be patterned through a separate exposure, resulting in a potential odd-cycle conflict for certain topologies. It must be also noted that for SAMP, shapes with different colors do not correspond to separate lithography masks, as deposition of other colored shapes is through spacer-deposition and the number of masks employed is lower than the number of decomposition colors.

1.3.3.3 Multiple Patterning Approach Comparison

Before selecting a particular MP technique for layer patterning, consideration must be given to the variability, complexity, and the cost associated with it, and the way in which it differs from another MP technique based on these metrics. Being the simplest and most commonly used multiple patterning techniques, LELE and SADP are contrasted here to determine their favorability as the preferred MP technique. In terms of cost, the SADP process is more expensive than LELE due to sequential etch and deposition steps (Vandeweyer et al. 2010). Liebmann et al. (2015) put the 193i LELE and 193i SADP normalized wafer costs at 2.5× and 3×, respectively, of the 193i SE cost. However, the lower variability and resulting smaller values for similar design rules that determine design density weigh in favor of SADP.

When the LELE steps define the line CD and as the two line populations are distinct, their CDU are generally uncorrelated (Arnold 2008). Any overlay between the two masks affects the space CD. When LELE steps define the space CD instead, the CDU between the two space populations is uncorrelated and overlay error affects the line CD. Thus, notwithstanding its use to define either the line or space CD, the CDU is entangled with overlay in the LELE process. The edge placement error (EPE) in LELE can be calculated as

$$EPE_{LELE} = \sqrt{\left(\frac{3\sigma CDU_{line1}}{2}\right)^2 + \left(\frac{3\sigma CDU_{line2}}{2}\right)^2 + \left(3\sigma Overlay_{line1-2}\right)^2}. \qquad (1.6)$$

for metal lines. On the other hand, in an SADP process, as the line or space CD is mostly defined through a single mask and spacer, the overlay or misalignment does not play a significant role in CD determination unless the target shape edges that determine CD are defined by the block mask. Consequently, block mask edge definition should be avoided. The edge placement error (EPE) in SADP, for spacer defined features, can thus be given as

$$EPE_{SADP} = \sqrt{\left(3\sigma CDU_{litho}\right)^2 + \left(3\sigma spacer_{left\_edge} + 3\sigma spacer_{right\_edge}\right)^2}. \qquad (1.7)$$

The spacer edge related terms in Equation (1.7) are associated with the spacer 3σ CDU, which can be as small as 1 nm (Ma et al. 2010). The absence of overlay due to the absence of block mask defined edges, together with the small spacer CDU, gives SADP an advantage over LELE in terms of EPE control for spacer-defined edges, assuming greater than 3 nm LELE overlay and well controlled SADP spacer CDU of about 1 nm. However, if LELE overlay can be made small enough (~ 2.5 nm), then the EPE for the LELE overlay influenced edges approach that for SADP spacer-defined edges, assuming that the remaining EPE constituent terms remain unchanged (Jung et al. 2007). As mentioned earlier, ASML's TWINSCAN NXT:1980Di 193i scanner has a 3σ MMO of 2.5 nm, but this is accompanied by a change in the CDU value, so that the advantage still lies with SADP. Finally, the spacers do not suffer from much line edge roughness (LER) (Oyama et al. 2012), which is another advantage that SADP offers over LELE. From an electrical perspective, the larger lithography-related variations for LELE patterned layers translate into larger RC variations, which can be nearly twice as much as that for SADP patterned layers employed for critical net routing, primarily due to capacitance variations (Ma et al. 2012). Thus, a choice of LELE creates a potential plethora of metal corners for designers to deal with.

The impact of topology-imposed design rule constraints must also be considered when selecting a MP technique, so as to limit any density penalty. Double patterning can typically overcome the large tip-to-tip (T2T) or tip-to-side (T2S) spacing requirements inherent to single exposure lithography by avoiding these features from being assigned to the same color. The SE T2T or T2S spacing values are larger than the minimum SE-defined width to spatially accommodate the hammerheads used for optical proximity correction (OPC) applied to ensure pattern fidelity (Wong et al. 2008). Large spacing also prevents tips from shortening due to bridging by ensuring sufficient contrast, as the regions between tips have low image contrast. However, even with double patterning, certain feature topologies, such as gridded metal routes with discontinuities, can result in T2T or T2S features on the same mask (Ma et al. 2010). This occurs because shapes on adjacent routing tracks must be colored alternatingly, which forces the shapes along the same track to be the same color. For LELE patterned layer as illustrated in Figure 1.3a, this color assignment necessitates the use of DR values related to SE T2T spacing, x, that are even larger than the minimum SE-defined width and nullify the advantage of double exposure. By comparison, the T2T and T2S features in SADP are block/cut mask defined (see Figure 1.3b). Although also SE-defined, the block/cut mask width is similar to the SE-defined line width, y, instead of the larger T2T or T2S SE spacing requirement. As identified by Ma et al. (2012), this enables SADP patterned metal routes on the same track to connect to pins on a lower metal interdigitating another lower metal route as shown in Figure 1.4a, while different tracks must be used as in Figure 1.4b to realize the same connections for LELE patterned metal routes. Thus, SADP can lower the density penalty in certain design scenarios.

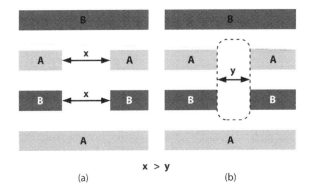

FIGURE 1.3
Comparison of tip-to-tip (T2T) spacing cases for different double patterning approaches. (a) Scenario where T2T spacing, x, is single exposure limited for LELE patterned layers are not too uncommon for designs with 1-D, gridded routing. (b) T2T spacing, y, for SADP patterned layer is defined by the block or cut mask (dotted polygon). Being a line-like feature, the block/cut mask width is similar to the single exposure-defined line width, which is smaller than the spacing required for single exposure-defined tips.

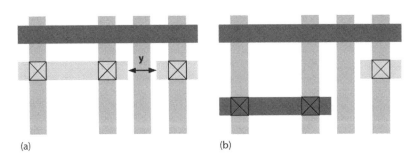

FIGURE 1.4
Pin connection scenarios for double patterned layers. (a) Connections to next neighboring pins can be made through metals on the same track by leveraging the smaller block mask defined T2T spacing for SADP patterned metal layer. (b) The same becomes impossible with LELE patterned metal layer due to larger T2T spacing requirements for same color shapes and different tracks must be utilized, which results in density penalty.

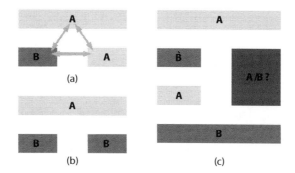

FIGURE 1.5
(a) Even the most common topologies suffer from LELE odd cycle conflicts. (b) SADP is largely free of such conflicts, given correct coloring and considering 1-D equal width metals. (c) Conflicts with SADP may arise for wide metals used in power or clock routing.

Furthermore, as shown in Figure 1.5a, LELE can produce odd cycle conflicts during mask decomposition into two colors. Such a conflict requires increasing the spacing between the features to the SE T2T value (x) in order to resolve the conflict. In contrast, SADP is largely free of odd cycle conflicts for 1-D topologies with equal metal width along routing tracks, as evident in Figure 1.5b. However, odd cycle conflicts may arise during mask decomposition into two colors for SADP, when unequal metal widths or 2-D features are used as shown in Figure 1.5c. Thus, the simpler SADP is adopted for BEOL in the ASAP7 PDK. Some simple DRs, such as limiting line widths to specific values and pitches, makes automatic decomposition possible. This is handled by the Calibre design rule check (DRC) flows, automatically, to simplify usage.

1.3.4 Extreme Ultra Violet Lithography (EUVL)

1.3.4.1 EUVL Necessity

Keeping in line with the area scaling trends requires manufacturing capability enhancement through technological advancement, with pitch scaling through photolithographic improvements constituting the major effort. This increases the wafer cost initially by nearly 20%–25% at a new technology node (Mallik et al. 2014). As a particular technology node matures, the subsequent process and yield optimization bring down the wafer cost. These improvements, together with the increased transistor density, eventually result in the overall cost reduction per transistor. 193i MP adoption to pattern increasingly small FEOL and BEOL pitches has led to a larger than conventional wafer cost increment. This is due to an increased number of photomasks and process steps required to pattern a given layer with MP—hitherto patterned through a single LE step. MOL layer and finFET introduction has led to further cost escalation as they constitute critical layers that require MP. The transistor cost reduction trend has slowed down over the previous few nodes, to which MP has contributed to some extent. This contribution will become even larger with an increase in layers patterned through MP techniques with each new node, resulting in a further slow-down in the per transistor cost-reduction trend. Thus, MP can potentially undermine the cost-effectiveness associated with transitioning to a new node.

In addition to cost, MP is also challenging from a variability perspective, which leads to stringent CDU and overlay considerations, as discussed in previous sections. Accommodating these requires guard-bending the projected pitch/spacing targets for relaxed process margins. Patterning a critical BEOL layer through MP is one such case. Patterning a 2-D shape using LE^x necessitates stitching, which complicates overlay requirements and requires that the pitch target be relaxed to ensure sufficient shape overlap. An alternative is using SADP patterning, which is not very amenable to 2-D shapes, and a 1-D patterning approach using SADP becomes the other choice (Ryckaert et al. 2014). This detrimentally affects the cell circuit density, but is easier to manufacture (Vaidyanathan et al. 2014). Mallik et al. (2015) estimate a 16% and 5%–15% lower area penalty for 2-D, instead of 1-D, metal layers for SRAM cell and standard cells, respectively. A pure 1-D approach makes layout of even relatively simple logic gates more difficult and results in poor input pin accessibility.

EUV lithography can mitigate some of the MP related issues. It uses a 13.5 nm light wavelength, instead of 193 nm used for ArF immersion lithography. This enables patterning the features at a much smaller pitch and resolution, so that a single EUV exposure suffices for patterning at the target pitch only attainable through multiple exposure with 193i; thus, greatly simplifying the design rules. Since academic use was a primary goal of the PDK, opting for simpler EUV rules was a key consideration, even at the risk of optimism in EUV availability. However, we are presently working on additions and libraries to support pure 1-D metallization.

1.3.4.2 EUVL Description and Challenges

ASML's EUVL scanners have a 0.33 NA and can operate with a processing factor k_1 of nearly 0.4 (VanSetten et al. 2015; van de Kerkhof et al. 2017). Using these values, together with a 13.5 nm wavelength, in Equation (1.4) gives a CD of approximately 16 nm. Improvements have led the CD to be reduced to 13 nm (ASML 2017; van de Kerkhof et al. 2017). An NA improvement to 0.5 can lead to a further CD reduction to 8 nm (van Schoot et al. 2015).

EUVL differs from 193i in a number of ways that make it a more challenging patterning approach and have contributed to the delay in EUVL being production-ready. Light at EUV wavelength is generated as follows. A droplet generator releases tin droplets, which are irradiated by a laser to create plasma containing highly ionized tin that emits light at 13.5 nm wavelength, which is gathered by a collector for further transmission (Tallents et al. 2010). Light in the EUV spectrum is absorbed in the air, which necessitates manufacturing under vacuum condition. It is also absorbed by nearly all materials, which precludes optical lens use to prevent high energy loss. Instead, reflective optics (i.e., mirrors) are used. These mirrors have a reflectivity of around 70% and an EUV system can contain over ten such mirrors, resulting in only around 2% of the optical transmission to reach the wafer (Tallents et al. 2010). Inefficiency in the power source reduces the transmitted optical power even further; thus, creating a demand for a high-power light source. It is also desirable for the photoresist to have a high sensitivity to EUV light.

Production throughput for a photolithography system, given in wafers per hour (WPH), is closely associated with cost-effectiveness. To a considerable extent, it relies on the optical power transmitted to the wafer and, thus, on the EUV light source power. It also depends on the amount of time the system is available for production (i.e., system availability). System downtime adversely affects the cost. For EUVL patterning to be cost-effective, a source power exceeding 250 W is desired for over 100 WPH throughput at a 15 mJ/cm^2 photoresist sensitivity (Mallik et al. 2014). Currently, the source power is around 205 W and droplet generator, hitherto a major factor in EUVL system unavailability, has become a smaller concern in more modern EUVL systems (Kim et al. 2017). These and other improvements have brought EUVL system close to the HVM production goals by increasing the throughput to 125 WPH (van de Kerkhof et al. 2017). Collector lifetime is the biggest contributor to the system unavailability at the moment and a number of other issues must be surmounted for further cost-effectiveness (Kim et al. 2017). Overall, EUVL systems continue to improve and are slated to be deployed by some foundries for production at N7 (Xie et al. 2016; Ha et al. 2017).

1.3.4.3 EUVL Advantages

The decision to choose single exposure EUV over 193i MP approaches comes down to both cost and complexity concerns. The EUV mask cost alone is approximately 1.5× that of a 193i mask (Mallik et al. 2015). Other operational expenses bring up the EUVL cost to nearly 3× of 193i single exposure (Liebmann et al. 2015). The number of masks used in each technology node have increased almost linearly up until N10, but the continuation of MP use will result in an abrupt departure from this trend at N7 (Dicker et al. 2015). The issue is compounded by an increase in the associated process steps, which further adds to the cost. Liebmann et al. (2015) estimate the normalized LE^2, LE^3, SADP, and SAQP cost to be 2.5×, 3.5×, 3×, and 4.5× that of 193i SE, respectively. Dicker et al. estimate a 50% patterning cost reduction with EUV as opposed to SAQP (Dicker et al. 2015). They also estimate a faster time to yield and time to market with EUVL due to cycle time reduction as compared to 193i MP approaches that suffer from large learning cycle time—as large as 30% compared to 2-D EUV, thus improving EUVL cost-effectiveness.

EUVL single exposure also reduces the process complexity by virtue of reduced overlay. The small EUV wavelength allows the processing factor k_1 to be relatively large—in the range 0.4–0.5. This enables EUVL to have a high contrast, given by normalized aerial image slope (NILS), than 193i and allows features to be printed with higher fidelity (Ha et al. 2017; van de Kerkhof et al. 2017). The high feature fidelity, better corner rounding, and a single exposure use with EUVL cause fewer line and space CD variations. Consequently, metals and vias patterned through EUVL have more uniform sheet resistance (Ha et al. 2017). They also have lower capacitance as compared to SADP patterned shapes, as EUV single exposure obviates dummy fills and metals cuts. These improvements contribute to improved scalability and better performance (Kim et al. 2017).

Mallik et al. estimate the normalized wafer cost to increase by 32% N10 and by a further 14% at N7 without EUV insertion at these nodes (Mallik et al. 2014). They also estimate a 27% cost reduction, as compared to the latter case, due to EUVL use at N7 for critical BEOL layer patterning with a 150 WPH throughput as a best-case scenario. Ha et al. put the number of mask reduction at N7 due to EUVL use at 25% (Ha et al. 2017). Dicker et al. estimate over 40% cost per function reduction in moving from N10 to N7, and further to N5 as a consequence of EUV insertion for critical BEOL layers. Thus, EUVL deployment at these nodes will likely help ensuring the economic viability of process node transition.

1.3.5 Patterning Cliffs

Patterning cliffs mark the pitch limits for a given lithography technique or MP approach. Table 1.3 summarizes these pitch limits for the metal layers (van Schoot et al. 2015; Sherazi et al. 2016).

It must be noted that EUV scanners are being continuously refined and their capabilities may vary, resulting in different the final patterning cliffs. However, the table gives good rule of thumb values.

TABLE 1.3

Patterning Cliffs, i.e., Minimum Feature Pitch for a Particular
Patterning Technique

Patterning Technique	Minimum Pitch (nm)
193i	80
193i LELE	64
193i LELELE	45
193i SADP	40
EUV SE (2-D, NA=0.33)	36
EUV SE (1-D, NA=0.33)	26
EUV SE (2-D, NA=0.55)	22
193i SAQP	20
EUV SE (1-D, NA=0.55)	16

1.3.6 Design Technology Co-Optimization (DTCO)

When a process is still in development, designers are faced with "what if?" scenarios, where the process developers naturally wish to limit the process complexity, but excessive limitation may make design overly difficult or lacking the needed density to make a new process node worthwhile. Those who must make decisions regarding cell architecture for future processes face significant challenges, as the target process is not fully defined. Specifically, as bends in diffusions and gates have become increasingly untenable, the MOL layers have been introduced to connect source drain regions and replace or augment some structures such as poly cross overs. Decisions that were once purely up to the technology developers increasingly affect design possibilities. Consequently, DTCO is used to feed the impact of such process structure support decisions on the actual designs back into the technology decision making process (Aitken et al. 2014; Liebmann et al. 2014b; Chava et al. 2015; Liebmann et al. 2015). It is increasingly important as finFET width discretization and MP constrain the possible layouts. Consequently, determining design rules progressed in this predictive PDK development by setting rules based on the equipment capabilities, designing cell layouts to use them, and iterating the rules based on the outcomes. This includes the APR and SRAM array aspects as described below.

1.4 Front End of Line (EOL) and Middle of Line (MOL) Layers

The transistors are assumed to be fabricated using a standard finFET type process: a high-K metal gate replaces an initial polysilicon gate, allowing different work functions for NMOS and PMOS, as well as different threshold voltages (V_{th}) (Vandeweyer et al. 2010; Schuegraf et al. 2013; Lin et al. 2014; Seo et al. 2014). Fins are assumed to be patterned at a 27 nm pitch and have a 7 nm drawn (6.5 nm actual) thickness. The layer active is drawn so as to be analogous to the diffusion in a conventional process and encloses the fins—over which raised source-drain is grown—by 10 nm on either side along the direction perpendicular to the fin run length (Clark et al. 2016). The drawn active layer differs from the actual active layer, which is derived by extending it halfway underneath the gates—perpendicular to the fins. The actual active layer, therefore, corresponds to the fin "keep" mask, with its horizontal extent marking the place where fins are cut and its vertical extent denoting the raised source-drain (RSD) regions.

Gates are uniformly spaced on a grid with a relatively conservative 54 nm CPP. Gates are 20 nm wide (21 nm actual). Spacer formation follows poly gate deposition (Hody et al. 2015). Cutting gate polysilicon with the gate cut mask, in a manner that keeps the spacers intact with a dielectric deposition following, ensures that fin cuts are buried under gates or the gate cut fill dielectric, so source/drain growth is on full fins. A double diffusion break (DDB) is assumed to be required to keep fin cuts under the gate. Recently announced processes have removed the DDB requirement, improving standard cell density. Adding this into the PDK as an option is under consideration for a future release. A 20 nm gate cap layer thickness is assumed. This thickness provides adequate distance to avoid TDDB after self-aligned contact etch sidewall spacer erosion, accounting for gate metal thickness non-uniformity (Demuynck et al. 2014). Dual spacer width is 9 nm.

The resulting FEOL and MOL process cross section comprises Figure 1.6. Figure 1.6a shows the (trapezoidal) source/drains grown on the fins. The MOL layers can be used for functions typically reserved for the first interconnect metal (M1) layer (Lin et al. 2014) and serve to lower M1 routing congestion, thereby improving standard cell pin accessibility (Ye et al. 2015). The connection to the MOL local interconnect source–drain (LISD) layer is through the source drain trench (SDT) contact layer. The minimum SDT vertical width of 17 nm is required in an SRAM cell, which necessitates patterning using EUVL. It has a 24 nm drawn horizontal width, with the actual width being 25 nm. This width is larger than the 15 nm gap between the spacers, so as to ensure complete gap coverage and contact with RSD, despite the 5 nm 3σ edge placement error for EUV.

LISD provides the means of connecting RSD regions to power rails and other equipotential RSDs within a standard cell. It is drawn at the same horizontal width as SDT for lower resistance when connecting to RSD through SDT. LISD may also be used for routing purpose within a standard cell at 18 nm width and 36 nm pitch, as it can pass over gates—further lowering M1 usage. The layer thicknesses are defined by different dielectric layers, so that appropriate etch stops can be used. This implies bidirectional (i.e., 2-D) LISD routing, which, when combined with width and pitch assumptions, means that EUVL must be used for patterning. LISD connects to M1 through via 0 (V0), which is another MOL layer.

The local interconnect gate (LIG) layer is used for connecting gates to M1 through V0 and for power delivery to standard cells by connection to LISD upon contact. The minimum LIG width of 16 nm is dictated by the LIG power rail spacing from the gate. The width value implies that LIG is also patterned using EUVL. Recent advances in EUVL have demonstrated resolutions lower than 18 nm (van Schoot et al. 2015; Neumann et al. 2015) even for 2-D patterns, but the PDK restricts this value as the minimum line width for 2-D layers, in accordance with a more conservative 36 nm pitch for bidirectional EUVL (Mallik et al. 2015). However, smaller line width is permitted for LIG as its usage are limited to unidirectional (i.e., 1-D) patterns. LIG connects to the gate through the cap layer in Figure 1.6b, which cuts through a standard cell between the NMOS and PMOS devices. Figure 1.6c shows the same view, but at the fins, so the source/drains are illustrated perpendicular to that in Figure 1.6a. As mentioned, the fin cuts occur under a dummy gate, which comprises ½ of a DDB. The SDT is self-aligned to the gate spacer as shown. MOL rules turn out to be limiting for both standard cells and SRAM. The details are described in Section 1.6.3 and Section 1.8.1. More details and the transistor electrical behavior are presented in Clark et al. (2016).

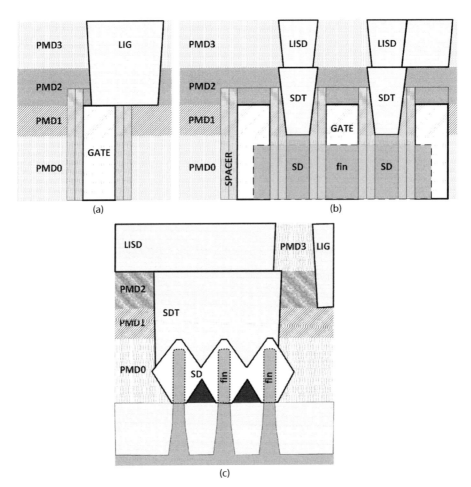

FIGURE 1.6
FEOL and MOL cross sections. (a) LIG connection to the gate. (b) LISD to SDT and SDT to source-drain (SD) connection. LISD location in the stack allows it to cross over gates, so as to be used for routing. (c) Fin and SD cross section. LIG is shown here to illustrate its necessary separation from LISD and SDT. Sub-fin and shallow trench isolation (STI) are evident underneath PMD0.

1.5 Back End of Line (BEOL) Layers

The ASAP7 PDK assumes nine interconnect metal layers (M1-M9) for routing purpose and corresponding vias (V1-V8) to connect these metals. Figure 1.7 shows a representative BEOL stack cross section, comprising of a lower metal layer (Mx), via (Vx), and an upper metal layer (Mx + 1). Following the industry trend (Lin et al. 2014), all BEOL layers assume copper (Cu) interconnects. Metal and vias have a 2:1 aspect ratio, in line with the ITRS roadmap (ITRS 2015). Table 1.4 enumerates thickness of the interconnect layers. Figure 1.7 also shows the barrier layers that increasingly consume the damascene trench, increasing resistance.

In the ASAP7 PDK, metals have the same thickness as the corresponding inter-layer dielectric (ILD), but the vias are thicker than the ILD by 10%—an amount corresponding to the assumed HM thickness. Actual processes at N7 include as many or more than 14 metals,

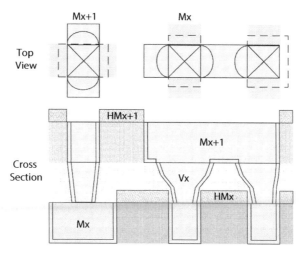

FIGURE 1.7
Representative BEOL cross section. The dotted lines represent the actual self-aligned via (SAV) masks top view. Vias with (left two) and without end-cap (rightmost) are shown. Arcs along Mx + 1 length denote via widening at the non-hard mask edges, evident in the cross section for the two vias at the right.

TABLE 1.4

BEOL Layer Thickness and Metal Pitches

Metal/Via Layer	Thickness (nm)
M1-M3	36
V1-V3	39.6
M4-M5	48
V4-V5	52.8
M6-M7	64
V6-V7	70.4
M8-M9	80
V8	88

with more metal layers at each thickness and pitch value, culminating in 2 layers that are much thicker and wider at the top than in our PDK. The layers here are representative of all but the very thick top layers, which are primarily for power distribution. We had initially not foreseen PDK use for large die power analysis, but are considering adding layers to make the PDK more amenable to full die analysis.

1.5.1 Self-Aligned Via (SAV) and Barrier Layer

In a typical via-first flow, the ILD corresponding to a via sustains damage and erosion—caused by dry etch and cleaning steps, respectively, during both via and metal patterning. The metal width itself is also affected, since via formation goes through the metal damascene trench. This results in via widening that can cause shorts to adjacent, non-equipotential metal lines (Brain et al. 2009; Baklanov et al. 2012). This issue is exacerbated by small metal pitches at lower technology nodes. Consequently, in the ASAP7 design rules, a self-aligned via (SAV) formation flow as described by Brain et al. (2009), is assumed, whereby vias are patterned after the upper interconnect metal layer is patterned on a hard mask (HM) that is relatively unaffected by the via etch. The HM greatly limits via widening perpendicular to the its edges, although some widening occurs along the metal as evident in Figure 1.7. The resulting via edges are delineated by the upper metal HM (i.e., via self-alignment). Via mask edges tend to extend outward where they are defined by the via rather than hard mask. Nonetheless, the via is perfectly aligned perpendicular to the upper metal direction despite via and upper metal overlay errors. The dashed lines in the figure show that the actual via mask overlaps so that the HM defines the width even with misalignment. However, for simplicity the vias are drawn conventionally in the PDK. They are sized as part of the design rule checks (DRC) flows.

Barrier materials, such as tantalum nitride (TaN), are required at the Cu and ILD interface to prevent Cu diffusion. Thickness of the barrier—composed of more resistive TaN—does not scale commensurately with the interconnect scaling. This, together with the diffuse electron scattering at interfaces, causes a greater increase in line resistance than is expected as a consequence of scaling (Im et al. 2005). Additionally, the presence of TaN at via interface has the undesirable effect of increased resistance (Schuegraf et al. 2013). For this reason, the PDK assumes the use of manganese-based self-forming barriers (SFBs) that assuage the shortcoming of TaN by virtue of their conformity, surface smoothness, smaller thickness in the total interconnect fraction, and high diffusivity in Cu (Au et al. 2010; Schuegraf et al. 2013).

The metal resistivity, as specified for extraction purpose, is calculated based on Pyzyna et al. (2015),

$$\rho = \rho_0 \frac{3}{8} C (1-p) \left(\frac{1}{h} + \frac{h}{A} \right) \lambda + \rho_0 \left[1 - \frac{3}{2}\alpha + 3\alpha^2 - 3\alpha^3 \ln(1 + \frac{1}{\alpha}) \right]^{-1} \qquad (1.8)$$

where the ρ_0 denotes the bulk resistivity and C is a geometry based constant. p is the electron collision specularity with surfaces, λ is the bulk electron mean free path, h is the line height, and A is the cross-section area. α is given as $\lambda R/[G(1-R)]$, where R is the electron reflection coefficient at the grain boundaries and G is the average grain size. The first and second terms are the resistivity due to surface electron scattering and grain boundary scattering, respectively. The latter dominates for the 7 nm node (Pyzyna et al. 2015).

1.5.2 EUV Lithography Assumptions and Design Rules

In addition to some of the FEOL (i.e., fin cut) and MOL layers, in the PDK, EUVL is also assumed for patterning M1-M3 and vias corresponding to these metals (i.e., V1 through V3). The choice of an EUVL assumption and the accompanying M1-M3 pitch of 36 nm is based on the premise that this pitch may be attained using single EUV exposure (Mallik et al. 2014). Meeting the same target using optical immersion lithography requires the use of MP techniques, such as SAQP with LE or LELE block mask, which also pushes toward all 1-D topologies. While EUVL is costlier than optical immersion lithography when considering a single exposure, the use of multiple masks in SAQP with block means that the MP approach becomes nearly as expensive as EUVL due to expensive mask tooling and associated processing steps (Liebmann et al. 2015). It is noteworthy that several ITRS target pitch values at N7 are at or over the EUV single exposure cliff. As a result, beyond N7, MP will be required even for EUV lithography defined layers. With N7 at the cut-off, we felt EUVL would be appropriate.

As EUVL permits the single mask use, employing it simplifies the design process by circumventing the issues related to MP, such as complicated cell pin optimization issues due to SADP and odd-cycle conflicts as a consequence of LELE use (Xu et al. 2015). As mentioned earlier, EUVL use at a slightly relaxed pitch of 36 nm also permits 2-D routing, which has the effect of further simplifying both standard cell and SRAM cell design. Thus, given similar cost and reduced design complexity, the PDK assumed EUVL over 193i MP schemes for a number of layers that complicate standard cell design, viz. SDT, LISD, LIG, V0, and M1. However, subsequent iterations of the PDK will revise this choice, so as to use SAQP for 1-D M1-M3, as a consequence of further delays in EUV readiness for high-volume manufacturing. While the aggressive pitch value of 36 nm that EUV affords are not required for M2 and M3, we considered these layers to be EUVL patterned due to the ease that this assumption lends when routing to standard cells during APR. Having the vertical M3 match the M2 also allows better routing density vs. a choice of say, the gate pitch for vertical M3. This also allows relative flexibility in metal directions. We foresee vertical M2 as a better choice for 1-D cells.

A 3σ edge placement error of 5 nm for two EUV layers, when determining inter-layer DRs, was calculated assuming a 3σ mixed machine overlay (MMO) of 1.7 nm for the EUV scanner, a 3σ error in placement of 2 nm due to process variations and through additional guard-banding (van Setten et al. 2014). Patterned M1-M3 lines have a 36 nm 2-D pitch and minimum line width of 18 nm is enforced by the DRs. T2T spacing for narrow lines is 31 nm following van Setten et al. (2015), while wider lines can have a smaller tip-to-tip spacing at 27 nm (Mulkens et al. 2015). As per the PDK DRs, lines narrower than 24 nm are considered thin lines and those wider than this value are considered as wide lines. This threshold value was determined based on the minimum LISD width, since LISD routes near the power rails become the limiting cases for T2T spacing. A moderate T2S spacing of 25 nm follows the results demonstrated by van Setten et al. (2014). Corner-to-corner EUV metal spacing of 20 nm enables via placement to metals on parallel tracks at the minimum possible via spacing of 26 nm with 5 nm EUV upper metal end-cap to allow full enclosure.

1.5.3 Multiple Patterning (MP) Optical Lithography Assumptions and Design Rules

Metal interconnect layers above the intermediate metal layers (i.e., M3) are assumed to be patterned using 193 nm optical immersion lithography and MP. The metal pitches are the same as the thickness values described in Table 1.3 (the 2:1 aspect ratio). The pitch values for M4-M5, and M6-M7 correspond to the targets defined by Liebmann et al. (2015) for 1.5× and 2× metal, respectively. The same metal pitch ratios do not apply to our PDK since the 1× metal pitch is 36 nm to ensure 2-D routing instead of the roadmap value of 32 nm. Moreover, recent foundry releases seem to be tending to the more conservative 36 nm pitch as well. Incrementing the pitch in multiples other than 0.5× does not have any design implication as proven by our DTCO APR experiments. The metal and via aspect ratios are in line with ITRS roadmap projections.

1.5.3.1 Patterning Choice

The M4-M5 pitch target of 48 nm can be attained using either SADP or LE[3] and that of 64 nm for M6-M7 using either SADP or LELE. LE[3] was dismissed outright for M4-M5 patterning, since it is costlier than SADP (Liebmann et al. 2015), and even though the lower LELE cost is appealing for M6-M7 patterning, SADP has lower EPE, LER, and hence RC variations as mentioned in Section 1.3.3.3. Therefore, we chose SADP over LELE for patterning the layers M4-M6. Furthermore, Ma et al. also found that LELE fails TDDB reliability test at 64 nm pitch with 6 nm overlay while SADP appears reliable (Ma et al. 2010).

Although we used a 3.5 nm overlay for 193i patterned layers that would allay the TDDB severity, we still assumed SADP patterning for M6-M7 so as to be more cautious, given the TDDB concern in addition to the other aforementioned SADP advantages. We chose SID over SIM as it permits multiple metal widths and spaces, which is beneficial for clock and most importantly, power routing. Patterning 2-D shapes is possible in SADP but presents challenges, as certain topologies, such as the odd-pitch U and Z constructs, may either not be patterned altogether or contain block mask defined metal edges that result in overlay issues and adversely affect metal CD (Ma et al. 2012). Consequently, we chose to restrict M4 through M7 1-D (straight line) routing in the DRs.

The ASAP7 PDK supports DRCs based on actual mask decomposition into two different masks (colors) for the purpose of SADP. The mask decomposition is performed as part of the rule decks using Mentor Graphics Calibre MP solution. An automated decomposition methodology (Pikus 2016) is employed and does not require coloring by the designer, which greatly simplifies the design effort. As mentioned, we consider this key in an academic environment. To the best of our knowledge, this is the first educational PDK that offers design rules based on such a decomposition flow for SADP. In contrast, the FreePDK15 also used multicolored DRs (Bhanushali et al. 2015) but required decomposition by a designer by employing different colored metals for the same layer.

1.5.3.2 SADP Design Rules and Derivations

DRs must ensure that shapes patterned using the two photomasks, viz. the block and the mandrel, can be resolved. This entails writing DRCs in terms of the derived photomasks and perhaps even showing these masks to the designer. However, fixing DRCs by looking at these masks, and the spacer, is both non-intuitive and confusing. In the flows here, the colors are generated automatically, and the flows can produce the block and mandrel masks created by metal layer decomposition. We formulated restrictive design rules (RDRs) that ensure correct-by-construction metal topologies and guarantee resolvable shapes created using mandrel and block mask referring only to the metals as drawn, so that the rules are color agnostic. Nonetheless, to validate the DRCs, the mandrel, block mask, and spacer shapes are completely derived in our separate validation rule decks (Vashishtha et al. 2017a).

The decomposition criteria for assigning shapes to separate colors for further SADP masks and spacer derivation is as follows. In addition to assigning shapes at a pitch below the 193i pitch limit of 80 nm to separate colors, a single CD routing grid based decomposition criterion is also used. The latter prevents isolated shapes, which do not share a common run length with another shape, from causing coloring conflict with a continuous mandrel through the design extent. Any off-track single CD shapes are marked invalid to enforce automated APR compatibility. Metals wider than a single CD are checked against the single CD metal grid to prevent them from causing incorrect decomposition. Such cases are essentially odd cycle conflicts between wide metals and single CD metal grid.

The ASAP7 PDK support metals wider than a single CD, but the design rules stipulate that the width be such that the metal span an odd number of routing tracks. This prevents odd cycle coloring conflicts between wide metals and single CD metal grids with a pre-assigned color. Wide metals are particularly useful for power distribution purpose. The PDK also supports rules to ensure correct block mask patterning. The minimum block mask width is considered as 40 nm, which corresponds to the minimum 193i resolution. The same T2T spacing value between two SADP patterned metals on the same routing tracks, as well as on adjacent the adjacent routing tracks, prevents minimum block mask width violation. A 44 nm minimum parallel run length between metals on adjacent tracks is also enforced to ensure sufficient block mask T2T spacing. These rules are described in greater detail in (Vashishtha et al. 2017a).

1.6 Cell Library Architecture

A standard cell library has become the usual way to construct general digital circuits, by synthesizing an HDL (typically Verilog) behavioral description and then performing APR of the gates and interconnections. This section describes the cell library. We also explain the cells provided and discuss changes to the design rules based on APR results.

1.6.1 Gear Ratio and Cell Height

As in planar CMOS with gridded gates, which was introduced by Intel at the 45 nm node and other foundries at 28 nm, the selection of standard cell height in a finFET process is also application specific. Low power application-specific integrated circuits (ASICs) tend to use short cells for density, and high performance systems, such as microprocessors, have historically used taller cells. The cell height constrains the number of usable fins per transistor and thus the drive strength as well as the number of M1 tracks for cell internal routing. Assuming horizontal M2, an adequate number of M2 pitches, with sufficient M2 track access to cell pins, are also required. Thus, the horizontal metal (here M2) to fin pitch ratio, known as the gear ratio, becomes an important factor.

The ¾ M2 to fin pitch ratio in the ASAP7 predictive PDK used for our libraries allows designing standard cells at 6, 7.5, 9, or 12 M2 tracks and with 8, 10, 12, or 16 fins fitting in the cell vertical height, respectively. Multiple foundries have mentioned "fin depopulation" on finFET processes; that is, using fewer fins per gate over time. This appears primarily driven by diminishing need for higher speed, increasingly power constrained designs, high drive per fin, MP lithography considerations, and the near 1:1 PMOS:NMOS drive ratio, as well as power density considerations. Moreover, for academic use and teaching cell design issues, a large number of tracks makes tall cells trivial and their excessive drive capability makes them unsuitable for low power applications. The two fin wide transistors in 6-track library cells have limited M1 routes and pin accessibility, but the best density. A 1-D layout, with two metals layers for intra-cell routing, is well suited for 6-track cells, but is not very amenable for classroom use. Therefore, we chose a 7.5-track cell library, which is in line with other publications targeted at N7 (Liebmann et al. 2015). This choice also allows wider M2 power follow rails at the APR stage, which are preferred for robust power delivery. However, wide M2, while supportable (and indeed required for SAQP M2) requires non SAV V1 on the power follow rails. This feature is currently not supported.

Utilizing all of the 10 fins in a 7.5 track library is not possible as sufficient separation is required between the layers SDT and LISD that connect to the transistor source-drain. Also, transistors must have sufficient enclosure by the select layers used for doping. The minimum transistor separation thus dictates that the middle two fins in a standard cell may not be used, as apparent in Figure 1.8, that shows a NAND2 and inverter cell adjacent to each other. This figure also illustrates the cell boundaries producing a DDB between the cells. The fin closest to the power rails cannot be used either. Thus, each transistor has three fins and

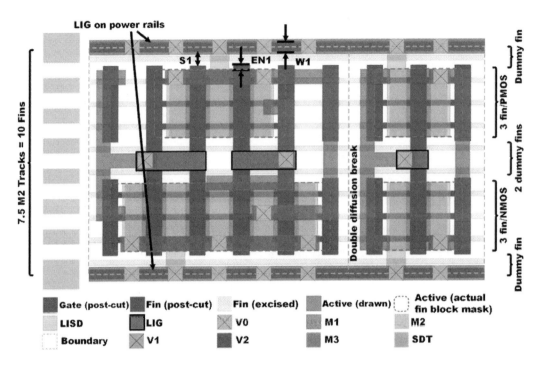

FIGURE 1.8

A 7.5 M2 track standard cell template, with 3-fins per device type, shows the FEOL, MOL, and M1 in adjacent NAND2 and inverter. A double diffusion break is required between the cells. Fins are tucked underneath the gate when breaking the diffusion. LIG, V0, and M1 are not shown here. S1 = minimum LIG to GATE spacing, EN1 = minimum GATE endcap, W1 = minimum LIG width.

this choice is aided by the nearly equal PMOS-to-NMOS ratio. Notwithstanding the number of fins, the transistors have sufficient drive capability due to the large per-fin drive current value (Clark et al. 2016). ASAP7 standard cells are classified based on their drive strengths, so that those with three fin transistors are said to have a 1× drive. Those will less drives strength have fraction values, e.g., a one fin transistor corresponds to a p33× (0.33×) drive strength.

1.6.2 Fin Cut Implications

Again referring to Figure 1.8, the vertical extent of the drawn active layer in the PDK denotes the raised source-drain region. As evident in Figures 1.6, 1.8, and 1.9, the actual active (fin "keep" mask) extends horizontally halfway underneath the non-channel forming (dummy) gates, where fins are actually cut. Implementing the fin cuts in such a manner necessitates a DDB requirement for the cases where diffusions at disparate voltages must be separated. Therefore, all standard cells end in a single diffusion break (SDB), resulting in a DDB upon abutment with other cells to ensure fin separation. Apart from the cell boundaries, an SDB may exist elsewhere in a cell if equipotential diffusion regions are present on either side of a dummy gate and is equivalent to shorted fins underneath such a gate. Nonetheless, this is a useful structure, particularly in the latch and flip-flop CMOS pass-gates.

As shown in Figure 1.9, block mask rounding effects, when implementing fin cuts, create sharp fin edges that have high charge density and, consequently, a high electric field

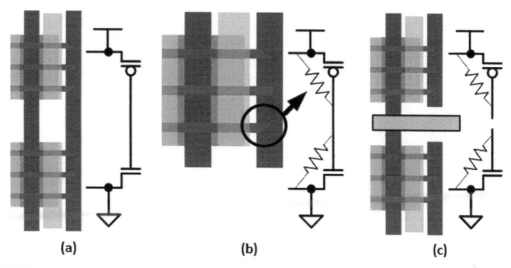

FIGURE 1.9
Dummy gate at DDB (a). Sharp fin edges arise due to mask rounding when cutting fins. This creates a TDDB scenario between the fins and dummy gate (b). This is avoided where possible by cutting the dummy gates (c).

(Vashishtha et al. 2017d). This creates a possible TDDB issue between the fin at the cut edge and the dummy gate, potentially causing leakage through the gate oxide. To mitigate this, in the libraries provided with the PDK, the dummy gates are cut at the middle in the standard cells to preclude an electrical path between the PMOS and NMOS transistors through the dummy gate. The gate cut also allows longer LIG for nearby routing without attaching to the cut dummy gate, as evident in Figure 1.9c.

SAQP requires fixed width and constrains the fin spacing of fins (Figure 1.10). Assuming an EUVL fin cut/keep mask enables the 7.5T library design in its present form (Figure 1.10a). If MP is assumed, LELE cut/keep is complicated as in Figure 1.10b and c, where the middle fins are excised. An even number of fins for PMOS and NMOS transistors allows the SAQP mandrel to completely define the fin locations, greatly easing the subsequent fin masking as illustrated in Figure 1.10d. A 6T (or 6.5T) or 9T library, which has an even number of fins per device at two or four, is more amenable to SAQP fin patterning than the 7.5T library we primarily used for DTCO.

1.6.3 Standard Cell Middle of Line (MOL) Usage

LISD is primarily used to connect diffusion regions to power rails and other equipotential diffusions within the cell through M1, somewhat relieving M1 routing congestion. However, connections to the M1 power rails cannot be completed using LISD alone, since that would require extending LISD past the cell boundary so as to allow sufficient V0 landing on LISD, or 2-D layout that would have issues with the T2S spacing. Larger LISD overlap (past the cell boundary) would violate the 27 nm LISD tip-to-tip (T2T) spacing by interaction with the abutting cell LISD that is not connected to the power rails. Therefore, an LIG power rail is used within the cells to connect LISD to M1, as LISD and LIG short upon contact, but being different layers, do not present 2-D MEEF issues. The PDK DRs require a 4 nm vertical gate endcap past active and a 14 nm LIG to gate spacing (S1).

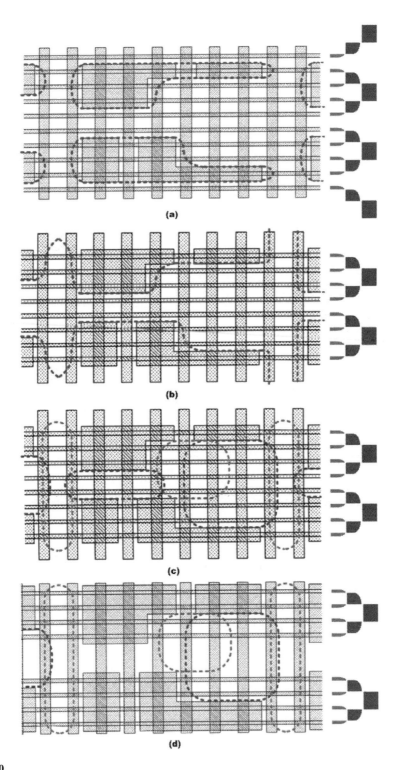

FIGURE 1.10
The original EUV fin keep mask assumptions and SAQP spacer/mandrels (a). Another fin keep option that allows larger patterns (b). LELE fin cut (c) may be problematic, but an even number of fins eases this (d).

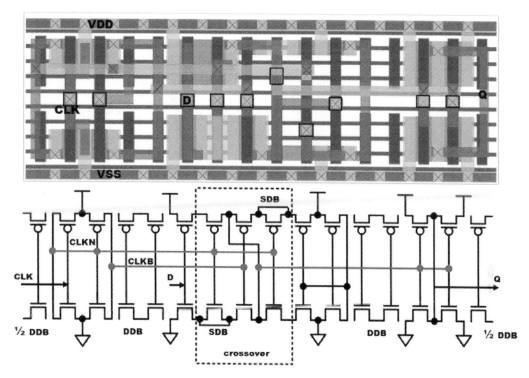

FIGURE 1.11

Cell layout of a transparent high D-latch. Double and single diffusion breaks are shown. The latter require LISD crossovers. Fins shown are prior to the cut.

These rules restrict the LIG power rail to its 16 nm minimum width (W1). This width does not provide fully landed V0, but the LIG power rail is fully populated with V0 to ensure robustness and low resistance.

LISD can also serve a secondary role of a routing layer within a cell, but is used sparingly, as local interconnects favor tungsten over copper, and therefore suffer from higher resistivity (Sherazi et al. 2016). However, LISD routing is helpful in complex cells, especially sequential cells, due to M1 routing congestion and allows us to limit M2 track usage. Figure 1.11 shows a D-latch, in which LISD use to connect diffusions across an SDB of a constituent tristate inverter becomes necessary, since M1 cannot be used due to intervening M1 routes or their large T2T and T2S spacing requirements. The SDBs arise due to the inability to accommodate two gate contacts along a single gate track, which also necessitates a M1 crossover. SDBs used in the cell CMOS pass-gate (crossover) comprised of a CLKN-CLKB-CLKN gate combination evident in the schematic at the bottom of Figure 1.11. Double diffusion breaks are used in the cell, as all diffusions cannot be shared.

1.6.4 Standard Cell Pin and Signal Routing

The cell library architecture emphasizes maximizing pin accessibility, since it has a direct impact on the block density after APR. Where possible, standard cell pins are extended, so as to span at least two, and preferably three M2 routing tracks. This ensures that even standard cells with a large number of pins, such as the AOI333 in Figure 1.12,

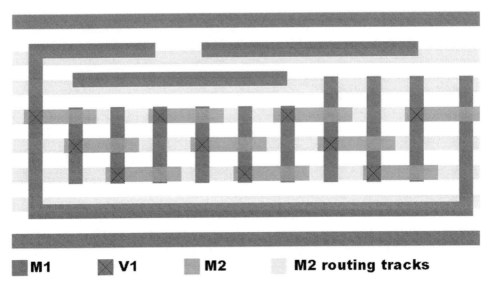

M1   **V1**   **M2**   **M2 routing tracks**

FIGURE 1.12
An AOI333 cell with all of its input M1 pins connected to M2. The staggered M2 allows further connections to M3. Note that this version does not use the bends on M1, following the post-APR DTCO changes. These vias may be slightly unlanded.

have all of their M1 pins accessible to multiple M2 tracks. The figure demonstrates that the use of staggered M2 routes to allow accessing the cell pins by M3. Good pin access is forced by our use of the M1 templates for cell design (Vashishtha et al. 2017d). The M1 template was employed for the rapid cell library development, as the pre-delineated metal constructs take the M1 design rules into consideration, which aids in DRC error free M1 placement. Most, but not all of the M1 routing in the standard cells is template based.

Referring to Figure 1.8, the non-power M1 lines closest to the cell boundary are horizontal, which alleviates the T2S DR spacing requirement for such routes as they constitute a side, rather than a tip feature. This approach was used for the initial cell version in the library, as it allowed better V1 landing from M2. However, this feature was removed after extensive APR validation, as mentioned below. The 7.5 track tall standard cell results in equidistant M2 routing tracks that are all 18 nm wide, except for wider spacing of 36 nm around the M2 power follow-rails. The cells accommodate seven horizontal M1 routes that are arranged as two groups of three equidistant tracks between the M1 power rails, with a larger spacing at the center of the cell where vertical M1 is required for pins. Accounting for the extra M1 spacing near the center, instead of the power rails, helps to maximize pin access through M1 pin extension past the M2 tracks adjacent to the power rails. The vertical M1 template locations, corresponding to the pins, match the CPP. Originally, they always ended at a spacing of 25 nm from the nearest horizontal M1 constructs that they do not overlap, so as to honor the M1 T2S spacing. However, L-shaped pins formed using solely vertical M1 constructs on either one of their ends, do not provide the stipulated 5 nm V1 enclosure necessary when connecting to an M2 track. Consequently, the final topology in Figure 1.12, which does not allow full V1 landing to input pins is adopted for releases after the 1st, as described in Section 1.6.6.

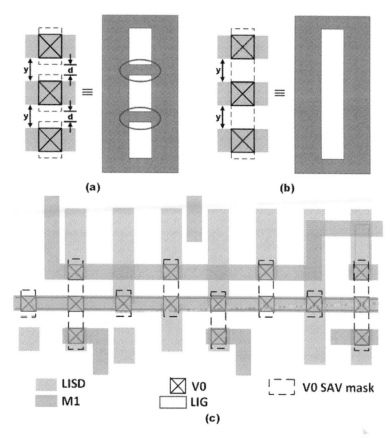

FIGURE 1.13

(a) Sub-lithographic features (ellipses) on SAV mask. (b) Merging SAV masks precludes such features. (c) Merging enables M1 signal routes to be at the minimum spacing from the power rails, which would otherwise have been not possible due to un-merged SAV mask shapes.

Connecting to M1 near the power rails is enabled by via merging. While the example presented here is for V0, all layers support via merging. The presence of V0 on the power rails may engender a case similar to that in Figure 1.13a, where the unmerged via mask spacing "d" may be smaller than the lithographically allowed minimum metal side-to-side spacing "y." Merging these vias (Figure 1.13b) in such cases permits the patterning of V0 containing M1 tracks at minimum spacing. It is enabled by keeping the vias on grid as evident in Figure 1.13c.

1.6.5 Library Collaterals

A cell library requires Liberty timing characterization files (compiled into .db for Synopsys tools) as well as Layout Exchange Format (LEF) of the cell pins and blockages (FRAM for Synopsys). We have focused primarily on Cadence collateral, since we are most familiar with those tools. The technology LEF provides basic design rule and via construction information to the APR tool. This is a key enabling file for high quality; that is, low DRC count APR results. Additional files are the .cdl (spice) cell netlists. We provide Calibre PEX extracted, and LVS versions. The former allow full gate level circuit simulation without needing to re-extract the cells. Data sheets of the library cells are also provided. These are generated automatically by the Liberate cell characterization tool. Mentor Graphics Calibre is used for DRC, LVS and parasitic RC extraction. The parasitic extraction decks allow accurate circuit performance evaluation.

1.6.6 DTCO Driven DR Changes Based on APR Results

The original M1 standard cell template shown in Figure 1.14a below was meant to provide fully landed V1 on M1 in all cases. However, this meant that where an L shape could not be provided, the track over that M1 portion of the pin was lost. APR tools generally complete the routing task, resulting in DRC violations when the layout cannot be completed without producing one. The APR exercises illuminated some significant improvements that could be made, as well as pointing out some substandard cells, which have been revised. A key change to the library templates is illustrated here.

Based on reviewing APR results described in Section 1.7.3, the V1 landing design rule was changed to 2 nm, which does not provide full landing with worst-case misalignment. However, most standard cell pins are inputs and thus drive gates. We decided that gaining the useful pin locations was desirable for multiple reasons. Firstly, a single gate load does not present much capacitance, so the extra RC delay from a misaligned (partially landed) via will not significantly affect delay. Secondly, some gates, notably those with a large number of inputs such as the AOI22 cell, have considerably better pin access, as shown in Figure 1.14b. Initially, two of the pins only had one usable M2-M1 intersection at the middle. These also conflicted for the same M2 track. The improved layout is simpler, and has lower capacitance, as well as moving the pins to a regular horizontal grid identical to the 54 nm gate grid. The cell output nodes still require 2-D routing to reach the correct diffusions. This maintains the original full 5 nm landing for the output node that drives a much larger load and is subject to self-heating.

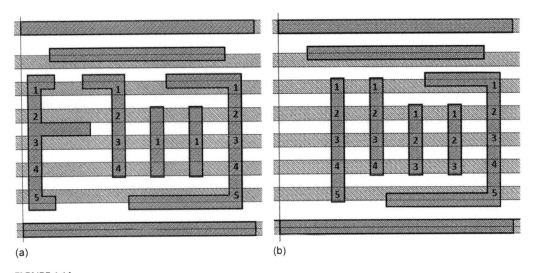

(a)   (b)

FIGURE 1.14
The original AOI22x1 M1 layout and M2 track overlay (a) and the new layout (b) showing the improved access by relaxing the V1 to M1 overlap rule from 5 nm (full landing) to 2 nm.

1.7 Automated Place and Route with ASAP7

1.7.1 Power Routing and Self-Aligned Via (SAV)

Wide power rails are important in their ability to provide a low resistance power grid. Low resistance alleviates IR drop, particularly the dynamic droop in each clock cycle. The ASAP7 PDK comprehends this by providing wide metal design rules that are compatible with the restrictive metal design rules. While the M2 and M3 are assumed EUV, they can easily be converted to SAQP assumptions and the place and route collateral keeps them on a grid that facilitates this change (Vashishtha et al. 2017b). Consequently, all M2 and M3 routes are, like M4 through M7, on grid with no bends in the provided APR flows. On the SADP M4 through M7, wide metals must be 5×, 9×, 13× of the minimum metal widths, so that the double patterning coloring is maintained. These rules are not enforced by DRCs on M2 and M3 as they are on M4-M7, but the APR flows follow the convention via the techLEF and appropriate power gridding commands in the APR tool.

A portion of the V_{SS} and V_{DD} power grid of fully routed digital design is shown in Figure 1.15. In Figure 1.15a the top down view shows the minimum width horizontal M2 power rails over cell top and bottom boundaries. Vertically oriented M3 is attached to each power rail. The V_{SS} on the left and V_{DD} on the right are connected by a single wide self-aligned V2 to each respective M2 rail. Horizontal M4 is connected to M3 with five self-aligned vias, which can be placed close together for minimum resistance due to the M3 EUV lithography assumption. The five vias fit perfectly in the 9× width M3 power routes. The similar V4 connections from M4 to M5 at the lower right of Figure 1.15a have only 3 vias, since these are assumed to be patterned by LELE. The technology LEF file provides via definitions at this spacing.

All vias are consistently SAV type, with the same minimum width as other vias on the layer and laid out on the routing grid. Ideally, wide M2 could be used for the follow rails to minimize the resistance. This, however, would require non-SAV V1 on the follow rails, since self-aligned vias would be too close to the underlying horizontal M1 tracks. The reader will also note that in an SAQP M3 and M2 scenario, all spaces are identical, so a wider M2 would be required. Figure 1.15b shows a 3-D extracted view of the V_{SS} and V_{DD} power rails. The perspective clearly shows the M2-V1-M1-V0-LIG follow-rail stack, with the LISD protruding into the cell gate areas to provide current paths to the NMOS and PMOS sources from V_{SS} and V_{DD}, respectively. The figure emphasizes the M2 over M1 follow-rails, but the M3 and M4 distribution grid is evident. There is substantial redundancy in the power scheme—if a break forms in M2, M1 or LIG due to electromigration or a defect, the other layers act as shunts.

1.7.2 Scaled Layout Exchange Format (LEF) and QRC TechFile

We use Cadence Innovus and Genus for the library APR and synthesis validation, respectively. Standard academic licensing did not allow features below 20 nm in 2016–2017, moving to support 14 nm in 2017. Neither is adequate for the 7 nm PDK. Synopsys academic licensing has the same issue as of this writing. To work around this, APR collateral is scaled 4× for APR, and the output.gds is scaled when streaming into Virtuoso to run Calibre LVS and DRC. Consequently, the technology LEF and macro LEF are scaled up by a factor of four, as is the technology LEF file. Since APR is performed at 4× size but Calibre parasitic extraction is at actual size, a scaled QRC technology file is used in APR. We ran APR on the EDAC design and constrained the

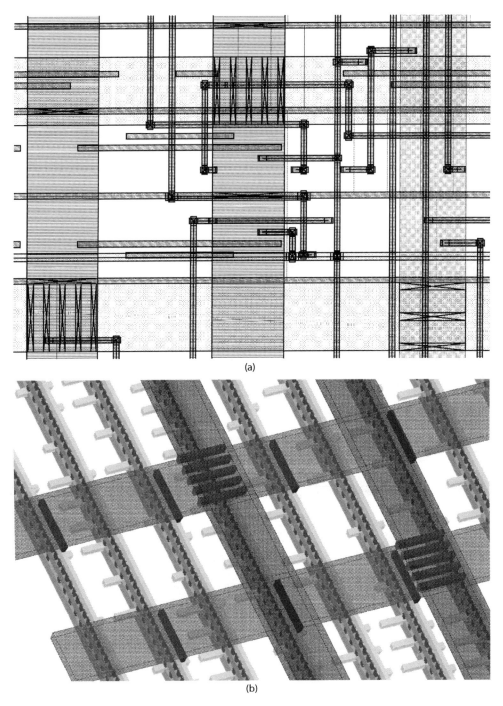

(a)

(b)

FIGURE 1.15
Layout portion of and APR block (a). The wide vias for power distribution are evident connecting M2, M3, and M4. A 3-D extraction of a section of a fully routed design, showing M4 down to the MOL layers (b). The way that LISD connects to LIG beneath the M1 power rails is evident.

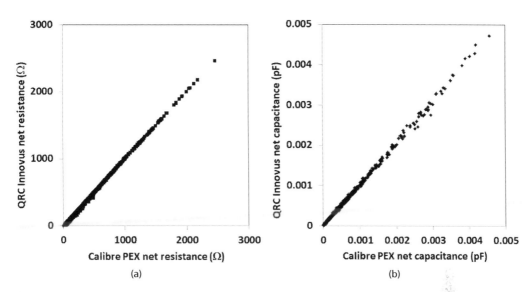

FIGURE 1.16
Correlation of Calibre and scaled QRC techFile resistance (a) and capacitance (b)

metal layers to provide high wire density. The QRC-based SPEF was then compared to the SPEF obtained from Calibre PEX after importing the design into Virtuoso. The QRC technology file was then iteratively "dialed in" to match Calibre. We had expected this to be a straightforward (linear) scaling, but it was not, presumably due to capacitance non-linearity from fringing fields. Figure 1.16 shows that good resistance and capacitance correlation was obtained, with nearly 98% total net capacitance correlation and better than 99% total net resistance correlation.

The technology LEF has fixed via definitions and via generation rule definitions for all layers. As mentioned, they are assumed SAV, allowing upper metal width vias. High-quality (low-resistance) multi-cut via generation for wide power stripes and rings, essential to low resistance power delivery, are also provided for one set of metal widths. User modification for different sizes should be straightforward, following the examples in the provided technology LEF file.

1.7.3 Design Experiments and Results

We performed numerous APR experiments in iterative DTCO on the library, to develop the technology LEF to drive routing and via generation in the Innovus APR tool, and to evaluate the library richness and performance. Three basic designs were used. The first is a small L2 cache error detection and correction (EDAC) block, which generates single error correct, double error correct (SEC-DEC) Hamming codes in the input and output pipelines. In the output direction, the syndrome is generated and in the event of incorrect data, the output is corrected by decoding the syndrome. This design has also been used in class laboratory exercises to use the APR (Clark et al. 2017). It has about 2k gates. The second design is a triple modular redundant (TMR) fully pipelined advanced encryption standard (AES), intended as a soft-error mitigation test vehicle (Chellappa et al. 2015;

Ramamurthy et al. 2015). This is a large design, with about 350k gates in most iterations. Finally, we used a MIPS M14k, with the Verilog adapted to SRAMs designed on the ASAP7 PDK (Vashishtha et al. 2017c) to test integration of SRAMs and their collateral (.lib and LEF). This design requires about 50k gates.

Versions of the designs are shown post-APR in Figure 1.17. In general, there are less than about 20 DRCs after importing these designs back into Virtuoso and running Calibre. Most, if not all of the DRCs are found by Innovus. Because they were the first arrays fully designed, 8kB SRAM arrays are used for the M14k instruction cache (left) and data cache (right) tag and data arrays. Ideally these would be smaller, but following field-programmable gate array (FPGA) convention, some address inputs and many storage locations are unused in this example. The EDAC (Figure 1.17a) allows fast debug of APR

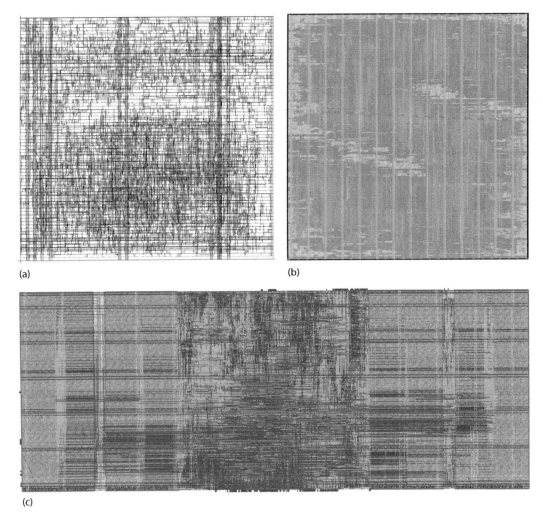

(a)   (b)

(c)

FIGURE 1.17
APR layouts of finished APR blocks. (a) L2 cache EDAC, which is approximately 22 μm by 20 μm. (b) the TMR AES engine, which is 215 μm by 215 μm. (c) The MIPS processor, which is 208 μm by 80 μm in this version.

problems, running in minutes. This design can route in three metal layers (Clark et al. 2017) but the figure includes seven. This design can reach 6 GHz at the TT process corner (25°C) with extensive use of the SLVT cells.

The AES (Figure 1.17b) has 14 pipeline stages with full loop unrolling for both key and data encryption/decryption and requires at least four routing layers (i.e., M2-M5). The design shown uses seven. It has 1596 input and output pins and contains three independent clock domains to support the TMR. TMR storage increases the number of flip-flops to over 15k. This design thus has large clock trees to ensure adequate library support of clock tree synthesis.

The MIPS M14k processor comprises Figure 1.17c and includes large SRAM arrays. As mentioned, the code was adapted from that provided for FPGA implementation by changing the cache arrays from BRAM to SRAM arrays designed for the ASAP7 PDK. The translation lookaside buffers (TLBs) and register file are synthesized, occupying about ¼ of the standard cell area. The 8kB SRAM arrays are apparent in the figure. They are too large for the design, which only needs 1kB tag arrays and 2kB data memory arrays for each cache, but confirmed the liberty and LEF files, as well as the APR routing over the arrays to the pins located in the sense/IO circuits at the center. The control logic, between the left and right storage arrays and the decoders, is laid out using Innovus APR, as are the pre-decoders.

1.8 SRAM Design

SRAMs are essential circuit components in modern digital integrated circuits. Due to their ubiquity, foundries provide special array rules that allow smaller geometries than in random logic for SRAM cells to minimize their area. SRAM addressability makes them ideal vehicles for defect analysis to improve yield in early production. Moreover, running early production validates the issues arising from the tighter design rules. One focus of this section is how SRAMs affected the DTCO analysis for the ASAP7 design rules. They turn out to be more limiting, and thus more important than the previously discussed cell library constructs.

Due to use of the smallest geometries possible, SRAMs are especially prone to random microscopic variations that affect SRAM cell leakage, static noise margin (SNM), read current (speed) and write margin. Consequently, they also provide a place to discuss the PDK use in statistical analysis, as well as our assumptions at N7. Historically, the 6-T SRAM transistor drive ratios required to ensure cell write-ability and read stability have been provided by very careful sizing of the constituent transistors; that is, the pulldown is largest, providing a favorable ratio with the access transistor to provide read stability, while the pull up PMOS transistor is smallest, so that the access transistor can overpower it when writing. Improving PMOS vs. NMOS strain has led to near identical, or in some literature greater PMOS drive strengths, which in combination with discrete finFET sizing, requires read or write assist techniques for a robust design. The yield limiting cases occur for cells that are far out on the tail of the statistical distribution, due to the large number of SRAM cells used in a modern device. Consequently, statistical analysis is required.

1.8.1 FinFET Implications and Fin Patterning

Besides the constraint on transistor width to discrete fin count, MP techniques, e.g., SADP or SAQP, further complicate the allowed cell geometries. On finFET processes, SRAM cells are divided into classes based on the ratios of the pullup, pass gate (access), and pulldown ratios, represented by PU, PG, and PD fin counts. The different sized cells that we used for DTCO in the PDK comprise Figure 1.18. For instance, the smallest cell is 111 (Figure 1.18a) and has nominally equal drive strength for each device. The cell that most easily meets the read stability and write-ability requirements outlined above is thus the 123 cell whose layout is illustrated in Figure 1.18d. As in the standard cells, there must be adequate spacing between the NMOS and PMOS devices for well boundaries, as well as active region separation. Thus, at least one fin spacing is lost between adjacent NMOS transistors in separate cells and between the NMOS and PMOS devices. The 112 and 122 cells comprise Figure 1.18b and c, respectively.

The SRAM DR active mask spacing, which is optimistically set at a single fin, is evident in Figure 1.19a and c. The single fin spacing allows the SAQP fin patterning to be uniform across the ASAP7 die. However, the fin patterning can be changed (and often is on foundry processes) by adjusting the mandrels. The EUV assumption drives some of the metal patterning. Referring to Figure 1.19b, note that the M1 is not 1-D, in that the cell V_{DD} connection zigzags through the array on M1. This M1 is redundant with the straight M2 route. This also provides full M1 landing for the via 1 (V1) connection in each cell. The zigzag is forced by the MEEF constraints described earlier (the M1 lines vertical in the figure that connect the V_{SS} and BL to the MOL). Since the MOL provides a great deal of the connections, M1 BL designs are also possible and given the slow actual roll out of EUV, are probably dominant at the 10 nm and possibly the 7 nm nodes. However, they are not compatible with the horizontal M2 direction outlined above for ASAP7. Consequently, we focus here on M2 BL designs, which maintain the same metal directions that are used in the standard cell areas across the SRAM arrays. The cell gate layers through metallization are illustrated in Figure 1.19d which emphasizes the high aspect ratios of modern metallization. The fins are omitted from the figure.

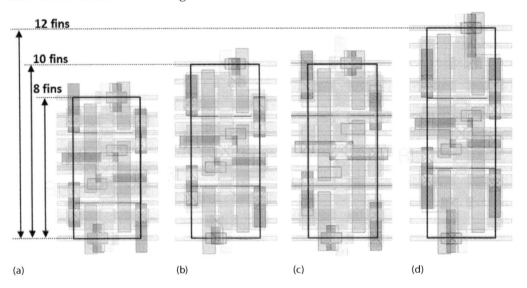

(a)   (b)   (c)   (d)

FIGURE 1.18
Layouts for 111 (a), 112 (b), 122 (c), and 123 (d) ASAP7 SRAM cells.

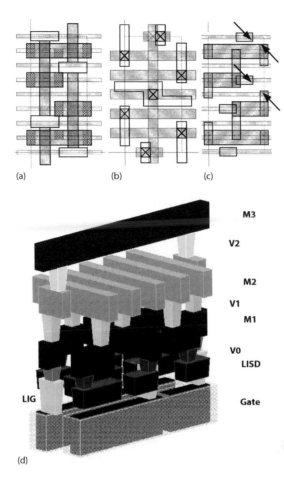

FIGURE 1.19
ASAP7 111 SRAM cell layout showing fin, active, SDT, gate and gate cut layers (a), M1 and M2 as well as via 1 (b) and active and MOL layers (c). 3-D cell view with the correct metal aspect ratios (d). The gate spacer is shown as transparent, but fins are not shown.

1.8.2 Statistical Analysis

AVT has long been used as a measure of local mismatch. AVT is a transistor channel area normalized $\sigma\Delta V_T$, where $\sigma\Delta V_T$ is the V_T difference variance as measured between nearby, identical transistors (Pelgrom et al. 1989; Kuhn et al. 2011). Thus,

$$AVT = \sigma\Delta V_T \times \sqrt{A} \tag{1.9}$$

$$\sigma\Delta V_T = \sigma(|V_{T1} - V_{T2}|). \tag{1.10}$$

In non-fully depleted devices, the dopant atoms are proportional to the area but vary statistically as random dopant fluctuations that in aggregate affect the transistor V_{th}, and dominate the mismatch. Other parameters affect the matching, but can often be driven out by improvements in the manufacturing processes and design. SRAM memories use the smallest devices that can be fabricated (in this case a single fin). Consequently SRAMs are strongly affected and statistical yield analysis is critical to the overall design (Liu et al. 2009).

As a result, SRAM yield falls off rapidly as V_{DD} is lowered toward V_{DDmin}, the minimum yielding SRAM V_{DD}. This is due to increasing variability in the drive ratios as the transistor gate overdrive $V_{GS}-V_{th}$ diminishes, due to V_{th} variations. In metal gate and finFET devices, the variability is primarily due to fin roughness and metal gate grain size variations (Liu et al. 2009; Matsukawa et al. 2009). Nonetheless, due to its ease of measurement and historical significance, AVT is still used to characterize mismatch (Kuhn et al. 2011). FinFET devices, while having different sources of variability since they are fully depleted, continue to use the AVT rubric for analysis. In general AVT tracks the inversion layer thickness, including the effective oxide thickness. It improved by nearly ½ with the advent of high-K metal gate processes, which eliminated the poly depletion effect, and again with finFETs, as they are fully depleted (Kuhn et al. 2011). The primary sources of variability in finFETs turn out to be metal gate grain size and distribution, followed by fin roughness and oxide thickness and work function variations. We use a 1.1 mV.μm AVT, which we chose as a compromise between the NMOS and PMOS values published for a finFET 14-nm process (Giles et al. 2015). Literature has reported pretty steady values across key device transistor fabrication technologies (i.e., poly gate, metal gate, and finFET). Of course, the overall variability increases in our 7 nm SRAM transistors vs. the 14 nm devices, since the channel area is less.

For sense amplifier analysis, we use a conventional Monte-Carlo (MC) approach, applying different differentials to a fixed variability case to find the input referred offset, then applying the next fixed variability, to reach the desired statistics. 1k MC points are used to determine the offset mean and sigma. For SRAM cell analysis, we use a combination of Monte-Carlo and stratified sampling (Clark et al. 2013). In general, unless huge numbers of simulations are used, Monte-Carlo is not adequate for estimations at the tails of the distributions. It is however, adequate to estimate the standard deviation σ and mean μ of the distributions. We confirm the needed σ/μ using the stratified sampling approach, which applies all the statistical variability at each circuit (sigma) strata, to all the possible combinations of the transistors within that strata, using a full factorial. Thus, 728 simulations are performed, one for each possible 1, 0, or −1 impact on each combination of devices at each strata, attempting to find failures in that circuit strata (at that circuit sigma). As an example, if the circuit sigma is distributed evenly across six devices, then there is minimal variation in the circuit. However, if all of the circuit level variability is applied to just the access and pulldown transistors, and in opposite fashion, then the read stability may be jeopardized. In this manner the technique evaluates the tail of the circuit distribution directly and efficiently. At each strata, the transistor variation applied changes with the number of non-zero variations, following the circuit σ_{DEVICE} as

$$\sigma_{DEVICE} = \frac{\sigma_{CELL}}{\sqrt{\Sigma a_i^2}}. \tag{1.11}$$

The technique has been successful at predicting the SRAM yield for foundry processes (Clark et al. 2013).

1.8.3 SRAM Cell Design and DTCO Considerations

1.8.3.1 MOL Patterning

Referring to Figure 1.19c the key DTCO limitation is the corner of LISD and SDT to LIG spacing as indicated by the arrows at the upper right. EUV single patterning would result in some rounding, which helps by reducing the peak electric field and increasing the spacing slightly. As before, we use conservative assumptions. Another result of maintaining adequate spacing for TDDB with worst-case misalignment of the LISD/SDT and LIG layers is that the SDT does not fully cover the active areas, evident at the top and bottom of the NMOS stacks (Figure 1.19a). This led us to separate the LISD and LIG drawn layers. We believe that given the late introduction of EUV, it will most likely be used for MOL layer patterning, as well as vias and cuts. The former is due to projections that MOL layers may require as many as five block masks for SRAM at N7, which makes the expense of EUV more favorable (Sakhare et al. 2015). Foundry presentations have also shown MOL experimental EUV results.

1.8.3.2 1-D Cell Metallization

The 111 cell (Figure 1.19) has 1-D M2 and M3, potentially making it very amenable with a 1-D cell library, and as mentioned, M1 BL designs are compatible with horizontal M1 1-D cell library architectures. The 2-D cell layouts of the 122 cell are shown in Figure 1.20a–d. The similarity to the overall architecture of the 111 cell is apparent. The same MOL limitations exist, but the two fin NMOS devices make the overall source/drain connections to the NMOS stacks considerably better, as they have more coverage and lower resistance at worst-case layer misalignments. The EUV M2 assumption allows narrow metals with wide spacing. This in turn reduces line to line coupling and overall capacitance. Other widths are possible.

Figure 1.20f–g show 1-D metal impact on the cell design at the M2 through MOL layers. Here, M1 and M2 are produced by 1-D stripes using SADP or SAQP, respectively, and then cut to produce separate metal segments. The cuts are assumed the same as SAV (i.e., 16 nm). Cuts are aligned to each other where possible. Wider M2 is required given the SID assumptions. The SAQP mandrel and 1st and 2nd level spacers to produce this layout is shown in Figure 1.20e. The spacers walk across the cells, but the wide, thin, wide, thin repetitive pattern is easily produced. The lines/cuts 1-D metal approach requires a dummy M1 in the middle of the cell, evident in Figure 1.20f and g. The SAV V1 are wider, but still aligned by the M2 hard mask, following the convention used for the APR power routes above. Note that these vias are un-landed on M1. Figure 1.20h is provided as an aid to the transistor layout pattern for readers who are not readily familiar with theses modern, standard SRAM layouts.

1.8.3.3 Stability and Yield Analysis

Read mode static noise margin (SNM) analysis following (Seevinck et al. 1978) at the typical process corner is remarkably similar for the four cells (Vashishtha et al. 2017c). The 112 cell has the best SNM due to the 1:2 PG to PD ratio, followed by the 123 cell with its 2:3 ratio. At TT, the 122 cell has the lowest value, but it is very close to the 111 cell. Referring to Table 1.5, the read SNM σ is inversely proportional to the number of fins as expected. The read SNM overall quality can be ranked by the standard μ/σ of each, since it gives an indication of the sigma level at which margin vanishes. Using this metric, the largest (123) cell is the best, as expected. The 111 cell is the worst, also as expected, entirely due to greater variability. The 112 and 122 cells are nearly even, with the latter making up for a lower baseline SNM with lower variability. The lithography for the 122 cell is easier, and as shown in Table 1.5, it has better write margins, so it becomes the preferred cell. With the possible exception of the 111 cell, all cells have good read margins.

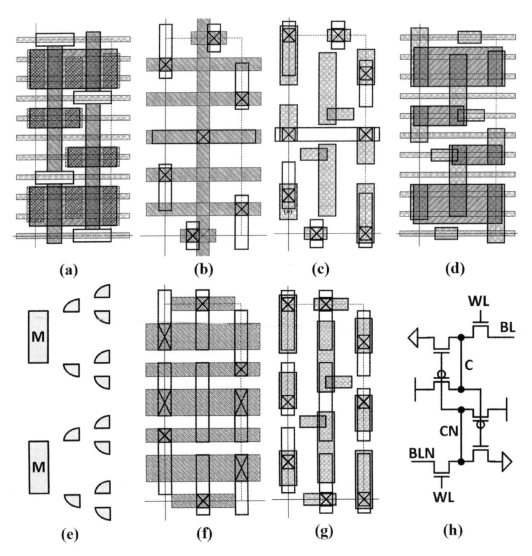

FIGURE 1.20
ASAP7 122 SRAM cell layout showing fin, active, SDT, gate and gate cut, active and SDT layers (a) M3, V2, M2, V1, and M1 layers (b), M1, V0 and MOL (c). Fin, active and MOL (d). The SAQP mandrel and spacers produced for SID M2 (e) and the resulting M2, V1, and M1 layers (f). M1 through MOL (g) and the schematic (h).

In the arrays presented here, a CMOS Y-multiplexer allows a single tristate write circuit per column and eliminates the need for separate write enables (Vashishtha et al. 2017c). Write margin is essentially the BL voltage at which the cell writes under the worst-case voltage and variability conditions. For this analysis, we use an approach that includes the write driver and Y-multiplexer variability, since the write signal passes through it. The analysis thus includes not only the write-ability of the cell, but also the ability of the column circuits to drive the BL low. MC generated V_{th} offsets are applied to the cell, Y-mux, and write driver transistors. The write driver input swings from V_{SS} to V_{DD}, sweeping the driver output voltage low. We tie the gate of the opposing PMOS transistor in the SRAM cell low,

TABLE 1.5

Mean, Variation, and Mean/Sigma of Read SNM, Hold SNM, and Write Margin Simulations for 111, 112, 122, and 123 SRAM Cells at Nominal $V_{DD} = 0.5$ V (hold at $V_{DD} = 350$ mV) and 25°C

	Cell Type				
Marg in Type	Quantity	111	112	122	1233
Read SNM	μ	105.2	116	100.4	106.6
	σ	11.2	9.48	8.3	7.5
	μ/σ	9.4	12.2	12.1	14.2
Hold SNM	μ	119.0	126.1	126.1	128.8
	σ	8.2	6.2	6.2	5.4
	μ/σ	14.4	20.3	20.3	23.9
Write Margin	μ	68.1	69.3	103.6	98.4
	σ	20.9	23.8	18.3	15.8
	μ/σ	3.3	2.9	5.7	6.2

so that the circuit ratio varies only with the write driver voltage. The margin is the BL difference between the point at which the far SRAM node (CN) rises past $V_{DD}/2$ and the lowest BL voltage that can be driven. Table 1.5 shows that the 122 cell has good margin. The high sigma margins are confirmed by the stratified sampling approach as shown in Figure 1.21. The first fails occur at the sigma predicted in Table 1.5, after 2.9 μ/σ for the 112 cell, and after 5.7 (with some added margin) for the 122 cell. While five sigma margins have been suggested in academic papers (Qazi et al. 2011) and this value has been used in commercial designs at the column group level, we consider it inadequate at the cell level since a column group can have as many as 8 × 256 cells. The conclusion is that the preferred 122 cell, which has good density and read stability, would benefit from write assist.

The hold margin controls V_{DDmin}, as well as limiting write assist using reduced V_{DD} or raised V_{SS}. Our target V_{DDmin} is 0.5 V, so we use V_{DDcol} of 350 mV as the hold SNM evaluation point to provide guard band. The results are also shown in Table 1.5. Low SRAM transistor I_{off} that allows good I_{on} to I_{off} ratio at low voltage, providing large hold SNM at 350 mV V_{DD} μ/σ.

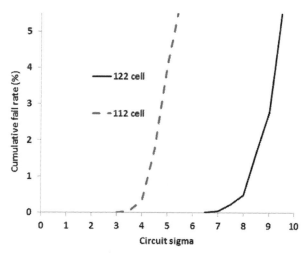

FIGURE 1.21

Comparing the failure rates of the 112 and 122 SRAM cell write margins using the stratified sampling approach.

1.8.4 Array Organization and Column Design

Further analysis comprehends more than one SRAM cell, so we proceed by describing the column group, which is the unit that contains the write, sense, and column multiplexing circuits, as well as multiple columns of SRAM above and below it. Typical choices are four or eight columns of SRAM per sense amplifier, so we (somewhat arbitrarily) chose four. The basic circuit is shown in Figure 1.22, including the "DEC" style sense amplifier. This sense circuit is chosen since it is naturally isolated from the BLs during writes and one of the authors prior experience has shown it to have similar mismatch, but simpler timing than the sense amplifier using just cross-coupled inverters. Note that changing the array size merely requires

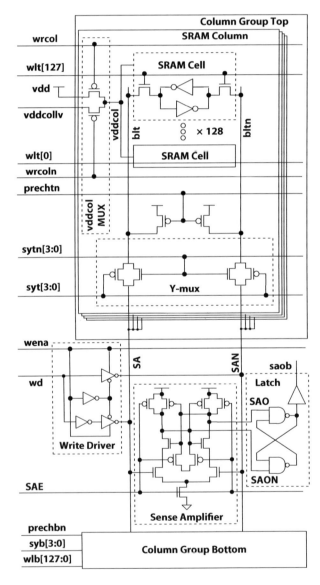

FIGURE 1.22

Column group circuits. Four SRAM columns are attached at the top and bottom of the sense and I/O circuits, sharing a common CMOS pass gate Y-mux for reads and writes.

changing the height of the SRAM columns and the number of column groups included in the array. As shown in the figure, the differential sense amplifier drives a simple set-reset (SR) latch to provide a pseudo-static output from the array. The SR latch has a differential input, so it will be well behaved, and does not need a delayed clock that a conventional D-latch would in this function. There is thus no race condition on the sense precharge.

The sense amplifier input referred offset voltage was determined by SPICE MC simulation. As noted above, the simulations apply between −100 and 100 mV to V(SA)−V(SAN) in 1 mV increments for each random (MC chosen) amplifier transistor mismatch selection. The point at which the amplifier output changes direction is the input-referred offset for that mismatch selection. The results produce a mean offset less than 1 mV, which is near the expectation of zero, also indicating no serious systematic offsets due to the layout. The distribution is Gaussian. The input referred offset standard deviation is 16.5 mV. Using the aforementioned five sigma offset for the sense, 82.6 mV of voltage difference is required at the sense nodes for correct operation. We guard band this up to 100 mV since extra signal is required to provide adequate speed—the sense operation with no residual offset starts the circuit in the metastable state and can produce significant output delays.

1.8.5 Write Assist

The limitations of transistor width and lengths possible given the MP and other limitations essentially requires the need for read or write assist techniques in the ASAP7 SRAM memory designs. Thus, circuit design techniques provide the needed statistical yield margins instead of the cell geometries (Chandra et al. 2010). While many assist approaches have been published, the impact on the overall SRAM size, as well as the variability of the assist techniques themselves, drove the choice for our ASAP7 SRAM arrays. We evaluated two approaches. First, lowering the VDD of the column being written. Previous designs have used timed pull-downs to V_{SS}. In the SRAMs here, we use charge sharing V_{DDcol} voltage generation as in Chandra et al. (2010). The resulting column V_{DD} voltage is capacitively matched to the SRAM columns and thus tracks corners well. However, this scheme adds eight poly tracks to the top and bottom of the column sense/write circuits.

This area is recovered by using a negative BL write assist that provides the write driver a low supply of less than V_{SS}. The same charge sharing circuit is used, with minor polarity changes, to drive a negative BL We see excellent margin improvement with −150 mV on the low driven BL. The CMOS BL multiplexer is unchanged and passes the negative voltage well. Leakage increases at lower voltages. For either scheme, sufficient capacitance is provided by eight SRAM width columns, which are integrated into the dummy cells at the left and right of each array. The circuit occupies the same columns in the read/write/IO circuit height. The reader is referred to (Vashishtha et al. 2017c) for details.

1.9 Chapter Summary

This chapter described the ASAP7 PDK and its development. It discussed the ASAP7 device electrical performance characteristics. The chapter also provided a basic overview of lithography variability concerns, double patterning approaches, and how they compare with each other. Furthermore, it covered the EUV lithography basics and the challenges associated with it. PDK details, such as patterning choices for the various layers, cell library architecture, and DTCO considerations for developing the architecture were discussed as well. The chapter concluded with the APR experiments and SRAM designs based on the PDK, which demonstrated the PDKs suitability for research into various VLSI circuit and system design related aspects.

The ASAP7 PDK has been deployed in graduate-level VLSI courses at the Arizona State University since 2015. The PDK is available for free to universities. It has received attention from research groups and faculty at a number of other universities. This indicates a high-likelihood of its adoption in the classroom. Thus, we hope that the ASAP7 PDK will fulfill its development intent of enabling sub-10 nm CMOS research in academia at a much larger scale and beyond a few university research groups with access to advanced foundry PDKs. We intend to continue our work for other process nodes, and ASAP5 PDK development for the 5 nm technology node is currently underway.

References

Aitken, R. et al., "Physical design and FinFETs," *Proc. ISPD*, pp. 65–68, 2014.

Arnold, W., "Toward 3 nm overlay and critical dimension uniformity: an integrated error budget for double patterning lithography", *Proc. SPIE*, vol. 6924, pp. 692404-1 9, 2008.

ASML. 2015. https://www.asml.com/asml/show.do?lang=EN&ctx=46772&dfp_product_id=8036.

ASML. 2016. https://www.asml.com/products/systems/twinscan-nxt/en/s46772?dfp_product_id=10567.

ASML. 2017. https://www.asml.com/products/systems/twinscan-nxe/twinscan-nxe3400b/en/s46772?dfp_product_id=10850.

Au, Y. et al., "Selective chemical vapor deposition of manganese self-aligned capping layer for Cu interconnections in microelectronics," *J. Electrochem. Soc.*, vol. 157, no. 6, p. D341-345, 2010.

Baklanov, M., P. S. Ho, and E. Zschech, editors, *Advanced Interconnects for ULSI Technology*, John Wiley & Sons, Chichester, UK, 2012.

Bhanushali, K. et al., "FreePDK15: An open-source predictive process design kit for 15nm FinFET technology," *Proc. ISPD*, pp. 165–170, 2015.

Brain, R. et al., "Low-k interconnect stack with a novel self-aligned via patterning process for 32nm high volume manufacturing," *Proc. IITC*, vol. M, pp. 249–251, 2009.

Chandra, V. et al., "On the efficacy of write-assist techniques in low voltage nanoscale SRAMs," *Proc. DATE*, vol. 1, pp. 345–350, 2010.

Chava, B. et al., "Standard cell design in N7: EUV vs. immersion," *Proc. SPIE*, vol. 9427, 94270E-1-9, 2015.

Chellappa, S. et al. "Advanced encryption system with dynamic pipeline reconfiguration for minimum energy operation," *Proc. ISQED*, pp. 201–206, 2015.

Chiou, T. et al., "Lithographic challenges and their solutions for critical layers in sub-14 nm node logic devices," *Proc. SPIE*, vol. 8683, pp. 86830R-1-15, 2013.

Clark, L.T. et al., "ASAP7: A 7-nm finFET predictive process design kit," *Microelectronics Journal*, vol. 53, pp. 105–115, 2016.

Clark, L.T. et al., "Design flows and collateral for the ASAP7 7nm FinFET predictive process design kit," *Proc. MSE*, pp. 1–4, 2017.

Clark, L.T. et al., "SRAM cell optimization for low AVT transistors," *Proc. ISLPED*, pp. 57–63, 2013.

de Graaf, R. et al., "NXT:1980Di immersion scanner for 7 nm and 5 nm production nodes," *Proc. SPIE*, vol. 9780, pp. 978011-1-9, 2016.

Demuynck, S. et al., "Contact module at dense gate pitch technology challenges," *Proc. IITC*, pp. 307–310, 2014.

Dicker, G. et al., "Getting ready for EUV in HVM," *Proc. SPIE*, vol. 9661, pp. 96610F-1-7, 2015.

Giles, M.D. et al., "High sigma measurement of random threshold voltage variation in 14nm Logic FinFET technology," *IEEE Symp. VLSI Tech.*, pp. 150–151, 2015.

Goldman, R. et al., "32/28nm educational design kit: Capabilities, deployment and future," *Proc. PrimeAsia*, pp. 284–288, 2013.

Ha, D. et al., "Highly manufacturable 7nm FinFET technology featuring EUV lithography for low power and high performance applications," *Proc. VLSIT*, pp. T68–T69, 2017.

Hody, H. et al., "Gate double patterning strategies for 10-nm node FinFET devices," *Proc. SPIE*, vol. 9054, pp. 905407-1-7, 2015.

Im, S. et al., "Scaling analysis of multilevel interconnect temperatures for high-performance ICs," *IEEE Trans. Electron Devices*, vol. 52, no. 12, pp. 2710–2719, 2005.

Ito, T., and S. Okazaki, "No Title," *Nature*, vol. 406, no. 6799, pp. 1027–1031, 2000.

ITRS. 2015. http://www.itrs2.net/

Jung, W. et al., "Patterning with amorphous carbon spacer for expanding the resolution limit of current lithography tool," *Proc. SPIE*, vol. 6520, no. 2007, pp. 1–9, 2007.

Kim, S.-S. et al., "Progress in EUV lithography toward manufacturing," *Proc. SPIE*, vol. 10143, pp. 1014306-1-10, 2017.

Kuhn, K.J. et al., "Process technology variation," *IEEE Trans. Electron Devices*, vol. 58, no. 8, pp. 2197–2208, 2011.

Liebmann, L. et al., "Demonstrating production quality multiple exposure patterning aware routing for the 10 nm node," *Proc. SPIE*, vol. 9053, pp. 905309–1-10, 2014.

Liebmann, L. et al., "The daunting complexity of scaling to 7NM without EUV: Pushing DTCO to the extreme," *Proc. SPIE*, vol. 9427, pp. 942701-1–12, 2015.

Liebmann, L. W. et al., "Design and technology co-optimization near single-digit nodes," *Proc. ICCAD*, pp. 582–585, 2014.

Lin, B.J. "Optical lithography with and without NGL for single-digit nanometer nodes," *Proc. SPIE* 9426, pp. 942602-1-10, 2015.

Lin, C. et al., "High Performance 14nm SOI FinFET CMOS Technology with 0.0174 μm² embedded DRAM and 15 Levels of Cu Metallization," *IEDM*, pp. 74–76, 2014.

Liu, Y., K. Endo, and O. Shinichi, "On the gate-stack origin of threshold voltage variability in scaled FinFETs and multi-FinFETs," *IEEE Symp. VLSI Tech.*, pp. 101–102, 2009.

Ma, Y. et al., "Decomposition strategies for self-aligned double patterning", *Proc. SPIE*, vol. 7641, pp. 76410T–1-13, 2010.

Ma, Y. et al., "Self-Aligned Double Patterning (SADP) compliant design flow," *Proc. SPIE*, vol. 8327, pp. 832706-1-13, 2012.

Mallik, A. et al., "Maintaining Moore's law: Enabling cost-friendly dimensional scaling," *Proc. SPIE*, vol. 9422, pp. 94221N–94221N-12, 2015.

Mallik, A. et al., "The economic impact of EUV lithography on critical process modules," *Proc. SPIE*, vol. 9048, pp. 90481R-1-12, 2014.

Martins, M. et al., "Open cell library in 15nm FreePDK technology," *Proc. ISPD*, pp. 171–178, 2015.

Matsukawa, T. et al., "Comprehensive analysis of variability sources of FinFET characteristics," *IEEE Symp. VLSI Tech.*, pp. 159–160, 2009.

Mulkens, J. et al., "Overlay and edge placement control strategies for the 7nm node using EUV and ArF lithography," *Proc. SPIE*, vol. 9422, pp. 94221Q-1-13, 2015.

Neumann, J.T. et al., "Imaging performance of EUV lithography optics configuration for sub-9nm resolution," *Proc. SPIE*, vol. 9422, pp. 94221H–1-9, 2015.

Oyama, K. et al. "CD error budget analysis for self-aligned multiple patterning," *Proc. SPIE*, vol. 8325, pp. 832517-1-8, 2012.

Oyama, K. et al., "Sustainability and applicability of spacer-related patterning towards 7nm node," *Proc. SPIE*, vol. 9425, pp. 942514-1-10, 2015.

Paydavosi, N. et al., "BSIM—SPICE models enable FinFET and UTB IC designs," *IEEE Access*, vol. 1, pp. 201–215, 2013.

Pelgrom, M.J.M. et al., "Matching properties of MOS transistors," *IEEE J. Solid-state Circuits*, vol. 24, no. 5, 1433-1439, 1989.

Pikus, F.G., "Decomposition technologies for advanced nodes," *Proc. ISQED*, pp. 284–288, 2016.

Pyzyna, A., R. Bruce, M. Lofaro, H. Tsai, C. Witt, L. Gignac, M. Brink, and M. Guillorn, "Resistivity of copper interconnects beyond the 7 nm node," *Symp. VLSI Circuits*, vol. 1, no. 1, pp. 120–121, 2015.

Qazi, M., K. Stawiasz, L. Chang and A. P. Chandrakasan, "A 512kb 8T SRAM macro operating down to 0.57 V with an AC-coupled sense amplifier and embedded data-retention-voltage sensor in 45 nm SOI CMOS," *IEEE J. Solid-State Circuits*, vol. 46, no. 1, pp. 85–96, 2011.

Ramamurthy, C. et al., "High performance low power pulse-clocked TMR circuits for soft-error hardness," *IEEE Trans. Nucl. Sci.*, pp. 3040–3048, vol. 6, 2015.

Ryckaert, J. et al., "Design technology co-optimization for N10," *Proc. CICC, pp. 1-8* 2014.

Sakhare, S. et al., "Layout optimization and trade-off between 193i and EUV-based patterning for SRAM cells to improve performance and process variability at 7 nm technology node," *Proc. SPIE*, vol. 9427, 94270O–1-10, 2015.

Schuegraf, K. et al., "Semiconductor logic technology innovation to achieve sub-10 nm manufacturing," *IEEE J. Electron Devices Soc.*, vol. 1, no. 3, pp. 66–75, 2013.

Seevinck, E. et al., "Static noise margin analysis of MOS SRAM cells," *IEEE J. Solid-State Circuits*, vol. SC-22, no. 5, pp. 748–754, 1978.

Seo, K. et al, "A 10 nm platform technology for low power and high performance application featuring FINFET devices with multi workfunction gate stack on bulk and SOI," *Proc. VLSIT*, pp. 1–2, 2014.

Servin, I. et al., "Mask contribution on CD and OVL errors budgets for double patterning lithography," *Proc. SPIE*, vol. 7470, pp 747009-1-13, 2009.

Sherazi, S.M.Y. et al., "Architectural strategies in standard-cell design for the 7 nm and beyond technology node," *Proc. SPIE*, vol. 15, no. 1, pp. 13507-1-11, 2016.

Standiford, K., C. Bürgel, "A new mask linearity specification for EUV masks based on time dependent dielectric breakdown requirements," *Proc. SPIE*, vol. 8880, pp. 88801M-1-7, 2013.

Stine, J.E. et al., "FreePDK: An open-source variation aware design kit," *Proc. MSE*, pp. 173–174, 2007.

Tallents, G., E. Wagenaars and G. Pert, "Optical lithography: Lithography at EUV wavelengths," *Nat. Photon.*, vol. 4, no. 12, pp. 809–811, 2010.

Vaidyanathan, K. et al., "Design implications of extremely restricted patterning," *J. Micro/Nanolith. MEMS MOEMS*, vol. 13, pp. 031309-1-13, 2014.

van de Kerkhof, M. et al., "Enabling sub-10nm node lithography: Presenting the NXE:3400B EUV scanner with improved overlay, imaging, and throughput," *Proc. SPIE*, vol. 10143, pp. 101430D-1-14, 2017.

van Schoot, J. et al., "EUV lithography scanner for sub-8nm resolution," *Proc. SPIE*, vol. 9422, pp. 94221F-1-12, 2015.

van Setten, E. et al., "Imaging performance and challenges of 10 nm and 7 nm logic nodes with 0.33 NA EUV," *Proc. SPIE*, vol. 9231, pp. 923108-1-14, 2014.

van Setten, E. et al., "Patterning options for N7 logic: Prospects and challenges for EUV," *Proc. SPIE*, vol. 9661, pp. 96610G-1-13, 2015.

Vandeweyer, T. et al., "Immersion lithography and double patterning in advanced microelectronics," *Proc. SPIE*, vol. 7521, pp. 752102-1-11, 2010.

Vashishtha, V. et al., "ASAP7 predictive design kit development and cell design technology co-optimization", *Proc. ICCAD*, pp. 992-998, 2017d.

Vashishtha, V. et al., "Design technology co-optimization of back end of line design rules for a 7 nm predictive process design kit," *Proc. ISQED*, pp. 149–154, 2017a.

Vashishtha, V. et al., "Robust 7-nm SRAM Design on a Predictive PDK," *Proc. ISCAS*, pp. 360–363, 2017c.

Vashishtha, V. et al., "Systematic analysis of the timing and power impact of pure lines and cuts routing for multiple patterning," *Proc. SPIE*, vol. 10148, pp. 101480P-1-8, 2017b.

Wong, B.P. et al., *Nano-CMOS Design for Manufacturability: Robust Circuit and Physical Design for Sub-65nm Technology Nodes*, John Wiley & Sons, Hoboken, NJ, 2008.

Wu, S.Y. et al., "A 16nm FinFET CMOS technology for mobile SoC and computing applications," *Proc. IEDM*, pp. 224–227, 2013.

Xie, R. et al., "A 7nm FinFET technology featuring EUV patterning and dual strained high mobility channels," *Proc. IEDM*, vol. 12, no. c, p. 2.7.1-2.7.4, 2016.

Xu, X. et al., "Self-aligned double patterning aware pin access and standard cell layout co-optimization," *IEEE Trans. Comput. Des. Integr. Circuits Syst.*, vol. 34, no. 5, pp. 699–712, 2015.

Ye, W. et al., "Standard cell layout regularity and pin access optimization considering middle-of-line," *Proc. GLSVLSI*, pp. 289–294, 2015.

Yeh, K., W. Loong, "Simulations of mask error enhancement factor in 193 nm immersion lithography," *Jpn. J. Appl. Phys.*, vol. 45, pp. 2481–2496, 2006.

2

Molecular Phenomena in MOSFET Gate Dielectrics and Interfaces

S. Arash Sheikholeslam, Hegoi Manzano, Cristian Grecu, and Andre Ivanov

CONTENTS

2.1 Introduction

Molecular phenomena at the Metal Oxide Semiconductor Field Effect Transistor (MOSFET) channel/dielectric interface are of considerable importance in understanding the kinetics of various degrading mechanisms in transistors. Most of the aging mechanisms impact MOSFETs over time; therefore, designers decrease the clock speed of their chips such that they do not fail during their intended lifetime [1]. This type of techniques is commonly known as guardbanding. Therefore, reliability is a limiting factor to high-speed computation. The main observable effect of aging at transistor level is the shift in threshold voltage. Device scaling is another contributor to the reliability issues. Scaling the MOSFET devices increases the heat density. More heat can accelerate the aging process. The most important aging mechanism are Negative/Positive Bias Temperature Instability (N/PBTI), Hot Carrier Injection (HCI) and Time Dependent Dielectric Breakdown (TDDB). Both NBTI and TDDB have the potential to be studied through classical molecular dynamics due to their nature. NBTI is an interface phenomenon that takes place when a hydrogen atom dissociates from the Si/Oxide interface and diffuses into the oxide due to the electric field caused by the gate potential (Figure 2.1). What remains are the positively charged dangling bonds at the interface [2]. The hydrogen atoms are originally placed at the interface to passivate the Si interfacial dangling bonds, which are known to degrade the threshold voltage. TDDB occurs when the gate dielectric irreversibly breaks down

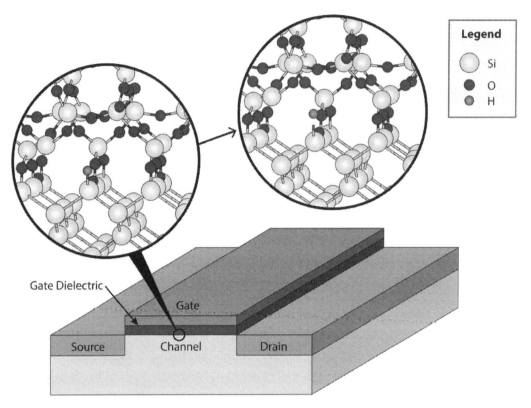

FIGURE 2.1
Hydrogen dissociation at the gate dielectric/channel interface.

when it is polarized with a high enough electric field such that a conductive path of a few kilo ohms is formed through the dielectric. In case of SiO_2, this conductive path will be made of Si.

The gate dielectric of MOSFET transistors is approaching dimensions below 1 nm [3], and therefore there has been significant amount of research on applications of high-k dielectrics, Hf and Zr oxides, in particular, as they can maintain the standards on current leakage while allowing for further scaling. While Zr and Hf are in the same column in the periodic table and have very similar characteristics, they did not receive similar attention in the research community, leaving ZrO_2 as the less studied of the two. ZrO_2 is considered as one of the more recent candidates for alternative gate dielectrics [3]. Amorphous ZrO_2 (a-ZrO_2) has been used in some memory applications and in transistors with high electron mobility as well as a candidate for Ge channel transistors [4–6]. Another interesting application of ZrO_2 is for fuel cell electrolytes [7]. Understanding the chemical and structural characteristics as well as the hydrogen diffusion behavior of ZrO_2 are very important due to their impact on device aging. However, only a handful of experimental studies of a-ZrO_2 characteristics are available in References [8] and [9]. To that end, we have dedicated parts of this chapter to the study of ZrO_2 systems.

The current understanding of hydrogen diffusion in amorphous materials is very limited and is based on experimental data. Such experiments are expensive and time consuming, which limits the scientific community's ability to explore new materials and processes for semiconductor industry's ever-growing demand. Molecular dynamics-based methods allow for quick, accurate, and effective design space exploration.

2.2 Methodology

Classical molecular dynamics (MD) is a method to investigate the movements and interactions of atoms within a system based on the laws of classical physics (as opposed to quantum MD). Classical simulation methods are invariably faster than their quantum physics counterparts, but they suffer from a lack of accuracy when it comes to explaining the details of chemical reactions and they are generally incapable of predicting electronic structure of various materials. Classical MD considers atoms to be hard balls with a charge assigned to them. Therefore, the evolution dynamics for every atom in a classical MD system can be obtained from the following nonlinear differential equation:

$$m\frac{\partial^2 r}{\partial t^2} = -\frac{\partial U}{\partial r} \tag{2.1}$$

where, in Equation 2.1, m represents the particle mass, r is the location and, U is the potential energy of the particle. The main job of a molecular dynamics simulation tool is to predict each particle's dynamical characteristics, such as speed ($\partial r/\partial t$), location (r), and force. The main input to an MD simulator is the initial location of the atoms, their type, electric charge, and a protocol to construct the potential energy (U) spatially. This protocol is called **force field**. Force fields are often designed based on experimental or theoretical data [10–13]. A force field structure and parameters depend on the application. For instance, there is the Lennard-Jones potential, which approximates the interactions between a pair of neutral molecules [14] and has a simple form: $U = 4\varepsilon(\sigma/r)^{12} - (\sigma/r)^6$, or the more complicated Stillinger-Weber potential, which is uniquely made for semiconductor applications as it favors tetrahedral structures that are common in semiconductor materials [15,16], and finally the bond order potentials that are designed to correctly model the structural characteristics of various systems [17,18]. In our case, Reax force fields [13] Molecular Dynamics have been used as the main tool to model the kinetics of hydrogen diffusion. We first explain why Reax is chosen over quantum chemical methods, such as Density Functional Theory (DFT), or other reactive force fields, such as the Parametrized Model number 3 (PM3) [19]. For systems in the order of 1000 atoms, a Reax force field was shown to be 100 times faster than PM3 while maintaining the accuracy [13]. A Reax force field is also known to be significantly faster than DFT [20].

While the full mathematical details of reactive force fields are beyond the scope of this chapter, we give a qualitative explanation of how the force field is used to calculate the potential energy of a system of atoms. In a Reax force field, the potential energy is calculated via the following equation [13]:

$$U = E_{Bond} + E_{Overcoordination} + E_{Undercoordination} + E_{Angle} + E_{Torsion} + E_{VDWaals}$$
$$+ E_{Coulomb} + E_{Conjugated} + E_{Penalty} + E_{Hydrogen-bond} \tag{2.2}$$

In order to take into account various possible carbonic, metallic and ionic bonds, E_{Bond} includes π, σ, and ππ bonds. The **over/under coordination** energy $E_{Overcoordination}$ regulates every atom's bond formation based on their known total bond order. The **angle** energy E_{Angle} sets the equilibrium valance angle for three bonded atoms and vanishes as the bond between two of the bonding atoms breaks. The VDWaals (**Van Der Waals**) energy $E_{VDWaals}$ accounts for the dispersion relations, and the **Coulomb** energy $E_{Coulomb}$ computes the electrostatic energy between two bonded atoms. This force field also has a dedicated energy parameter for the **hydrogen bonds**. Reax force fields are parametrized via a three-step process: (1) Multiple simulations on small molecules are performed using accurate quantum chemistry methods, such as DFT, and the equilibrium structures and energies of these small systems are recorded. (2) The Reax force field parameters are optimized such that classical MD with Reax force fields can predict the same equilibrium structures and energies as in step 1. (3) The force field is then validated against experimental data. If this step does not achieve the desired accuracy, the computation reiterates through steps 1 and 2 with a more complete set of small molecular systems to cover a larger set of reaction between the atomic species in the system.

All the simulations throughout this chapter were performed using the Sandia Labs molecular dynamics suite known as **Large-scale Atomic/Molecular Massively Parallel Simulator** (LAMMPS) [21]. The simulations were performed with a Nose-Hoover thermostat and barostat, also known as NVT and NPT ensembles [22–24]. Further details regarding the thermostat usage are discussed where applicable. The temperature and pressure relaxation were performed over 50 and 500 time-steps, respectively. Verlet algorithm was used to perform the integration of motion equations, with a time-step of 0.1 fs unless specified otherwise. The pressure in all the NPT ensembles is fixed at 1 atm (room pressure). We have also calculated the charges on every atom in the simulation. The charge equilibration was performed using the QEq method at every step. QEq method is based on the overlap of Slatter determinants and can predict the atomic charges for equilibrium and near equilibrium distances [25,26].

As mentioned earlier we focus on molecular phenomena at the gate dielectric/channel interface because the effect of such phenomena on the MOSFET threshold voltage. To that end, we considered two known dielectric oxides, namely silicon dioxide and zirconium dioxide. We explained the technical simulation details of each oxide in the following sub sections. Note that the focus of this chapter is proton dissociation and diffusion as a degrading agent in MOSFETs.

2.2.1 Silicon Oxide

As will be presented later, hydrogen diffusion in silicon dioxide mainly takes place through dissociation and formation of partially covalent OH bonds. This process cannot be described by nonreactive force fields [27]. Reactive force fields were shown to model diffusion of protons in oxides very accurately [13] and were also successful in modeling the structure of amorphous silica [28]. Therefore, we decided to use a variation of the reactive force field reported in [29].

Silicon dioxide is studied in both bulk and thin film forms (at the gate dielectric interface) by the authors. The bulk systems were prepared through an annealing method in accordance with [30]. The simulated oxide should resemble that of a MOSFET that has a high quality and higher breakdown voltage [31]. Higher quality in this case means less porous material with fewer defects. This in turn means a slightly denser oxide. Therefore, the simulation boxes were prepared by randomly placing SiO_2 molecules in the simulation box such that the box's initial density is 3.2 gr/cm^3, which is between the Stishovite [32] and low pressure [33] forms of quartz. The following procedure was used to prepare the amorphous oxide:

1. The temperature was increased to 4000 K using an NVT ensemble and was maintained constant for 150 ps, then cooled down to 300 K. This step randomizes the atoms in the initial simulation box.

2. The system was heated up to 4000 K using an NPT ensemble and maintained at 4000 K for 75 ps until the oxide was fully melted.

3. The system was cooled back down to 300 K (room temperature) at the rate of 10^{13} K/s. This is to guarantee a smooth transition from liquid to solid amorphous structure [34].

4. The system was further relaxed at 300 K.

The modifications employed lead to a higher density/quality oxide, which closely resembles that of the gate dielectric material fabricated by thermal oxidation. The calculated density is 2.28 gr/cm^3 which is in close agreement with the density obtained by thermal oxidation [35]. The following table compares the quality of our oxide and that of the real glass as well of those of other simulations (Table 2.1).

Structural comparisons between the SiO$_2$ prepared by our method and those of real data do not show any noticeable difference. There is only a 3% mismatch in the Si-O-Si angle data. Other than the short-range characteristics, we also have studied the global characteristics, such as the ring statistics. To that end, we have analyzed the occurrence of Guttmann's rings [39] of various lengths. The occurrence frequency of rings of various length is also consistent with the existing experimental data.

2.2.2 Zirconium Oxide

As mentioned earlier in the text, ZrO$_2$ is among high-k alternatives for gate dielectric material. Here we explain how it was prepared for simulation. Very similar to the silicon dioxide, an annealing process was used to prepare amorphous ZrO$_2$ from crystalline form (in this case monoclinic crystal [40]). A simulation cell containing 1500 atoms was prepared, melted at different temperatures, then cooled down with various cooling rates to find the best match with the real material. This process is necessary because within the simulation, cooling takes places at rates that are much shorter than experimental values. The monoclinic crystal has a density of 5.68 gr/cm^3, while various crystalline forms of ZrO$_2$ are known to be denser than its amorphous form. Throughout the ZrO$_2$ annealing process, the NPT ensemble has been employed extensively. NPT ensemble allowed for expansion and contraction of the system under constant pressure. The pressure was fixed at 1 atm that is the normal atmospheric pressure. Also, since annealing takes place over a very short time span, the annealing temperature should be much higher than the actual melting point of ZrO$_2$ at 2988.15 K. Therefore, our annealing temperatures were chosen to be 4000 K, 6000 K,

TABLE 2.1

Structural Characteristics of Silicon Dioxide

	Our Simulation [36]	Real Glass [37,35]	ReaxFF [30]	MD [38]
Density (gr/cm^3)	2.28	2.27	2.14	2.24–2.38
Si-O bond length (Å)	1.62	1.62	1.59	—
Si-O-Si Angle	139.8	144	150	141–152
O-Si-O Angle	108.3	109.5	109.2	108.3

and 8000 K. We used a systematic approach to find an annealing process that can guarantee an amorphous system with our desired density. The annealing recipe is as follows:

1. Multiple monoclinic simulation cells were prepared at 4000 K and were cooled down to 300 K using an NVT ensemble. The cooling rate at this stage was chosen arbitrarily as the sole purpose of this stage is to randomize the molecular structure.

2. The simulation cells were split into three groups at this stage and each group was heated up to 4000 K, 6000 K, and 8000 K using an NPT ensemble for 1 ps.

3. The simulation cells were kept at the annealing temperatures above for 40 ps. This duration is chosen based on the time needed to guarantee the melting of the original simulation cell. The evolution of Zr and O coordination numbers guarantees that melting takes place at this stage [34].

4. Each group of the previous annealing groups was split into another five groups for different cooling rates at this stage and cooled back down to the room temperature using an NPT ensemble.

5. The simulation cells were then relaxed at 300 K for a few ps.

The above process leads to Figure 2.2, which is a plot for the amorphous structure's density vs the cooling rate. Using this plot, one can decide on their desired amorphous ZrO_2 density and pick the cooling rate accordingly.

Based on the above plot, to achieve various possible amorphous ZrO_2 densities we need to tweak the cooling rate, but the relevant annealing temperatures do not change the outcome. Note that the results in this section were obtained by choosing a value of 4000 K for the annealing temperature.

2.2.3 Diffusion Calculations

The Hydrogen diffusion coefficients in this work were calculated based on the Mean Square Displacement (MSD) and Einstein formula [13,41]. The MSD per trajectory was computed using the following equation:

$$MSD = \left\langle \left| r\left(t + t_0\right) - r\left(t\right) \right|^2 \right\rangle$$

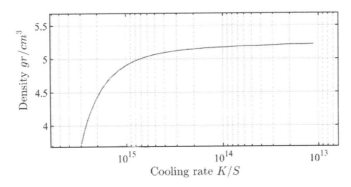

FIGURE 2.2
ZrO_2 density vs cooling rate. (From Sheikholeslam, S.A. et al., *Thin Solid Films*, 594, 172–177, 2015.)

Where $r(t)$ is the location of a hydrogen atom at time t. The data for the first 50 ps of every molecular trajectory was discarded and the rest were used in the MSD equation, which was calculated over trajectories of 200 ps ($t = 200$ ps). The calculations were performed on over 500 trajectories per hydrogen atom. The total MSD (MSD_{Tot}) per hydrogen atom was computed by averaging over these 500 trajectories. The MSD_{Tot} was later used along with the Einstein formula:

$$D(T) = \frac{1}{6Nt_0} MSD_{Tot}$$

Where N is the total number hydrogen atoms. Substituting the diffusion coefficient into the Arrhenius relationship [42] we could compute the diffusion activation energy.

$$D(T) = D_0 e^{-E_a/KT}$$

Where E_a is the activation energy, K is the Boltzmann constant, and T is the temperature. To quantify E_a, diffusion coefficients should be calculated at multiple different temperatures, and an Arrhenius fit should be used to calculate the activation energies.

2.2.4 Diffusion Path Identification

Barrier Climbing Nudged Elastic Band (BC-NEB) method was used to determine the diffusion paths in this work. BC-NEB is a method to calculate the minimum reaction paths and transition state barriers in chemical reactions [43–46]. While activation energy is statistically relevant and shows the general trend in transitions, it does not reveal any information about various types of diffusion path. The latter requires the analysis of individual hydrogen trajectories from one potential minimum to the other. The method works by optimizing a fixed number of transition images along a transition path in a constrained optimization scheme where these images are laid along the transition path and are connected to each other via spring forces.

2.3 Results

In this section, the outcomes of our simulations are explained in full detail. It was found that hydrogen diffusion in silicon dioxide has two regimes and has implications as far the NBTI aging is concerned. We also explained various modes of hydrogen transport in zirconium dioxide, which is a more complex material compared to silicon oxide. While amorphous silicon oxide is made of a single type of building block (SiO_4 tetrahedral), zirconium oxide is made of various type of building blocks with Zr participating in the structure with various coordination numbers (ZrO_x where $4 < x < 8$.) This is also due to the fact that zirconium oxide has five different types of crystal structures ([40,47–49]) with various energies [50], while silicon oxide has only one. In the final part of this chapter, we will discuss some other predictions and characterizations on new materials that can be performed using reactive forcefields. To that end we will explain, using an example, how these forcefields can be used to predict the $(SiO_2)_{1-x}(ZrO_2)_x$ dielectric constant for various values of x.

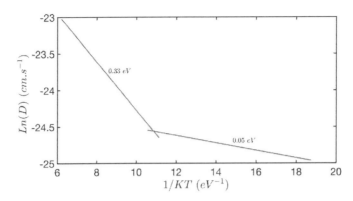

FIGURE 2.3
SiO$_2$ diffusion regimes.

2.3.1 Silicon Dioxide Diffusion Coefficients and Paths

As explained earlier an Arrhenius plot was used to quantify the activation energy of the zirconium oxide. Figure 2.3 depicts the Hydrogen diffusion Arrhenius plot reproduced from [36]. As one can observe, two different diffusion regimes are present. The regime with lower activation energy corresponds to hydrogen atom vibrations around their equilibrium position while the high activation energy represents the actual diffusion. Basically, hydrogen presence in the bulk SiO$_2$ is in the form of hydroxyl group (–OH) and the diffusion takes place through dissociation and formation of bonds between O and H. This diffusion mechanism was previously observed in crystalline and amorphous silicates, as well as phosphates [51–57]. It was also observed that the H atom is on the middle plane formed by adjacent Si atoms. This result agrees with DFT simulations [53]. Along with other structural characteristics of an OH bond, this suggests that the OH bond points toward inside the ring and the diffusion takes place via cross-ring hydrogen hopping as opposed to hopping between neighboring oxygen atoms, as depicted in Figure 2.4.

The above analysis brings forth the possibility of achieving higher diffusion activation energies (which is desirable when dealing with NBTI) via introducing mechanical changes in SiO$_2$ to change the cross-ring OH distances. One such possibility is one-dimensional straining of the amorphous structure. In our simulations, straining was achieved via deforming the simulation cell by applying various amounts of tensile strain along the z direction.

The stress strain relationships and the Young modulus match with the experimental values reported for silica glass [58] and [59]. Table 2.2 shows the structural results of straining the amorphous SiO$_2$ in z direction [36]. As seen in the table, straining has almost no effect on O-Si-O angle and Si-O bond length, but it slightly increases the O-O/Si-O distances on the strain direction. The Si-O-Si angle increases as well.

These characteristics suggest that while the internal structure of SiO$_4$ tetrahedrals are impervious to straining, their orientations and arrangement within the amorphous structure change significantly.

The above analysis shows that increasing the diffusion activation energy can be achieved by straining the SiO$_2$. Therefore, we computed the activation energy of strained systems with 3%, 4%, 5%, and 10% strains and depicted the outcome in Figure 2.5. While the vibration activation energy remains almost constant, the higher (diffusion) activation energy significantly increases from 0.33 eV to almost 1.1 eV.

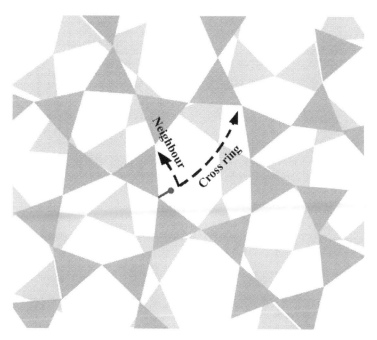

FIGURE 2.4
Cross ring vs. neighbor hydrogen diffusion. The triangles represent SiO_4 tetrahedral.

In NBTI degradation the hydrogen atoms diffuse away from the interface or else they have a chance to repassivate the dangling Si bond (the degradation will be reversed.) Therefore, straining the oxide will make it costly (from an energy standpoint) for the hydrogen atom to diffuse away from the interface and, in that sense, will increase the chance for hydrogen atoms to repassivate the dangling bond. This means that while the bond dissociation still takes place, the process remains practically reversible because the hydrogen atoms stay close to the Si dangling bond.

2.3.2 Zirconium Dioxide Diffusion Coefficients and Paths

As mentioned earlier, ZrO_2 is one of the high dielectric constant replacements (high permittivity, also referred to as high-k) for SiO_2. We have calculated the corresponding

TABLE 2.2

Structural Comparisons between SiO_2 with Various Strain Percentages

Strain Percentage	0%	5%	10%
Si-O-Si Angle	139.8	142.3	145.5
O-Si-O Angle	108.3	108.3	108.3
Si-O bond length	1.62	1.62	1.62
O-O distance (z projection)	1.32	1.35	1.38
Si-O distance (z projection)	0.82	0.84	0.87

Source: Sheikholeslam, S.A. et al., *J. Mater. Chem. C*, 4, 8104–8110, 2016.

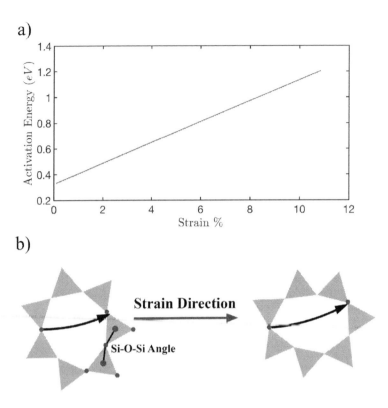

FIGURE 2.5
(a) Effect of straining on the higher activation energy. (b) Effect of strain on the cross-ring O-O distance.

hydrogen diffusion activation energies and the diffusion paths as a first step in characterizing this material for what is relevant when NBTI is concerned.

Amorphous ZrO_2 comes in a range of densities [9,60]. In order to take the entire spectrum into account, the hydrogen diffusion was calculated for two types of amorphous ZrO_2: one with a density of $4.2/cm^3$ and another with a density of 5.1 gr/cm^3. The cooling rate to achieve the desired density with MD simulations was obtained from the following equation [34]:

$$Density = -4.229\left(\log_{10}\left(\frac{1}{cooling\,rate} \right) + 16.7 \right)^{-5.22} + 5.149$$

As a result of performing the hydrogen diffusion simulations we found three major mechanisms of H transport in ZrO_2. One mechanism is through angular vibrations of H in a hydroxyl group. A second type occurs through dissociation and formation of bonds with oxygen atoms. A third type manifests through metastable bonding between a hydrogen and three oxygen atoms. The second mode is similar to the typical hydrogen diffusion mechanism in SiO_2. However, there is one major difference between these three modes of transport and those in SiO_2. Hydrogen modes of transport in ZrO_2 are complementary to each other and are part of the transport process while the vibrations in SiO_2 were local.

The angular vibration mode of transport is presented in the following, where the OH bond vibrates between inside and outside a ring (centered at O):

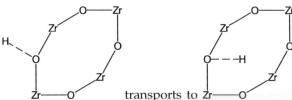

transports to 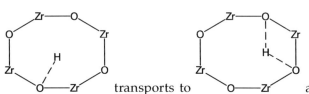 and vice versa. This mode of transport also has the lowest activation energy at less than 0.5 eV (the activation energies were calculated using the NEB method.) The angular transport displaces the hydrogen atom around 2 Å. The other (nonlocal) transport mechanisms complete the above as discussed earlier. The metastable transport mechanism is depicted in the following:

transports to and vice versa. The activation energies from metastable bonds to the hydroxyl group are directional. The activation energy is around 0.3 eV from the metastable bond to a hydroxyl group and around 1.6 eV from the hydroxyl group to the metastable bond [61]. The OH group formation and dissociation mechanism has a range of activation energies based on the coordination number of the oxygen atoms involved. Based on our observations, this range spans from 0.5 eV to 2.5 eV. The hydrogen atom forms stronger bonds with undercoordinated oxygen atoms. This leads to a higher overall diffusion activation energy for lower density oxides (as the simulations also confirm) because lower density oxides have more undercoordinated oxygen atoms.

To complete this section, we present the overall diffusion activations energies and diffusivities in Table 2.3.

One can conclude that controlling the hydrogen diffusion activation energy is possible through introducing mechanical changes into the ZrO_2 structure (similar to SiO_2.)

TABLE 2.3

Diffusion Activation Energies and Diffusivities of Amorphous Zirconium Oxide

Oxide Type	High Density	Low Density
Activation Energy	0.45 eV	0.57 eV
Diffusivity	1.14×10^{-2} cm^2/s	4.57×10^{-1} cm^2/s

Our results might be considered counter intuitive in that a higher activation energy is observed (both in ZrO_2 and SiO_2) at lower concentration of transport medium. The reason is the kinetics of hydrogen transport that were discussed throughout the chapter. The molecular resolution in studying the hydrogen transport mechanics was achieved through the reactive nature of the force fields we have used. In the next section, we will stretch the Reax forcefield to its limits beyond which some direct modeling of free electrons becomes necessary.

2.3.3 Other Characterizations

In this section we will show how Reax force fields are capable of characterizing material features as long as they don't need to involve some description of free electrons. Basically, the Reax forcefield is incapable of modeling different electronic energy levels. There have been attempts at including free electrons into the force field explicitly, but it will increase the computation cost and takes away from the simplicity of the model.

For the simulations presented in this section, we developed a Reax force field that includes Silicon, Oxygen, Zirconium, and Hydrogen. To that end, we merged two existing force fields [30,62] and optimized the missing parameters as explained briefly in the Methodology section. Since we were interested in $(SiO_2)_{1-x}(ZrO_2)_x$ with $0 < x < 0.1$, we parametrized the force field on clusters of molecules with a Si-O-Zr angle because we do not expect any Si-Zr bonds in such system. We used this force field to predict the permittivity of $(SiO_2)_{1-x}(ZrO_2)_x$ as a function of x. Dielectric constant was shown to have the following relation with the total dipole moment [63]:

$$\varepsilon_r = 1 + \frac{4\pi}{3Vk_BT}\left(\left\langle M^2 \right\rangle - \left\langle M \right\rangle^2\right) \tag{2.3}$$

Where V is the volume of the system, k_B is the Boltzmann constant, and M is the total dipole moment, which is calculated as a sum over all the atoms present in the system. Equation 2.3 along with simple potentials, such as SPC, was used to calculate the permittivity of water and ethanol [64,65]. Here we extended their work to study a more complex system using a more complete force field. We performed our simulations over a 25 ns time period, which is considered to be very long for a molecular dynamic simulation. However, when working with complex systems, such extended time frames are needed for Equation 2.3 to converge. We calculated the dielectric constant of SiO_2 to be 3.6, which is very close to experimental value (between 3.7 and 3.9). The calculated dielectric constant for ZrO_2 is 22.6, which is within the range of experimental values for this material (between 18 and 27.) We also found that there is no linear relation between the percentage of ZrO_2 in $(SiO_2)_{1-x}(ZrO_2)_x$ and the dielectric constant.

X	ε
0	3.6
0.02	4.5
0.05	5.4
0.7	6
0.1	13.1
1	22.6

Our simulations show that the dielectric constant relates to the number of six-coordinated zirconium atoms rather than the percentage of zirconium in silica glass. These findings back up earlier work that made similar assessments [66].

In this chapter we showed how molecular dynamics with reactive force fields can be used to explore new dielectric material for semiconductor application. We explored the capabilities and limitations to classical models through various examples.

References

1. V. Huard, M. Denais, F. Perrier, N. Revil, C. Parthasarathy, A. Bravaix and E. Vincent, "A thorough investigation of MOSFETs NBTI degradation," *Microelectronics Reliability*, vol. 45, pp. 83–98, 2005.
2. G. Pobegen, M. Nelhiebel and T. Grasser, "Detrimental impact of hydrogen passivation on NBTI and HC degradation," in *Reliability Physics Symposium (IRPS), 2013 IEEE International*, 2013.
3. M. Neisser and S. Wurm, "ITRS lithography roadmap: 2015 challenges," *Advanced Optical Technologies*, vol. 4, pp. 235–240, 2015.
4. D. Panda and T.-Y. Tseng, "Growth, dielectric properties, and memory device applications of $ZrO2$ thin films," *Thin Solid Films*, vol. 531, pp. 1–20, 2013.
5. A. Salaun, H. Grampeix, J. Buckley, C. Mannequin, C. Vallée, P. Gonon, S. Jeannot, C. Gaumer, M. Gros-Jean and V. Jousseaume, "Investigation of $HfO2$ and $ZrO2$ for resistive random access memory applications," *Thin Solid Films*, vol. 525, pp. 20–27, 2012.
6. G. Ye, H. Wang, S. Arulkumaran, G. I. Ng, R. Hofstetter, Y. Li, M. J. Anand, K. S. Ang, Y. K. T. Maung and S. C. Foo, "Atomic layer deposition of $ZrO2$ as gate dielectrics for AlGaN/GaN metal-insulator-semiconductor high electron mobility transistors on silicon," *Applied Physics Letters*, vol. 103, p. 142109, 2013.
7. M. Youssef and B. Yildiz, "Hydrogen defects in tetragonal $ZrO2$ studied using density functional theory," *Physical Chemistry Chemical Physics*, vol. 16, pp. 1354–1365, 2014.
8. J. Li Vage, C. Mazieres et al., "Nature and thermal evolution of amorphous hvdrated zirconium oxide," *Journal of the American Ceramic Society*, vol. 51, pp. 349–353, 1968.
9. M. Winterer, "Reverse Monte Carlo analysis of extended x-ray absorption fine structure spectra of monoclinic and amorphous zirconia," *Journal of Applied Physics*, vol. 88, pp. 5635–5644, 2000.
10. T. Aichinger, M. Nelhiebel and T. Grasser, "On the temperature dependence of NBTI recovery," *Microelectronics Reliability*, vol. 48, pp. 1178–1184, 2008.
11. B. W. Dodson, "Development of a many-body Tersoff-type potential for silicon," *Physical Review B*, vol. 35, p. 2795, 1987.
12. S. L. Price, "Toward more accurate model intermolecular potentials for organic molecules," *Reviews in Computational Chemistry*, vol. 14, pp. 225–289, 2000.
13. A. C. T. Van Duin, S. Dasgupta, F. Lorant and W. A. Goddard, "ReaxFF: A reactive force field for hydrocarbons," *The Journal of Physical Chemistry A*, vol. 105, pp. 9396–9409, 2001.
14. J. E. Jones, "On the determination of molecular fields. —II. From the equation of state of a gas," *Proceedings of the Royal Society A*, vol. 106, no. 738, pp. 463–477, 1924.
15. R. L. C. Vink, G. T. Barkema, W. F. Van der Weg and N. Mousseau, "Fitting the Stillinger—Weber potential to amorphous silicon," *Journal of Non-Crystalline Solids*, vol. 282, pp. 248–255, 2001.
16. B. Leimkuhler and C. Matthews, *Molecular Dynamics*, Springer, Cham, Switzerland, 2016.
17. M. Bauchy and M. Micoulaut, "Atomic scale foundation of temperature-dependent bonding constraints in network glasses and liquids," *Journal of Non-Crystalline Solids*, vol. 357, pp. 2530–2537, 2011.

18. D. Conrad and K. Scheerschmidt, "Empirical bond-order potential for semiconductors," *Physical Review B*, vol. 58, p. 4538, 1998.
19. J. J. P. Stewart, "Optimization of parameters for semiempirical methods II. Applications," *Journal of Computational Chemistry*, vol. 10, pp. 221–264, 1989.
20. A. Van Duin, "Reactive force fields: Concepts of ReaxFF," *Computational Methods in Catalysis and Materials Science: An Introduction for Scientists and Engineers*, pp. 167–181, 2009.
21. S. Plimpton, P. Crozier and A. Thompson, "LAMMPS-large-scale atomic/molecular massively parallel simulator," *Sandia National Laboratories*, vol. 18, 2007.
22. W. G. Hoover, "Canonical dynamics: Equilibrium phase-space distributions," *Physical Review A*, vol. 31, p. 1695, 1985.
23. S. Nose, "A unified formulation of the constant temperature molecular dynamics methods," *The Journal of Chemical Physics*, vol. 81, pp. 511–519, 1984.
24. S. Nose, "A molecular dynamics method for simulations in the canonical ensemble," *Molecular Physics*, vol. 52, pp. 255–268, 1984.
25. A. K. Rappe and W. A. Goddard III, "Charge equilibration for molecular dynamics simulations," *The Journal of Physical Chemistry*, vol. 95, pp. 3358–3363, 1991.
26. A. Nakano, "Parallel multilevel preconditioned conjugate-gradient approach to variable-charge molecular dynamics," *Computer Physics Communications*, vol. 104, pp. 59–69, 1997.
27. P. Raiteri, J. D. Gale and G. Bussi, "Reactive force field simulation of proton diffusion in BaZrO3 using an empirical valence bond approach," *Journal of Physics: Condensed Matter*, vol. 23, p. 334213, 2011.
28. Y. Yu, B. Wang, M. Wang, G. Sant and M. Bauchy, "Revisiting silica with ReaxFF: towards improved predictions of glass structure and properties via reactive molecular dynamics," *Journal of Non-Crystalline Solids*, vol. 443, pp. 148–154, 2016.
29. H. M. Aktulga, J. C. Fogarty, S. A. Pandit and A. Y. Grama, "Parallel reactive molecular dynamics: Numerical methods and algorithmic techniques," *Parallel Computing*, vol. 38, pp. 245–259, 2012.
30. J. C. Fogarty, H. M. Aktulga, A. Y. Grama, A. C. T. Van Duin and S. A. Pandit, "A reactive molecular dynamics simulation of the silica-water interface," *The Journal of Chemical Physics*, vol. 132, p. 174704, 2010.
31. R. C. Jaeger, *Introduction to Microelectronic Fabrication*, Prentice Hall, Upper Sadle River, NJ, 2002.
32. N. L. Ross, J. Shu and R. M. Hazen, "High-pressure crystal chemistry of stishovite," *American Mineralogist*, vol. 75, pp. 739–747, 1990.
33. R. T. Downs, "The pressure behavior of a cristobalite," *American Mineralogist*, vol. 79, pp. 9–14, 1994.
34. S. A. Sheikholeslam, G. M. Xia, C. Grecu and A. Ivanov, "Generation and properties of bulk a-ZrO2 by molecular dynamics simulations with a reactive force field," *Thin Solid Films*, vol. 594, pp. 172–177, 2015.
35. G. E. McGuire, *Semiconductor Materials and Process Technology Handbook: For Very Large Scale Integration (VLSI) and Ultra Large Scale Integration (ULSI)*, Noyes Publications, Park Ridge, IL, 1988.
36. S. A. Sheikholeslam, H. Manzano, C. Grecu and A. Ivanov, "Reduced hydrogen diffusion in strained amorphous SiO2: Understanding ageing in MOSFET devices," *Journal of Materials Chemistry C*, vol. 4, no. 34, pp. 8104–8110, 2016.
37. R. L. Mozzi and B. E. Warren, "The structure of vitreous silica," *Journal of Applied Crystallography*, vol. 2, pp. 164–172, 1969.
38. K. Vollmayr, W. Kob and K. Binder, "Cooling-rate effects in amorphous silica: A computer-simulation study," *Physical Review B*, vol. 54, p. 15808, 1996.
39. L. Guttman, "Ring structure of the crystalline and amorphous forms of silicon dioxide," *Journal of Non-Crystalline Solids*, vol. 116, pp. 145–147, 1990.
40. C. J. Howard, R. J. Hill and B. E. Reichert, "Structures of ZrO2 polymorphs at room temperature by high-resolution neutron powder diffraction," *Acta Crystallographica Section B: Structural Science*, vol. 44, pp. 116–120, 1988.

41. S. M. Wood, C. Eames, E. Kendrick and M. S. Islam, "Sodium ion diffusion and voltage trends in phosphates Na4M3 (PO4) 2P2O7 (M= Fe, Mn, Co, Ni) for possible high-rate cathodes," *The Journal of Physical Chemistry C*, vol. 119, pp. 15935–15941, 2015.

42. T. Bauer, P. Lunkenheimer and A. Loidl, "Cooperativity and the freezing of molecular motion at the glass transition," *Physical Review Letters*, vol. 111, no. 22, p. 225702, 2013.

43. G. Henkelman and H. Jónsson, "Improved tangent estimate in the nudged elastic band method for finding minimum energy paths and saddle points," *The Journal of Chemical Physics*, vol. 113, pp. 9978–9985, 2000.

44. G. Henkelman, B. P. Uberuaga and H. Jónsson, "A climbing image nudged elastic band method for finding saddle points and minimum energy paths," *The Journal of Chemical Physics*, vol. 113, pp. 9901–9904, 2000.

45. S. D. Schwartz, *Theoretical Methods in Condensed Phase Chemistry*, vol. 5, Springer Science & Business Media, Dordrecht, the Netherlands, 2002.

46. D. Sheppard, R. Terrell and G. Henkelman, "Optimization methods for finding minimum energy paths," *The Journal of Chemical Physics*, vol. 128, p. 134106, 2008.

47. E. H. Kisi, C. J. Howard and R. J. Hill, "Crystal structure of orthorhombic zirconia in partially stabilized zirconia," *Journal of the American Ceramic Society*, vol. 72, pp. 1757–1760, 1989.

48. L.-G. Liu, "New high pressure phases of ZrO2 and HfO2," *Journal of Physics and Chemistry of Solids*, vol. 41, pp. 331–334, 1980.

49. G. Teufer, "The crystal structure of tetragonal ZrO2," *Acta Crystallographica*, vol. 15, pp. 1187–1187, 1962.

50. G. Jomard, T. Petit, A. Pasturel, L. Magaud, G. Kresse and J. Hafner, "First-principles calculations to describe zirconia pseudopolymorphs," *Physical Review B*, vol. 59, p. 4044, 1999.

51. J. D. Kubicki, J. O. Sofo, A. A. Skelton and A. V. Bandura, "A new hypothesis for the dissolution mechanism of silicates," *The Journal of Physical Chemistry C*, vol. 116, pp. 17479–17491, 2012.

52. H. Manzano, E. Durgun, I. López-Arbeloa and J. C. Grossman, "Insight on tricalcium silicate hydration and dissolution mechanism from molecular simulations," *ACS Applied Materials & Interfaces*, vol. 7, pp. 14726–14733, 2015.

53. J. Godet and A. Pasquarello, "Proton diffusion mechanism in amorphous SiO2," *Physical Review Letters*, vol. 97, p. 155901, 2006.

54. T. S. S. Mahadevan and S. H. H. Garofalini, "Dissociative chemisorption of water onto silica surfaces and formation of hydronium ions," *Journal of Physical Chemistry C*, vol. 112, pp. 1507–1515, 2008.

55. M. S. Islam, R. A. Davies and J. D. Gale, "Proton migration and defect interactions in the CaZrO3 orthorhombic perovskite: A quantum mechanical study," *Chemistry of Materials*, vol. 13, pp. 2049–2055, 2001.

56. I. M. Markus, N. Adelstein, M. Asta and L. C. De Jonghe, "Ab Initio calculation of proton transport in DyPO4," *The Journal of Physical Chemistry C*, vol. 118, pp. 5073–5080, 2014.

57. K. Toyoura, N. Hatada, Y. Nose, I. Tanaka, K. Matsunaga and T. Uda, "Proton-conducting network in lanthanum orthophosphate," *The Journal of Physical Chemistry C*, vol. 116, pp. 19117–19124, 2012.

58. A. Pedone, G. Malavasi, M. C. Menziani, U. Segre and A. N. Cormack, "Molecular dynamics studies of stress-strain behavior of silica glass under a tensile load," *Chemistry of Materials*, vol. 20, pp. 4356–4366, 2008.

59. P. K. Gupta and C. R. Kurkjian, "Intrinsic failure and non-linear elastic behavior of glasses," *Journal of Non-Crystalline Solids*, vol. 351, pp. 2324–2328, 2005.

60. X. Zhao, D. Ceresoli and D. Vanderbilt, "Structural, electronic, and dielectric properties of amorphous ZrO2 from ab initio molecular dynamics," *Physical Review B*, vol. 71, p. 085107, 2005.

61. S. A. Sheikholeslam, W. Luo, C. Grecu, G. Xia and A. Ivanov, "Hydrogen diffusion in amorphous ZrO2," *Journal of Non-Crystalline Solids*, vol. 440, pp. 7–11, 2016.

62. A. C. van Duin, B. V. Merinov, S. S. Han, C. O. Dorso, & W. A. Goddard Iii, "ReaxFF reactive force field for the Y-doped BaZrO3 proton conductor with applications to diffusion rates for multigranular systems." *The Journal of Physical Chemistry A*, vol. 112, no. 45, p. 11414–11422, 2008.
63. M. Neumann, "Dipole moment fluctuation formulas in computer simulations of polar systems," *Molecular Physics*, vol. 50, no. 4, pp. 841–858, 1983.
64. M. B. Roberto Olmi, "Can molecular dynamics help in understanding dielectric phenomena?" *Measurement Science and Technology*, vol. 27, p. 014003, 2017.
65. R. J. S. Gabriele Raabe, "Molecular dynamics simulation of the dielectric constant of water: The effect of bond flexibility," *The Journal of Chemical Physics*, vol. 134, p. 234501, 2011.
66. G. M. Rignanese, F. Detraux, X. Gonze, A. Bongiorno, A. Pasquarello, "Dielectric constants of Zr silicates: a first-principles study," *Physical Review Letters*, vol. 89, no. 11, p. 117601, 2002.
67. K. D. Nielson, A. C. T. Duin, J. Oxgaard, W.-Q. Deng and W. A. Goddard, "Development of the ReaxFF reactive force field for describing transition metal catalyzed reactions, with application to the initial stages of the catalytic formation of carbon nanotubes," *The Journal of Physical Chemistry A*, vol. 109, pp. 493–499, 2005.

3

Tunneling Field Effect Transistors

Amir N. Hanna and Muhammad Mustafa Hussain

CONTENTS

3.1 Introduction

Tunneling field effect transistors (TFETs) offer interesting opportunities to address two major challenges faced by aggressively scaled conventional CMOS technology:

1. Scaling the supply voltage (V_{DD}).
2. Minimizing the leakage currents that degrade the I_{ON}/I_{OFF} switching ratio.

Both challenges aim at lowering power consumption in devices where we need more ultra-mobile computation capability [1]. As the transistor gate length is reduced, improving performance requires that both the supply voltage, V_{DD}, and the threshold voltage, V_T, be lowered to maintain a high overdrive factor ($V_{DD} - V_T$). However, by doing this, it exponentially increases the "OFF" state leakage current (I_{OFF}) due to a physical limitation commonly referred to as the 60 mV/dec subthreshold slope (SS) bottleneck. This is inherent in all current generation electronics that utilizes CMOS transistors with over-the-barrier charge transport physics. Additionally, transistor off-state power dissipation is considered to be semi-empirically proportional to [1]:

$$P \alpha I_{OFF} V_{DD}^3 \qquad (3.1)$$

To reduce power, reducing V_{DD} is absolutely critical, which in turn demands devices with steep SS for enabling faster turn on at low supply voltages. Contrary to classical MOSFETs, where charge carriers are thermally injected by lowering an energy barrier, the primary transport mechanism in a TFET is inter-band tunneling, where charge carriers transfer from one energy band into another at a heavily doped p^+/n^+ junction. In a TFET, inter-band tunneling can be switched "ON" and "OFF" abruptly by controlling the band bending in the channel region using gate-to-source bias, V_{gs}. This can be realized in a reverse-biased p-i-n structure, where asymmetric doping is used to suppress ambipolar transport [1]. While all-silicon TFETs have been studied rigorously using technology drive current boosters, such as the use of a high-κ gate dielectric, abrupt doping profiles at the tunnel junction, ultra-thin body, higher source doping, a double gate, a gate oxide aligned with the intrinsic region, and a shorter intrinsic region (and gate) length, I_{ON} of ~120 μA/μm have been achieved [2]. As for steep SS devices, sub 60mV/dec SS all-Si-based devices have been demonstrated by using dopant segregation followed by silicidation process, also known as the "green FET," as shown in Figure 3.1 [3]. The 46 mV/dec point SS was attributed to carrier tunneling through Schottky barrier [3].

However, the sub 60 mV/dec SS was only achieved for less than 3 decades of current, which does not fulfill requirement of sustaining an average sub 60 mV/dec SS for 5 decades of current, put forth in the literature [1], to have a reasonable I_{ON} at low V_{DD} values.

On the other hand, two-dimensional (2D) materials-based TFET have recently shown impressive results for subthermionic-limit conduction with an average SS of 31.1 millivolts per decade for over 4 decades of drain current at room temperature, as shown in Figure 3.2 [4]. This was achieved through using heavily doped Ge source and exfoliated bi-layer MoS_2 as the channel material. The use of the van der Waals bonded 2D materials allows for a strain-free hetero-interface between the heavily doped Ge source and MoS_2 channel; thus, allowing for the engineering of the band off-sets and the tunnel barrier width and height [4].

Also, the atomically thin 2D material intrinsically allows for excellent electrostatics, which made the fabrication of a vertical Ge/MoS2 hetero-structure feasible [4]. However, the need to use exfoliated 2D materials poses a challenge regarding the large-scale production of this technology. In addition, the large contact resistance, due to Schottky contact formation, of 2D materials poses a question on the scalability of such technology to smaller gate lengths [5].

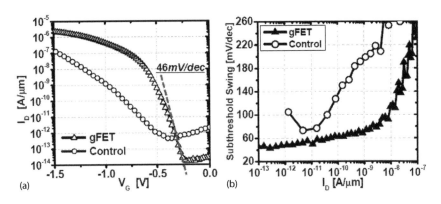

FIGURE 3.1
(a) I_D-V_G of measured and simulated green transistor FET showing SS of 46mV/dec. (L_G = 20 μm, V_{DS} = −1.0 V). (b) SS < 60 mV/dec over almost 3 decades of I_D for the green transistor FET, but not seen in the control TFET. (Reprinted with permission from Jeon, K. et al., *Symp. VLSI Tech. (VLSI)*, 121–122, 2010.)

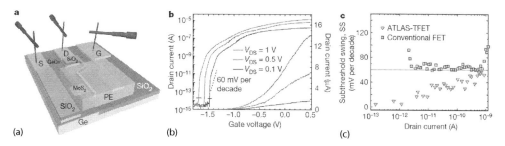

(a)   (b)   (c)

FIGURE 3.2

(a) Schematic diagram showing the probing configuration for measurement of the characteristics of the atomically thin and layered semiconducting-channel tunnel-FET (ATLAS-TFET). (b) Drain current as a function of gate voltage for three different drain voltages of 0.1 V, 0.5V and 1V. (c) SS as a function of drain current for an ATLAS-TFET (part c triangles) as well as a conventional MOSFET (part c squares) at V_{DS} = 0.5 V. The horizontal line in part c demarcates the fundamental lower limit of SS of conventional FETs. (Reprinted with permission from Sarkar, D. et al., *Nature*, 526, 91–95, 2015.)

This makes the use of low band gap, small effective mass, and CMOS-compatible materials, as the source charge injectors, a promising alternative to achieve high I_{ON} while scaling down V_{DD}. Here, the smaller effective mass of charge carriers increases the tunneling probability, according to the triangular Wentzel–Kramer–Brillouin (WKB) approximation, this will be discussed in detail in the Section 3.4 [6,7]. Potential source material candidates for N/PMOS TFETs are Germanium (Ge) and Indium Arsenide (InAs), respectively. Simulation studies using the above materials in a hetero-structure have corresponded to I_{ON} enhancements by factors > 400x and > 100x for the N- and P-type TFETs, respectively, over their all-Si planar counterparts for a planar single gate architecture [1]. Combining this with new and unique non-planar architectures opens up new opportunities for TFETs that are on-par with traditional Boltzmann transistors. For example, a recent demonstration of sub 60 mV/dec SS for almost 4 orders of magnitude and a minimum SS of 21 mV/dec shown in Figure 3.3 InAs Nanowires (NWs) based TFETs grown on Si(111) substrate [8].

Table 3.1 summarizes some recent state-of-the-art TFET employing group IV and III-V source materials that are integrated on Si and benchmarks their results against state-of-the

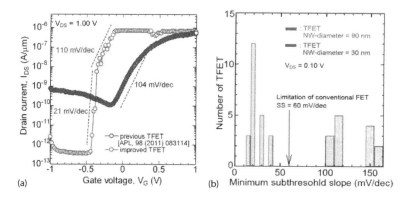

(a)   (b)

FIGURE 3.3

(a) Experimental transfer characteristics of optimized TFET with a NW-diameter of 30 nm (curve) V_{DS} = 1.00 V. (b) Distribution of minimum SS. The average SS for TFETs with NW-diameter of 30 nm is 25 mV/dec, which is much lower than the physical limit of SS in conventional MOSFETs. (Reprinted with permission from Tomioka, L. et al., *Symp. VLSI Tech. (VLSI)*, 47–48, 2012.)

TABLE 3.1

Recent State-of-the-Art TFET Employing Group IV and III-V Source Materials that are Integrated on Si Substrate

Affiliation	References	Year	Technology	Source	Channel	Smallest SS (mV/dec)	I_{on} (A) per NW	I_{on} (μA/μm)	V_{ds} (V)	I_{on}/I_{off}
UC Berkeley	[10]	2007	Planar	Si (100)	Si (100)	52.8	N/A	12	1	10^5
Stanford	[11]	2008	Planar	Ge (100)	Ge (100)	50–60	N/A	10	1	10^7
UC Berkeley/ SEMATECH	[12]	2009	Planar	Ge (100)	Si (100)	40	N/A	0.42	0.5	$>10^6$
SEMATECH	[3]	2010	Planar	Si (100)	Si (100)	46	N/A	1.2	−1	$\sim10^8$
NUS	[13]	2011	Vertical GAA NW	Si (100)	Si (100)	30	8×10^{-7}	1.2	−2	10^5
NUS	[14]	2011	Vertical GAA NW	Si (100)	Si (100)	30–50	1.5×10^{-9}	0.031	2	10^5
IBM	[15]	2011	Vertical GAA NW	InAs	Si (111)	220	10^{-7}	0.4	1	10^5
Peking University	[16]	2012	Planar	Si (100)	Si (100)	36	N/A	0.14	1	10^5
IBM	[17]	2012	Vertical GAA NW	InAs	Si (111)	150	Not reported	2.4	−0.5	10^6
Hokkaido Univ	[8]	2012	Vertical GAA NW	InAs	Si (111)	21/114	Not reported	~0.005	0.1	$>10^6$
Hokkaido Univ	[18]	2013	Vertical hetero NW	InAs	Si (111)	21	Not reported	500	1	$>10^4$
IBM	[9]	2015	Vertical GAA NW	InAs	Si (100)	160	Not reported	6	1	$\sim10^5$

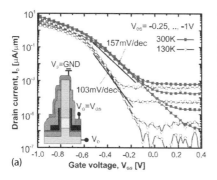
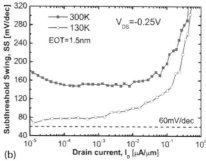

FIGURE 3.4

(a) I_D-V_{GS} at 300 K and 130 K of 100-nm-diameter GAA NW p-type TFET with EOT ~1.5 nm. The source (n-InAs) is grounded, while the drain (p-Si) is swept with −0.25 V bias step, see inset for biasing. Ion maximum of 6 µA/µm at $V_{GS} = V_{DS} = −1$ V is measured at 300 K. (b) SS versus ID for the device shown in (a), at 300 K and 130 K. At 130 K, the average SS is reduced to 75 mV/dec over the exponential tail. (Reprinted with permission from Cutaia, D. et al., *IEEE J. Electron Devices Soc.*, 3, 176–183, 2015.)

art results for all-Si and 2D based TFETs. The rest of the chapter would focus primarily on P-type TFETs based on InAs/Si heterojunction, which has shown a lot of promise recently for the high normalized I_{ON} and the possibility of growth on industry standard Si(100) substrate, as shown in Figure 3.4 [9].

3.2 Growth Techniques

Integration of III-V NWs on Si using a bottom-up approach could be achieved using both metal catalyzed growth and selective-area growth techniques. For the former technique, self-catalyzed NWs are grown using a vapor–liquid–solid technique using in-situ self-forming metal droplets. Both GaAs [19] and $GaAs_xSb_{1-x}$ [20] NWs were grown on Si (111) substrate using this technique.

As for the latter approach, CMOS compatible non-catalyzed selective-area growth is achieved by using lithographically-defined openings in a hard mask, typically SiO_2, for III-V NW growth. Recent experimental demonstration of InAs NWs on Si(111) has shown the possibility of growing sub 20 nm NW using selective-area metal-organic vapor epitaxy (SA-MOVPE) using SiO_2 as a hard mask [21]. Authors reported NW growth without misfit dislocation for diameters less than 20 nm, which is attributed to the reduction of strain field owing to the nanometer-scale footprint of NW [21]. The authors found that the hetero-structure with small diameter NW possesses fewer misfit dislocations, which quantitatively suppress trap-assisted tunneling via dislocation levels and has pure band-to-band tunneling as the dominant tunneling process. This showed the potential of wafer-scale production of high performance InAs/Si TFETs showing SS of less than 60 mV/dec for more than 4 orders of magnitude of drain current with high yield [16].

Another recent demonstration has shown the possibility of growing InAs NWs on the industry standard Si (100) using Metal Organic Chemical Vapor Deposition (MOCVD) [22]. The authors use SiO_2 nanotube templates to both guide the growth within the template, independent of substrate orientation, and to control the diameter of the grown NW [22]. NWs of diameters down to 25 nm were achieved, which demonstrated the scalability of this technique. This shows the potential path of growing high-quality CMOS-compatible III-V materials on Si (100) substrates.

3.3 Physics of Tunneling

Tunneling in TFET devices is governed by the inter-band tunneling rate across the tunneling barrier, which is typically calculated using WKB tunneling probability approximation [23]:

$$T_{WKB} \approx \exp\left(-\frac{4\lambda\sqrt{2m^*}\sqrt{E_g^3}}{3q\hbar\left(E_g + \Delta\Phi\right)} \right) \tag{3.2}$$

where m* is the effective mass, E_g is the band gap, λ is the screening tunneling length, and $\Delta\Phi$ is the potential difference between the source valence band and channel conduction bands. From this simple triangular approximation, we can see that the band gap (E_g), the effective carrier mass (m*), and the screening tunneling length (λ) should be minimized to increase the tunneling probability. A proper derivation of the WKB approximation will be shown Section 3.2. While E_g and m* are material dependent parameters, λ depends on other parameters, such as the device geometry, doping profiles, and gate capacitance. A small λ value would result in a strong modulation of the channel bands by the gate. It has been shown that the highest tunneling rate and, hence, lowest λ values were found for the gate-all-around (GAA) architecture for a 10 nm diameter nanowire, while ultra-thin body (UTB) double gate FETs has shown comparatively higher λ values. Planar UTBs have the highest λ values [24]. Because λ is also sensitive to gate capacitance, tunneling probability can also be enhanced by using high-κ gate dielectrics, as well as, small channel body thickness. Also, the abruptness of the doping profile at the tunnel junction is also important to control $\Delta\Phi$. In order to minimize the tunneling barrier, the high source doping level must fall off to the intrinsic channel in as short a width as possible. Typically, this requires a change in the doping concentration of about 4–5 orders of magnitude within a distance of only a few nanometers [1]. The following Section, 3.3.1, discussed the two governing modes of tunneling, namely point tunneling and vertical tunneling, and which one of them dominates for the InAs/Si TFET.

3.3.1 Point Tunneling versus Line Tunneling

In designing a TFET based on BTBT, there are two different tunneling modes that can be employed depending on the device structure. The first is lateral tunneling, also known as point tunneling, in which the electrons are injected from the source to the channel region in a direction parallel to the semiconductor/gate-dielectric interface, which is shown schematically by Figure 3.5a [25]. Point tunneling requires heavy doping of the TFET source to induce steep band bending of the source to channel junction, as dictated by $\Delta\Phi$ in equation (3.2). The source region should also perfectly align with the gate edge without gate-to-source overlap since gate-field induced depletion at the edge of the source region would result in degraded BTBT due to reduction in the abruptness of the energy band profile [25]. Also, gate-to-source underlap would also result in a degraded tunneling rate [26,27].

The other type of BTBT is vertical tunneling in which the electrons tunnel within the source region in the direction perpendicular to the semiconductor/gate-dielectric interface, as illustrated by Figure 3.5b. Similar to Gate-Induced Drain Leakage (GIDL) [28], the existence of a sufficient gate to source overlap region allows for the electrons to

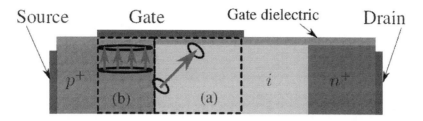

FIGURE 3.5

General nTFET configuration with the region of point tunneling (a) and line tunneling (b) schematically indicated. The direction of BTBT in the semiconductor is indicated by arrows and the regions of highest tunneling efficiency are circumscribed by an ellipsoid. (Reprinted with permission from Vandenberghe, W. et al., *Int. Conf. Simul. Semicond. Processes Devices*, 137–140 2008.)

be injected from within the source to the inverted surface region of the source. Unlike the requirement for point tunneling, the doping of the source must be moderate to allow for significant energy band bending within the source for electrons to tunnel from the conduction band to the valence band of the source material, or vice versa [25].

According to recent results in the literature, point tunneling was shown to dominate the BTBT current due to the small effective "tunnel gap" between heavily doped InAs source and the intrinsic silicon channel [29]. For the InAs/Si material system no experimental band offsets are available to date [29]; however, Anderson's rule gives a VB offset of 80 meV, as shown in Figure 3.6 [30]. This is also known as type II band alignment, or staggered band alignment, which is also used for is known to occur in the $In_xGa_{1-x}As/GaAs_{1-y}Sb_y$ system [31].

3.3.2 WKB Tunneling Probability Derivation

For Band-To-Band-Tunneling (BTBT) is a quantum mechanical phenomenon, where electrons tunnel through the forbidden gap of a semiconductor. It is experimentally evaluated, typically, in degenerately doped PN junctions, where negative differential resistance (NDR) behavior is observed [23].

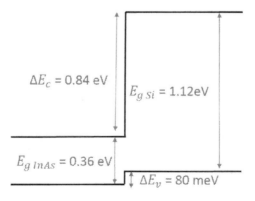

FIGURE 3.6

Band alignment at the InAs/Si hetero-interface. (From Schenk, A. et al., *ECS Trans.*, 66, 157–169, 2015; Anderson, R., *IBM J. Res. Dev.*, 4, 283–287, 1960.)

For tunneling to occur a PN junction, the following conditions should be met [23]:

1. Occupied energy states occur on one side of the tunnel barrier from which the electron would tunnel.
2. Unoccupied energy states at the same energy level on the other side.
3. Low tunneling potential barrier height and narrow barrier width to allow for a finite tunneling probability.
4. Conservation of momentum in the tunneling process.

While in classical mechanics an electron of energy E cannot enter into a region with a potential energy V > E because it would have to possess a negative kinetic energy. For example, for an electron traveling in the positive x-direction, which possess energy E > V, the wave function can be expressed as [32]:

$$\psi = Ae^{(-iKx)} \tag{3.3}$$

$$K = \sqrt{2m^* \frac{E-V}{h}} \tag{3.4}$$

Where A is the amplitude of the wave function, m^* is the electron effective mass, k is the wave vector, and h is Planck's constant. If E < V, which is a case for a particle with energy E incident upon a potential barrier with height V, then K becomes an imaginary number inside the potential barrier; thus, ik is real, such that [32]:

$$\psi = Ae^{(-Kx)} \tag{3.5}$$

$$|K| = \sqrt{2m^* \frac{V-E}{h}} \tag{3.6}$$

Thus, in quantum mechanics, the wave function of an electron with energy E, incident upon a potential barrier with height V, where E < V is represented as a wave with an attenuating amplitude.

In a degenerately doped PN junction, the tunnel probability across the potential barrier formed within the depletion region, as shown in Figure 3.1a, is given by the WKB approximation, which expresses the probability as [23,32]:

$$T_t \approx \exp\left(-2\int_0^{x_2} |k(x)| dx \right) \tag{3.7}$$

Where the limits are between the classical turning points, 0 and x_2, of the tunneling barrier shown schematically in Figure 3.1b, which are the points at which the particle's energy E equals the barrier's potential energy V, and $|k(x)|$ is the absolute value of the wave vector of the electron inside the barrier. $E_{Fn}-E_C$ and E_v-E_{FP}, shown in Figure 3.7a, quantify the amount of degeneracy on the n and p sides, respectively, due to the high doping levels on both sides [23]. The tunnel barrier for this case has a triangular shape, as shown in Figure 3.7b. Thus, the problem reduces to tunneling across a triangular

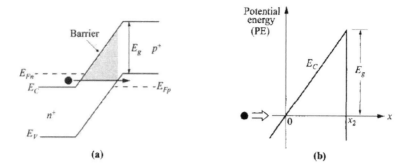

FIGURE 3.7
(a) Tunneling in a tunnel diode can be analyzed by (b) a triangular potential barrier. (Reprinted with permission from Sze, S., *Physics of Semiconductor Devices*, Chapter 8 "Tunnel Devices," Wiley, New York, 1981.)

barrier of width x_2 and height equal to the bandgap of the material, E_g, as shown in Figure 3.7a and b [23].

The E-k dispersion relationship for electron inside the triangular barrier, in Figure 3.7b, is given by [23]:

$$k(x) = \sqrt{\frac{2m^*(PE - E_c)}{h}} \tag{3.8}$$

Where m^* is the electron effective mass, PE is the potential energy of the incoming electron, E_c is the conduction band edge, and h is Planck's constant. Since the incoming electron has a PE equal to the bottom of the energy gap, the left-hand side of Figure 3.7a, which is lower than the conduction band edge, E_c, values for the triangular barrier and makes k an imaginary number. Also, since we can express the conduction-band edge, E_c, in terms of the applied electric field, ε, in the tunneling direction, the dispersion relationship becomes [23]:

$$k(x) = \sqrt{\frac{2m^*(-q\varepsilon)}{h^2}} \tag{3.9}$$

Where q is the electron charge.

Substituting (3.9) into (3.7) gives [23]:

$$T_t \approx \exp\left(-2\int_0^{x_2}\sqrt{\frac{2m^*(q\varepsilon x)}{h^2}}\, dx\right) \tag{3.10}$$

For a triangular barrier in a uniform field, x_2 can be expressed as $x_2 = E_g/\varepsilon q$, and thus the tunneling probability becomes [23]:

$$T_t \approx \exp\left(-\frac{4\sqrt{2m^*}\, E_g^{3/2}}{3qh\varepsilon}\right) \tag{3.11}$$

This is the expression for BTBT tunneling probability under uniform field limit, which is very useful for calibrating experimental data as will be shown in the Section 3.3.

3.3.3 Direct and Indirect Tunneling

When considering BTBT tunneling in semiconductors, and since tunneling requires conservation of mementum, we have to distinguish between two cases [23]:

1. Direct tunneling case, where the both the initial and final states have the same momentum, i.e.: $K_f = K_i$.
2. Indirect tunneling case, where the final state has a different momentum than the initial state, i.e.: $K_f \neq K_i$.

Figure 3.8a and b show E-K relationship superimposed on the band diagram, for the direct, Figure 3.8a, and indirect tunneling, Figure 3.8b, respectively [23].

Figure 3.8a shows direct tunneling case where electrons tunnel from the vicinity of the conduction-band minimum to the vicinity of the valence-band maximum, without a change of momentum, $k_i = k_f$. In other words, for direct tunneling to occur, the conduction-band minimum and the valence-band maximum must have the same momentum. This is typically fulfilled by direct bandgap semiconductors, such as InAs and GaAs [23]. Figure 3.8b shows the indirect tunneling case, where the conduction band minimum does not align with valence band maximum in the E-K diagram. This is the case for indirect band gap materials, such as Si and Ge, shown schematically in Figure 3.8b as a π/a difference between the direct and indirect band gaps [23].

For this case, in order to conserve momentum, the difference in momentum between the initial and final states must be compensated by a scattering agent, such as lattice vibration (phonons) or impurities [23]. For the case of phonon-assisted tunneling, the sum of the initial electron momentum, and the phonon momentum, is equal to the final electron momentum after it has tunneled, i.e: $k_f = k_i + k_{phonon}$. For indirect tunneling, conservation of energy also requires that the sum of the phonon energy and the initial electron energy is equal to the final electron energy after it has tunneled, i.e.: $E_f = E_i + E_{phonon}$.

However, to obtain a closed form solution for the BTBT tunneling rate, G_{BTBT}, from the tunneling probability expression in (3.7), the following assumptions have to be made [33]:

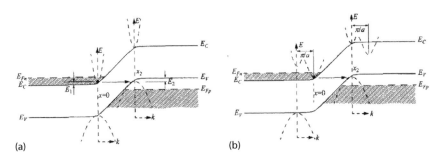

(a)   (b)

FIGURE 3.8
Direct and indirect tunneling processes demonstrated by E-k relationship superimposed on the classical turning points (x = 0 and x_2) of the tunnel junction. (a) Direct tunneling process with $k_{min} = k_{max}$. (b) Indirect tunneling process with $k_{min} \neq k_{max}$. (Reprinted with permission from Sze, S., *Physics of Semiconductor Devices*, Chapter 8 "Tunnel Devices," Wiley, New York, 1981.)

1. Assuming a direct band-gap semiconductor, ignoring phonon-scattering contribution, and thus conserving momentum along the tunneling path; that is: $k_f = k_i$.

2. Assuming that the electron's and hole's quasi Fermi levels are equal to the conduction band and valence band edges, respectively; that is: $E_{Fn} = E_c$ and $E_{Fp} = E_v$.

3. Assuming that the valence band states are completely occupied by electrons and conduction band states completely empty so that the carrier statistics (Fermi distribution) are simplified (i.e., $f_V - f_C = 1$).

4. Assume a constant electric field, ε, across the p-n junction in the tunneling direction.

5. Assume symmetric 2-band relation to model the imaginary wave vector dispersion relation within the band gap.

Under these assumptions, G_{BTBT} can evaluated by [33]:

$$G_{BTBT} = \frac{qm^*}{2\pi^2 h^3}\left(\frac{1}{q}\frac{\partial E_x}{\partial x}\right)\int_0^\infty T_t\left(E_x, E_T\right)dE_T \qquad (3.12)$$

Where, $E_x = q\varepsilon x$ is the electron energy component due to the momentum in the tunneling direction, and E_T is the energy associated with transverse momentum, i.e.: $E = E_x + E_T$. Under the above assumptions, G_{BTBT} can be written as [33]:

$$G_{BTBT} = \frac{\sqrt{2m^*}\,q^2}{2\pi^3 h^2 \sqrt{E_g}}\varepsilon^2 \exp\left(-\frac{\pi\sqrt{m^*}\,E_g^{3/2}}{2\sqrt{2}\,q\varepsilon h}\right) = A\varepsilon^2\exp(-\frac{B}{\varepsilon}) \qquad (3.13)$$

Where,

$$A = \frac{\sqrt{2m^*}\,q^2}{2\pi^3 h^2 \sqrt{E_g}}\ , B = \frac{\pi\sqrt{m^*}\,E_g^{3/2}}{2\sqrt{2}\,qh}$$

A and B are the tunneling parameters and they are typically used for calibration of Technology Computer Aided Design (TCAD) models, such as the widely used dynamic nonlocal BTBT model, to experimental data [34]. The final sheds light on why using a small E_g and m^* materials could increase the tunneling probability and G_{BTBT}. It also shows the dependency of G_{BTBT} on the electric field in the tuneling direction, ε.

3.3.4 TCAD Models

The most commonly used model in TFET literature is the dynamic nonlocal BTBT model [34].

The model assumes that tunneling occurs between two parabolic bands, representing dispersion at the bottom of conduction and top of valence bands, respectively [35]. As for the magnitude of the imaginary wavevector, k, used for calculation of the tunneling path length, it is obtained from Kane's two-band model dispersion relation, which is a "simple two-band model capable of including one conduction band and one valence band and it is formulated as two coupled Schrodinger-like equations for the conduction-band and valence-band envelope functions. The coupling term is treated by the **k·p** perturbation method, which gives the solutions of the single electron Schrodinger equation in the neighborhood of the bottom of the conduction band and the top of the valence bands, where most of the electrons and holes, respectively, are concentrated" [35]. The expression for the tunneling probability used in the dynamic nonlocal BTBT was derived by Kane using time-dependent perturbation theory and Fermi's Golden Rule. Since BTBT can take place between either heavy hole (HH) band and conduction band (CB) or between light hole (LH) band and CB, and since the effective mass of holes in the HH band is much larger compared to that in the LH band, tunneling from the HH band is completely suppressed [29]. Therefore, it is neglected in the simulations.

The generation rate is obtained from the nonlocal path integration, and electrons and holes are generated non-locally at the ends of the tunneling path. The tunnel path can belong to either direct OR phonon-assisted band-to-band tunneling processes. The model has the following assumptions [34]:

1. The tunneling path starts from the valence band in a region where the nonlocal path model is active.

2. The tunneling path is a straight line with its direction opposite to the gradient of the valence band at the starting position.

3. The tunneling path ends at the conduction band.

4. The tunneling energy is equal to the valence band energy at the starting position and is equal to the conduction band energy plus band offset at the ending position.

The Kane model has been used in calibration of experimental results for hetero-structure TFETs for both direct [36] and indirect band gap [12] source materials. The model have been used to simulate Si/InAs hetero-structure devices in the literature [6,7,37].

3.3.5 BTBT Model in InAs/Si Hetero-Interfaces

Since a tunnel path in dynamic nonlocal BTBT model in hetero-interface can either belong to a direct (zero-phonon) or to a phonon-assisted tunnel process, combinations are excluded in accordance with the missing theoretical understanding of the case of tunneling from a direct band-gap to an indirect band-gap semiconductor [37].

So, as in TCAD simulation literature of InAs/Si hetero-interface, there are two possible workarounds [6,7,37]:

1. The Kane model for direct material is also used on the silicon side after fitting it to experimental data in the literature [38] and using calibrated data for InAs. The A, B parameters for InAs have been calculated according to [36], where the pre-exponential factor A was used as a fitting parameter, and B was in agreement for experimental data of InAs diode and TFET.

2. Use the calibrated model for Si [39] based on indirect tunneling and apply it to the InAs side, but set the phonon energy to a very small value and adjust the pre-exponential factor for the best fit to other techniques, such as Non-Equilibrium Green's Function (NEGF) based quantum transport characteristics [6,7,37].

For model calibration and under uniform field limit, the A and B parameters are calculated according to [34]:

$$R_{net} = A \left(\frac{F}{F_0} \right)^P \exp\left(-\frac{B}{F} \right). \tag{3.14}$$

Where $F_0 = 1$ and F is applied electric field, P = 2.5 for indirect tunneling process [34].

3.3.6 Midgap States Trap-Assisted Tunneling

Trap-Assisted Tunneling (TAT), due to midgap states, can be included into TCAD Sentaurus using the Hurkx Model, since the silicon recombination parameters have been calibrated to this model according to the paper by G.A. Hurkx et al. [39], according to Sentaurus manual [34]. This model is typically used to account for high electric fields that are typically in excess of 3×10^5 V/cm at reverse biased PN junctions, which are extremely sensitive to defect-assisted tunneling, which causes electron–hole pair generation before band-to-band tunneling sets in. The model reduces to SRH recombination rate under low electric fields.

The Hurkx model TAT recombination rate is expressed as [34]:

$$R_{trap} = \frac{pn - n_{ie}^2}{\dfrac{\tau_p}{1+\Gamma_p} \left[n + n_{ie} \exp\left(\dfrac{\bar{E}}{kT} \right) \right] + \dfrac{\tau_n}{1+\Gamma_n} \left[p + n_{ie} \exp\left(\dfrac{-\bar{E}}{kT} \right) \right]} \tag{3.15}$$

Where, $\bar{E} = E_T - E_i$, and E_T and E_i are the trap level and intrinsic level, respectively. τ_p and τ_n are the hole and electron recombination lifetimes, and n_{ie} is the intrinsic carrier concentration. $\Gamma_{n,p}$ is a field effect function related to the electron/hole trap emission through

$\Gamma_n = \dfrac{n_t}{n_0} - 1$, $\Gamma_p = \dfrac{p_t}{p_0} - 1$, where n_0 and p_0 are equilibrium electron and hole carrier concentrations, $\Gamma_{n,p}$ is equal to

$$\Gamma_{n,p} = \frac{\Delta E_{n,p}}{kT} \int_0^1 \exp\left(\frac{\Delta E_{n,p}}{kT} u - K_{n,p} u^{1.5} \right) du \tag{3.16}$$

And

$$K_{n,p} = \frac{4}{3} \frac{\sqrt{2m^* \Delta E_{n,p}^3}}{qh|F|} \tag{3.17}$$

Where F is the local electric field, and $\Delta E_{n,p}$ are the integration intervals

So, if $\bar{E} = E_T - E_i = 0$ for intrinsic level the expression would reduce to the Shockley-Read-Hall (SRH) recombination formula.

3.3.7 Metal-Semiconductor Interfacial States Trap-Assisted Tunneling

Simulations were performed to test the sensitivity of the TFET transfer characteristics to interface traps. A density-of-states of interface traps (D_{it}) of 2×10^{13} cm^{-2}eV^{-1} was used [29]. The energetic trap distribution in the band gap was assumed to be uniform.

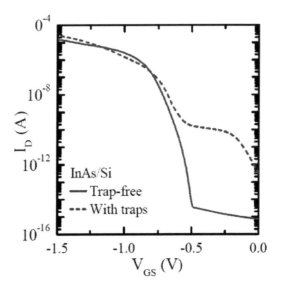

FIGURE 3.9
InAs/Si pTFETs with and without traps at the semiconductor interface. A uniform trap distribution was assumed in the band gap InAs and the D_{it} was set to 2×10^{13} cm^{-2}eV^{-1}. (Reprinted with permission from Schenk, A. et al., *ECS Trans.*, 66, 157–169, 2015.)

The "Dynamic nonlocal path TAT model" in the SDevice was activated in the simulations [34]. TAT acts as an additional electron-hole pair generation mechanism in TFETs which dominates conduction prior to BTBT due to the lower tunnel barrier, as shown in Figure 3.9, and is responsible for the degraded SS demonstrated in InAs/Si TFETs literature [14,15,18]. The small tunnel gap at the InAs/Si interface favors "point tunneling." Therefore, it is obvious that point tunneling (in which the tunnel paths cross the interface) is more sensitive to interface traps than line tunneling.

3.3.8 SRH Recombination for Si/InAs Interface

In addition to the BTBT model, SRH generation is typically used to account for the thermal generation current ("leakage") in the off-state. The expression for SRH recombination is similar to equation (3.15), and the life time of both the electron and holes were chosen to be equal to 1 n sec, i.e.: $\tau_p = \tau_n = 1$ n sec, similar to other reports in the literature [29]. SRH contribution to total current will be discussed in more details in Section 4.3.

3.4 Effect of Transistor Architecture on TFET Performance

GAA NW architecture is believed to be the best architecture for vertical NW TFET since it provides the best electrostatic control and the smallest λ according to equation (3.2) [24]. However, the need for scaled NW diameter, the wide NW pitch needed for fabrication of arrays of NW, typically in the range of 400 nm, and the variability in device parameters, such as the threshold voltage, of the individual NW TFETs within the array impose a restriction on the current per unit chip area that could be achieved using vertical GAA NW architecture [40]. Thus, an alternative architecture for vertical TFETs has been recently proposed, namely the nanotube architecture, which mimics the gate all-around nanowire (GAA NW) devices by having an outer (shell) gate, as well as, an inner (core) gate inside the nanowire making it a hollow cylindrical structure. When compared to arrays of nanowires, the nanotube architecture outperforms in terms of drive current capability, CV/I metric (i.e., intrinsic gate delay), power consumption, and area efficiency [41–44].

In this section, a hetero-structure Si/InAs p-channel TFET device concept is presented that combines the advantages of a low band-gap InAs source injector and inherent high drive current advantage in NTFET (Figure 3.10). The nanotube TFET's excellent electrostatic control enables steep turn on characteristics, while maintaining low I_{OFF} values comparable to NW TFET. This transistor architecture in conjunction with a low band gap source injector enables a higher inter-band tunneling rate, when compared to all-silicon TFET structure [42].

3.4.1 TCAD Models Used

To study the benefits of a nanotube architecture over a nanowire on a hetero-structure Si/InAs TFET platform, 3D simulations of a nanotube (NT) (Figure 3.10) and GAA NW TFET using Synopsys™ using the dynamic nonlocal path BTBT model [36,37]. For this case, indirect BTBT model was assumed as Si is an indirect bandgap material. Both devices are compared for a gate length (L_g) of 20 nm. Silicon drain is p-doped with acceptor active concentration $N_A = 1 \times 10^{20}$ cm^{-3}, while an intrinsic channel is used. The InAs source was used with n-doping with donor active concentration $N_D = 1 \times 10^{18}$ cm^{-3}, both typical to the previously demonstrated device [38]. Both the nanotube thickness and nanowire diameter are kept at 10 nm to allow for a fair comparison of both architectures. The gate metal in both devices has a work function of 4.53 eV, and a nitride gate dielectric is assumed with an effective oxide thickness (EOT) of 0.5 nm. A dynamic nonlocal band-to-band tunneling model is utilized in conjunction with SRH recombination and drift–diffusion physics. The BTBT parameters, "A" and "B" for silicon are 4×10^{14} cm^{-3} s^{-1} and 1.9×10^7 V cm^{-1}, respectively [34,39]. While for InAs, the BTBT parameters were taken as 9×10^{19} cm^{-3} s^{-1} and 1.3×10^6 V cm^{-1}, respectively [36].

This comparative simulation study does not include any gate overlap with the source, assumes an ideal interface with no defects due to strain, and takes into account

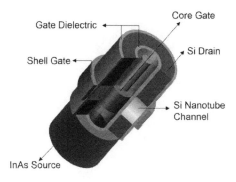

FIGURE 3.10
Schematic of the Nanotube (NT) architecture. (Reprinted with permission from Hanna, A. et al., *Sci. Rep.*, 5, 9843, 2015.)

82 *Energy Efficient Computing & Electronics*

trap-assisted tunneling due to dopants induced defect levels. However, it does not take into account bands offset due to strain, quantum confinement effects, and multiple valley BTBT effects. The valence band and conduction band dispersion relationship are assumed to be bulk-like. This simulation framework has been previously utilized in the literature for simulating tunneling in hetero-structures for both Si/InAs Esaki diodes and TFETs [36,37]. The nanotube TFET has a silicon channel thickness of 10 nm and an inner core-gate diameter (CG_{dia}) of 100 nm. All contacts are assumed to be Ohmic with zero contact resistance.

3.4.2 Results

Figure 3.11 shows the energy band diagram of the simulated p-channel nanotube TFET. In the "ON" state (i.e., $V_{gs} = V_{ds} = -1V$), electrons tunnel from the Si (channel) valence band to InAs(source) conduction band due to the small width of the tunnel barrier, thus generating holes in the channel's valence band that, under the lateral drain's electric field, gets swept toward the drain contact. In the "OFF" state, the barrier for tunneling widens, thus suppressing inter-band tunneling, and only the SRH generation/recombination and trap-assisted tunneling terms contribute to the total current.

Figure 3.12a compares the normalized I_{ds}–V_{gs} characteristics of a 10 nm thin NTFET (with 100 nm inner core-gate diameter, CG_{dia}) and 10 nm diameter NWFET. We have used the NW circumference (π d), where d is the NW diameter, and, in the case of the NT, we have used average circumference $\left(\pi \times (CG_{dia} + NT_w)\right)$, where CG_{dia} and NT_w are the nanotube core-gate diameter and thickness, respectively. As it can be seen the nanotube architecture has 5x higher normalized current output than that of the GAA NW architecture. The NT TFET also has 3x lower I_{OFF} compared to the NW TFET. Both architectures provide I_{ON}/I_{OFF} of more than 10^5. Figure 3.12b compares the SS values of the 10 nm NT and NW FETs. Both architectures show SS values less than 60 mV/dec over 5 decades. However, the NW architecture TFET shows lower point SS values as low as

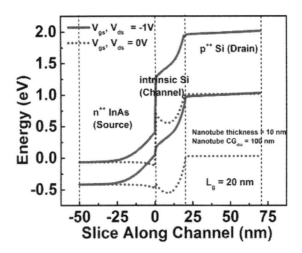

FIGURE 3.11
Band diagram of the P-type nanotube architecture TFET showing both ON state OFF states. (Reprinted with permission from Hanna, A. et al., *Sci. Rep.*, 5, 9843, 2015.)

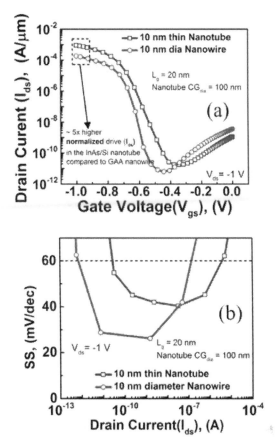

FIGURE 3.12
(a) Normalized I_{ds}-V_{gs} characteristics of a 10 nm diameter NW and 10 nm thin NT p-channel TFETs and (b) Subthreshold Slope (SS) for the NT and NW TFET showing sub 60 mV/dec for more than 5 orders of magnitude of current. (Reprinted with permission from Hanna, A. et al., *Sci. Rep.*, 5, 9843, 2015.)

25 mV/dec, while the lowest SS values for the NTFET is ~40 mV/dec. We have shown before that this is due to the ultimate electrostatic control in the GAA architecture TFET [24,42]. The ambipolar conduction could be suppressed by using a high work function p++ gate, as shown in Figure 3.13 for the NW TFET, thus allowing V_{DD} scaling. However, we chose silicon midgap work function in our comparison for the sake of generality.

However, when since current per unit chip area is a valuable metric for circuit design, the non-normalized current of both architectures was compared in Figure 3.14. We have also compared the non-normalized "ON" current of the 10, 20, 30 nm NW TFETs with the 10 nm NT TFET in Figure 3.14a. The 10 nm NT TFET shows a non-normalized I_{ON} ~ 0.32 mA, while the NW TFETs show I_{ON} ~ 5.9×10^{-6} A, 2.51×10^{-5} A, 5.31×10^{-5} A for the 10 nm, 20 nm, and 30 nm NWs, respectively. This means that the 10 nm NTFET shows **54X, 13X**, and **6X** increase in drive current over that of the 10 nm, 20 nm, and 30 nm diameter NW TFETs. As for SS, Figure 3.14b shows that only 10 nm and 20 nm NW TFET and the 10 nm NT TFET can achieve sub 60 mV/dec SS. So, having a small diameter NW is essential to maintaining a low SS.

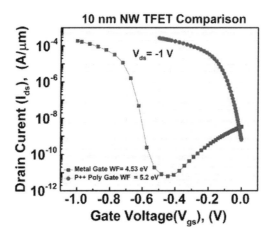

FIGURE 3.13
Comparison of two different metal work functions, 4.53 and 5.2 eV, showing the possibility of both suppressing ambipolar transport and V_{DD} scaling.

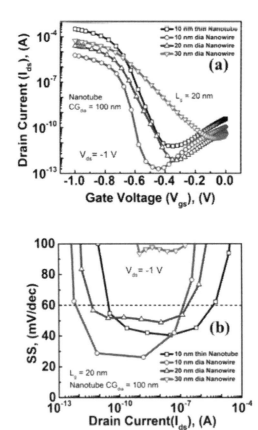

FIGURE 3.14
(a) Transfer (I_{ds}-V_{gs}) characteristics of NWs of 10, 20, 30, 40, and 50 nm diameter and 10 nm NT, and (b) SS comparison between 10 nm thick NT and 10, 20, and 30 nm diameter NW TFETs. (Reprinted with permission from Hanna, A. et al., *Sci. Rep.*, 5, 9843, 2015.)

3.4.3 Discussion

To supply high drive current while maintaining small SS, the arraying of small diameter NWs is inevitable. However, this would come at the expense of chip area and "OFF" state leakage, as will be seen the following sections. Additionally, the low I_{OFF} current characteristic of the TFET would be lost as arraying would cause, at the least, multiplying the "OFF" state leakage current by the number of NWs in the array needed to supply the same "ON" current as one NT. So, in the case of 54 NW array of 10 nm diameter NWs, the leakage would be at least 13X higher compared to a single 10 nm NT. Finally, although it could be argued that arraying could boost the "ON" current value, sensitivity of parameters like threshold voltage to, for example, variations in NW width could lead to degradation of the SS swing for a large array of devices [40,45,46]. Recent demonstrations of sub 60 mV/dec of have been for all silicon single NW N- and P-type TFET of diameter < 20 nm, which supplies a maximum "ON" current in the nA regime and a normalized "ON" current of 1.2 µA/µm [13,14]. However, when an array of TFETs is tested, sub 60 mV/dec SS have been demonstrated for currents as low as 0.01 µA/µm [46]. Also, degraded SS slope was noticed for higher drain current giving a maximum $I_{ON} = 64$ µA/µm at $V_{DD} = 1.0$ V and at a higher gate bias $V_{GS} = 2$ V. On the other hand, a SS of 52 mV/dec have been shown at an even higher $I_{ON} = 100$ µA/µm at $V_{DD} = 1.0$ V and $V_{gs} = 1$ V for all-silicon single gated SOI-based TFETs with vertical self-aligned top gate structure supplying and for 70 nm thick SOI with 2 nm EOT [47]. Even higher drain currents have been shown for double gate strained-Ge hetero-structure TFET with a drive current of 300 µA/µm at a SS of 50 mV/dec [11]. That is why the NT architecture could be an excellent candidate for a vertical structure that resembles double gate structure and provide a higher integration density compared to the NW structure.

To consider the scalability of the nanotube architecture, we studied the non-normalized drain current as a function of the inner core-gate diameter as shown in Figure 3.15a. An important observation here is that the core-gate contact scaling tunes the on-state drive performance without compromising the subthreshold swing, as can be seen from Figure 3.15b. This becomes a competitive technology option compared to GAA NWFETs. To fully comprehend this concept, consider the top-down plan-view chip layout perspective of a single nanotube and an array of GAA nanowires in Figure 3.16a, p. 87. In vertical GAA nanowire technology, maintaining a small nanowire pitch (NW_{pitch}) ensures high integration density and drivability. In order to compete with this, the core-gate diameter (CG_{dia}) of the nanotube should be highly scalable just like the nanowire pitch. One pragmatic approach to investigate this is by studying the device dimension scalability effects on chip area. Using the ITRS overall roadmap technology characteristics (ORTC), scalable parameters for current generation FinFET technology are adapted here, which are summarized in Table 3.2.

In this comparison, the nanowire pitch is assumed to be equal to the fin half-pitch, and the nanotube core-gate diameter is assumed to be equal to contact/via size as specified in the 2013 ORTC target for future technology nodes. The contact/via size scaling is considered equal to the M1 metal half-pitch. The other parameters in the comparison are the nanowire diameter (NW_{dia}) and the nanotube thickness (NT_w), both of which are assumed equal to the fin width target given in Table 3.2. From Figure 3.15a, a 100 nm core-gate diameter, 10 nm thin InAs/Si nanotube has greater than 10x higher non-normalized drive capability compared to a single 10 nm diameter GAA InAs/Si nanowire. To achieve similar performance levels as the nanotube, more than 10x nanowires need to be stacked in an array. Using the above parameters and assuming that the normalized drive current scales linearly with channel thickness (NT_w and NW_{dia}) in both the nanotube and nanowire architecture, chip-area estimates of the single nanotube and a 10x nanowire array are

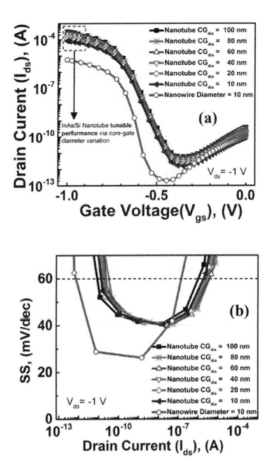

FIGURE 3.15
Non-normalized NT drive current (a) and SS (b) as a function of inner core-gate diameter, reprinted with permission from [42].

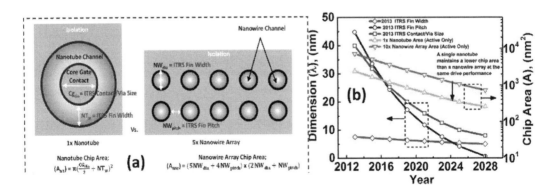

FIGURE 3.16
(a) Illustrated top-down plan-view comparison of between and (b) chip area comparison using ITRS predicted parameters between a single nanotube device and an array of 10x GAA nanowires. (Reprinted with permission from Hanna, A. et al., *Sci. Rep.*, 5, 9843, 2015.)

TABLE 3.2

2013 ITRS ORTC FinFET Scaling Parameters

Technology Node	Year	MPU/ASIC M1 ½-Pitch (nm)	Fin ½-Pitch (nm)	Fin Width (nm)
16/14	2013	40	30	7.6
10	2015	31.8	24	7.2
7	2017	25.3	19	6.8
5	2019	20	15	6.4
3.5	2021	15.9	12	6.1
2.5	2023	12.6	9.5	5.7
1.8	2025	10	7.5	5.4

carried out at different technology nodes using simple analytical calculations. As it can be seen from Figure 3.16b, the contact/via scaling at the moment is comparable to the nanowire-pitch. But at around the 7 nm technology node (2017), the pitch scaling will start to become more aggressive. However, even after this and considering the fact that large nanowire arrays are required to achieve similar drivability as a single nanotube, the InAs/Si nanotube architecture will outperform InAs/Si GAA nanowire arrays at the extreme scaling limit in terms of chip area consumed.

3.4.4 BTBT and SRH Color Map Comparison

Cross-sectional color maps of BTBT generation rates and SRH Recombination rates of the 10 nm NT, 10 nm NW, and 20 nm NW TFETs are shown in Figures 3.17 through 3.19, respectively.

In order to compare the BTBT generation and SRH recombination rates of the various devices, measurements along X-marked lines in Figures 3.17 through 3.20 at the middle of the channel were plotted in Figure 3.20a and b on p. 89. Figure 3.20a compared the BTBT

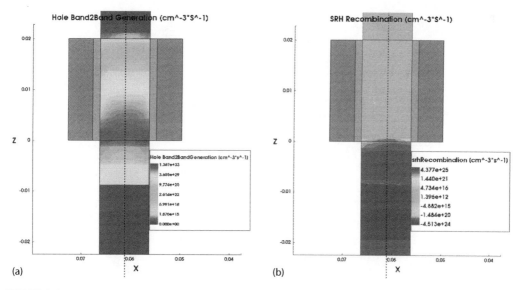

FIGURE 3.17
Color maps of 10 nm NT TFET (a) Hole Band-to-Band Generation rate, and (b) SRH Recombination rate. X sections are indicated where the BTBT generation and SRH recombination rates are measured in Figure 3.20. (Reprinted with permission from Hanna, A. et al., *Sci. Rep.*, 5, 9843, 2015.)

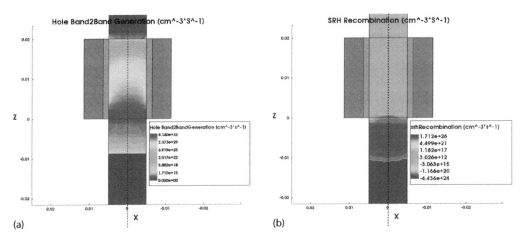

FIGURE 3.18
Color maps of 10 nm NW TFET (a) Hole Band-to-Band Generation rate, and (b) SRH Recombination rate. X sections are indicated where the BTBT generation and SRH recombination rates are measured in Figure 3.20. (Reprinted with permission from Hanna, A. et al., *Sci. Rep.*, 5, 9843, 2015.)

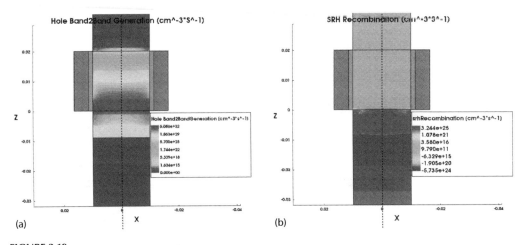

FIGURE 3.19
Color maps of 20 nm NW TFET (a) Hole BTBT Generation rate, and (b) SRH Recombination rate. X sections are indicated where the BTB generation and SRH recombination rates are measured in Figure 3.20. (Reprinted with permission from Hanna, A. et al., *Sci. Rep.*, 5, 9843, 2015.)

generation rate and showed that the 10 nm diameter NW has slightly higher peak BTBT generation of 1.6×10^{32} (cm^{-3}s^{-1}), when compared to the 10 nm thick nanotube, showing ~1.5×10^{32} (cm^{-3}s^{-1}), due to the shorter tunneling length, λ, for the nanowire architecture. Similar peak BTBT generation rate is expected as it has been theoretically shown that the differences in the scaling tunneling length, λ, between the GAA and double gate architectures, which resembles the NT Core-Shell gates architecture, reduces for body thickness ≤ 10 nm and almost diminishes for body thickness of 5 nm [24]. Thus, it shows the potential of the NT architecture at scaled body thicknesses to outperform the GAA NW architecture.

On the other hand, the larger diameter nanowire (20 nm) shows lower peak tunneling rate of 5.53×10^{31} (cm^{-3}s^{-1}) due to larger body thickness leading to higher tunneling length, λ, values. However, one major difference between the two architectures is the distance

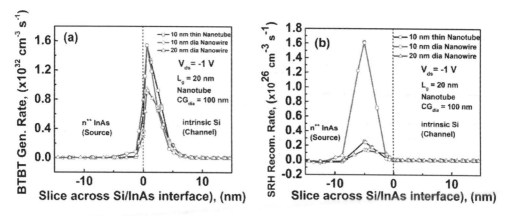

FIGURE 3.20
(a) BTBT Generation Rate and (b) SRH Recombination Rate for the 10 nm NT and 10, 20 nm NWs as function of the distance from the Si/InAs interface. (Reprinted with permission from Hanna, A. et al., *Sci. Rep.*, 5, 9843, 2015.)

over which BTBT generation is significant away from the Si/InAs interface. For the case of the NT, we observe that BTBT generation rate is higher over a larger distance, almost 7 nm into the Silicon channel, when compared to both the 10 nm and 20 nm NW, as shown in Figure 3.20a. This is an indication of higher lateral tunneling across the Si-InAs interface for the NT architecture when compared to the NW architecture, which could explain the higher normalized current, as lateral (point) tunneling is the dominant tunneling mechanism for InAs/Si band diagram as mentioned in Section 3.1 [29].

However, the 10 nm NW TFET is shows higher BTBT generation rate when moving away from the interface into the InAs source, indicating higher vertical tunneling within the source, when compared to both the 10 nm NT and 20 nm NW TFETs, as shown from Figure 3.20a. Although higher vertical tunneling is desired in hetero-structure TFETs for increasing the drive current [12], it could also lead to higher SRH recombination in the small direct bandgap InAs source. When analyzing the SRH recombination rate in Figure 3.20b, it was found that the peak SRH recombination rate for the 10 nm NW is almost an order of magnitude higher than the 10 nm NT TFET, as well as, the 20 nm NW TFET. The high SRH recombination rate for the 10 nm GAA NW TFET could explain the higher normalized I_{OFF} observed in Figure 3.12a for the NW versus the NT TFETs when compared at a body thickness of 10 nm.

To get a quantitative sense of the effect of BTBT generation and SRH recombination rates on the drive current, we analyzed the integrated area under the curve in (Figure 3.20a and b). For the 10 nm NT BTBT generation curve, the area under the curve was found to be 8.4% higher than that of the 10 nm NW curve, and 45% higher than that of the 20 nm NW curve. This can partially account for the larger non-normalized current seen for 10 NT TFET, 54X, compared to that of the 10 nm nanowire TFET. The nanotube has larger available cross-sectional area for tunneling, when compared to the 10 nm nanowire. Since a 100 nm core-gate diameter nanotube with 10 nm thickness has 44x the cross-sectional area of a 10 nm nanowire. Doing a back-of-the-envelope multiplication, the extra 8.4% in volumetric band-band generation by the additional cross-sectional area gives 47.7X the anticipated increase in the current. The area under the SRH recombination curve for the 10 nm NW 8.7X of that of the 10 nm NT, which could explain the higher I_{OFF} observed in Figure 3.12a.

3.5 Conclusion and Future Outlook

The potential of the nanotube transistor architecture to radically enhance the drive current capability and lower the "OFF" current of tunnel FETs to values comparable to state-of-the-art CMOS was demonstrated due to both higher BTBT generation rate and lower SRH recombination rate when compared to the GAA NW architecture at the same body thickness. 3D device simulations have shown that a p-channel nanotube TFET is able to outperform nanowire arrays, while preserving chip area and at comparable SS values. We believe the nanotube architecture combined with hetero-structure III-V/IV material systems hold a great promise for high performance, ultra-low power consumer computing applications [48].

References

1. A. Ionescu, and H. Riel, Tunnel field-effect transistors as energy-efficient electronic switches, *Nature* **479**, 329–337 (2011).
2. W. Loh et al., Sub-60nm Si tunnel field effect transistors with I_{on} >100 µA/µm, *IEEE Euro. Sol. Stat. Dev. Res. Conf.*, 162–165 (2010).
3. K. Jeon et al., Si tunnel transistors with a novel silicided source and 46mV/dec swing, *Symp. VLSI Tech. (VLSI)*, 121–122 (2010).
4. D. Sarkar, X. Xie, W. Liu, W. Cao, J. Kang, Y. Gong, S. Kraemer, P. Ajayan and K. Banerjee, A subthermionic tunnel field-effect transistor with an atomically thin channel, *Nature* **526**, 91–95 (2015).
5. D. Jena, K. Banerjee, and G. Xing, 2D crystal semiconductors: Intimate contacts, *Nat. Mater.* **13**, 1076–1078 (2014).
6. A. Verhulst, W. Vandenberghe, K. Maex, and G. Groeseneken, Boosting the oncurrent of a n-channel nanowire tunnel field-effect, *J. Appl. Phys.* **104**, 064514 (2008).
7. A. Verhulst et al., Complementary silicon-based heterostructure tunnel-FETs with high tunnel rates, *IEEE Electron Device Lett.* **29**, 1398–1401 (2008).
8. K. Tomioka, M. Yoshimura, and T. Fukui, Steep-slope tunnel field-effect transistors using III-V nanowire/Si heterojunction, *Symp. VLSI Tech. (VLSI)*, 47–48 (2012).
9. D. Cutaia, K. Moselund, M. Borg, H. Schmid, L. Gignac, C. Breslin, S. Karg, E. Uccelli, and H. Riel, Vertical InAs-Si gate-all-around tunnel FETs integrated on Si using selective epitaxy in nanotube templates, *IEEE J. Electron Devices Soc.* **3**, 176–183 (2015).
10. W. Y. Choi, B. G. Park, J. D. Lee, and T. -J. King Liu, Tunneling field-effect transistors (TFETs) with subthreshold swing (SS) less than 60 mV/dec, *IEEE Elect. Dev. Lett.* **28**, 743–745 (2007).
11. T. Krishnamohan, K. Donghyun, S. Raghunathan, and K. Saraswat, Double gate strained-Ge heterostructure tunneling FET (TFET) with record high drive currents and 60 mV/dec subthreshold slope, *IEEE Int. Electron Devices Meet.*, 1–3 (2008).
12. S. Kim, H. Kam, C. Hu, and T. King Liu, Germanium-source tunnel field effect transistors with record high I_{ON}/I_{OFF}, *Symp. VLSI Tech. (VLSI)*, 178–179 (2009).
13. R. Gandhi, Z. Chen, N. Singh, K. Banerjee, and S. Lee, CMOS-Compatible Vertical-Silicon-Nanowire Gate-All-Around p-Type Tunneling FETs With ≤ 50-mV/decade Subthreshold Swing, *IEEE Elect. Dev. Lett.* **32**, 1504–1506 (2011).
14. R. Gandhi, Z. Chen, N. Singh, K. Banerjee, and S. Lee, Vertical Si-nanowire n-type tunneling FETs with low subthreshold swing (≤ 50 mV/decade) at room temperature, *IEEE Elect. Dev. Lett.* **32**, 437–439 (2011).

15. H. Schmid et al., Fabrication of vertical InAs-Si heterojunction tunnel field effect transistors, *IEEE. Device Res. Conf.*, 181–182 (2011).

16. Q. Huang et al., A novel Si tunnel FET with 36mV/dec subthreshold slope based on junction depleted-nodulation through striped gate configuration, *IEEE Int. Electron Devices Meet.*, 8.5.1–8.5.4 (2012).

17. H. Riel et al., InAs-Si heterojunction nanowire tunnel diodes and tunnel FETs, *IEEE Int. Electron Devices Meet.*, 16.6.1–16.6.4 (2012).

18. K. Tomioka, M. Yoshimura and T. Fukui, Sub 60 mV/decade switch using an InAs nanowire–Si heterojunction and turn-on voltage shift with a pulsed doping technique, *Nano Lett.* **13**, 5822–5826 (2013).

19. S. Plissard, G. Larrieu, X. Wallart, and P. Caroff, High yield of self-catalyzed GaAs nanowire arrays grown on silicon via gallium droplet positioning, *Nanotechnology* **22**, 275602 (2011).

20. E. Alarcon-Llado, S. Conesa-Boj, X. Wallart, P. Caroff, and A. Morral, Raman spectroscopy of self-catalyzed $GaAs_{1-x}Sb_x$ nanowires grown on silicon, *Nanotechnology* **24**, 405707 (2013).

21. K. Tomioka, M. Yoshimura, and T. Fukui, Sub 60 mV/decade switch using an InAs nanowire–Si heterojunction and turn-on voltage shift with a pulsed doping technique, *Nano Lett.* **13**, 5822–5826 (2013).

22. M. Borg et al., Vertical III–V nanowire device integration on Si(100), *Nano Lett.* **14**, 1914–1920 (2014).

23. S. Sze, *Physics of Semiconductor Devices* Ch. 8 "Tunnel Devices" (Wiley, New York, 1981).

24. Y. Lu et al., Geometry dependent tunnel FET performance-dilemma of electrostatics vs. quantum confinement, *IEEE. Device Res. Conf.*, 17–18 (2010).

25. W. Vandenberghe, A. Verhulst, G. Groeseneken, B. Sorée, and W. Magnus, Analytical model for point and line tunneling in a tunnel field-effect transistor, *Int. Conf. Simul. Semicond. Processes Devices*, 137–140 (2008).

26. P. Wang, T. Nirschl, D. Schmitt-Landsiedel, and W. Hansch, Simulation of Esaki-tunneling FET, *Solid-State Electron.* **47**, 1187–1192 (2003).

27. Q. Zhang, S. Sutar, T. Ksel, and A. Seabaugh, Fully-depleted Ge interband tunnel transistor: Modeling and junction formation, *Solid-State Electron.* **53**, 30–35 (2008).

28. T. Chan, J. Chen, P. K. Ko, and C. Hu, The impact of gate-drain leakage current on MOSFET scaling, *IEEE Int. Electron Devices Meet.*, 718–721 (1987).

29. A. Schenk, S. Santa, K. Moselund, and H. Riel, Comparative simulation study of InAs/Si and All-III-V hetero tunnel FETs, *ECS Trans.* **66**, 157–169 (2015).

30. R. Anderson, Germanium-gallium arsenide heterojunctions, *IBM J. Res. Dev.* **4**, 283–287 (1960).

31. H. Sakaki, L. Chang, R. Ludeke, C. Chang, G. Sai-Halasz, and L. Esaki, $In_{1-x}Ga_x$ As- $GaSb_{1-y}As_y$ heterojunctions by molecular beam epitaxy, *Appl. Phys. Lett.* **31**, 211–213 (1977).

32. D. Griffith, *Introduction to Quantum Mechanics*, Ch.8 "WKB approximation" (Prentice Hall, Upper Saddle River, NJ, 1995).

33. S. Kim, "Germanium-source tunnel field effect transistors for ultra-low power digital logic", PhD thesis, UC Berkeley, Berkeley, CA, 2012.

34. Sentaurus Device User Guide, Synopsys, Version H-2013.3. March (2013).

35. L. Barletti, L. Demeio, G. Frosali, "Multiband quantum transport models for semiconductor devices," *Transport Phenomena and Kinetic Theory*, 55–89, 2007.

36. A. Ford, C. Yeung, S. Chuang, H. Kim, E. Plis, S. Krishna, C. Hu, A. Javey, Ultrathin body InAs tunneling field-effect transistors on Si substrates, *Appl. Phys. Lett.* **98**, 113105 (1–3) (2011).

37. A. Schenk, R. Rhyner, M. Luisier, and C. Bessire, Analysis of Si, InAs, and Si-InAs tunnel diodes and tunnel FETs using different transport models, *Int. Conf. Simul. Semicond. Processes Devices*, 263–266 (2011).

38. P. Solomon, J. Jopling, D. Frank, C. Emic, O. Dokumaci, P. Ronsheim, and W. Haensch, Universal tunneling behavior in technologically relevant P/N junction diodes, *J. Appl. Phys.* **95**, 5800–5812 (2004).

39. G. Hurkx, D. Klaassen, and M. Knuvers, A new recombination model for device simulation including tunneling, *IEEE Trans. Electron Devices* **39**, 331–338 (1992).

40. G. Larrieu, and X. Han, Vertical nanowire array-based field effect transistors for ultimate scaling, *Nanoscale* **5**, 2437–2441 (2013).

41. A. Hanna, H. Fahad, and M. Hussain, InAs/Si hetero-junction nanotube tunnel transistors, *Sci. Rep.* **5**, 9843(1–7) (2015).

42. H. Fahad, and M. Hussain, High-performance silicon nanotube tunneling FET for ultralow-power logic applications, *IEEE Trans. Electron Devices* **60**, 1034–1039 (2013).

43. H. Fahad, and M. Hussain, Are nanotube architectures more advantageous than nanowire architectures for field effect transistors? *Sci. Rep.* **2**, 475(1–7) (2012).

44. H. Fahad, C. Smith, J. Rojas, and M. Hussain, Silicon nanotube field effect transistor with core–shell gate stacks for enhanced high-performance operation and area scaling benefits, *Nano Lett.* **11**, 4393–4399 (2011).

45. S. Choi, D. Moon, S. Kim, J. Duarte, and Y. Choi, Sensitivity of threshold voltage to nanowire width variation in junctionless transistors, *IEEE Electron Device Lett.* **32**, 125–127 (2011).

46. L. Knoll et al., Demonstration of improved transient response of inverters with steep slope strained Si NW TFETs by reduction of TAT with pulsed IV and NW scaling, *IEEE Int. Electron Devices Meet.*, 4.4.1–4.4.4 (2013).

47. C. Young, B. Park, L. Jong-Duk, and T. King Liu, Tunneling field-effect transistors (TFETs) with subthreshold swing (SS) less than 60 mV/dec, *IEEE Electron Device Lett.* **28**, 743–745 (2007).

48. A. Hanna, and M. Hussain, Si/Ge hetero-structure nanotube tunnel field effect transistor, *J. Appl. Phys.* **117**, 014310–014317 (2015).

4

The Exploitation of the Spin-Transfer
Torque Effect for CMOS Compatible
Beyond Von Neumann Computing

Thomas Windbacher, Alexander Makarov, Siegfried Selberherr,
Hiwa Mahmoudi, B. Gunnar Malm, Mattias Ekström, and Mikael Östling

CONTENTS

The exponential growth in (affordable) computational power over the last decades was only sustainable due to continuous successful scaling of CMOS devices. The shrinking of the CMOS transistors allowed not only an increase in the speed and performance of circuits, but also ensured that the costs per transistor dropped for every technology generation. However, with each technology generation, new and ever harder to resolve obstacles appeared. Currently, out of the multitude of potential showstoppers in charge-based CMOS technology, the dissipated power and the energy associated with the transport of information are major concerns. The fast evolving field of spintronics offers a potential remedy for these problems by introducing "More than Moore" devices. The quest for the future universal memory candidate not only led to spin-based magnetoresistive random-access memory (MRAM), but also culminated in the first off-the-shelf MRAM products. Nevertheless, the core of the MRAM, the magnetic tunnel junction (MTJ), is not limited to memory applications. It can also be exploited for building logic-in-memory circuits with nonvolatile storage elements, as well as very compact on-chip oscillators with low power consumption. In general, the advent of nonvolatile elements, and especially spintronics in circuits, gives the unique opportunity to rethink how information is processed and moved. The concept of continuous information exchange between physically separated memory and processing units—also known as the Von Neumann architecture—has become a performance limiting bottleneck. The transition towards beyond Von Neumann architectures obviously also requires a redesign of all basic computational building blocks. In the this chapter, we will give an overview about the ideas and concepts for such beyond Von Neumann systems. First, we will present a short introduction into the physics necessary to understand the spintronic effects, like the magnetoresistance effect, spin-transfer torque (STT), spin Hall effect, and the magnetoelectric effect. Then we will move towards spintronic devices and circuits and their different concepts and architecture levels, where they introduce nonvolatility, such as thermally-assisted (TA)-MRAM, STT-MRAM, domain wall (DW)-MRAM, spin-orbit torque (SOT)-MRAM, spin-transfer torque and spin Hall oscillators, logic-in-memory, all-spin logic, buffered magnetic logic gate grid, ternary content addressable memory (TCAM), and random number generators. From our point of view, there will be no disruptive transition from pure CMOS to pure spintronic circuits. Instead, there will be a gradual introduction and substitution of existing CMOS devices by spintronic devices, where they outperform CMOS devices in one or more aspects. Therefore, we will concentrate on and emphasize concepts and devices that are CMOS compatible and present possibilities for different levels of integration into CMOS technology.

Finally, we summarize the current state-of-the-art and extrapolate an outlook regarding future development of the field and prospective devices from our point of view.

4.1 Introduction

The persistence and ingenuity of scientists and engineers made it possible to maintain the miniaturization of electronic components and interconnects for many decades. This still ongoing strategy led to the current 14 nm node with multi-gate three-dimensional transistors [3] and culminated in the announcement of the mass production of 10 nm node products for 2017 [4–6]. In principle, devices with a few nanometers gate length are feasible [7], but their introduction into large scale manufacturing is rather challenging due to fabrication and control issues that translate into reliability problems. In conjunction with their broad variability, which manifests in high integration costs, it is clear that in the foreseeable future scaling will come to a halt.

However, looking at the very core of the MOSFET operation, the interaction between the electrons' charge and an electric field, reveals that there is another intrinsic electron property, which can be harnessed as an alternative degree of freedom—the electron spin. It not only holds the potential to complement, but to substitute the currently omnipresent charge degree of freedom for future electronic devices [8,9]. The electron spin is the angular momentum of the electron due to its intrinsic rotation and is commonly measured by its projection along a given axis. The introduction of the axis results in two possible projections (parallel and antiparallel to the axis), which can be facilitated for digital information processing. A further advantage of exploiting spin as a degree of freedom is the very small amount of energy, which is required to invert its orientation. All spin-based technologies share advantageous features like a low supply voltage, small device count, and zero static power [10]. An essential aspect for the realization of all-spin-based computing is the understanding and control of the injection, propagation, and detection of spin signals, which has been achieved only recently. The difficulties to demonstrate spin injection from a ferromagnetic layer into a semiconductor origin from the inherent spin impedance mismatch between these materials [11]. This problem can be solved by the introduction of a potential barrier between the metal and the semiconductor [12]. Another obstacle on the way towards all-spin computation is the growth of contacts with low resistivity per area for good spin injection. In [13], it has been shown that spin injection through single layer graphene contacts are a promising close to optimal solution [14].

One of the major differences between spin and charge injected into a semiconductor is that the spin signal is not conserved. During the diffusion of the spin information carrying electrons, their net spin relaxes through scattering events to the equilibrium value of nonmagnetic semiconductors—zero. Even though Huang et al. [15] successfully demonstrated spin injection and propagation over 350 µm through a silicon wafer at 77 K, the diffusion length is reduced to approximately 200 nm at room temperature [14].

Unfortunately, this length reduces even further in CMOS technology mainly due to the increased number of scattering events at the interfaces [16]. However, there is a trick to boost the spin lifetime in such systems. In (001) silicon films, the governing scattering mechanism that reduces the spin lifetime is the intervalley scattering between equivalent valleys. If one introduces uniaxial stress along the (110) direction, the degeneracy is lifted and the respective intervalley scattering is significantly reduced, which leads to a large increase in the spin lifetime [17,18]. Strain has been used for many years in the semiconductor industry to boost the electron mobility; thus, it is easy to exploit the same well established methods for enhancing the spin lifetime.

Furthermore, it has been shown that purely electrical spin manipulation in InGaAs heterostructures with point contacts is possible at low temperatures [19]. The down side of this is the very poor control of the spin signal by voltage-dependent spin-orbit interaction in silicon channels. Therefore, the only feasible way to introduce spin into nano-scale CMOS technology is to add ferromagnetic source and drain contacts [20]. Such structures exhibit different currents depending on the relative orientation of the magnetization orientation of source and drain, which can be exploited for the realization of reprogrammable nonvolatile logic. However, this is quite unsatisfying due to the rather low magnetoresistance ratios in comparison to MTJs. Therefore, the most promising way for the introduction of practical spin-driven applications within the next few years will likely be an MTJ-based solution.

An MTJ comprises two magnetic layers that sandwich a nonmagnetic thin insulating layer (cf. Figure 4.1). Depending on the relative orientation between the magnetizations of the two magnetic layers, MTJs either exhibit a low resistance state (LRS, parallel) or a high resistance state (HRS, antiparallel). The two resistance states LRS and HRS are assigned to logic "0" and "1," respectively [21,22].

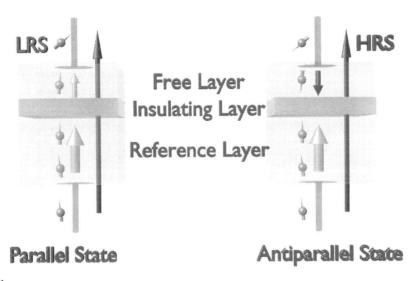

FIGURE 4.1
An MTJ consists of two magnetic layers separated by a nonmagnetic insulating layer. Depending on the magnetization orientation of the free and the reference magnetic layer with respect to each other, the electrons traversing through the layer stack experience more (antiparallel) or less (parallel) scattering, which is reflected in a high (HRS) and a low resistance state (LRS), respectively.

A universal memory that squares the circle of simultaneously being fast, nonvolatile, small in size, allows high integration density, and is CMOS compatible is spin-transfer-torque-based MRAM, one of the most promising candidates so far [22–25].

But the emerging spin-based technology has much more to offer. For instance, it can be used to build very compact versatile on-chip oscillators with low power consumption for consumer electronics and telecommunication applications. MRAM is also exploitable for logic-in-memory architectures, where the memory elements sit on top of the CMOS logic circuits. The combination of the nonvolatility of the memory elements and the considerably shorter interconnects guarantee low power losses and fast operation. There are already spin-based solutions able to compete with pure CMOS with respect to energy consumption and speed; however, one of the key aspects to be competitive in the market—the integration density—is still worse than in pure CMOS. Therefore, we will also look into ideas and technologies that have the potential for high integration density.

In the following section, we will first give an overview about the physical fundamentals of spintronics to allow the reader to concentrate on the devices and circuits in later sections. Since the peculiarities of the employed materials and their processing are essential to understand the current limitations for designing and manufacturing spintronic devices, the subsequent section is dedicated to these aspects. Then the different types of spintronic memory will be elucidated, followed by a spintronic logic section where different possibilities to implement logic will be explained. Afterwards the applications section will highlight some spintronic solutions to demonstrate the potential of spintronics in future applications. Finally, we will conclude the chapter and try to extrapolate how spintronics will develop in the future.

4.2 Fundamentals of Spintronics

In order to enable the reader to concentrate on the spintronic devices and circuits without the need to take breaks to look up physics details, a short section that will help to grasp the most relevant basic physical effects is provided here.

4.2.1 Magnetoresistance

The discovery of first the giant magnetoresistance (GMR) and later the tunneling magnetoresistance (TMR) were essential for the development of widely usable spintronic devices.

4.2.1.1 Giant Magnetoresistance

The GMR has been observed for the first time in Fe/Cr superlattices in the late 1980s by two independent researchers Baibich et al. [26] and Binasch et al. [27].

The GMR effect is observed when a current is passed through a stack of two or more magnetic layers that are separated by nonmagnetic conducting spacer layers. The measured resistance depends on the magnetization orientation of the magnetic layers with respect to each other. Commonly the strength of the GMR effect is expressed as the ratio between the high and low resistance states [28]:

$$GMR = \frac{R_{AP} - R_P}{R_P} = \frac{\rho_{AP} - \rho_P}{\rho_P} = \frac{\sigma_P}{\sigma_{AP}} - 1 \qquad (4.1)$$

R_{AP} (high resistance) and R_P (low resistance) denote the resistances for antiparallel and parallel layer magnetization orientations, ρ_{AP} and ρ_P are the associated resistivities, and σ_{AP} and σ_P the corresponding conductivities, respectively.

According to the definition in Equation 4.1 the GMR can become larger than 1, if $\rho_{AP} > \rho_P$. To avoid confusion, there is also an alternative definition where the GMR is never larger than 1 ($\rho_{AP} > \rho_P$) [28]:

$$GMR' = \frac{\rho_{AP} - \rho_P}{\rho_{AP}} = 1 - \frac{\sigma_{AP}}{\sigma_P} \qquad (4.2)$$

The simplest explanation for the GMR effect assumes that the electrons, which are traveling through the magnetic stack, can be described by two independent conduction channels. One channel describes electrons with a certain direction; for example, "Up," while the other channel describes electrons with opposite direction "Down" (see Figure 4.2) [29]. The sum of these two spin currents (I_{Up} and I_{Down}) forms the total charge current that passes through the stack. If these two spin currents flow through a ferromagnetic layer with a fixed magnetization direction, the electrons with "Up" and "Down" orientation experience different scattering rates depending on their orientation with respect to the orientation of the magnetic layer. This difference is reflected in different resistances for the two groups of electrons. For instance, if the magnetization orientations of the spin valve stack from Figure 4.2 are parallel, there is always one electron channel whose spin is antiparallel (electron spin and magnetic moment are antiparallel) and, thus, able to travel through the stack with only little scattering. On the contrary, if the magnetization orientation of the layers is antiparallel one of the channels always experiences enhanced scattering. As a result, the total resistance of the spin valve is lower for parallel magnetization ("Down"

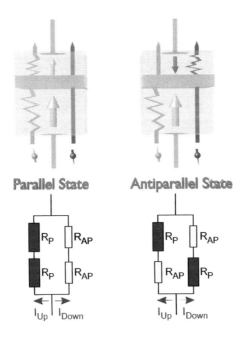

FIGURE 4.2
The GMR effect can be explained by assuming that the charge current can be split into two spin currents (I_{Up} and I_{Down}), which experience different scattering rates during their travel through the magnetic layers.

channel experiences only little scattering) than for antiparallel (both "Up" and "Down" channels exhibit a zone with increased scattering) [30].

This effect opened up the path for the development of today's hard disk drive read heads and encouraged research in GMR-based MRAM [31].

4.2.1.2 Tunnel Magnetoresistance

Another important effect that has great significance in current MRAM applications is the TMR. It was discovered by Julliere et al. [32] in a Fe/Ge/Co junction at temperatures below 4.2K in 1975. Similar to the GMR effect, TMR can be observed, where two magnetic layers sandwich a nonmagnetic insulating layer and the measured resistance depends on the magnetization orientation of the two magnetic layers with respect to each other. However, in this case the separating nonmagnetic layer is a thin metal-oxide (e.g., Al_2O_3 or MgO) and forms a MTJ in contrast to the GMR, where the nonmagnetic layer is composed of a metal (e.g., Cu) and forms a spin valve.

The TMR effect is quantified as the relative ratio between the parallel and antiparallel resistance states of the stack [33,34]:

$$TMR = \frac{R_{AP} - R_P}{R_P} \tag{4.3}$$

Analog to before, R_{AP} and R_P denote the high (antiparallel) and low (parallel) resistance states of the stack.

The source of this effect can be attributed to the difference in tunneling probabilities for the electrons with certain orientation (e.g., "Up" or "Down") from one ferromagnetic layer (reference layer) to the other ferromagnetic layer (free layer) through the oxide for a given magnetization state. Figure 4.3 depicts the energy bands and their respective occupation for the parallel (left) and the antiparallel (right) magnetization state. If the magnetizations of both layers are parallel (e.g., both point "Up"), the majority of the electrons occupies "Up" states and the minority "Down" states in the reference layer as well as in the free layer. Therefore, the bands and their occupation match, which makes it easier for the electrons to tunnel through the thin nonconducting layer. This state has a higher conductance (lower resistivity). In the case of antiparallel magnetization orientations (e.g., reference layer→"Up" and free layer→"Down"), the majority of the electrons in the free layer are in "Down" states and the minority of the electrons in "Up" states. Therefore, there are far more electrons in the reference layer with "Up" than matching states available in the free layer, which leads to a strongly reduced tunneling probability. Even though the spin "Down" electrons from the reference layer find plenty of available states in the reference layer, their total number is much smaller than the amount of "Up" electrons. Thus, they can only contribute little to the total conductance of the stack. In summary, the overall conductance is strongly decreased and an increase in the stack resistance is observed.

Although the TMR effect was found earlier than the GMR effect, its practical use was limited due to poor TMR values, until the advent of stacks with amorphous Al_2O_3 as tunnel barrier. Moodera et al. [35] and Miyazaki et al. [36] where the first who developed independently such structures. The largest TMR ratio for an MTJ with amorphous Al_2O_3 tunnel barrier at room temperature so far was demonstrated by Wang et al. [37] in 2004 and amounts to 70.4%.

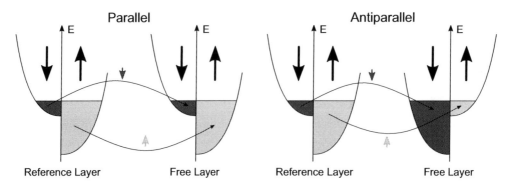

FIGURE 4.3
The energy bands and their respective occupation is different for parallel (left) and antiparallel (right) magnetization orientations. For parallel magnetization, the available states in the free layer match with the reference layer. Therefore, the electrons with "Up" and "Down" orientation are able to tunnel into matching states. For antiparallel orientation, there are far more electrons with spin "Up" than states available in the free layer. This reduces the tunneling probability considerably and causes an increase in resistance.

The next leap towards the realization of MRAM exploiting the TMR effect was the discovery of a giant TMR in an MTJ with an epitaxially grown MgO barrier. Again two scientists, Butler et al. [38] and Mathon et al. [39], predicted independently a giant TMR in MTJs with MgO tunnel barrier in 2001. Furthermore, Mathon predicted a TMR ratio >1000% for an MgO barrier [39].

This exceedingly large TMR ratio can be explained by a symmetry-based spin filtering that occurs in the MgO tunnel barrier [34]. Bowen et al. [40] were the first to measure TMR in Fe/MgO/FeCo (001) single-crystal epitaxial junctions. These measurements showed a much smaller TMR (27% at 300 K, 60% at 30 K) than predicted previously. In 2004, it was possible to increase the TMR ratio considerably in single-crystal Fe/MgO/Fe/MTJs, to a level of 220% [41] and 180% [42] at room temperature. Thanks to the rapid progress in the epitaxially growth techniques of MTJ stacks, the TMR increased swiftly [33]. By 2006, TMR values up to 410% could be demonstrated [43], followed by 604% at room temperature and 1144% at 4.2 K in $Ta/Co_{20}Fe_{60}B_{20}/MgO\ Co_{20}Fe_{60}B_{20}/Ta$ junctions [44] in 2008.

4.2.2 Spin-Transfer Torque

Before the discovery of the spin-transfer torque (STT), the free layers of MRAMs were switched by the application of magnetic fields (cf. Figure 4.4). The magnetic fields were created by passing currents through adjacent wires. In order to protect the free layers from accidental switching, the memory cells must be designed in a way that two magnetic fields generated by two physically separated wires add up to switch the memory cells without unintentional switching events. The field-based switching method has the disadvantage of increasing current densities, when the structures are scaled down. This stems from the fact that the current must not change to ensure sufficient switching field strength, while at the same time the cross section of the wires decreases, when the structures are shrunk. This counteracting prerequisites made the shrinking of field-based MRAM below 90 nm unfeasible [45].

This field related limit was circumvented, when Slonczewski's [46] and Berger's [47] theoretical work predicted the existence of the STT effect in 1996. The exploitation of the STT effect represents a technological breakthrough, which allows the direct manipulation

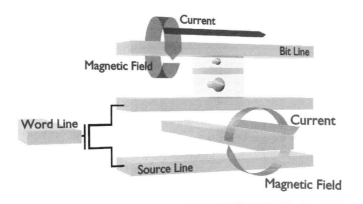

FIGURE 4.4

Field-based MRAM requires two wires for the generation of the writing field. Only when the magnetic fields created by both wires add up, the free layer will switch its magnetization. This design is deliberate to protect neighboring memory cells from accidental switching.

of the magnetization of a layer through a spin polarized current and renders the previously employed indirect switching via Ørsted fields superfluous.

When electrons move through a (thick) fixed reference layer, their magnetic moment aligns with the local magnetization (see Figure 4.5). If these spin-polarized electrons subsequently enter the free layer, they align again to the local magnetization orientation within a few Ångström. During the relaxation of the electrons to the local magnetization, not only the electrons experience a torque, but also the local magnetic moments (total sum of torques must be zero). This STT is able to excite precessions in the free layer and, if strong enough to overcome the damping, eventually switches the whole free layer. The precessions are carried out around the effective field H_{eff}. Changing the polarity of the

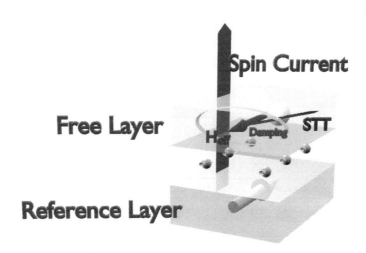

FIGURE 4.5

The electrons traversing through the stack, first pass the reference layer, where they align parallel to the reference layers' magnetization orientation (bottom). Then they pass the nonmagnetic layer (transparent gap) and finally enter the free layer, where they relax to the free layers magnetization orientation. This relaxation creates a spin-transfer torque that drives magnetization precessions. If the torque is strong enough to overcome the damping, the free layer is switched.

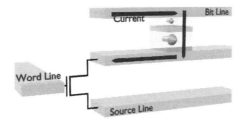

FIGURE 4.6
In contrast to the field-based MRAM (cf. Figure 4.4) STT-MRAM does not require an extra wire to prevent switching failures. Instead the magnetization is manipulated by a spin polarized current, which allows a considerably simplified memory cell design.

applied current flips the orientation of the exerted STT and, thus, allows to repeatably switch the free layer between antiparallel and parallel orientation with respect to the reference layer (see Figure 4.6).

Nevertheless, it took years until STT-induced switching could be demonstrated experimentally on all-metallic stacks [45]. Co/Cu/Co was the first GMR-based stack to proof the concept of STT-induced switching [48–52]. The first working STT-switched MTJ memory cells based on AlO_x were shown in 2004 [53] and based on MgO in 2005 [54].

4.2.3 Spin Hall/Spin-Orbit Effect

Another effect that has attracted a lot of attention, is the Spin Hall effect (SHE). It also generates a spin current capable of switching the magnetization of a layer and was predictedby D'yakonov and Perel in 1971 [55]. Driving a charge current through a metal line with strong spin orbit interaction generates a spin current perpendicular to the current's flow direction (see Figure 4.7). Vorob'ev et al. were the first to confirm the spin Hall effect experimentally in 1979 by observing a change in the rotation rate of the polarization plane for light propagating through a Te crystal [56]. Kato et al. [57] were able to demonstrate and confirm the same effect in 2004. The first direct electronic measurements were carried out

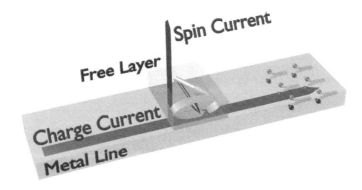

FIGURE 4.7
When a charge current flows through a metal line with strong spin-orbit interaction, a spin current perpendicular to the current flow is generated. The spin polarized electrons accumulate at the wire's surface and diffuse into the neighboring free layer, where they relax to the local magnetization and exert a spin torque on the free layer's magnetic moments.

by Valenzuela and Tinkham [58]. As it turned out later, they actually observed the inverse spin Hall effect (ISHE), since in their case they created a spin current, which generated a perpendicular charge current that accumulated at the edges of the sample exploited for electrical measurement [59]. For their work, they used a ferromagnetic electrode to generate a spin current and subsequently injected it into a nonmagnetic metal strip, where they took advantage of the ISHE as well as the nonlocal spin valve effect with the aid of a ferromagnetic probe electrode for the spin signal detection.

Further work regarding the SHE and the ISHE effect was carried out by Kimura et al. [60,61] and is based on NiFe/Cu/Pt structures. The spin current was measured by exploiting a nonlocal spin signal and the ISHE. Their work paved the way for the exploitation of the SHE and the ISHE as spin injection and detection tools.

4.2.4 Magnetoelectric Effect

Analog to the initially employed current controlled bipolar junction transistors, also the STT-based spintronic devices always require some kind of charge flow and thus also exhibit Joule heating as an energy dissipation mechanism during switching. This problem drove the transition from bipolar junction transistors to first N/P-MOS devices and eventually to the state-of-the-art CMOS technology. Therefore, ideally one could switch in spintronics from current-based to voltage-based magnetization dynamics manipulation in order to benefit the same way from the significant reduction in power dissipation [62].

Weisheit et al. [63] showed that the magnetocrystalline anisotropies of FePt and FePd compounds can be reversibly switched by an externally applied electric field. It was also demonstrated that a relatively small electric field can induce a large ~40% change in the magnetic anisotropy of a bcc Fe(001)/MgO(001) junction [64]. Furthermore, it was demonstrated that the magnetocrystalline anisotropy of $Fe_{80}Co_{20}(001)/MgO(001)$ cannot only be changed by an electric field, but actually voltage-assisted switched [65]. Nozaki et al. [66] showed high-frequency voltage-assisted magnetization reversal in MgO-MTJs in 2014. They could demonstrate a switching field reduction of >80% at a radio frequency of 3 dBm. Recently, Li et al. [67] could show that the introduction of a thin Mg layer at the CoFeB/MgO interface causes a 3× increase in the voltage controlled anisotropy coefficient (from commonly ~30 fJ/Vm to ~100 fJ/Vm). This is very encouraging, because it allows to reduce the write voltage below 0.6 V, which allows to employ advanced CMOS transistors.

The drastic change in the magnetocrystalline anisotropy strength of ultra-thin layers under the application of an electric field can be attributed to a change in the occupation of the atomic orbitals at the CoFeB/MgO interface, which together with the spin-orbit interaction, alters the anisotropy [62,67,68]. However, it can be also explained by the interfacial Rashba effect [62,69].

4.3 Materials and Their Processing

Since the peculiarities of the employed materials and their processing are essential to understand the current limits for designing and manufacturing spintronic devices, this section is dedicated to these aspects. A recurring theme of discussion is the integration with advanced CMOS process nodes, since a complete MRAM cell features a controlling transistor in combination with the MTJ element. MRAM technology has a few distinct reliability issues and the large interest for MRAM technology has prompted tool vendors to develop dedicated tools. Acceptance of MRAM technology is manifested by its adoption into foundry process lines.

4.3.1 Back End of Line Integration

In integrated circuits the back end of line (BEOL) process refers to the fabrication of metal interconnects and the intermetal dielectrics (IMD) layers. Using successive deposition of metal (Cu), patterning of metal lines, IMD deposition, and planarization of the IMD layers, more than 10 layers of interconnecting Cu-lines can be realized. This is sufficient for the routing of signal and power supply lines in very complex circuits, with 100 millions of integrated transistors. All BEOL process steps are performed at low temperature, typically in the range 350°C–400°C. Therefore, the integration of spintronic memory and logic based on multilayer ferromagnetic metallic stacks with thin metal-oxide tunneling barriers is feasible. The MTJ stacks will not suffer from interdiffusion and the integrity of the tunneling barrier can be maintained [2,44]. Specifically the MgO barrier must be annealed under controlled conditions to obtain a proper crystallographic reorientation epitaxially along the (001) direction. More importantly, annealing is also necessary to induce the interfacial perpendicular magnetic anisotropy (PMA) effect for stacks based on CoFeB, which is intrinsically an in-plane material [70]. A comprehensive review of the PMA and its applications in [71]. The PMA can also be strengthened by using, for example, multilayer Co/Pt with inherent PMA or synthetic antiferromagnetic (SAF) layers in the MTJ stack [72]. Capping layers (e.g., silicon nitride) are used to protect the MTJ from unintentional reoxidation during later stages of processing. The MTJs are typically inserted close to the top metal layers. The MTJ bottom electrode is connected to an already available Cu-line in, for example, metal level 5 (M5), [1]. Subsequent MTJ layers are deposited without breaking the vacuum, patterned by lithography and etching and then embedded in the subsequent IMD layer. The IMD thickness depends on the layer and is chosen to minimize the interconnect capacitances. The MTJ stack total thickness is less than the IMD thickness so that the MTJ becomes fully embedded. For an illustration of production near embedded MRAM, see Figure 4.8.

To implement MTJs in the BEOL process flow the minimum additional lithographic mask count is three or four. To put this into perspective, a 14 nm advanced CMOS process node uses close to 70 mask steps. Also for comparison it is interesting to note that embedded flash nonvolatile memory has an added mask count as high as a dozen. Embedded static random-access memory (SRAM) has a significantly larger footprint or cell area, while embedded dynamic random-access memory (DRAM) is a quite complex process module, including high aspect ratio etching and filling steps for the storage capacitors.

4.3.1.1 MRAM Cell Density

The metal pitch in advanced technology nodes is compatible with the size of an MTJ element and the area of the complete MRAM cell, including one controlling CMOS transistor, follows the standard CMOS design rules. The width of the CMOS transistor must be chosen so that enough drive current can be supplied in order to reach the critical current density for STT switching. This has led some researchers to pursue devices that are voltage controlled (VC) MRAM and consume less current, allowing smaller transistors to be used [73]. A 4Gbit MRAM density has been demonstrated with 90 nm pitch [74]. The minimum pitch is used in the lower metal layer while the metal pitch increases for higher layers [75].

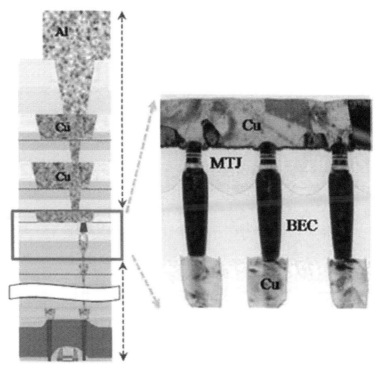

FIGURE 4.8
Example of production-near embedded MRAM [1]. Left panel showing schematic vertical structure of 8 Mb STT-MRAM cell array embedded in 28 nm logic process. Right panel showing transmission electron micros-copy (TEM) picture of MTJ module inserted between Cu BEOL lines.

A via needs to be opened in the IMD and aligned to the MTJ top contact area. There are several methods to achieve this, including self-aligned process schemes [76]. A combina-tion of chemical mechanical polishing (CMP) and deposition of sacrificial or etch stop layer on top of the actual MTJ are examples for such self-alignment solutions.

4.3.1.2 MTJ Multilayer Stack Deposition

Metal layers in the BEOL flow are deposited by sputtering tools, also known as physical vapor deposition (PVD) tools. The particular requirements for MTJ stacks include the pos-sibility to deposit a large number of elements, for example, Ta, Ru, Co, Ni, Fe, Cu, Pt, B, Mg, and Al to name the most common. The tools operate under ultra-high-vacuum conditions (UHV) corresponding to 10^{-8} Torr or better and feature in-situ annealing capability. The UHV condition is a key requirement, for growing sub-nm atomically abrupt layers. In research, molecular beam epitaxy is sometimes used for abrupt layers, but for production purposes PVD tools are the only choice in terms of throughput, wafer scale uniformity and metal targets available. Dedicated PVD tools for MRAM fabrication are offered in multi-cathode configuration, able to handle the large number of elements. Examples of deposition and etching tool vendors include Applied Materials, Singulus, Canon Anelva, Oxford Instruments, and LAM. It should be noted that several of these companies already have a strong presence in microelectronics fabrication.

4.3.1.3 Two-Dimensional Materials in the MTJ Stack

Using emerging two-dimensional (2D) materials either graphene, boron-nitride (BN), MoS_2, or WS_2 offers an interesting path to improve the MTJ stack [77]. There have been successful demonstrations of using graphene as a tunneling barrier [78]. However, in this case the TMR is too low to consider applications. On the other hand, graphene and other monolayer materials are excellent diffusion barriers and can be used in combination with oxide tunneling barriers, since they alleviate interdiffusion issues during high temperature process steps [79].

4.3.1.4 MTJ Shape, Patterning, and Etching

MTJs are patterned and etched into pillars with their material stack sandwiched in between nonmagnetic top and bottom metal contacts. First and second generation MRAM cells relied on in-plane magnetization and shape anisotropy to stabilize the magnetization of the fixed or reference layer. Therefore, elliptic shapes were mandatory. This requirement put very stringent boundaries on the pattering process since variability in shape could be detrimental to the switching energy barrier. Basically, elliptic shapes are not optimal from a patterning perspective. In standard CMOS foundry design rules, circular contacts are patterned at minimum lithographic dimensions. In current generation MRAM, the use of materials with perpendicular anisotropy removed this constraint of having elliptically shaped MTJ elements and hence significantly eased the process integration. As discussed above, the MTJs are comparable to the pitches used in the BEOL and deep-UV optical lithography provides the necessary resolution and alignment. It should be mentioned that the scientific community relies almost exclusively on electron beam lithography, which has nanometer resolution but suffers from long writing times, and is impractical to use for alignment of multiple layers with critical dimensions.

Regardless of the patterning technique, the etching of MTJ stacks is known to be challenging, because the etching residues are not very volatile. This becomes an issue in reactive ion etching (RIE), where the substrate temperature must be raised to achieve enough etching rate for the removal of residues and to avoid redeposition. The temperature the metal stack can tolerate is limited, so other solutions must be considered. The main technique is physical etching by sputtering with low energy Ar ion beams. For this technique, there are also redeposition issues. Furthermore, since the etch is typically performed at glancing angles, the area of the patterned element will be reduced by lateral etching of the pillar sidewalls [80]. It could be advantageous to shrink the lithographic pattern size [81], but it is generally considered as a drawback of these etching tools. Many ion beam tools are equipped with in-situ analysis capabilities of the etching residues, which is highly useful for controlled etching of monolayers. Alternative methods include atomic layer deposition (ALD)/atomic layer etching (ALE), where the volatility of the etching residues is increased by controlled deposition of selected elements on the metallic surfaces [82]. For etching of multilayer stacks, the chemical reactions provide a degree of selectivity to the different materials being etched. In contrast, ion beam etching has virtually no selectivity due to its purely physical nature of material removal.

4.3.1.5 Nonvolatile Logic

Processing of nonvolatile logic based on MRAM cells is a straightforward adaption of the standard MRAM blocks (also known as macros). While these highly regular memory matrices are based on a 1T/1MTJ configuration, the relative number of transistors as compared to MTJs increases in the hybrid realization of nonvolatile logic [83]. As an example, one would use at least six transistors and two MTJs in a flip-flop. To reduce this area and computational energy overhead, it would be preferable to move all computation to the spin domain and to use electronics only for interfacing purposes (see Section 4.5).

4.3.1.6 Access to Foundry Process Flow

All main foundry-based actors in the semiconductor business, such as TSMC, UMC, Global Foundries, and Samsung, have announced embedded MRAM (eMRAM) options by end of 2017 (see [84]). The intellectual property (IP) needed to get started in the MRAM field has been transferred to the foundry partners from start up companies, closely connected to academia. Some companies have entered manufacturing agreements and continue to develop their own IP. Finally, there are some companies with strong in-house activities, notably Toshiba.

4.3.2 Reliability and Yield Issues

During qualification, all memory devices are subjected to thorough cycling at various operating conditions, including elevated temperatures and increased humidity. Typical specifications require that a nonvolatile cell retains its state over a specified time (10 years) [84] and that a cell can be cycled (10^7 times) without any penalty in read or write voltage margin [85]. In fact, current MRAM offering surpass this significantly. In particular, the number of cycles is often given as close to unlimited [86].

4.3.2.1 Time Dependent Dielectric Breakdown

For MTJs one can identify several reliability issues. The main one is the relatively high current density passing through the MgO tunneling barrier during switching. This might lead to time dependent dielectric breakdown (TDDB), which is a well known issue in advanced CMOS with thin gate oxides. The TDDB is strongly affected by the temperature, as discussed further below.

4.3.2.2 Electromigration and Self-Heating

The relatively high critical current densities posed a serious threat due to possible electromigration in early generation spintronic devices [87]. For MTJ devices based on STT, the current densities are orders of magnitude lower. Since MTJs are embedded in isolating material with relatively low thermal conductance, there might be a significant temperature increase during switching. Increased temperature is known to accelerate electromigration by a power law [88].

4.3.2.3 Shorting of the Tunnel Junction and Etch Damage

During ion beam etching of the MTJ pillars, redeposition of metal can potentially cause an electrical short along the pillar sidewall and hence short the tunneling barrier [84]. This type of defect must be avoided since a parallel resistive path forms and effectively eliminates the difference between high and low resistance states of the MRAM cell. Careful tailoring of etch process, including good control of the sidewall slope is key to obtain high yield. In practice, the MTJs in production-near MRAM cells feature a small intentional sidewall slope (see Figure 4.9).

FIGURE 4.9
Cross-sectional transmission electron microscopy image of a fully functional device integrated on 90 nm CMOS
[2]. The diameter of this device is about 50 nm.

Vertical sidewalls cannot be controlled with sufficient accuracy in wafer scale production. In case of RIE tools, redeposition on the MTJ sidewalls also occurs, but in this case of polymer residues. These can be removed by proper post-etch cleaning steps. The chemical species used in RIE tools are very corrosive and could damage the sensitive MgO tunneling barrier. Again, post process cleaning is essential to hinder any further corrosion due to remaining etchant species.

4.3.2.4 Voids/Open Failures

The MTJs are sputter deposited on bottom electrodes, which are part of the standard BEOL process flow. As in any contact opening the surface should be free of residues of such polymers, which remain from previous process steps [84]. Good via filling is essential particularly for the top contact, so that all of the MTJ area is contacted. Having a partial void at either the bottom or top contact will degrade the relative changes in resistance during switching and increase the absolute value of the MTJ resistance, so that the voltage drop becomes too high.

4.3.2.5 Disturbance by Internal and External Fields

Both read and write operations of MRAM can be disturbed by external fields. As the first generation MRAM cells were field-switched, current generations rely on spin transfer torque or voltage controlled anisotropy (cf. Sections 4.2.2 and 4.2.4). As an example of the field sensitivity, data sheets for commercial products give a limit of 8000 A/m. In addition, in a scenario where MRAM cells are placed at minimum design rules in advanced CMOS technology nodes, neighboring cells could affect each other due to their internal stray fields.

4.4 Spintronic Memory

Memory can be distinguished into two categories: volatile and nonvolatile. The volatile memories, such as SRAM and DRAM, retain their data as long as they are supplied with power. The nonvolatile memories, such as EEPROM and Flash, retain the data when powered off.

Conventional computers are organized in a memory hierarchy to improve their performance and optimize the cost [89]. The hierarchy is illustrated in Figure 4.10. The fastest, highest performance memory technology is placed at the top of the pyramid. The high performance memory is expensive so its size is kept small. At the top of the hierarchy is the so called Level 1 cache (L1), which is typically a small volume memory placed on the same chip as the microprocessor. L1 cache is realized through SRAM and fabricated with the same CMOS technology as the microprocessor. Further levels of cache (L2 and L3) are also SRAM, but typically on dedicated stand-alone chips. Below the cache is the main or primary memory, with considerably larger size than the cache. DRAM is employed for the main memory and its size is a tradeoff between cost and required performance. All data in the cache is also present in the main memory in order to avoid accessing the relatively slow main memory as much as possible. Below the main memory is the nonvolatile storage or secondary memory, where volume is more important than performance. The storage memory was for a long time occupied by HDDs, but now faces competition from the

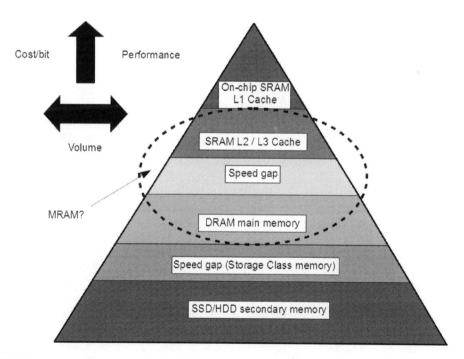

FIGURE 4.10
Pyramidal representation of the memory hierarchy. MRAM is a suitable candidate for L2/L3 cache and main memory.

NAND-flash-based solid state drives (SSD). The main memory mirrors the data from the comparatively very slow storage memory in order to speed up the access times.

Currently, there are two speed-gaps in the hierarchy: Between the cache and the primary memory and between the primary memory and the secondary memory [90,91]. For the gap between the primary memory and the high volume secondary memory, a new hierarchy level—the storage class memory (SCM)—has been proposed. To fill the gap the employed memory must exhibit, a density higher than DRAM, an access time shorter than NAND-Flash, and nonvolatility. Right now there is a competition between several nonvolatile random access memory (NVRAM) types that have potential for SCM applications, such as phase change memory, conductive bridge memory, resistive memory, and MRAM.

MRAM is a high performance NVRAM suitable for SCM applications, but currently not used in the SRAM, DRAM and HDD/SSD dominated memory hierarchy. MRAM has been proposed as a universal memory that can fill all levels of the memory hierarchy. However, the up to now rather low density prohibits any serious competition with the well established HDD/SSD technology. Due to the demand of a high density for an SCM, other NVRAMs are better suitable [90,91]. Especially, a three-dimensional monolithic integration of cross-bar memory arrays is more likely to succeed. These require memory cells that use a 1D-1R memory cell architecture (see Section 4.4.3). A more realistic application for MRAM is to replace SRAM and DRAM in L2/L3 cache and primary memory, respectively. It can bridge the speed gap between the cache and the primary memory. The required high endurance has been successfully demonstrated [92] and Kitagawa et al. [93] showed that a simulated mobile CPU would use less power, if it employs MRAM instead of SRAM as L2 cache. Other examples of MRAM for cache-applications can be found in [94]. MRAM is available on the market for main memory applications (DDR3 DRAM compatible) [95]. The major benefit of replacing DRAM with a NVRAM is the removal of the refresh action, the reduction of the overall power consumption and the simplification of the circuit design.

In this section, the MTJ, the core of the spintronic memory, will be discussed in depth. Its properties and trade-offs will be presented. The different varieties of spintronic memories and their peculiarities will be shown and the different memory cell architectures compared.

4.4.1 Magnetic Layer Design

4.4.1.1 Free and Reference Layer

A basic MTJ is composed of three elements: The reference layer, the tunneling barrier, and the storage layer. The tunneling barrier was covered in Section 4.2.1.2.

Storage layer: The storage layer, or free layer, is the layer that stores information as magnetization direction. A necessary requirement is that the free-layer material possess an energetically favorable nonzero magnetization in the absence of an external magnetic field or a remanent magnetization. There are elemental ferromagnetic materials (e.g., Fe, Co, Ni), ferrimagnetic half-metal oxides (Fe_3O_4, $La_{1-x}Sr_xMnO_3$), and various ferromagnetic alloys ($Ni_xFe_yCo_z$, Heusler alloys). Today, CoFeB ($Co_{0.20}Fe_{0.60}B_{0.20}$) is the material of choice as it has low damping [96] and provides high TMR in combination with MgO as tunneling barrier [44,97]. The free layer is a planar thin-film and can be further differentiated in films with in-plane and perpendicular magnetization direction. In order to achieve a free layer with in-plane direction, the free layer commonly exhibits an elongated shape (such as elliptic or rectangular). This form creates a shape anisotropy with two well-defined stable magnetization states along the major axis. The effective anisotropy field H_k for in-plane design is given by [96,98]:

$$H_k = 2(M_s d_F) \times \left(\frac{l_F - w_F}{w_F \times l_F} \right) = 2(M_s d_F) \times \left(\frac{AR-1}{w_F AR} \right) \tag{4.4}$$

l_F and w_F is the length and width of the free layer ($l_F > w_F$), respectively. AR denotes the aspect ratio ($= l_F / w_F > 1$), d_F the layer thickness, and M_s the magnetization saturation. The largest value the effective anisotropy field can reach is $2(M_s d_F)/w_F$. The magnetic moment $m = M_s A_F \times d_F$ is accessible through measurements with a vibrating sample magnetometer (VSM), and, if the area $A_F = l_F \times w_F$ is known, $M_s d_F$ can be determined ($= m/A_F = M_s d_F$). A typical ferromagnet (Co, Fe, Ni) exhibits an M_s of $\sim 10^6$ A/m. The two stable states of the magnetization are separated by an energy barrier that prevents the magnetization from freely switching the direction. The energy barrier determines the retention and switching properties, as will be discussed in Section 4.4.1.2. The general model for the energy barrier is found in [96,99]:

$$E_b = \mu_0 M_s H_k V_F / 2 = \mu_0 (M_s d_F) H_k A_F / 2 = K_u V_F \tag{4.5}$$

μ_0 describes the vacuum permeability, V_F the volume of the free layer ($= A_F \times d_F$), and K_u is the magnetic anisotropy energy density. The model for the energy barrier can with the aid of Equation 4.4 be further refined to analyze implications for in-plane designs:

$$E_b = \frac{\mu_0 (M_s d_F) H_k A_F}{2} \sim \mu_0 (M_s d_F)^2 w_F \times (AR-1) \tag{4.6}$$

Assuming a constant aspect ratio, one can see from Equation 4.6 that the barrier depends quadratically on the layer thickness and linearly on the layer width.

In contrast, for the perpendicular design, the effective anisotropy field H_k is governed by three terms [96]:

$$H_k = \frac{2M_s}{\mu_0} K_b + \frac{2}{\mu_0 d_F} \sigma - \frac{1}{2} M_s \tag{4.7}$$

The first term corresponds to the perpendicular bulk anisotropy (given by K_b, J/m^3), the second term to the perpendicular surface anisotropy (given by σ_i, J/m^2), and the third term to the demagnetization energy. The perpendicular design becomes unstable, if the demagnetization energy dominates. Ferromagnetic materials behave differently in different nonequivalent crystal directions, which manifests in the magnetocrystalline anisotropy. As an example, Co, which has hexagonal symmetry, prefers a magnetization along c-axis instead of lying in the a-plane, and has an anisotropy energy of 4.5×10^5 J/m^3, or 2.8×10^{-3} eV/nm^3 [100]. The magnetocrystalline anisotropy can be used as source for bulk anisotropy. If this bulk term dominates, the anisotropy field is independent of the film thickness and the energy barrier is given as

$$E_b = \frac{\mu_0 (M_s d_F) H_k A_F}{2} \sim \left(K_b - \frac{1}{2} \mu_0 M_s^2 \right) d_F \times A_F \tag{4.8}$$

In this case, the energy barrier increases with increasing volume. Both magnetocrystalline anisotropy and damping are correlated functions of spin-orbit coupling [100,101]; thus, systems using intentionally high magnetocrystalline anisotropy (Pt/Pd systems, FePt [102], or Co/Pt-multilayers [101]) exhibit large damping, which will be discussed in relation with switching current in Section 4.4.1.2.

If the interface term dominates, the barrier is given as

$$E_b = \frac{\mu_0 (M_s d_F) H_k A_F}{2} \sim \left(\sigma_i - \frac{1}{2} \mu_0 \frac{(M_s d_F)^2}{d_F} \right) \times A_F. \tag{4.9}$$

The barrier increases with increasing area, but decreases with increasing thickness. The demagnetization term dominates, if the thickness is larger than a critical value, $d_C = 2\sigma_i / \mu_0 M_s^2$. The interface anisotropy for a CoFeB film sandwiched between Ta (bottom layer) and MgO (top layer) is $\sim 1.8 \times 10^{-3} \, J/m^2$ and has a critical thickness of 1.1 nm [103].

Reference layer: Unlike the storage layer, the magnetization of the reference layer is not supposed to switch. For in-plane and bulk anisotropy designs, a simple solution is to have a relatively thick film in comparison to the storage layer. For interface anisotropy designs, it is the opposite. High anisotropy is achieved by having a relatively thin film [104]. A more sophisticated solution is to couple the ferromagnet to an antiferromagnetic material (AFM). For an AFM, it is energetically more favorable to be in a state with net-zero magnetization. This is achieved by arranging the magnetic moments in a regular pattern with neighboring moments (on different sublattices) pointing in opposite directions. At the interface between a ferromagnet and an AFM, the moments can align across the interface and couple the two layers. Reversing the magnetization of the coupled ferromagnet requires that the coupling energy is overcome, as the AFM will resist reversal. Thus, the ferromagnetic layer is pinned into a magnetic state. This exchange bias acts like an additional anisotropy field that forces the magnetization of the ferromagnetic layer into a specific state [105]. PtMn is an example for an antiferromagnetic material used in devices, with an interface anisotropy energy of $3.2 \times 10^{-4} \, J/m^2$ [105].

If the reference layer comprises only one ferromagnetic layer, the reference layer exerts a fringe field on the free layer. This fringe field biases the free layer towards the anti-parallel state. A remedy is to use a synthetic antiferromagnet (SAF). The SAF has two ferromagnetic layers forced into antiparallel state with net-zero magnetization. The two ferromagnetic layers are coupled through a very thin (typically less than 1 nm) nonmagnetic metal (like Ru or Cu). Depending on the thickness of the spacer, it may be favorable to have ferromagnetic coupling (both align) or antiferromagnetic coupling (anti-parallel). The interlayer exchange coupling can be described by the Ruderman-Kittel-Kasuya-Yosida (RKKY) model [106–108].

State-of-the-art MTJs are complex multilayer devices. Examples of MTJs include (from bottom to top): seed layer/PtMn/CoFeB/Ru/CoFeB/MgO/CoFeB/capping layer (an in-plane design with AFM and SAF) [109], Si wafer /Ta/Ru/Ta/CoFeB/MgO/CoFeB/ Ta/Ru (an in-plane design without AFM or SAF) [44] and Ta/Ru/Ta/CoFeB/MgO/CoFeB/ Ta/CoFeB/Ta/CoFeB/ MgO/Ta/Ru (interface perpendicular design, the MgO/CoFeB/Ta/ CoFeB/ MgO cleverly doubles the energy barrier by doubling the interface anisotropy without increasing the switching current) [110]. The various designs are illustrated in Figure 4.11.

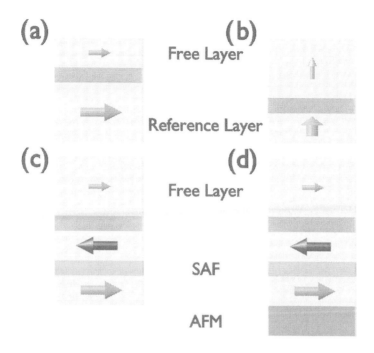

FIGURE 4.11
MTJ designs, where the reference layers are represented by big arrows and the free layers by small arrows. (a) An in-plane design, with the reference layer thicker than the free layer. (b) An interface perpendicular design, with the reference layer thinner than the free layer. (c) An in-plane design with a SAF. (d) An in-plane design with both a SAF and an AFM.

4.4.1.2 MTJ Properties

Memory devices are characterized by several key characteristics, such as endurance, retention, power consumption, read/write time, and density.

4.4.1.2.1 Endurance

Assuming an access interval of 1 ns to 10 ns for a high performance memory, the memory will experience about 3×10^{16} operations for an expected operational lifetime of 10 years. Assuming further a 256 kiB memory density/64 B cache line (L2 cache [92]), an individual memory cell is accessed about 10^{13} times. The number of times the memory is read is usually not a problem, but changing the state of a memory device can degrade the storage mechanism. As such, memories can wear out by repeated writing, and a cache memory should, therefore, withstand 10^{13} write operations. The endurance describes how many times a memory can be rewritten before the memory states become indistinguishable, which is a special reliability concern of memory devices. In a magnetic memory, the storage mechanism is the direction of magnetization in a ferromagnetic metal. There is no known degradation mechanism for the magnetization—the direction can be switched an infinite number of times. Of more concern is the degradation of the tunneling barrier in MTJs, which can be degraded by current injection during writing. However, for MRAM it has been demonstrated that it has "practically" unlimited endurance ($>10^{12}$) [86,92,111].

4.4.1.2.2 Retention, Volatility, and Thermal Stability

The distinction between a volatile and a nonvolatile memory is that the retention time, the time which the data is retained after power off, for nonvolatile memories is more than 10 years. MRAM belongs to the thermodynamically stable nonvolatile memory category, which means that the two possible states are approximately equally stable. Nevertheless, the memory is still susceptible to thermal fluctuations. The failure mechanism is modeled as [99]

$$\frac{1}{\tau} = \frac{1}{\tau_0} \exp(-E_b / k_B T). \tag{4.10}$$

τ is the mean time, τ_0 the attempt period (approximately equal to the gyromagnetic resonance period [112], 10^{-10}–10^{-9} s), E_b the energy barrier (cf. Section 4.4.1.1), k_B the Boltzmann constant, and T the absolute temperature. The ratio $\Delta_H = E_b / k_B T$ is called the thermal stability factor and dimensionless. The thermally activated process is stochastic and the probability is modeled by the cumulative exponential distribution function:

$$p(t) = 1 - \exp(-t / \tau) \tag{4.11}$$

Nonvolatile memories require less than one bit-flip during 10 years ($\sim 10^8$ s), or $p(t_{10a}) < 1 / N$, where N is the number of bits in the memory. The required thermal stability factor can be estimated as

$$\Delta = \ln\left(\frac{t_{10a}}{\tau_0} \times \frac{1}{\ln(N / (N-1))} \right). \tag{4.12}$$

For 64 MebiByte memory and an attempt period of 1 ns, the thermal stability factor must be larger than 60 to qualify as a nonvolatile memory. The thermal energy at room temperature is approximately 26 meV; thus, the energy barrier must be larger than 1.56 eV. In practice, the minimum energy barrier is taken at the maximum expected working temperature ($\sim 80°C$), in this case the energy barrier must be larger than 1.84 eV, or the thermal stability factor $\Delta > 71$ at room temperature.

4.4.1.2.3 Writing and Critical Current

MRAM dissipates energy during writing in the form of Joule heating,

$$P_J = RI^2. \tag{4.13}$$

P_J describes the dissipated power, R the resistance of the MTJ, and I the current. For STT-MTJs, the major dissipation occurs during writing, when the current is large. Thermal stability is modified by the STT current [99,113,114]:

$$\Delta_I(I) = \Delta_I(I = 0) \times \left(1 - \frac{I}{I_c} \right) \tag{4.14}$$

I_c is the critical current density for switching. In microscopic models, it corresponds to the minimum spin torque required to reverse magnetization at absolute zero [113]. Δ_I is not necessarily identical with Δ_H; it has been shown experimentally that Δ_I is smaller than Δ_H [109]. If the temperature is nonzero, the MTJ can switch by thermal fluctuations, even if $I < I_c$. Obviously, also reducing the barrier leads to an increase in switching probability. If one bit is written once every 100 ns for 10 years (3×10^{15} write operations), the probability of *not switching* must be below 3×10^{-16} ($= 1 / 3 \times 10^{15}$). The necessary current as a function of pulse time can be estimated from

$$I(t_p) = I_c \times \left(1 - \frac{1}{\Delta_I(I = 0)} \ln\left(\frac{t_p}{\tau_0} \times \frac{1}{\ln(1 / p)} \right) \right). \tag{4.15}$$

Considering the assumptions from before, I / I_c must be larger than 0.986 ($\Delta_I = 71$). The above formular is valid for thermally activated switching regimes (a few ns upwards). For shorter switching times, there is not sufficient time for thermal excitations to aid the switching process and the switching changes towards the purely STT driven precessional switching regime with a steeper current increase for shorter switching times [114]. Fast, reliable switching requires that the current is close to or larger than the critical current. The critical current depends on the layer design, such as in-plane or perpendicular. For in-plane design, the critical current [21,96] is given by:

$$I_{c\text{in-plane}} = \left(\frac{2q}{\hbar} \right) \times \left(\frac{\alpha}{\eta} \right) \times (2E_b + \mu_0 M_s^2 V_F / 2). \tag{4.16}$$

q describes the elementary charge, $\hbar$ denotes the reduced Planck constant, α is the phenomenological Gilbert damping constant, and η the polarizing factor or spin-transfer efficiency. η depends on the spin polarization P and direction [115]:

$$\eta(P, \theta) = \frac{P}{2(1 + P^2 \cos(\theta))} \tag{4.17}$$

$\theta = 0$ describes the parallel and $\theta = \pi$ the anti-parallel state. η assumes its smallest value for the parallel state (large I_c). For the perpendicular design, the critical current is given by

$$I_{c\text{perpendicular}} = \left(\frac{2q}{\hbar} \right) \times \left(\frac{\alpha}{\eta} \right) \times 2E_b. \tag{4.18}$$

Comparing the critical currents for in-plane and perpendicular designs shows that in-plane designs have an additional energy contribution that must be overcome. This term originates from the fact that the STT, which switches the in-plane magnetization, must move the magnetization out of the layer plane. The related energy barrier is $\mu_0 M_s^2 V_F / 2$ higher than the switching barrier E_b between the major and minor ellipses axis of the layer. Perpendicular designs do not need to overcome this extra energy barrier and, thus, exhibit lower critical currents.

Since both the retention time (Equation 4.10) and the critical current (Equations 4.16 and 4.18) depend on the energy barrier, there is a trade-off between high retention time and low critical current. A corresponding figure of merit for an MTJ design is Δ_H / I_c (the higher the better). The perpendicular design has a higher figure of merit than the in-plane design, if all parameters are the same, since the perpendicular design, unlike the in-plane design, does not have to overcome the demagnetization field.

For a perpendicular layer CoFeB with a damping of 0.005 [93,96], an assumed spin polarization of 60% [104] in parallel state and an energy barrier of 1.8 eV, the critical current is 40 µA and its figure of merit at room temperature (Δ_H / I_c) is 1.7 µA^{-1}. The resistance of an MTJ is on the order of 1–10 kΩ [93]; thus, the voltage and power dissipation when writing is ~100 mV and ~1 µW, respectively. Given the switching time (~10–100 ns), the energy consumed during writing is on the order of 10–100 fJ. Low energy operation (90 fJ write energy) has been experimentally demonstrated [93]. The most important material parameters are damping and polarization.

4.4.1.2.4 Reading and TMR

Reading involves determining if the state is in a LRS or a HRS. The relevant metric is the TMR, given by Equation 4.3. The TMR is microscopically connected to the spin polarization P through Julliere's model, which assumes that spin is conserved during tunneling [32].

$$\frac{\Delta G}{G_p} = \frac{2P^2}{1 + P^2} \qquad (4.19)$$

G is the conductance ($= 1 / R$). Using Equation 4.3, Equation 4.19, and some algebra, the polarization can be estimated from the device TMR as

$$P = \sqrt{TMR / (TMR + 2)}. \qquad (4.20)$$

An infinitely high TMR corresponds to an ideal polarization of 1; thus, high polarization is important for material consideration. Polarization also improves the writability, as previously discussed. The state is read by a sense current that develops a voltage drop across the MTJ. The magnitude of the voltage is used to determine the resistance state. Since a current passes through the MTJ during reading, power is dissipated and its magnetization is excited, which can lead to a read disturb error. A read disturb error is an accidental bit-flip of a memory during the read operation. If one bit is read every 100 ns for 10 years (3×10^{15} read operations), the probability of *switching* must be below 3×10^{-16} ($= 1 / 3 \times 10^{15}$). The read disturb error is a switching event due to thermal activation over the current reduced barrier. The maximum allowed sense current can be estimated from

$$I(t_p) = I_c \times \left(1 - \frac{1}{\Delta_I(I = 0)} \ln\left(\frac{t_p}{\tau_0} \times \frac{1}{\ln(1 / (1 - p))} \right) \right). \qquad (4.21)$$

For a 10 ns read, I / I_c must be smaller than 0.47, for a 1 ns read 0.5 ($\Delta_I = 71$). Using the same values as for writing, the estimated sense voltage and dissipated power is on the order of 10–100 mV and 100 nW–1 µW. The time required to determine the state does not intrinsically depend on the MTJ but on the CMOS sense amplifier. A large difference in resistance between HRS and LRS (corresponds to high TMR) allows trading-off sense amplifier sensitivity to faster reading [96]. If the signal compared to noise is small, then the sense amplifier must be more sensitive and will be comparatively slow [116]. It should be noted that for the most common memory architecture (1T-1R, see Section 4.4.3), the transistor is connected in series with the MTJ when reading. For instance, assuming a transistor impedance of 1 kΩ, an LRS of 1 kΩ and an HRS of 7 kΩ (TMR = 600%), the resistance ratio of the memory cell is 300%.

The reading operation is unipolar for all spintronic designs—only the magnitude of the voltage/current matters, not the sign.

4.4.1.2.5 Density and Scaling

Memory density is arguably the most important metric for commercial products. The density is a measure of the number of bits per area, although it is often given by the number of bits per chip.

The density is not only determined by the size of the MTJ alone, but also by the size of the access device, typically a transistor (see Section 4.4.3). The size of the transistor is primarily determined by its drive current capability, which must be large enough to switch the MTJ (10–100 µA). The transistor width-normalized on-current for a low-power design transistor is about 600 µA/µm or A/m [116,117]. To provide 10–100 µA, the width has to be between 17 and 170 nm. For a hypothetical minimum sized memory cell (1T-1R) with a 20 nm memory half-pitch, the gate length would also be 20 nm (low-power logic transistor) [118]. The respective MTJ must be smaller than 20 nm and its switching current lower than 12 µA.

It is not trivial to take an MTJ design and scale it to smaller size. Looking at Equation 4.5 shows that decreasing the volume will cause a reduction in the energy barrier. To account for this decrease, the magnetic anisotropy strength has to be increased without degrading other parameters, for example like the damping by adding a second interface anisotropy [110]. The energy barrier and the switching current also depend on how the magnetization reversal takes place. Above a certain size (~40–70 nm [21,96]), it is easiest to reverse the magnetization by first nucleating a new domain and then having it grow. However, for decreasing size the magnetization reversal becomes nucleation dominated and the thermal stability factor almost independent of size. The figure of merit Δ_H / I_c improves with decreasing size. Below ~40–70 nm, the magnetic film prefers single-domain states and the entire domain switches instead of first growing a new domain. This is reflected by a saturation in Δ_H / I_c, as predicted by Equations 4.16 and 4.18 [96].

Most reports of MRAM circuits are on the order of Mib [94,95,109,119,120], and a few up to Gib [74,111].

4.4.1.2.6 Harsh Environment

There are applications where the electronics must be able to operate in harsh environments, such as military, vehicular, aerospace, space, and nuclear technology. The demands could be operation at low temperature (−40°C), high temperature (125°C), thermal cycling, and high radiation environment. Low temperature is not an issue for MTJ devices, as functional devices have been demonstrated to operate at liquid helium temperature (4 K). The spin-polarization through the tunneling layer is a function of temperature, and degrades with increasing temperature [121,122]. Nonvolatile operation at high temperature can be maintained as long as the thermal stability factor still exceeds about 60 at operation temperature. Or in other words the energy barrier must be about 30% larger as compared to room temperature ($\approx$2eV or $\Delta_H \approx 78$). MTJs are exceptionally radiation hard. Ionizing radiation cannot cause the magnetization to switch direction; thus, there are no single event upsets (SEU) [123] or loss of information. High enough radiation doses could cause displacement damage in the tunneling layer, which would degrade the overall memory cell. Nevertheless, the real radiation vulnerability lies in the CMOS circuit. CMOS electronics would break at doses below the doses necessary to damage the MTJs [120]. But an SEU can cause transient currents in the CMOS periphery during reading the MTJ, which can lead to bit-flip [124].

4.4.2 Magnetic Random Access Memory

This section covers different designs for spintronic memories. All of them share the same method of reading, the tunneling magnetoresistance effect. What distinguishes the designs is the way the memory state is switched. The designs covered here are thermally-assisted, STT, spin-Hall/spin-orbit, domain wall and finally voltage controlled magnetic anisotropy (VCMA).

4.4.2.1 Thermally-Assisted MRAM

Thermally-assisted (TA) MRAM adds temperature as a controllable variable. The principle is quite simple—the thermal stability factor is smaller at a higher temperature and, therefore, it is easier to switch the state. This allows to design very stable devices at room temperature, without any penalty for writing, because the energy barrier for writing and the energy barrier responsible for the stability are decoupled by temperature.

Taken to the extreme, the device can undergo phase-changes during the writing procedure. If the temperature is higher than the Curie temperature of the storage layer, it becomes paramagnetic. If the storage layer is cooled below the Curie temperature while biased into a state, it magnetizes easily into the biased state. Another design possibility is to couple the storage layer to an AFM. The storage layer can easily be switched, if the AFM is heated above its blocking temperature or Néel temperature (the AFM becomes paramagnetic), but is otherwise very difficult to switch [125]. Such devices are typically heated up to ~200°C.

The devices proposed in [125] are for field-written MRAM, but TA-writing has also been demonstrated for STT-MTJs. The Joule dissipation is used to heat the storage layer (above 150°C). The perpendicular anisotropy of the storage layer is reduced, allowing the STT to bias the storage layer into a state. The anisotropy recovers as it cools down and the storage layer settles into the state it was biased into by the STT [126].

4.4.2.2 Spin-Transfer Torque MRAM

STT-based MRAM is the most mainstream design and was extensively covered in Section 4.4.1. A typical scaled memory device uses a perpendicular CoFeB/MgO/CoFeB stack, thanks to its low damping, high spin polarization, high TMR and low switching current. Gib-density has been demonstrated [74,111] and STT-MRAM is commercially available in density of 32Mib × 8 [95].

4.4.2.3 Spin-Hall/Spin-Orbit MRAM

The two-terminal MTJ suffers from its shared read- and write-path. A large write current can cause degradation in the tunneling layer, while reading can cause read disturb errors. The three-terminal Spin-Hall or Spin-Orbit MRAM offers a way to decouple the writing and reading path. The major benefit is that the properties that determine reading and writing can be optimized independently of each other. The reading is still carried out by the TMR-effect, but the writing is performed by spin injection from a heavy-metal film by utilizing the SHE. The reference layer has one terminal, and the heavy-metal film has two terminals.

The SHE details have been explained in Section 4.2.3. Metals generating a spin-current are Ta [127,128], W [129], Ir-doped Cu [130], and several others [131]. There are differences between the metals in SHE strength and polarization efficiency. The performance metric is given by the spin Hall angle Θ_{SH}, which is the ratio of spin-current density J_s to charge-current density J_q. For example, Ta has a value of 0.12 [127] and 0.33 for W [129].

Geometric effects can significantly amplify the SHE [131]. The spin-current I_s to charge-current I_q ratio is given by $\Theta_{SH} A_s / A_q$, where A_s is the cross-section of the MTJ ($w_F l_F$) and A_q is the cross-section of the metal film ($d_{Metal} l_F$). Thus, $\Theta_{SH} A_s / A_q = \Theta_{SH} w_F / d_{Metal}$. The metal thickness d_{Metal} is usually several times thinner than the MTJs width w_F. Functional memory devices have been demonstrated in [127–130].

4.4.2.4 Domain Wall MRAM

Like the SHE-MRAM, domain wall MRAM is a three-terminal MRAM device that decouples the writing and reading paths. The reference layer has also one terminal, like SHE-MRAM, but the other two terminals are not connected to a metal film under the MTJ. Instead, they are connected to the storage layer that extends laterally out of the MTJ stack (see Figure 4.12). The storage layer is connected to spin-polarizers at each end. The spin-polarizers comprise ferromagnetic films pinned into opposite spin-states by AFMs. Since the two spin-polarizers have fixed spin-states, two domains form in the magnetic film, which are separated by a DW. When current is injected from a polarizer into the storage layer, the domain grows and the DW moves towards the other end of the layer. Depending on the current direction through the storage layer, the DW can be moved repeatedly back and forth, allowing to deliberately set the magnetization orientation below the MTJ stack either in a parallel or antiparallel orientation with respect to the reference layer [132].

The formation and design of domain walls is a complex topic, depends on the material parameters as well as on the geometry, and is the result of the energy minimization of the magnetic film, where several energy contributions compete with each other (e.g., exchange energy, anisotropy energy, and demagnetization energy). For DW-MRAM, the storage layer's material properties and geometry have been chosen so that the layer sustains stable domain walls. Memory devices based on DWs have been demonstrated [133,134] and there are also designs with more than three terminals [135,136].

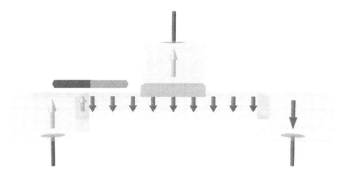

FIGURE 4.12
The DW memory has three leads, one connected to the MTJ reference layer (spin-up) and one to each polarizer. The DW is at the left hand side of the ferromagnetic layer. The spin-down domain encompasses both the right hand side polarizer and the storage layer. The memory is in antiparallel state. If a spin current is injected from the left hand side polarizer into the storage layer, the DW moves to the right and changes the state of the memory.

4.4.2.5 Voltage Controlled Magnetic Anisotropy MRAM

The voltage controlled magnetic anisotropy was discussed in Section 4.2.4, and it offers a new way to control an MTJ. As the free layer thickness scales down, the interface dominates over the bulk. Thin enough films can exhibit a perpendicular net magnetization due to interface effects, as describe in Section 4.4.1. The VCMA is controlled by the application of a voltage or an electric field. It can be described as [62]

$$\sigma_i(V) = \sigma_i(V=0) - \xi V / d. \tag{4.22}$$

ξ describes the VCMA coefficient, V the applied voltage, and d the thickness of the tunneling barrier. The ξ values for the CoFeB/MgO systems range from 30–100 fJ/Vm [62,67]. The VCMA effect can be used either in conjunction with STT, where it reduces the energy barrier and thus the switching current, or by pure voltage switching by removing the barrier. When the barrier is removed, the magnetization freely precesses between the states, allowing STT-free switching at the physically fastest switching rate [62]. The precessional switching, while extremely fast and energy efficient, is circuit-wise complicated by the nondeterministic end-state (it keeps precessing between the states as long as the voltage pulse is on). The pulse must be very precise to ensure that it toggles into the desired state. Since there is no force driving it into a certain state, the memory must be read after writing to determine if the writing was successful or must be performed again.

4.4.3 Memory Cell Architecture

Memory matrices are typically active matrices, where the memory cell is composed of a selector device and a memory functional device [90]. The selector is a non-linear device that decouples the bit line from the memory functional device, unless selected (cf. Figure 4.13).

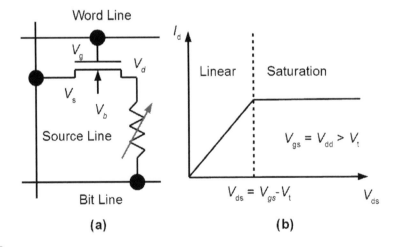

FIGURE 4.13
(a) A 1T-1R memory cell. V_g, V_s, and V_d are connected to the word line, source line, and an MTJ, respectively. The body connection V_b is not connected in this figure, but is usually tied to ground. (b) The simplified transistor output current characteristics. For small V_{ds}, the transistor behaves as a resistor, and, for large V_{ds}, it sinks the maximum amount of current.

Examples of selector devices are diodes (PN, Schottky) and transistors (NPN, NMOS, and PMOS). Diodes are preferred for their smaller size over transistors, but their rectifying property limits the memory functional device to the unipolar class. NMOS and PMOS transistors are used for bipolar class memories. NMOS transistors are preferred over PMOS transistors because they deliver larger currents for a given size due to their higher charge carrier mobility.

The access transistor is a four terminal (gate, source, drain, and body) device for a bulk CMOS process and a three terminal device for a silicon on insulator (SOI) process (gate, source, and drain). The body of the bulk transistor is connected to the lowest potential for an NMOS device (typically ground) and the highest potential for a PMOS device (typically V_{dd}). The transistor is characterized by its threshold voltage V_t. The NMOS transistor is on, when the source-gate voltage $V_{gs} = V_g - V_s$ is larger than V_t and off if smaller than V_t (for the PMOS it is the opposite, and V_t is negative). The maximum current through a long-channel transistor occurs, when the transistor is on ($V_{gs} - V_t = V_{ov} > 0$) and the source-drain voltage $V_{ds} = V_d - V_s$ is larger than the overdrive voltage V_{ov}. The transistor operates in saturation and is approximately independent of V_{ds} as long as the saturation condition is met. Since STT-switched MTJs require large currents, the transistor may operate in saturation during programming. For small V_{ds}, the transistor operates in the linear region and behaves as a resistor with an impedance on the order 100 Ω–1 kΩ. The transistor is operated in the linear region during reading. Figure 4.13 shows the transistor connected to an MTJ and the transistor output current characteristics.

The most basic CMOS circuit has only two potentials, ground and V_{dd}. The power supply depends on the CMOS technology and its intended application, but is in the range of 1.2–3.3 V for submicron transistors. A necessary but not sufficient condition for turning the NMOS transistor on is that the gate potential is V_{dd}. The NMOS transistor can still be off, if the source potential is larger than $V_{dd} - V_t$, since $V_{gs} = V_g - V_s < V_{dd} - (V_{dd} - V_t) = V_t$. This condition can occur for the access NMOS transistor, if it tries to source a large current to a large resistor, such as in the case of programming an MTJ with high resistance. The equivalent case for a PMOS transistor solution is when it tries to sink a current. This condition is known as source degeneration. There are several ways to avoid this. A charge pump can boost the gate voltage to $V_{dd} + V_t$, in such case, a full V_{dd} voltage-drop can be applied to the MTJ. But the charge pump solution is less preferably, as it consumes extra overhead area, power, and adds circuit complexity [99]. The largest current occurs during writing P→AP, when the current is sourced through the reference layer. For this case, it is best to connect the highest potential to the reference layer, or in other words, to connect the reference layer to the bit line. Consequently, the free layer is then connected to the drain of the NMOS transistor, which sinks current to the source line during the P→AP transition. Another solution is to use a PMOS transistor [91], where the drain is connected to the reference layer. The PMOS transistor will not have any source degeneration, when it sources the current during the P→AP transition, since the source is connected to the highest potential.

MRAM devices can be separated into devices with unipolar and bipolar switching. Most, but not all, MRAM devices are two-terminal devices. If there are multiple terminals, they are typically connected to different transistors. As such, MRAM can be realized by several memory cell architectures, such as 1D-1R, 1T-1R, 2T-1R, or 2T-2R (D = Diode, T = Transistor, R = Resistor, sometimes referred to as MTJ in literature). There are even more memory cell architectures, like 6T-2R, which will be briefly covered in Section 4.7.

1D-1R: This is the smallest memory cell for spin-based memories, with a minimum cell size of $4F^2$. F refers to the half-pitch in DRAM memory (2F = width + spacing = pitch), and is connected to the minimum lithographic feature size. It requires that the MTJ itself is not larger than $4F^2$ and can be placed on-top of the diode. A requirement of this memory cell is that the MTJ allows unipolar switching, which is possible for the orthogonal MTJ design (cf. [137,138]) and devices using VCMA [62]. The cell is selected (selector in low-impedance state), when there is a voltage drop across the diode (about 0.2 V for a Schottky and 0.7 V for a PN).

1T-1R: This is the standard memory cell for spin-based memories, shown in Figure 4.13. The 1T-1R is superficially similar to the 1T-1C of the DRAM. The densest DRAM cell uses an open architecture with a transistor size of $6F^2$. The size is minimized by having one contact shared between two transistors. It has been argued that the 1T-1R cell can be as small as $6F^2$ at the 90 nm node by [99], $9F^2$ at 45 nm feature size [74], and $22F^2$ was demonstrated at 28 nm feature size [111].

2T-1R: This cell design is used for three-terminal spintronic devices, like SHE-MRAM (Section 4.4.2.3) and DW-MRAM (Section 4.4.2.4). The size is larger or equal to $12F^2$, as transistors cannot share their contacts [134]. These devices target SRAM replacement instead of DRAM replacement [135].

2T-2R: This cell should be interpreted as 2×1T-1R. It dedicates "two" 1T-1R cells to store the data and their complement. The minimum size is larger or equal to $12F^2$. The memory density is about half of designs with1T-1R cells. This is a major disadvantage, but the design offers some performance advantages. The storing of both the data and its complement effectively doubles the read signal and allows to trade sensitivity for fast differential reading (see Section 4.4.1.2). It is also very robust to process variations. The performance of individual MTJs is strongly correlated to their locality, because MTJs near each other are more likely to show the same performance compared to MTJs further apart. As such, the R_P and R_{AP} of the two MTJs are very likely a close match, which almost guarantees correct reading. For example, it is highly unlikely that the R_P of one MTJ is as high as the R_{AP} of the adjacent MTJ, or conversely that the R_{AP} is as low as the R_P of the adjacent MTJ. For large arrays, the spread of values of R_P and R_{AP} in 1T-1R designs must be very small (i.e., $20\sigma < \bar{R}_{AP} - \bar{R}_P$ [96,99]) to ensure that $\min(R_{AP})$ of the array can never be misinterpreted as parallel state, and that $\max(R_P)$ of the array is not misinterpreted as antiparallel state. An example of a 2T-2R design can be found in [94], using a 90 nm CMOS process.

4.5 Spintronic Logic

As explained in the previous sections spintronic devices, in particular MTJs, are very promising for memory applications due to their nonvolatility, CMOS-compatibility, fast operation, and (nearly) unlimited endurance. But they are by no means limited to pure memory applications, which is reflected in the ITRS [118], where it is suggested to exploit nonvolatile devices to circumvent current limits of state-of-the-art logic circuits. Among the many challenges in current CMOS technology development, the introduction of nonvolatile elements into logic circuits, allows to tackle the issue of the exponentially growing standby power dissipation [139]. Within the smorgasbord of available memory technologies STT-switched devices are especially appealing for logic applications [25,140–151].

4.5.1 Logic-in-Memory

Logic-in-Memory supplements the logic circuit plane by adding nonvolatile elements [152–155]. This way instant-on and more importantly an energy efficient transition between the shut down state and the active state are possible, but at the same time also the circuit complexity and the layout footprint increases. Typical application scenarios are in microprocessors and field programmable gate arrays [156]. Microprocessors are already incorporating various power reducing technologies and operation schemes (e.g., reduced operation voltage, clock gating, and power gating), but the power reduction commonly comes at the price of reduced performance. The more energy is saved, the longer it takes to enter into and to exit from the power saving mode. Introducing nonvolatile flip flops [157], or STT-MRAM [158], speeds up this transitions considerably and allows more frequent transitions into the power saving states as well as into deeper sleep states, resulting in a reduction of the total system power without degrading performance [159,160]. The second application area are field programmable gate arrays (FPGAs). They belong to the most popular reconfigurable hardware platforms and are employed for rapid prototyping and as a generic hardware for mapping arbitrary applications [161]. Commonly, they consist of elementary logic functions (lookup tables), which are connected through wire segments and programmable switches. The content of the lookup tables and the states of the programmable switches are fully defined by the bits stored in the configuration memory. Currently, there are two main groups of FPGAs, SRAM-based that store the configuration in SRAM memory cells and flash or anti-fuse FPGAs that employ nonvolatile memory for storing the configuration [162]. SRAM-based FPGAs are very fast, but need an additional nonvolatile off-chip storage for storing the configuration. Therefore, their startup is rather slow and takes around 100 ms. Additionally, SRAM cells have a big footprint in comparison to other memroy cells. In contrast, anti-fuse and flash-based FPGAs have a smaller footprint and startup faster, but anti-fuse FPGAs can only be programmed once and flash-based FPGAs have a very slow and energy consuming writing. MTJ/CMOS hybrid FPGA designs combine the advantageous features of both technologies without their drawbacks. They make the off-chip nonvolatile memory superfluous, which allows a very fast boot process and reduces the design complexity at the chip level, while featuring at the same time the speed of SRAM-based FPGAs and (partial) run-time reconfigurability. Another benefit of the transition to MTJs is the improved single event upset reliability of the resulting FPGAs. Especially for the deep-submicron technology nodes, this has become a concern [163]. Since the first proposal to use 100 nm thermally-assisted MTJs in combination with 130 nm CMOs technology to build a nonvolatile FPGA by Bruchon et al. [164] in 2006, a wide variety of publications picking up the idea and trying to enhance FPGAs from circuits up to the architectural level has followed [165–170]. The naive approach to simply replace the SRAM cells and flip flops in the FPGAs with nonvolatile counter parts suffers under area and leakage current increase due to the additional required components to read and write the MTJs [171,172]. Therefore, Suzuki et al. [170,173,174] proposed a six-input lookup table circuit with shared write driver and sense amplifiers in combination with redundant MTJs to decrease the influence of resistance variations and improve the programmability. In the remainder of this section, we focus on the spintronic logic proposals that use the spin-based device as the main logic element and replace a CMOS-based logic block rather than acting as a complement. This paves the way for realizing intrinsic logic-in-memory architectures.

4.5.2 Spin-Transfer Logic

The dissipated power and the interconnection delay are central issues that have far-reaching implications throughout the digital ecosystem [118,175]. Nowadays, the static power consumption is minimized by shutting down unused circuit parts. Even though this strategy is simple and effective, it bears the disadvantage of loosing all the information stored in the circuit through dissipation. Therefore, it must be copied back into the circuit when it is brought online again, which adds delay and power consumption. A way to avoid this, is to use nonvolatile elements in the circuits. Spintronics exhibits a number of features that make it very attractive for such nonvolatile elements and circuits. Among the currently available multitude of ideas the technology readiness level for commercialization strongly varies [25,176] and despite the availability of many candidates for potential CMOS successors, CMOS will be around forever. Even more the upcoming generation of widespread commercially available products within the next few years will be nonvolatile CMOS MTJ hybrids [25,177–180]. In fact, STT-MRAM is already available on the market and many more applications will very likely follow soon [95,181]. Even though the CMOS MTJ hybrid solutions progress fast and that they are already competitive with respect to speed and power consumption in comparison to pure CMOS, at one of the essential features that guaranteed CMOS success, its integration density, the current solutions are still inferior. In principle, STT-MRAM is suited for high integration density and it is already three-dimensionally integrated at the BEOL, but it still suffers under relatively high switching currents for the MTJs, which limits the minimum useable transistor size. This led to the investigation of alternative switching mechanisms, such as the spin Hall effect, to surpass this limit [182,183]. The exploitation of STT-MRAMs for Compute-in-Memory (see Section 4.5.7) applications is very appealing due to their potential for high integration densities as well as the exploration of novel computation concepts, but they all remain limited by the same boundaries as the state-of-the-art STT-MRAM. To overcome these obstacles, researchers are also investigating alternatives to push the achievable integration densities beyond todays limits, i.g. [25].

A way to increase density is to put as much as possible of the CMOS functionality into the spintronic devices. The result of such efforts are the proposal of a nonvolatile magnetic flip flop (NVMFF) and a nonvolatile magnetic shift register [184]. The nonvolatile flip flop exploits spin-transfer torques and magnetic exchange coupling within its free layer to perform the actual computation instead of relying on external CMOS transistors. Thereby it is possible to reduce the number of required transistors, reduce the structural complexity, and take advantage of the resulting very dense layout footprint. Rigorous simulation studies were carried out to explore the capabilities as well as the limits of the NVMFF [185–189]. Additionally, the NVMFF can be combined with a STT majority gate in order to create a novel nonvolatile buffered magnetic gate grid.

4.5.2.1 Nonvolatile Magnetic Flip Flop

Flip flops belong to the group of sequential logic circuits and are an essential part of modern digital electronics [190]. Thus, the nonvolatile magnetic flip flop is a fundamental building block required in the creation of a nonvolatile STT computation environment.

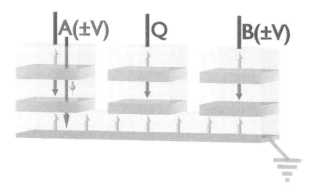

FIGURE 4.14

The nonvolatile magnetic flip flop comprises three MTJ or spin valve stacks that share a common magnetic free layer. It is operated by two simultaneously applied current pulses at its inputs A and B. Logic "0" and "1" are encoded via the pulse polarity and the result of each operation is stored in the common free layer as the layer's magnetization orientation. The information stored in the flip flop is accessible through its output Q as a high or low resistance state by exploiting either the GMR or the TMR effect.

Therefore, we will first explain how a single NVMFF works and use this knowledge later as basis for more complex applications, such as a nonvolatile shift register or a nonvolatile buffered gate grid. The NVMFF comprises three antiferromagnetically coupled polarizer stacks with perpendicular magnetization orientation (see Figure 4.14). The polarizer stacks are connected with each other through nonmagnetic interconnection layers (e.g., MgO, Al_2O_3 or Cu) and a common free layer with perpendicular magnetization orientation. One of the stacks is used for readout Q and the remaining two are dedicated to input A and B. Due to the anti-ferromagnetic nature of the polarizer stacks it is assumed that their stray field is negligible. The information is stored as magnetization orientation of the shared free layer and accessible via GMR or TMR effect. Depending on the relative orientation between the magnetization orientation of the shared free layer and the readout stack Q a high resistance state HRS (antiparallel) or a low resistance state LRS (parallel) is measured. The respective HRS and LRS are assigned to logic "0" and logic "1," respectively. For the operation of a single NVMFF, the polarity of the input pulses is mapped to logic "0" and "1." If now, a negative voltage is applied to one of the inputs (i.e., A), then electrons will flow from the leads through the polarizer stack, where they align with the local magnetization orientation, and eventually enter the common free layer before they get absorbed by the grounded bottom contact. During their time in the common free layer, the electrons relax to the free layers magnetization orientation. This creates a localized torque in the region where the electrons traverse, which depends on the pulse polarity and the relative orientation between the layers. Depending on the electrons polarization orientation the exerted torque either tries to push the magnetization into its other stable position or damps the magnetization precessions and stabilizes its current orientation. For the operation of the flip flop, two synchronous input pulses are applied to the two inputs A and B. Both pulses exert an STT on the common free layer. For fixed parallel magnetization orientations of the two input stacks and two input pulse polarities, four input combinations are feasible. Depending on the input pulse polarities the two created STTs either add up and accelerate the switching (same polarity) or counter act each other and damp the switching (opposing polarities). Translating this behavior to a logic table shows that the device can be SET/RESET (same polarity) as well as HOLD its current

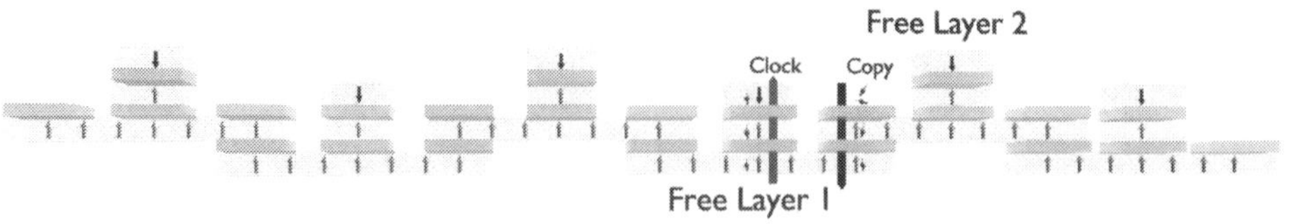

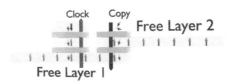

FIGURE 4.15
A shift register consists of flip flops that are connected in series in order to pass the information stored in one flip flop to its subsequent neighbor. To create this kind of functionality, the nonvolatile flip flops are arranged in two rows in two distinct levels. The free layer of every flip flop overlaps at its boundaries with its neighbor flip flops on the respective other level. The polarizer stack in the middle of the flip flops is exploited for the generation of an auxiliary (clocked) STT.

state, which is exactly required for sequential logic as flip flops and latches [190]. In comparison to other ideas in this field, the key advantage of the nonvolatile magnetic flip flops is its very small footprint. But, in general, a single flip flop by itself is of limited use. Only when it is seamlessly integratable into bigger circuits without compromising the overall integration density, it is of practical value. Therefore, it was investigated how one can build bigger structures like a shift register out of the flip flops without degrading the overall integration density. It is essential to keep as much as possible of the required functionality in the spintronic domain in order to sustain the achieved integration density and the key to this is to directly copy the information from one flip flop's free layer into a subsequent free layer (see Figure 4.15).

4.5.2.2 Nonvolatile Magnetic Shift Register

The copy operation is achieved by first traversing an unpolarized current through the layer that is read (cf. Figure 4.16, Free Layer 2 overlapping region) and exploit the afterwards orientation encoded spin polarized electrons in the subsequent layer (Free Layer 1) to exert a spin-transfer torque on the local magnetization in the region where they enter. This way it is possible to pass directly the information from one device to the next without complex external CMOS blocks. As explained in Section 4.5.3, there are always two torques when the electrons interact with the local magnetization. One acts on the electrons while the other acts on the magnetization. Therefore, when the electrons are polarized in the read layer (i.e., Free Layer 2), there is always a torque that destabilizes the magnetization of the free layer and might cause a read error. The solution for this problem is to speed up the

FIGURE 4.16
The n-Bit shift register from Figure 4.15 has been reduced to a 2-Bit shift register, in order to reduce the computational effort. During the copy operation, an unpolarized current is pushed through Free Layer 2. The with the orientation of Free Layer 2 encoded electrons enter Free Layer 1, where they exert a spin-transfer torque on the magnetization of the layer. The auxiliary pulse through the clock stack creates a second spin-transfer torque that speeds up the copy operation by either damping or enforcing the switching of the magnetization in Free Layer 1.

switching of the written layer (i.e., Free Layer 1) by adding a second auxiliary STT. Thereby, the copy operation will require less time than it takes to cause a read disturbance. The concept was tested by an extensive set of simulations. To keep the computational effort on a manageable level, the n-Bit shift register was reduced to a 2-Bit shift register and rigorous simulation studies were carried out to at first test the idea [191] and later check its limits with respect to manufacturing related misalignment [192,193]. It was not only found that the concept works, but that the structures are even capable of tolerating moderate in-plane as well as out-of-plane misalignment.

4.5.2.3 Nonvolatile Buffered Magnetic Gate Grid

A further example for the possible application of the flip flop is its use in a nonvolatile buff-ered gate grid [189], which, like the shift register, takes advantage of passing directly the stored information from one free layer to the next. To create the nonvolatile buffered gate grid, one needs an extra ingredient, the STT majority gate. The majority gate employs the same material stack for the free layer and the polarizers as the flip flop and is also based on the same information encoding principle via input polarity [185,194]. Therefore, the flip flop and the majority gate can be synergetically combined into bigger circuits. Since both devices are similar, the focus will be on explaining the differences between these two and how they interact. First of all the STT majority gate belongs to the class of combinational logic devices, while the flip flop belongs to the class of sequential logic devices. Both types are essential for building a computing environment and complement each other with their functionalities. The most obvious structural difference between them is that the free layer of the STT majority gate is cross shaped and features four instead of three polarizer stacks (cf. Figure 4.17). Three of the polarizer stacks A, B, and C are used as inputs and one polar-izer stack Q is used for readout. The STT majority gate is operated via three synchronous polarity encoded input pulses and the final orientation of the free layer is defined by the majority of the input signals. One must mention that it is crucial that the number of applied inputs is odd. Otherwise, it can happen that the number of "0" and "1" input signals are equal, and the created torques perfectly balance each other (assuming equal input cur-rents and equal torque strength), which leads to an undefined state after the operation. Only when an odd number of inputs is applied, there is one uncompensated torque during

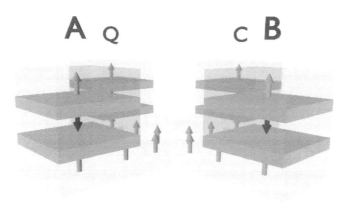

FIGURE 4.17
The STT majority gate comprises four equidistantly spaced legs. Each of the legs exhibits a polarizer stack and is connected through an interconnection layer to a common cross shaped free layer. One of the stacks is dedicated to readout Q and the remaining three A, B, and C are employed as inputs.

operation left, which will decide the final state. Another important feature for building arbitrary logic functions is functional completeness. In CMOS logic circuits, the NAND and NOR gates are widely employed due to their functional completeness. Looking at the truth table of the majority function shows that it consists of a two-input AND and a two-input OR gate, when one of the inputs is fixed to logic "0" and "1," respectively. Therefore, the NOT operation must be added in order to reach functional completeness. The easiest way to introduce the NOT operation is to invert the acting torque by inverting the polarity of the input signal.

By combining the STT majority gates and the nonvolatile flip flops into a buffered gate grid, the resulting circuit is not only CMOS compatible and able to complement CMOS logic, but also one achieves much more. Namely, the communication between the (external) memory and the logic is significantly decreased and the auxiliary CMOS circuits for the signal conversion between the CMOS and the spintronic domain become redundant, which in turn greatly improves the omnipresent leakage power and interconnection delay problems [10,139,176]. The devices are periodically distributed over the die plane and positioned in two separate levels with zones where they overlap with their neighbor devices in the respective other level (see Figure 4.18). Adding contacts at the top and the bottom of the overlapping regions allows to directly copy the stored information from one free layer to the next, the same way as for the shift register explained before. The resulting structure is highly regular, allows parallel execution of operations, and brings the advantage of a shared buffered between adjacent logic gates. The synergetic combination of all the features allows to keep the integration density high, while at the same time the energy and time spent for the information transport are minimized. Even more, it also enables the investigation of computing alternatives to the nowadays performance limiting Von Neumann architecture, where the computation units and the memory are physically separated and the information is continuously pushed back and forth between them. Furthermore, this structure gives considerable freedom in allocating the employed resources and it is very easy to reconfigure its logic; that is, the number of operating gates and buffers can be adjusted on the fly depending on the current computing task. To give an idea of how this nonvolatile buffered gate grid can be exploited in practice, the example of an easily concatenable one-bit full adder (cf. Figure 4.19) will be discussed in the following.

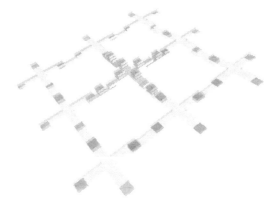

FIGURE 4.18
The nonvolatile buffered gate grid is formed by a periodic continuation of the STT majority gates and nonvolatile flip flops. The nonvolatile flip flops (rectangles) act as shared buffers and the STT majority gates (crosses) perform the calculations.

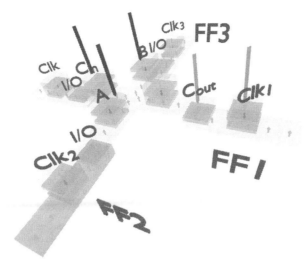

FIGURE 4.19
A single node of the buffered magnetic gate grid comprising a single majority gate and three flip flops is already capable to perform the calculations of a concatenable one-bit full adder. As for the shift register, the key for this is the exploitation of the devices' free layers as polarizers.

For the one-bit full adder three inputs, A, B, and C_{in} (carry in from a previous adder stage) and two outputs Sum and C_{out} are assumed. Sum is given by [190]:

$$\text{Sum} = A \ XOR \ B \ XOR \ C_{in}$$

$$= A \cdot B \cdot C_{in} + A \cdot \overline{B} \cdot \overline{C}_{in} + \overline{A} \cdot B \cdot \overline{C}_{in} + \overline{A} \cdot \overline{B} \cdot C_{in} \qquad (4.23)$$

C_{out} denotes the carry out and takes care of the overflow into the next digit for a multi-bit addition:

$$C_{out} = MAJORITY(A, B, C_{in})$$

$$= A \cdot B + A \cdot C_{in} + B \cdot C_{in} \qquad (4.24)$$

In order to perform both calculations on a single majority gate one has to translate them into a sequence of *MAJORITY*, *NOT*, and copy operations. Since the *MAJORITY* function and the *NOT* form a functional complete basis, there is a well-defined sequence that achieves the calculation of SUM and C_{out}. For instance, as first step $MAJORITY(A, B, C_{in})$ is performed and the result is copied into the first buffer $FF1$ (see Figure 4.19). Then $MAJORITY(A, B, NOT(C_{in}))$ is performed and its result is stored in the second buffer $FF2$. Now, in the final step, the information contained in $FF1$ and $FF2$ is combined through $MAJORITY(NOT(FF1), FF2, C_{in})$ to calculate the Sum and to copy it into $FF3$. At the end of this sequence, Sum is stored in $FF3$ and C_{out} is contained in $FF1$. Since the results are safely stored in the buffers $FF1$ and $FF3$, they can be exploited for further calculations in their neighbor gates. For example, the C_{out} stored in $FF1$ can be used as carry in the next one-bit adder stage (e.g., majority gate at the right side of the central gate), even before the Sum calculation in the initial stage has been finished. This illustrates the parallelization potential of the structure and how the expensive information transport over a common bus can be minimized.

Straightforwardly, one can build a one-bit full adder without the proposed buffering [195]. However, without the buffers, it is much harder to generalize the layout and fit it into large scale integration schemes and it suffers under further drawbacks, which led to the proposal of an alternative solution with in-plane magnetization [196]. This alterative approach employs an all-spin logic inspired hybrid shift register and a stacking scheme that is similar to a previously proposed shift register [184,186]. Their exploitation of in-plane instead of out-of-plane magnetization also requires a redesign of the majority gate, which was realized by a ring structure.

4.5.3 All-Spin Logic

The direction of the spin transport is often tied to the direction of the charge transport, but by nature spin and charge transport are independent. Behin-Aein et al. proposed a type of spintronic logic that takes advantage of this fact and coined the term "All-Spin Logic" (ASL) [141,197,198]. This type of logic employs magnets to store logic information and to create spin polarized signals. The magnets are connected via nonmagnetic conducting channels in order to propagate spin signals between them (see Figure 4.20). These structures feature pure spin signal transport and can be exploited for sequential as well as combinational logic [141,199–201].

The proposal of this very interesting idea triggered a broad spectrum of activities to explore the different aspects of the concept. For example, ways to create and operate sequential logic, like a shift register and a ring oscillator, have been investigated in [200]. The scaling properties and the energy delay of ASL devices are investigated in [198,202]. The optimization of the device structure including interconnect materials and their respective advantages and disadvantages has been studied in [203–211]. A further important step for the progress in the field was the development of a generalized framework for modeling spintronic devices on the circuit level [212]. One of the key features that makes ASL attractive is the consequent avoidance of charge transport and by that a significant reduction in the dissipated power.

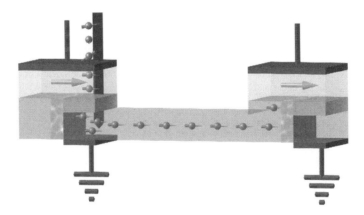

FIGURE 4.20
The magnets (boxes with white arrows) are used to store logic information (magnetization orientation) and to create spin polarized signals. Electrons, entering from the top electrode, are polarized when they move through the magnet and create a spin accumulation at the bottom electrode after they passed an oxide layer. This accumulation drives a spin diffusive current that reaches the neighbor magnet, where it relaxes and creates a spin-transfer torque able to switch the magnetization. This way a pure spin signal, encoded with the magnetization orientation of the magnet, is created and exploited to copy information from one magnet to another (left to right).

However, current estimates for the power consumption are worse than for state-of-the-art CMOS circuits [10,176,208]. This is not specific to ASL and shared by many other spintronic technologies, which can be easily explained by the head start of many decades for CMOS technology in research and development. Nevertheless, we feel confident that over time with growing knowledge and experience in the field of spintronics the gap will not only be diminished, but that spintronics will even substitute CMOS electronics for certain applications. In order to be able to study and explore ASL circuits larger than a few gates, one has to simplify their description and translate the device behavior into compact models. Many of these compact models employ the assumption that the magnets can be described by a single macro-spin [207,210,212–214]. It is obvious that the quality of the gained results on the circuit level crucially depends on the physical accuracy of the employed compact models. The findings of Verma et al. [215] show that the macro-spin assumption, which relies on a uniform precession and switching of the magnetization, is not valid. It has been further demonstrated that the current flow through the magnet and the related STT is strongly nonuniform. This translates into a significant influence on the overall switching behavior of the magnet and therefore must be incorporated in the compact models. An additional effect commonly ignored is the pairwise occurrence of the spin-transfer torques. If one pushes electrons through a magnetic layer to polarize them, not only the electrons experience a torque that aligns them to the local magnetization. There is always a simultaneously acting back torque that acts on the magnetic layer and destabilizes the magnet, thus causing a read error [216]. Therefore, the influence of these effects must be quantified and considered in the compact model description to gain meaningful ASL circuit modeling and analysis.

4.5.4 Domain Wall Logic

The creation and manipulation of magnetic domain walls via spin polarized currents for storing information and realizing logic has been a very hot topic for many years [134,217–227]. Since many of the domain wall related logic ideas are similar, we pick All-Metallic Logic (mLogic) [225,226] as a representative for this class of logic and discuss its basic features in the following. The basic mLogic device is shown in Figure 4.21 and comprises a free layer holding a domain wall sandwiched between two fixed oppositely magnetized polarizers (bottom of structure). The free layer is coupled to an adjacent free magnetic layer through a coupling layer (mottled layer) and its state can be accessed through the GMR stacks on the left and right side, which connect it to the leads R and R'. If one applies a current through the W^+ and W^-, the electrons will first pass a polarizer stack and subsequently exert a spin-transfer torque that pushes the domain wall along the layer. By changing the polarity of the applied current, the domain wall can be reversibly moved back and forth through the free layer. Due to the coupling between the upper and the lower free layer, the magnetization orientation of the upper layer follows accordingly. The logic state of the device is accessible through the GMR stacks, which are connected to the upper free layer, and their respective low and high resistance states depend on the free layers' orientation. The advantage of this gate is that it mimics CMOS behavior (n- and p-type by swapping input ports), allowing for the reuse of CMOS circuit design and replacing the n-type and p-type transistors by corresponding mLogic gates. The all-metallic structure also enables very small supply voltages. However, it also can cause high leakage currents and lead to degraded energy efficiency. Of course, it is also possible to create domain wall logic with simpler structures by cascading, for example DW memory cells (cf. Section 4.4.2.4), but it comes at the price of coupling the read and write paths of the devices. Thus, the control currents must pass the tunneling oxides, which has implications on wear and supply voltage [221,222,224].

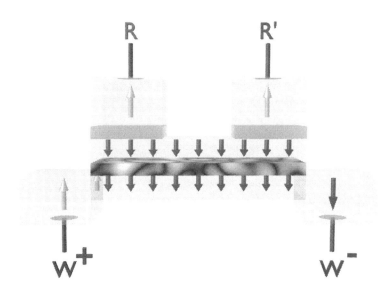

FIGURE 4.21

The mLogic device exhibits separate read and write paths. The writing path is formed by the free layer at the bottom of the structure and its two adjacent polarizers. Pushing a current through W^+ and W^- moves the domain wall (left corner) through the free layer. The read path is formed by the upper free layer and the two polarizer stacks at its ends. Upper and lower free layer interact through an insulating coupling layer (mottled layer).

4.5.5 Reprogrammable Logic

To exploit magnetic devices as computing elements and provide logic-in-memory, it has been shown that the use of direct communication between STT-MTJs can realize the basic Boolean logic operations. The experimental demonstration of two-input and three-input reprogrammable logic gates (Figure 4.22) to implement AND, OR, NAND, NOR, and the Majority operations is reported in [142] and [143]. A Boolean logic operation is executed

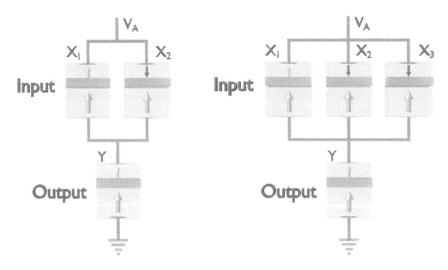

FIGURE 4.22

STT-MTJ-based two-input (left) and three-input (right) reprogrammable logic gates. X_i (Y) shows an input (output) MTJ.

in two sequential steps. These steps comprise an appropriate preset operation (parallel or antiparallel state) in the output MTJ. Then a voltage pulse (V_A) with a proper amplitude is applied to the gate. Depending on the logic states of the input MTJs (X_i), the preset in the output MTJ (Y), and the voltage level applied to the gate, a conditional switching behavior in the output MTJ is provided, which corresponds to a particular logic operation.

Tables 4.1 and 4.2 illustrate how the AND and OR operation and NAND and NOR operation, respectively, are performed using the two-input reprogrammable gate in two steps. The HRS and the LRS correspond to logical 0 and 1, respectively and the variable x_i and y represent the logic state of the input (X_i) and output (Y) MTJs. In order to perform a logic operation, first a preset of $y = 1$ or $y = 0$ is performed in the output MTJ and then in the following step a proper voltage level ($V_A < 0$ or $V_A > 0$ is applied to the gate to enforce the desired (high-to-low or low-to-high) resistance switching event in the output MTJ, which corresponds to a logic operation (i.e., AND/OR or NAND/NOR). The value of the voltage V_A has to be optimized to ensure a reliable conditional switching behavior of the output MTJ for any possible input pattern [228]. In fact, this optimization is required for any logic operation to maximize (minimize) the switching probability in the output MTJ ($P \rightarrow 1$ or $P \rightarrow 0$), when it is a desired (an undesired) switching event in Step 2. One should note that the switching probability of any input MTJ is negligible as the current flowing through the output MTJ splits between the input MTJs, and their currents are below the critical current required for the STT switching. Therefore, the logic state of the input MTJs is left unchanged.

TABLE 4.1

The realized conditional switching behavior is equivalent to the AND and OR operations with a preset of $y = 1$. Using the two-input reprogrammable gate.

	Input Patterns		$y' \leftarrow x_1$ AND x_2		$y' \leftarrow x_1$ OR x_2	
			Step 1	Step 2	Step 1	Step 2
State	x_1	x_2	y	y'	y	y'
1	LRS (0)	LRS (0)	HRS (1)	**LRS (0)**	HRS (1)	**LRS (0)**
2	LRS (0)	HRS (1)	HRS (1)	**LRS (0)**	HRS (1)	HRS (1)
3	HRS (1)	LRS (0)	HRS (1)	**LRS (0)**	HRS (1)	HRS (1)
4	HRS (1)	HRS (1)	HRS (1)	HRS (1)	HRS (1)	HRS (1)

Desired switching events in the output (y') are indicated by boldface type.

TABLE 4.2

The realized conditional switching behavior is equivalent to the NAND and NOR operations with a preset of $y = 0$. Using the two-input reprogrammable gate.

	Input Patterns		$y' \leftarrow x_1$ NAND x_2		$y' \leftarrow x_1$ NOR x_2	
			Step 1	Step 2	Step 1	Step 2
State	x_1	x_2	y	y'	y	y'
1	LRS (0)	LRS (0)	LRS (0)	**HRS (1)**	LRS (0)	**HRS (1)**
2	LRS (0)	HRS (1)	LRS (0)	**HRS (1)**	LRS (0)	LRS (0)
3	HRS (1)	LRS (0)	LRS (0)	**HRS (1)**	LRS (0)	LRS (0)
4	HRS (1)	HRS (1)	LRS (0)	LRS (0)	LRS (0)	LRS (0)

Desired switching events in the output (y') are indicated by boldface type.

Because of the easy integration with CMOS, the reprogrammable gates are generalizable to provide stateful logic arrays for large-scale logic circuit applications. In fact, unlike the ASL, which is based on spin-current in a spin-coherent channel, reprogrammable logic is based on an electric current to apply conditional switching at the output and, therefore, the logic operation is not limited to physically adjacent magnetic elements. This is an important feature for complex logic applications as discussed later.

4.5.6 Implication Logic

Material implication (IMP) is a fundamental two-input (e.g., s and t) Boolean logic operation ($s \rightarrow t$), which reads "s implies t" or "if s, then t," and is equivalent to "(NOT s) OR t." The IMP operation has been classified as one of the four basic logic operations by Whitehead and Russell [229], but has been ignored in digital electronics as Shannon founded modern digital electronics based on AND, OR, and NOT operations [230]. Only recently has it received renewed attention, when it was demonstrated that memristive switches intrinsically enable the IMP operation in a crossbar array of TiO_2 memristive switches [231]. Table 4.3 shows the truth tables of the basic implication operations, IMP and negated IMP (NIMP).

The MTJ-based realization of the IMP operation was demonstrated in [145] and it has been shown that the MTJ-based implication logic gate (Figure 4.23) significantly improves

TABLE 4.3

Truth table of IMP and NIMP operations.

State	$s\,t$	$s \rightarrow t$	$\overline{t \rightarrow s}$
1	0 0	1	0
2	0 1	1	1
3	1 0	0	0
4	1 1	1	0

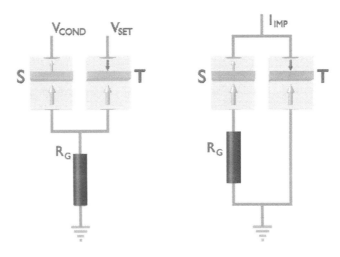

FIGURE 4.23
Voltage-controlled (left) and current-controlled (right) STT-MTJ-based implication logic gates.

the reliability of the MTJ-based logic as compared to the reprogrammable gates [145]. It has been shown that all three-input as well as two-input OR and NOR reprogrammable gates suffer from high error probability and, therefore, cannot provide reliable logic operation [228]. In fact, as reliability is an essential prerequisite to realize spin-based logic, the implication logic is a promising alternative.

Figure 4.23 shows the circuit topologies of the implication gates [145]. In both gates, two STT-MTJs are combined with a simple resistor R_G, where the initial resistance states of the source (S) and target (T) MTJs (logic variable s and t) act as the logic inputs of the gate. The final resistance state of T (t') is the logic output of the gate. The logic operation ($t' = s \rightarrow t$) is performed by simultaneous application of two voltage pulses (V_{SET} and V_{COND}) in the voltage-controlled gate or the application of the current I_{imp} in the current-controlled gate. When these voltage or current pulses are applied, a conditional switching behavior in T is provided, depending on the resistance state of the MTJs S and T (Table 4.4). Such a switching behavior corresponds to the IMP or NIMP (negated IMP) operation (Table 4.3). If we define HRS $\equiv 1$ and LRS $\equiv 0$ (the convention of Shannon), the logic output of the implication gate corresponds to the NIMP operation.

$$\{t' = t \text{ NIMP } s\} \equiv \overline{t \rightarrow s} \equiv \{t' = t.\bar{s} = t \text{ AND } \bar{s}\} \tag{4.25}$$

t' is the final state of the variable t after the operation. In combination with the TRUE operation (here low-to-high resistance switching), the NIMP operation forms a complete logic basis to compute any Boolean function. Therefore, stateful logic is enabled as MTJs are simultaneously used as nonvolatile memory and logic elements. The universal NOR and NAND operations, for example, can be performed in three and five sequential steps, respectively.

Step 1 (TRUE): $a = 1$

Step 2 (NIMP): $\overline{a \rightarrow b} \equiv \{a' = a.\bar{b} = \bar{b}\}$

Step 3 (NIMP): $\overline{a \rightarrow c} \equiv \{a' = a.\bar{c} = \bar{b}.\bar{c} = \overline{b + c} = b \text{ NOR } c\}$ $\qquad$ (4.26)

TABLE 4.4

The realized conditional switching behavior is equivalent to the operation IMP or NIMP. Depending on the definitions for the HRS and LRS as logical "0" and "1".

State	Implication Operation (Conditional Switching)		HRS $\equiv$ 0, LRS $\equiv$ 1 $t' = s \rightarrow t$			HRS $\equiv$ 1, LRS $\equiv$ 0 $t' = \bar{t} \rightarrow s$		
	$s\ t$	$s'\ t'$	s	t	t'	s	t	t'
1	HRS HRS	HRS **LRS**	0	0	1	1	1	0
2	HRS LRS	HRS LRS	0	1	1	1	0	0
3	LRS HRS	LRS HRS	1	0	0	0	1	1
4	LRS LRS	LRS LRS	1	1	1	0	0	0

Step 1 (TRUE): $a = 1$

Step 2 (NIMP): $\overline{a \to b} \equiv \{a' = a.\overline{b} = \overline{b}\}$

Step 3 (NIMP): $\overline{c \to a} \equiv \{c' = c.\overline{a} = c.b\}$

Step 4 (TRUE): $a = 1$

Step 5 (NIMP): $\overline{a \to c} \equiv \{a' = a.\overline{c} = \overline{c.b} = b \text{ NAND } c\}$ (4.27)

Here, a (a') indicates the initial (final) variable equivalent to the resistance state of an auxiliary MTJ storing the logic result of intermediary steps and the final result of stateful NAND and NOR operations.

It has been shown that the implication logic outperforms the conventional Boolean logic based on reprogrammable gates from both reliability and power consumption point of views [232]. In addition, a combination of implication logic and the reprogrammable logic reduces the number of required logic steps implementing complex logic functions [233]. Therefore, the total time and the energy consumption can be decreased at the cost of higher error probability [233].

4.5.7 Compute-in-Memory

An important issue of nonvolatile logic, the fan-out, needs to be addressed to generalize the intrinsic logic-in-memory proposals in order to perform complex logic functions and for large-scale logic circuits. When the input and output of memory elements are physically connected to form a logic gate, additional connection elements could disturb the correct logic operation (e.g., conditional switching behavior in MTJ gates). Therefore, the extension of the logic gates to provide more complex functions is problematic. In fact, highly localized computations limit the possibility of performing logic operations among data located in arbitrary parts of the circuit. Therefore, intermediate circuitry is usually required to perform additional read/write operations increasing the complexity, energy consumption, and delay. There is a lot of effort to offer compute-in-memory capabilities in large-scale implementing complex functions [234–239].

This issue appears unsolvable in all-spin logic as it is based on spin-current. However, the reprogrammable and implication gates are based on electric current and, therefore, extendable to stateful arrays without being limited to physically adjacent elements [234,235]. This makes MTJ-based logic very promising, especially when MTJs with high TMR are available to guarantee reliable operation with negligible error probabilities.

In previous sections, it has been described how direct communication between STT-MTJs via reprogrammable and implication logic gates realizes intrinsic logic-in-memory architectures and extends the functionality of nonvolatile memory circuits to incorporate logic computations.

It has been shown that by replacing the MTJ devices with one-transistor/one-MTJ cells (see Figure 4.6), the reprogrammable and implication logic gates can be realized in MRAM arrays [234,235]. Since the 1T/1MTJ cell is the basic building block for STT-MRAM structures [24,240], an STT-MRAM array can be used not only as memory, but also as magnetic logic circuit for the development of innovative nonvolatile large-scale logic architectures [234,235]. The realization of the MTJ gates in STT-MRAM arrays enables the extension of nonvolatile MRAM from memory to logical computing applications and eliminates the need for sensing amplifiers and intermediate circuitry as compared to other hybrid CMOS/MTJ nonvolatile logic proposals, where the MTJs are used only for nonvolatile storage.

4.6 Spin-Torque Oscillator

Oscillators are important devices ubiquitously needed for many applications [190] and the STT-effect can be exploited to build oscillators. Commonly a spin-torque oscillator (STO) is built as a GMR-pillar or an MTJ. Thus, the GMR or the TMR effect can be used to detect the magnetization oscillation as a high frequency voltage. The precession frequency of STOs is tunable over a wide range of frequencies 5–46 GHz by a DC current as well as by the application of an external magnetic field, which makes them very competitive in comparison to voltage controlled oscillators and Yttrium garnet oscillators [241–244]. They are also very small. STOs are over fifty times smaller than a standard LC-tank voltage controlled oscillator due to the very large required inductor footprint [245]. Their large operation frequency, small size, and low power consumption is very attractive for several microwave-based applications, like broadband oscillators [241–244], fast modulators [246–251], and sensitive field/current detectors [252].

There are several ways to categorize STOs. We focus here on their structure and distinguish between nano-contact oscillators, where the current enters though a nano-constriction into an extended magnetic structure and nano-pillar STT oscillators (spin valve or MTJ stack). Nano-contact STOs can be further differentiated by their number of contacts and have been demonstrated for different geometries [253–255]. The nano-pillar STOs can be subdivided into two categories, depending on the magnetization orientation of the free layer: "out-of-plane" with the magnetization perpendicular to the layer and "in-plane" with the magnetization parallel to the layer. Looking at STOs with nano-pillars and in-plane magnetization [256] reveals on one hand high frequency capabilities, but on the other hand the prerequisite of a large external magnetic field and low output power levels [257]. In contrast, oscillators with an out-of-plane magnetization of the free layer [258] are able to oscillate without an external magnetic field, but suffer under relatively low output power. Additionally, they typically feature lower operation frequencies (≤ 2 GHz), which limits their potential application as tunable oscillators [257].

Makarov et al. [259,260] proposed a bias-field-free STT oscillator with in-plane MgO MTJ, elliptical cross-section, and nonperfect overlap between the free and the fixed layers. This structure exhibits the drawbacks of a weak frequency dependence on the current density and a narrow range of frequencies. A way around these limitations is to use an alternative structure, which employs two MTJ stacks that share a common free layer (cf. Figure 4.24; similar to the NVMFF). This structure allows stable high frequency oscillations without the need of an external magnetic field and its operation frequency is widely tunable by varying the current density through the MTJs [261,262]. In [263], it could be demonstrated that in such a structure oscillations up to ≈ 30 GHz are achievable. As further investigations have shown, the structure based on two MTJs with a common shared free layer also exhibits stable oscillations with an out-of-plane magnetization orientation of the free layer [264].

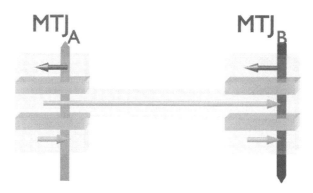

FIGURE 4.24
This structure exploits one MTJ for driving the oscillations, while the other prevents switching and relaxation into a stable state (opposite current direction). This way large stable oscillations of the common free layer can be not only sustained, but steered over a wide frequency range.

In general, the output power of current STOs is not sufficient for practical applications yet. Commonly the output power for GMR-based STOs lies in the sub-nW range. CoFeB/MgO/CoFeB based nano-pillar structures achieve higher output power, but still remain in the nW range [265–268]. In order to overcome this problem, the synchronization of several STOs has been proposed [248,254,269–276]. Another solution is to use an external microwave current or field to injection lock the synchronization [277–281]. Parametric synchronization with an external microwave field frequency close to twice the STO's free frequency has been reported, a bit more recently [281–284] and benefits from the advantage that the measurement is not interfered by the external signal. First experimental observation of parametric excitation in a nano-contact STO at cryogenic temperatures and an excitation frequency of twice the STO's free frequency was achieved by a separately manufactured strip line on top of the STO [285]. Bortolotti et al. [286] were able to parametrically excite vortex gyration by passing a microwave current through a vortex-based MTJ-STO with a sufficing Ørsted field strength at room temperature. A very encouraging result was shown by Sani et al. [254] in 2013. They were able to mutually synchronize three nano-contact STOs. Nevertheless, even under ideal conditions the parametric excitation only shows imperfect locking and the output power as well as the phase noise need further improvement [287].

The above mentioned STOs exploit magnetically hard layers to polarize the electrons before they interact with the free layer to drive precessions. But it is also possible to take advantage of the spin Hall effect for the polarization of the electrons to build spin Hall nano-oscillators (SHNOs). An SHNO comprises a nonmagnetic layer with a strong spin-orbit coupling adjacent to a magnetic layer. These devices are able to create microwave signals in the range of 2–10 GHz, which is appealing for applications in the telecommunication domain. These SHNOs work with pure spin signals (cf. Section 4.2.3), only require little power, operate in a wide range of frequencies, and are rather small (≤5 μm) in comparison to state-of-the-art technologies. Several SHNO devices in a variety of geometries have been manufactured and their operation was properly demonstrated, such as a disk with triangular contacts (nano-gap) [288,289], a nano-wire [290], and a nano-constriction [291]. It has been demonstrated that relatively large power and small auto-oscillation linewidth are possible for a localized spin current injection at cryogenic temperatures in 2013 [288]. Unfortunately, both features decrease considerably at elevated temperatures due to the availability of additional modes and the arising thermal mode hopping [291]. These problems can be avoided by deliberately exciting a single mode; that is, by adjusting the geometric area of the auto-oscillation zone. One would intuitively expect that the auto-oscillation area is correlated to the experimental setup, like the spin injection geometry, but as it turns out the control of the auto-oscillation characteristics is rather tricky [292]. The local injection of a spin signal into a continuous magnetic film causes the spontaneous excitation of the bullet auto-oscillation mode [293] and the dimensions of the "bullet" is governed by nonlinear self-localization effects and not the spin injection area [289,294].

4.7 Applications

In this final section we highlight a few showcases for spintronic computing, which are likely to be commercialized in the next few years.

4.7.1 Random Number Generator

Since the magnetization of MTJs experiences thermal excitations, their switching shows stochastic behavior and the switching probability of MTJs is governed by the applied current amplitude. This is commonly considered as an effect that has to be controlled by careful circuit design, but one can make a virtue out of necessity by exploiting it for the physical realization of a random number generator [295]. In 2013, a first spin-based random-number generator (spin dice) was built by employing a conventional in-plane MTJ and a current adjusted to achieve 50% switching probability [72]. Unfortunately, in-plane MTJs suffer of a rather small magnetic field range for bistable states and demand high switching current densities, which caused problems in the practical realization of the spin dice [296]. The next generation utilizes MTJs with perpendicular free layer and a perpendicular synthetic antiferromagnetic bottom reference layer [295,296]. Besides the well known random number generator applications, like Monte Carlo simulations and cryptography, they can also be used to improve analog-to-digital information conversion systems for low energy applications. Lee et al. [297] proposed a very interesting voltage-controlled stochastic oscillator for event-driven random sampling. Due to the exploitation of a VCMA and their deliberately reduced thermal barrier, they are able to reduce the power consumption by more than three times and improve the area efficiency even by a factor of 20 in comparison to the state-of-the-art.

4.7.2 Ternary Content-Addressable Memory

Content addressable memory (CAM) is the kind of technology that people use every day, but commonly are completely unaware of its existence. Even more, modern databases and search engines, like Google, could not offer high-speed access to data without them. In contrast to RAM, where the user sends an address to the RAM and gets a data word in return, CAMs get a data word from the user, search their entire memory for the data word and, if such entries exist, return a list of addresses where the data word is stored [298]. These memories are designed in a way that they can search their entire memory within one operation, which makes them very fast but at the same time expensive. The added search capability is realized by additional comparison circuitry in each memory cell, which causes a high power consumption and a considerably increased memory cell footprint. A ternary CAM (TCAM) extends the functionality of binary CAMs (BCAMs) by adding a third search option "X"—do not care, which gives more freedom for search queries but adds additional complexity to the circuit design [299]. The combination of rising demand for (T)CAMs together with their high energy consumption and large layout footprint makes them very attractive candidates for spintronic complementation. For instance, Govindaraj et al. [300] proposed a 6T 2 STT-MTJ-based NOR-TCAM in 2015 and lately another 9T 2 STT-MTJ based NAND TCAM in 2017 [301]. Considering that a typical CMOS only TCAM cell exhibits 16 transistors, this is a big step forward with respect to integration density and dissipated power. There are TCAMs with smaller memory cell size, like 4T-2MTJ TCAM [302], 3T-2DW BCAM [303], or even 2T-2MTJs [304]. They all feature significantly smaller memory cells and offer zero standby power, but depending on the application scenario and the overhead complexity one is able to afford, one or another of these designs will prevail [300,301].

4.7.3 Spin-Transfer Torque Compute-in-Memory (STT-CiM)

One very recent and interesting proposal for a STT-based Compute-in-Memory design was presented by Jain et al. [305]. They suggest to take advantage of the resistive nature of the STT-MRAM cells to perform a range of arithmetic, bitwise, and complex vector operations. The trick is to enable multiple word lines simultaneously and sense the effective resistance of all enabled cells in each bit-line in order to directly perform logic functions dependent on the values stored in the cells. Such a scheme is not feasible in SRAMs, because it would cause short circuit paths, but, since the STT-MRAM cells are intrinsically resistors, this problem does not appear. This idea is not new (Section 4.5.7), but the implementation simultaneously addresses process variation issues and allows arithmetic and complex vector operations without modifying either the bit-cell or the data array. The bundle of adjustments on different levels (sensing scheme, error correction scheme, and extension of the instruction set) cumulates in an average 4× performance improvement and simultaneous memory system energy reduction.

4.8 Conclusion and Outlook

In this chapter, we tried to give an overview about the many facets of CMOS compatible spintronics and its importance for future beyond Von Neumann computing. Especially STT-switched MTJs have not only become so mature that off-the-shelf MRAM is already available, but the technology as a whole including the process know-how reached a level that we are confident that first CMOS MTJ hybrids for logic applications will be brought to market very soon. However, MTJs of course also have issues like the still rather high switching currents, reduction of damping, increased thermal stability, and device variability. There are ideas to circumvent (some of) these problems via the spin Hall effect, domain wall motion, or voltage controlled magnetic anisotropies. But domain wall-based MRAM and spin Hall MRAM are by nature three terminal devices and, therefore, require more space, while the voltage controlled magnetic anisotropy coefficient needs a boost to make it compatible with advanced CMOS transistors. STT oscillators are an essential building block for digital electronics and show great potential due to their large operation frequencies, small size, and low power consumption. Currently, they suffer under too low output power for many applications, but there are niches where they already can shine like in STO-based random number generators and analog-to-digital information conversion systems for low energy applications.

Overall one can see a gradual evolution on all levels (materials, processing, devices, circuits, and architectures). This evolution drove the introduction of STT-MRAM into market and will also lead to first spintronic logic and later beyond Von Neumann products. Most likely, this will happen in high performance computing and database hardware, where reduced cooling power and less power consumption immediately brings a big advantage in operating expenses for high performance computation clusters and big server farms. TCAMs with their rather big and complex memory cells as well as their high power dissipation in combination with their crucial role in modern data base applications are perfect replacement candidates. Also FPGAs have a high potential to boost their performance with spintronic logic and will help to familiarize the current generation of developers and engineers with the next generations spintronic technology. FPGAs are very important due to their widespread application in aerospace, medical electronics, application-specific integratet circuit (ASIC) prototyping, digital signal processing, image processing, consumer electronics, high performance computing, scientific instruments, data mining, and many more. They also open up the next step towards the huge system-on-chip market. More disruptive approaches try to draw from the unique advantages of spintronics and break up the CMOS dominance. For example, ASL or STT logic will take off later, when the CMOS MTJ hybrid logic ecosystem has been established and the companies as well as the market are ready for the next more powerful technology generation. In summary, we can see that spintronics managed to become mature enough for first real products, like STT-MRAM, and we are confident that it will not stop there and a plethora of nonvolatile spin-based logic is going to appear within the next 5–10 years on the market.

Acknowledgment

This work is supported by the European Research Council through the grant #692653 NOVOFLOP. B. Gunnar Malm wants to acknowledge professor Johan Åkerman and the Applied Spintronics Group at KTH. Special thanks go to Anders Eklund, Sohrab Sani, Stefano Bonetti, and Sunjae Chung.

References

1. D. Shum, D. Houssameddine, S. T. Woo, Y. S. You, J. Wong, K. W. Wong, C. C. Wang et al. CMOS-embedded STT-MRAM arrays in 2x nm nodes for GP-MCU applications. In *2017 Symposium on VLSI Technology*, pages T208–T209, 2017.
2. L. Thomas, G. Jan, J. Zhu, H. Liu, Y.-J. Lee, S. Le, R.-Y. Tong et al. Perpendicular spin transfer torque magnetic random access memories with high spin torque efficiency and thermal stability for embedded applications. *Journal of Applied Physics*, 115(17):172615, 2014.

3. S. Natarajan, M. Agostinelli, S. Akbar, M. Bost, A. Bowonder, V. Chikarmane, S. Chouksey et al. A 14nm logic technology featuring 2nd-generation FinFET, air-gapped interconnects, self-aligned double patterning and a 0.0588 m² SRAM cell size. In *IEEE International Electron Devices Meeting (IEDM)*, pages 3.7.1–3.7.3, 2014.

4. Intel's 10 nm technology. https://newsroom.intel.com/newsroom/wp-content/uploads/sites/11/2017/03/10-nm-technology-fact-sheet.pdf. Accessed: July 17, 2017.

5. Samsung Electronics 10 nm ramp-up. https://news.samsung.com/global/samsung-electronics-on-track-for-10nm-finfet-process-technology-production-ramp-up. Accessed: July 17, 2017.

6. TSMC's 10 nm process for Apple's A10X fusion chip. http://www.idownloadblog.com/2017/06/30/apples-latest-a10x-fusion-chip-is-built-using-tsmcs-10nm-process/. Accessed: July 17, 2017.

7. B. Doris, M. Ieong, T. Kanarsky, Y. Zhang, R. A. Roy, O. Dokumaci, Z. Ren et al. Extreme scaling with ultra-thin Si channel MOSFETs. In *IEEE International Electron Devices Meeting (IEDM)*, pages 267–270, 2002.

8. I. Žutić, J. Fabian, and S.D. Sarma. Spintronics: Fundamentals and applications. *Reviews of Modern Physics*, 76(2):323–410, 2004.

9. J. Fabian, A. Matos-Abiague, C. Ertler, P. Stano, and I. Žutić. Semiconductor spintronics. *Acta Physica Slovaca*, 57(4–5):565–907, 2007.

10. J. Kim, A. Paul, P. A. Crowell, S. J. Koester, S. S. Sapatnekar, J. P. Wang, and C. H. Kim. Spin-based computing: Device concepts, current status, and a case study on a high-performance microprocessor. In *Proceedings of the IEEE*, Vol. 103, pages 106–130, 2015.

11. G. Schmidt, D. Ferrand, L. W. Molenkamp, A.T. Filip, and B.J. Van Wees. Fundamental obstacle for electrical spin injection from a ferromagnetic metal into a diffusive semiconductor. *Physical Review B*, 62(8):R4790–R4793, 2000.

12. E. I. Rashba. Theory of electrical spin injection: Tunnel contacts as a solution of the conductivity mismatch problem. *Physical Review B - Condensed Matter and Materials Physics*, 62(24):R16267–R16270, 2000.

13. O. M. J. Van't Erve, A. L. Friedman, E. Cobas, C. H. Li, J. T. Robinson, and B. T. Jonker. Low-resistance spin injection into silicon using graphene tunnel barriers. *Nature Nanotechnology*, 7(11):737–742, 2012.

14. R. Jansen. Silicon spintronics. *Nature Materials*, 11(5):400–408, 2012.

15. B. Huang, D. J. Monsma, and I. Appelbaum. Coherent spin transport through a 350 micron thick silicon wafer. *Physical Review Letters*, 99:177209, 2007.

16. J. Li and I. Appelbaum. Lateral spin transport through bulk silicon. *Applied Physics Letters*, 100(16):162408, 2012.

17. V. Sverdlov. *Strain-Induced Effects in Advanced MOSFETs*. Computational Microelectronics. Springer-Verlag, Wien, Austria, 2011.

18. V. Sverdlov and S. Selberherr. Silicon spintronics: Progress and challenges. *Physics Reports*, 585:1–40, 2015.

19. P. Chuang, S.-C. Ho, L. W. Smith, F. Sfigakis, M. Pepper, C.-H. Chen, J.-C. Fan et al. All-electric all-semiconductor spin field-effect transistors. *Nature Nanotechnology*, 10(1):35–39, 2015.

20. T. Tahara, H. Koike, M. Kameno, T. Sasaki, Y. Ando, K. Tanaka, S. Miwa, Y. Suzuki, and M. Shiraishi. Room-temperature operation of Si spin MOSFET with high on/off spin signal ratio. *Applied Physics Express*, 8(11):113004, 2015.

21. A. Makarov, T. Windbacher, V. Sverdlov, and S. Selberherr. CMOS-compatible spintronic devices: A review. *Semiconductor Science and Technology*, 31(11):113006, 2016.

22. A. Makarov, V. Sverdlov, and S. Selberherr. Emerging memory technologies: Trends, challenges, and modeling methods. *Microelectronics Reliability*, 52(4):628–634, 2012.

23. S. A. Wolf, J. Lu, M. R. Stan, F. Chen, and D. M. Treger. The promise of nanomagnetics and spintronics for future logic and universal memory. In *Proceedings of the IEEE*, Vol. 98, pages 2155–2168, 2010.

24. C. Augustine, N. Mojumder, X. Fong, H. Choday, S. P. Park, and K. Roy. STT-MRAMs for future universal memories: Perspective and prospective. In *Proceedings of the International Conference on Microelectronics (MIEL)*, pages 349–355, 2012.

25. W. Zhao and G. Prenat. *Spintronics-Based Computing*. Springer International Publishing, 2015.

26. M. N. Baibich, J. M. Broto, A. Fert, F. Nguyen Van Dau, F. Petroff, P. Etienne, G. Creuzet, A. Friederich, and J. Chazelas. Giant magnetoresistance of (001)Fe/(001)Cr magnetic superlattices. *Physical Review Letters*, 61:2472–2475, 1988.

27. G. Binasch, P. Grünberg, F. Saurenbach, and W. Zinn. Enhanced magnetoresistance in layered magnetic structures with antiferromagnetic interlayer exchange. *Physical Review B*, 39:4828–4830, 1989.

28. P. Zahn and I. Mertig. Enhanced magnetoresistance. In *Handbook of Magnetism and Advanced Magnetic Materials*, Vol. 1, Fundamentals and Theory. John Wiley & Sons, Hoboken, NJ, 2007.

29. G. A. Prinz. Spin-polarized transport. *Physics Today*, 48(4):58–63, 1995.

30. C. Chappert, A. Fert, and F. Nguyen Van Dau. The emergence of spin electronics in data storage. *Nature Materials*, 6(11):813–823, 2007.

31. S. Tehrani, Jon M. Slaughter, M. DeHerrera, B. N. Engel, N. D. Rizzo, J. Salter et al. Magnetoresistive random access memory using magnetic tunnel junctions. In *Proceedings of the IEEE*, Vol. 91, pages 703–714, 2003.

32. M. Julliere. Tunneling between ferromagnetic films. *Physics Letters A*, 54(3):225–226, 1975.

33. H. Imamura and S. Maekawa. Theory of spin-dependent tunneling. In *Handbook of Magnetism and Advanced Magnetic Materials*. John Wiley & Sons, Hoboken, NJ, 2007.

34. A. V. Khvalkovskiy, D. Apalkov, S. Watts, R. Chepulskii, R.S. Beach, A. Ong, X. Tang et al. Basic principles of STT-MRAM cell operation in memory arrays. *Journal of Physics D: Applied Physics*, 46(8):074001, 2013.

35. J. S. Moodera, Lisa R. Kinder, Terrilyn M. Wong, and R. Meservey. Large magnetoresistance at room temperature in ferromagnetic thin film tunnel junctions. *Physical Review Letters*, 74:3273–3276, 1995.

36. T. Miyazaki and N. Tezuka. Giant magnetic tunneling effect in $Fe/Al_2O_3/Fe$ junction. *Journal of Magnetism and Magnetic Materials*, 139(3):L231–L234, 1995.

37. D. Wang, C. Nordman, J.M. Daughton, Z. Qian, and J. Fink. 70% TMR at room temperature for SDT sandwich junctions with CoFeB as free and reference layers. *IEEE Transactions on Magnetics*, 40(4):2269–2271, 2004.

38. W. H. Butler, X.-G. Zhang, T. C. Schulthess, and J. M. MacLaren. Spin-dependent tunneling conductance of Fe/MgO/Fe sandwiches. *Physical Review B*, 63:054416, 2001.

39. J. Mathon and A. Umerski. Theory of tunneling magnetoresistance of an epitaxial Fe/MgO/Fe(001) junction. *Physical Review B*, 63:220403, 2001.

40. M. Bowen, V. Cros, F. Petroff, A. Fert, C. Martínez Boubeta, J. L. Costa-Krämer, J. V. Anguita et al. Large magnetoresistance in Fe/MgO/FeCo(001) epitaxial tunnel junctions on GaAs(001). *Applied Physics Letters*, 79(11):1655–1657, 2001.

41. S. S. P. Parkin, C. Kaiser, A. Panchula, P. M. Rice, B. Hughes, M. Samant, and S.-H. Yang. Giant tunnelling magnetoresistance at room temperature with MgO(100) tunnel barriers. *Nature Materials*, 3(12):862–867, 2004.

42. S. Yuasa, T. Nagahama, A. Fukushima, Y. Suzuki, and K. Ando. Giant room-temperature magnetoresistance in single-crystal Fe/MgO/Fe magnetic tunnel junctions. *Nature Materials*, 3(12):868–871, 2004.

43. S. Yuasa, A. Fukushima, H. Kubota, Y. Suzuki, and K. Ando. Giant tunneling magnetoresistance up to 410% at room temperature in fully epitaxial Co/MgO/Co magnetic tunnel junctions with BCC Co(001) electrodes. *Applied Physics Letters*, 89(4):042505, 2006.

44. S. Ikeda, J. Hayakawa, Y. Ashizawa, Y. M. Lee, K. Miura, H. Hasegawa, M. Tsunoda, F. Matsukura, and H. Ohno. Tunnel magnetoresistance of 604% at 300K by suppression of Ta diffusion in CoFeB/MgO/CoFeB pseudo-spin-valves annealed at high temperature. *Applied Physics Letters*, 93(8):082508, 2008.

45. R. Sbiaa, H. Meng, and S. N. Piramanayagam. Materials with perpendicular magnetic anisotropy for magnetic random access memory. *Physica Status Solidi - Rapid Research Letters*, 5(12):413–419, 2011.

46. J. C. Slonczewski. Current-driven excitation of magnetic multilayers. *Journal of Magnetism and Magnetic Materials*, 159(1–2):L1–L7, 1996.

47. L. Berger. Emission of spin waves by a magnetic multilayer traversed by a current. *Physical Review B*, 54:9353–9358, 1996.

48. J. Grollier, V. Cros, A. Hamzic, J. M. George, H. Jaffrès, A. Fert, G. Faini, J. Ben Youssef, and H. Legall. Spin-polarized current induced switching in Co/Cu/Co pillars. *Applied Physics Letters*, 78(23):3663–3665, 2001.

49. J. A. Katine, F. J. Albert, R. A. Buhrman, E. B. Myers, and D. C. Ralph. Current-driven magnetization reversal and spin-wave excitations in Co/Cu/Co pillars. *Physical Review Letters*, 84:3149–3152, 2000.

50. E. B. Myers, D. C. Ralph, J. A. Katine, R. N. Louie, and R. A. Buhrman. Current-induced switching of domains in magnetic multilayer devices. *Science*, 285(5429):867–870, 1999.

51. J.Z Sun. Current-driven magnetic switching in manganite trilayer junctions. *Journal of Magnetism and Magnetic Materials*, 202(1):157–162, 1999.

52. M. Tsoi, A. G. M. Jansen, J. Bass, W.-C. Chiang, M. Seck, V. Tsoi, and P. Wyder. Excitation of a magnetic multilayer by an electric current. *Physical Review Letters*, 80:4281–4284, 1998.

53. Y. Huai, F. Albert, P. Nguyen, M. Pakala, and T. Valet. Observation of spin-transfer switching in deep submicron-sized and low-resistance magnetic tunnel junctions. *Applied Physics Letters*, 84(16):3118–3120, 2004.

54. Z. Diao, D. Apalkov, M. Pakala, Y. Ding, A. Panchula, and Y. Huai. Spin transfer switching and spin polarization in magnetic tunnel junctions with MgO and AlO_x barriers. *Applied Physics Letters*, 87(23):232502, 2005.

55. M. I. Dyakonov and V. I. Perel. Spin relaxation of conduction electrons in noncentrosymmetric semiconductors. *Fizika Tverdogo Tela*, 13:1382–1397, 1971.

56. E. L. Ivchenko, G. E. Pikus, I. I. Farbshtein, V. A. Shalygin, A. V. Shturbin, and L. E. Vorob'ev. Optical activity in tellurium induced by a current. *JETP Letters*, 29(8):441–444, 1979.

57. Y. Kato, R. C. Myers, A. C. Gossard, and D. D. Awschalom. Observation of the spin Hall effect in semiconductors. *Science*, 306(5703):1910–1913, 2004.

58. S. O. Valenzuela and M. Tinkham. Direct electronic measurement of the spin Hall effect. *Nature*, 442(7099):176–179, 2006.

59. J. E. Hirsch. Spin Hall effect. *Physical Review Letters*, 83:1834–1837, 1999.

60. J. Wunderlich, B. Kaestner, J. Sinova, and T. Jungwirth. Experimental observation of the spin-Hall effect in a two-dimensional spin-orbit coupled semiconductor system. *Physical Review Letters*, 94:047204, 2005.

61. T. Kimura, Y. Otani, T. Sato, S. Takahashi, and S. Maekawa. Room-temperature reversible spin Hall effect. *Physical Review Letters*, 98:156601, 2007.

62. P. K. Amiri, J. G. Alzate, X. Q. Cai, F. Ebrahimi, Q. Hu, K. Wong, C. Grèzes et al. Electric-field-controlled magnetoelectric RAM: Progress, challenges, and scaling. *IEEE Transactions on Magnetics*, 51(11):1–7, 2015.

63. M. Weisheit, S. Fähler, A. Marty, Y. Souche, C. Poinsignon, and D. Givord. Electric field-induced modification of magnetism in thin-film ferromagnets. *Science*, 315(5810):349–351, 2007.

64. T. Maruyama, Y. Shiota, T. Nozaki, K. Ohta, N. Toda, M. Mizuguchi, A.A. Tulapurkar et al. Large voltage-induced magnetic anisotropy change in a few atomic layers of iron. *Nature Nanotechnology*, 4(3):158–161, 2009.

65. Y. Shiota, T. Maruyama, T. Nozaki, T. Shinjo, M. Shiraishi, and Y. Suzuki. Voltage-assisted magnetization switching in ultrathin $Fe_{80}Co_{20}$ alloy layers. *Applied Physics Express*, 2(6):063001, 2009.

66. T. Nozaki, H. Arai, K. Yakushiji, S. Tamaru, H. Kubota, H. Imamura, A. Fukushima, and S. Yuasa. Magnetization switching assisted by high-frequency-voltage-induced ferromagnetic resonance. *Applied Physics Express*, 7(7):073002, 2014.

67. X. Li, K. Fitzell, D. Wu, C. T. Karaba, A. Buditama, G. Yu, K. L. Wong et al. Enhancement of voltage-controlled magnetic anisotropy through precise control of Mg insertion thickness at CoFeB|MgO interface. *Applied Physics Letters*, 110(5):052401, 2017.

68. K. L. Wang, J. G. Alzate, and P. K. Amiri. Low-power non-volatile spintronic memory: STT-RAM and beyond. *Journal of Physics D: Applied Physics*, 46(7):074003, 2013.

69. S. E. Barnes, J. Ieda, and S. Maekawa. Rashba spin-orbit anisotropy and the electric field control of magnetism. *Scientific Reports*, 4:4105, 2014.

70. W. Kim, J. H. Jeong, Y. Kim, W. C. Lim, J. H. Kim, J. H. Park, H. J. Shin et al. Extended scalability of perpendicular STT-MRAM towards sub-20nm MTJ node. In *IEEE International Electron Devices Meeting (IEDM)*, pages 24.1.1–24.1.4, 2011.

71. B. Dieny and M. Chshiev. Perpendicular magnetic anisotropy at transition metal/oxide interfaces and applications. *Reviews of Modern Physics*, 89:025008, 2017.

72. S. Yuasa, A. Fukushima, K. Yakushiji, T. Nozaki, M. Konoto, H. Maehara, H. Kubota et al. Future prospects of MRAM technologies. In *IEEE International Electron Devices Meeting (IEDM)*, pages 3.1.1–3.1.4, 2013.

73. S. Wang, H. Lee, F. Ebrahimi, P. K. Amiri, K. L. Wang, and P. Gupta. Comparative evaluation of spin-transfer-torque and magnetoelectric random access memory. *IEEE Journal on Emerging and Selected Topics in Circuits and Systems*, 6(2):134–145, 2016.

74. S. W. Chung, T. Kishi, J. W. Park, M. Yoshikawa, K. S. Park, T. Nagase, K. Sunouchi et al. 4Gbit density STT-MRAM using perpendicular MTJ realized with compact cell structure. In *IEEE International Electron Devices Meeting (IEDM)*, pages 27.1.1–27.1.4, 2016.

75. D. Apalkov, A. Khvalkovskiy, S. Watts, V. Nikitin, X. Tang, D. Lottis, K. Moon et al. Spin-transfer torque magnetic random access memory (STT-MRAM). *Journal of Emerging Technology in Computer System*, 9(2):13:1–13:35, 2013.

76. Y. Lu, X. Li, S. H. KANG, and S. Gu. Self-aligned top contact for MRAM fabrication, 2016. EP Patent App. 3095144.

77. M. Piquemal-Banci, R. Galceran, M.-B. Martin, F. Godel, A. Anane, F. Petroff, B. Dlubak, and P. Seneor. 2D-MTJs: Introducing 2D materials in magnetic tunnel junctions. *Journal of Physics D: Applied Physics*, 50(20):203002, 2017.

78. E. Cobas, A. L. Friedman, O. M. J. van't Erve, J. T. Robinson, and B. T. Jonker. Graphene as a tunnel barrier: Graphene-based magnetic tunnel junctions. *Nano Letters*, 12(6):3000–3004, 2012.

79. M.-B. Martin, B. Dlubak, R. S. Weatherup, H. Yang, C. Deranlot, K. Bouzehouane, F. Petroff et al. Sub-nanometer atomic layer deposition for spintronics in magnetic tunnel junctions based on graphene spin-filtering membranes. *ACS Nano*, 8(8):7890–7895, 2014.

80. M. Gajek, J. J. Nowak, J. Z. Sun, P. L. Trouilloud, E. J. O'Sullivan, D. W. Abraham, M. C. Gaidis et al. Spin torque switching of 20nm magnetic tunnel junctions with perpendicular anisotropy. *Applied Physics Letters*, 100(13):132408, 2012.

81. H. Sato, E. C. I. Enobio, M. Yamanouchi, S. Ikeda, S. Fukami, S. Kanai, F. Matsukura, and H. Ohno. Properties of magnetic tunnel junctions with a MgO/CoFeB/Ta/CoFeB/MgO recording structure down to junction diameter of 11nm. *Applied Physics Letters*, 105(6):062403, 2014.

82. S. Tan, T. KIM, W. Yang, J. Marks, and T. Lill. Dry plasma etch method to pattern MRAM stack, 2016. US Patent App. 20160308112.

83. N. Sakimura, T. Sugibayashi, R. Nebashi, and N. Kasai. Nonvolatile magnetic flip-flop for standby-power-free SoCs. In *2008 IEEE Custom Integrated Circuits Conference*, pages 355–358, 2008.

84. Y. J. Song, J. H. Lee, H. C. Shin, K. H. Lee, K. Suh, J. R. Kang, S. S. Pyo et al. Highly functional and reliable 8Mb STT-MRAM embedded in 28nm logic. In *IEEE International Electron Devices Meeting (IEDM)*, pages 27.2.1–27.2.4, 2016.

85. J. M. Slaughter, N. D. Rizzo, J. Janesky, R. Whig, F. B. Mancoff, D. Houssameddine, J. J. Sun et al. High density ST-MRAM technology. In *IEEE International Electron Devices Meeting (IEDM)*, pages 29.3.1–29.3.4, 2012.

86. J. J. Kan, C. Park, C. Ching, J. Ahn, L. Xue, R. Wang, A. Kontos et al. Systematic validation of 2x nm diameter perpendicular MTJ arrays and MgO barrier for sub-10 nm embedded STT-MRAM with practically unlimited endurance. In *IEEE International Electron Devices Meeting (IEDM)*, pages 27.4.1–27.4.4, 2016.

87. M. C. Gaidis. *Magnetoresistive Random Access Memory*. Wiley-VCH Verlag GmbH & Co. KGaA, 2010.

88. J. R. Black. Mass transport of aluminum by momentum exchange with conducting electrons. In *2005 IEEE International Reliability Physics Symposium, 2005. Proceedings. 43rd Annual*, pages 1–6, 2005.

89. K. Itoh. *VLSI Memory Chip Design*. Advanced Microelectronics. Springer-Verlag, Berlin, Germany, 2001.

90. A. Chen. A review of emerging non-volatile memory (NVM) technologies and applications. *Solid-State Electronics*, 125:25–38, 2016.

91. T. Endoh, H. Koike, S. Ikeda, and H. Ohno. An overview of nonvolatile emerging memories – Spintronics for working memories. *IEEE Journal on Merging and Selected Topics in Circuits and Systems*, 6(2):109–119, 2016.

92. J. J. Kan, C. Park, C. Ching, J. Ahn, Y. Xie, M. Pakala, and S. H. Kang. A study on practically unlimited endurance of STT-MRAM. In *IEEE Transactions on Electron Devices (IEDM)*, Vol. 64, pages 3639–3646, 2017.

93. E. Kitagawa, S. Fujita, K. Nomura, H. Noguchi, K. Abe, K. Ikegami, T. Daibou et al. Impact of ultra low power and fast write operation of advaced perpendicular MTJ on power reduction for high-performance mobile CPU. In *IEEE International Electron Devices Meeting (IEDM)*, pages 29.4.1–29.4.4, 2012.

94. T. Ohsawa, S. Miura, H. Honjo, S. Ikeda, T. Hanyu, H. Ohno, and Endoh T. A 500ps/8.5ns array read/write latency 1Mb twin 1T1MTJ STT-MRAM designed in 90nm CMOS/40nm MTJ process with novel positive feedback S/A circuit. In *2014 International Conference on Solid State Devices and Materials*, pages 458–459, 2014.

95. Everspin's STT technology. https://www.everspin.com/ddr3-dram-compatible-mram-spin-torque-technology-0. Accessed: September 7, 2017.

96. D. Apalkov, B. Dieny, and J. M. Slaughter. Magnetoresistive Random Access Memory. *Proceedings of the IEEE*, 104(10):1796–1830, 2016.

97. Y. M. Lee, J. Hayakawa, S. Ikeda, F. Matsukura, and H. Ohno. Effect of electrode composition on the tunnel magnetoresistance of pseudo-spin-valve magnetic tunnel junction with a MgO tunnel barrier. *Applied Physics Letters*, 90:212507, 2007.

98. D. Apalkov, S. Watts, A. Driskill-Smith, E. Chen, Z. Diao, and V. Nikitin. Comparison of scaling of in-plane and perpendicular spin transfer switching technologies by micromagnetic simulation. *IEEE Transactions on Magnetics*, 46:2240–2243, 2010.

99. E. Chen, D. Apalkov, Z. Diao, A. Driskill-Smith, D. Druist, D. Lottis, V. Nikitin et al. Advances and future prospects of spin-transfer torque random access memory. *IEEE Transactions on Magnetics*, 46(6):1873–1878, 2010.

100. B. D. Cullity and C. D. Graham. *Introduction to Magnetic Materials*. John Wiley & Sons, Hoboken, NJ, 2009.

101. S. Mangin, D. Ravelosona, J. A. Katine, M. J. Carey, B. D. Terris, and E. E. Fullerton. Current-induced magnetization reversal in nanopillars with perpendicular anisotropy. *Nature Materials*, 5:210–215, 2006.

102. T. Seki, S. Mitani, K. Yakushiji, and K. Takanashi. Magnetization switching in nanopillars with FePt alloys by spin-polarized current. *Journal of Applied Physics*, 5:08G521, 2006.

103. D. C. Worledge, G. Hu, D. W. Abraham, J. Z. Sun, P. L. Trouilloud, J. Nowak, S. Brown, M. C. Gaidis, E. J. O'Sullivan, and R. P. Robertazzi. Spin torque switching of perpendicular Ta|CoFeB|MgO-based magnetic tunnel junctions. *Applied Physics Letters*, 98:022501, 2011.

104. H. Sato, M. Yamanouchi, K. Miura, S. Ikeda, H. D. Gan, K. Mizunuma, R. Koizumi, F. Matsukura, and H. Ohno. Junction size effect on switching current and thermal stability in CoFeB/MgO perpendicular magnetic tunnel junctions. *Applied Physics Letters*, 99:042501, 2011.

105. J. Nogués and I. K. Schuller. Exchange bias. *Journal of Magnetism and Magnetic Materials*, 192:203–232, 1999.

106. D. M. Edwards, J. Mathon, R. B. Muniz, and M. S. Phan. Oscillations of the exchange in magnetic multilayers as an analog of de Haas–van Alphen effect. *Physical Review Letters*, 67(4):493–496, 1991.

107. S. S. P. Parkin and D. Mauri. Spin engineering: Direct determination of the Ruderman-Kittel-Kasuya-Yosida far-field range function in ruthenium. *Physical Review B*, 44(13):7131–7134, 1991.

108. M. D. Stiles. Interlayer exchange coupling. *Journal of Magnetism and Magnetic Materials*, 200:322–337, 1999.

109. T. Min, Q. Chen, R. Beach, G. Jan, C. Horng, W. Kula, T. Torng et al. A study of write margin of spin torque transfer magnetic random access memory. *IEEE Transactions on Magnetics*, 46(6):2322–2327, 2010.

110. H. Sato, M. Yamanouchi, S. Ikeda, S. Fukami, F. Matsukura, and H. Ohno. Perpendicular-anisotropy CoFeB-MgO magnetic tunnel junctions with a MgO/CoFeB/Ta/CoFeB/MgO recording structure. *Applied Physics Letters*, 101:022414, 2012.

111. C. Park, J. J. Kan, C. Ching, J. Ahn, L. Xue, R. Wang, A. Kontos et al. Systematic optimization of 1 Gbit perpendicular magnetic tunnel junction arrays for 28 nm embedded STT-MRAM and beyond. In *IEEE International Electron Devices Meeting (IEDM)*, pages 26.2.1–26.2.4, 2015.

112. W. F. Brown Jr. Thermal fluctuations of a single-domain particle. *Physical Review*, 130(5):1677–1686, 1963.

113. Z. Li and S. Zhang. Thermally assisted magnetization reversal in the presence of a spin-transfer torque. *Physical Review B*, 69:134416, 2004.

114. R. H. Koch, J. A. Katine, and J. Z. Sun. Time-resolved reversal of spin-transfer switching in a nanomagnet. *Physical Review Letters*, 92(8):088302, 2004.

115. Z. Diao, M. Pakala, A. Panchula, Y. Ding, D. Apalkov, L.-C. Wang, E. Chen, and Y. Huai. Spin-transfer switching in MgO-based magnetic tunnel junctions. *Journal of Applied Physics*, 99:08G510, 2006.

116. R. J. Baker. *CMOS: Circuit Design, and Simulation*, 3rd ed. John Wiley & Sons, Hoboken, NJ, 2010.

117. Y. Taur and T. H. Ning. *Fundamentals of Modern VLSI Devices*, 2nd ed. Cambridge University Press, New York, 2009.

118. International Technology Roadmap for Semiconductors (ITRS). http://www.itrs2.net/. Accessed: October 6, 2017.

119. M. Durlam, P. J. Naji, A. Omair, M. DeHerra, J. Calder, J. M. Slaughter, B. N. Engel et al. A 1-Mbit MRAM based on 1T1MTJ bit cell integrated with copper interconnects. *IEEE Journal of Solid-State Circuits*, 38(5):769–773, 2003.

120. R. R. Katti, J. Lintz, L. Sundstrom, T. Marques, S. Scoppettuolo, and D. Martin. Heavy-ion and total ionizing dose (TID) performance of a 1 Mbit magnetoresistive random access memory (MRAM). In *2009 IEEE Radiation Effects Data Workshop*, pages 103–105, 2009.

121. C. H. Shang, J. Nowak, R. Jansen, and J. S. Moodera. Temperature dependence of magnetoresistance and surface magnetization in ferromagnetic tunnel junctions. *Physical Review B*, 58(6):R2917–R2920, 1998.

122. K. Ono, T. Kawahara, R. Takemura, K. Miura, M. Yamanouchi, J. Hayakawa, K. Ito, H. Takahashi, H. Matsuoka, S. Ikeda, and H. Ohno. SPRAM with large thermal stability for high immunity to read disturbance and long retention for high-temperature operation. In *2009 Symposium on VLSI Technology Digest of Technical Papers*, pages 228–229, 2009.

123. D. Kobayashi, Y. Kakehashi, K. Hirose, S. Onoda, T. Makino, T. Ohshima, S. Ikeda et al. Influence of heavy ion irradiation on perpendicular-anisotropy CoFeB-MgO magnetic tunnel junctions. *IEEE Transactions on Nuclear Science*, 61(4):1710–1716, 2014.

124. D. Chabi, W. Zhao, J.-O. Klein, and C. Chappert. Design and analysis of radiation hardened sensing circuits for spin transfer torque magnetic memory and logic. *IEEE Transactions on Nuclear Science*, 61(6):3258–3264, 2014.

125. J. M. Daughton and A. V. Pohm. Design of Curie point written magnetoresistance random access memory cells. *Journal of Applied Physics*, 93:7304–7306, 2003.

126. S. Bandiera, R. C. Sousa, M. Marins de Castro, C. Ducruet, C. Portemont, S. Auffret, L. Vila, I. L. Prejbeanu, B. Rodmacq, and B. Dieny. Spin transfer torque switching assisted by thermally induced anisotropy reorientation in perpendicular magnetic tunnel junctions. *Applied Physics Letters*, 99:202507, 2011.

127. L. Liu, C.-F. Pai, Y. Li, H. W. Tseng, D. C. Ralph, and R. A. Buhrman. Spin-torque switching with the giant spin Hall effect of tantalum. *Science*, 336:555–558, 2014.

128. C. Zhang, M. Yamanouchi, H. Sato, S. Fukami, S. Ikeda, F. Matsukura, and H. Ohno. Magnetization reversal induced by in-plane current in Ta/CoFeB/MgO structures with perpendicular magnetic easy axis. *Journal of Applied Physics*, 115:17C714, 2014.

129. C.-F. Pai, L. Liu, Y. Li, H. W. Tseng, D. C. Ralph, and R. A. Buhrman. Spin transfer torque devices utilizing the giant spin Hall effect of tungsten. *Applied Physics Letters*, 101:122404, 2012.

130. M. Yamanouchi, L. Chen, J. Kim, M. Hayashi, H. Sato, S. Fukami, S. Ikeda, F. Matsukura, and H. Ohno. Three terminal magnetic tunnel junction utilizing the spin Hall effect of iridium-doped copper. *Applied Physics Letters*, 102:212408, 2013.

131. A. Hoffmann. Spin Hall effects in metals. *IEEE Transactions on Magnetics*, 49(10):5172–5193, 2013.

132. A. Yamaguchi, T. Ono, S. Nasu, K. Miyake, K. Mibu, and T. Shinjo. Real-space observation of current-driven domain wall motion in submicron magnetic wires. *Physical Review Letters*, 92(7):077205, 2004.

133. H. Numata, T. Suzuki, N. Ohshima, S. Fukami, K. Nagahara, N. Ishiwata, and N. Kasai. Scalable cell technology utilizing domain wall motion for high-speed MRAM. In *2007 Symposium on VLSI Technology Digest of Technical Papers*, pages 232–233, 2007.

134. S. Fukami, T. Suzuki, K. Nagahara, N. Ohshima, Y. Ozaki, S. Saito, R. Nebashi et al. Low-current perpendicular domain wall motion cell for scalable high-speed MRAM. In *2009 Symposium on VLSI Technology*, pages 230–231, 2009.

135. S. Fukami, M. Yamanouchi, S. Ikeda, and H. Ohno. Domain wall motion device for nonvolatile memory and logic – Size dependence of device properties. *IEEE Transactions on Magnetics*, 50(11):3401006, 2014.

136. Y. Seo, X. Fong, and K. Roy. Domain wall coupling-based STT-MRAM for on-chip cache applications. *IEEE Transactions on Magnetics*, 62(2):554, 2015.

137. H. Liu, D. Bedau, D. Backes, J. A. Katine, J. Langer, and A. D. Kent. Ultrafast switching in magnetic tunnel junction based orthogonal spin transfer devices. *Applied Physics Letters*, 97:242510, 2010.

138. H. Liu, D. Bedau, D. Backes, J. A. Katine, J. Langer, and A. D. Kent. Precessional reversal in orthogonal spin transfer magnetic random access memory devices. *Applied Physics Letters*, 101:032403, 2012.

139. N. S. Kim, T. Austin, D. Baauw, T. Mudge, K. Flautner, J. S. Hu, M. J. Irwin, M. Kandemir, and V. Narayanan. Leakage current: Moore's law meets the static power. *Computer*, 36:68–75, 2003.

140. S. Ikeda, J. Hayakawa, Y. M. Lee, F. Matsukura, Y. Ohno, T. Hanyu, and H. Ohno. Magnetic tunnel junctions for spintronic memories and beyond. *IEEE Transactions on Electron Devices*, 54:991–1002, 2007.

141. B. Behin-Aein, D. Datta, S. Salahuddin, and S. Datta. Proposal for an all-spin logic device with built-in memory. *Nature Nanotechnology*, 5(4):266–270, 2010.

142. A. Lyle, J. Harms, S. Patil, X. Yao, D. Lilja, and J. P. Wang. Direct communication between magnetic tunnel junctions for nonvolatile logic fan-out architecture. *Applied Physics Letters*, 97:152504, 2010.

143. A. Lyle, S. Patil, J. Harms, B. Glass, X. Yao, D. Lilja, and J. P. Wang. Magnetic tunnel junction logic architecture for realization of simultaneous computation and communication. *IEEE Transactions on Magnetics*, 47:2970–2973, 2011.

144. J. Das, S. M. Alam, and S. Bhanja. Ultra-low power hybrid CMOS-magnetic logic architecture. *IEEE Transactions on Circuits and Systems I: Regular Papers*, 59:2008–2016, 2012.

145. H. Mahmoudi, T. Windbacher, V. Sverdlov, and S. Selberherr. Implication logic gates using spin-transfer-torque-operated magnetic tunnel junctions for intrinsic logic-in-memory. *Solid-State Electronics*, 84:191–197, 2013.

146. B. Dieny, R. Sousa, G. Prenat, L. Prejbeanu, and O. Redon. *Emerging non-volatile memories*, chapter Hybrid CMOS/Magnetic Memories (MRAMs) and Logic Circuits, pages 37–101. Springer US, 2014.

147. J. Butler, M. Shachar, B. Lee, D. Garcia, B. Hu, J. Hong, N. Amos, and S. Khizroev. Reconfigurable and non-volatile vertical magnetic logic gates. *Journal of Applied Physics*, 115(16):163903, 2014.

148. Cheol Seong Hwang. Prospective of semiconductor memory devices: from memory system to materials. *Advanced Electronic Materials*, 1(6), 2015.

149. S. Verma, A. A. Kulkarni, and B. K. Kaushik. Spintronics-based devices to circuits: Perspectives and challenges. *IEEE Nanotechnology Magazine*, 10(4):13–28, 2016.

150. V. Jamshidi, M. Fazeli, and A. Patooghy. Mgate: A universal magnetologic gate for design of energy efficient digital circuits. *IEEE Transactions on Magnetics*, 53:3400813, 2017.

151. V. Jamshidi and M. Fazeli. Design of ultra low power current mode logic gates using magnetic cells. *AEU-International Journal of Electronics and Communications*, 83:270–279, 2018.

152. S. Matsunaga, J. Hayakawa, S. Ikeda, K. Miura, T. Endoh, H. Ohno, and T. Hanyu. MTJ-based nonvolatile logic-in-memory circuit, future prospects and issues. In *2009 Design, Automation Test in Europe Conference Exhibition*, pages 433–435, 2009.

153. T. Hanyu. Challenge of MTJ/MOS-hybrid logic-in-memory architecture for nonvolatile VLSI processor. In *2013 IEEE International Symposium on Circuits and Systems (ISCAS2013)*, pages 117–120, 2013.

154. M. Natsui, D. Suzuki, N. Sakimura, R. Nebashi, Y. Tsuji, A. Morioka, T. Sugibayashi et al. Nonvolatile logic-in-memory LSI using cycle-based power gating and its application to motion-vector prediction. *IEEE Journal of Solid-State Circuits*, 50(2):476–489, 2015.

155. B. Jovanović, R.M. Brum, and L. Torres. Logic circuits design based on MRAM: From single to multi-states cells storage. In *Spintronics-Based Computing*. pages 179–200. Springer, Cham, Switzerland, 2015.

156. T. Hanyu, T. Endoh, D. Suzuki, H. Koike, Y. Ma, N. Onizawa, M. Natsui, S. Ikeda, and H. Ohno. Standby-power-free integrated circuits using MTJ-based VLSI computing. *Proceedings of the IEEE*, Vol. 104, pages 1844–1863, 2016.

157. T. Endoh, S. Togashi, F. Iga, Y. Yoshida, T. Ohsawa, H. Koike, S. Fukami et al. A 600 MHz MTJ-based nonvolatile latch making use of incubation time in MTJ switching. In *IEEE International Electron Devices Meeting (IEDM)*, pages 4.3.1–4.3.4, 2011.

158. H. Koike, S. Miura, H. Honjo, T. Watanabe, H. Sato, S. Sato, T. Nasuno et al. 1T1MTJ STT-MRAM cell array design with an adaptive reference voltage generator for improving device variation tolerance. In *2015 IEEE International Memory Workshop (IMW)*, pages 1–4, 2015.

159. H. Koike, T. Ohsawa, S. Ikeda, T. Hanyu, H. Ohno, T. Endoh, N. Sakimura et al. A power-gated MPU with 3-microsecond entry/exit delay using MTJ-based nonvolatile flip-flop. In *2013 IEEE Asian Solid-State Circuits Conference (A-SSCC)*, pages 317–320, 2013.

160. N. Sakimura, Y. Tsuji, R. Nebashi, H. Honjo, A. Morioka, K. Ishihara, K. Kinoshita et al. A 90nm 20MHz fully nonvolatile microcontroller for standby-power-critical applications. In *2014 IEEE International Solid-State Circuits Conference Digest of Technical Papers (ISSCC)*, pages 184–185, 2014.

161. P. Chow, S. O. Seo, J. Rose, K. Chung, G. Paez-Monzon, and I. Rahardja. The design of a SRAM-based field-programmable gate array-Part II: Circuit design and layout. *IEEE Transactions on Very Large Scale Integration (VLSI) Systems*, 7(3):321–330, 1999.

162. I. Kuon, R. Tessier, and J. Rose. *FPGA Architecture: Survey and challenges*, volume 2. Now Publishers, Hanover, MA, 2008.

163. L. V. Cargnini, Y. Guillemenet, L. Torres, and G. Sassatelli. Improving the reliability of a FPGA using fault-tolerance mechanism based on magnetic memory (MRAM). In *2010 International Conference on Reconfigurable Computing and FPGAs*, pages 150–155, 2010.

164. N. Bruchon, L. Torres, G. Sassatelli, and G. Cambon. New nonvolatile FPGA concept using magnetic tunneling junction. In *IEEE Computer Society Annual Symposium on Emerging VLSI Technologies and Architectures (ISVLSI'06)*, page 6, 2006.

165. W. Zhao, E. Belhaire, C. Chappert, and P. Mazoyer. Spin transfer torque (STT)-MRAM–based runtime reconfiguration FPGA circuit. *ACM Transactions on Embedded Computing Systems*, 9(2):14:1–14:16, 2009.

166. S. Paul, S. Mukhopadhyay, and S. Bhunia. A circuit and architecture codesign approach for a hybrid CMOS-STTRAM nonvolatile FPGA. *IEEE Transactions on Nanotechnology*, 10(3):385–394, 2011.

167. W. Zhao, E. Belhaire, C. Chappert, B. Dieny, and G. Prenat. TAS-MRAM-based low-power high-speed runtime reconfiguration (RTR) FPGA. *ACM Transactions on Reconfigurable Technology and Systems*, 2(2):8:1–8:19, 2009.

168. O. Goncalves, G. Prenat, G. Di Pendina, and B. Dieny. Non-volatile FPGAs based on spintronic devices. In *Design Automation Conference (DAC)*, pages 1–3, 2013.

169. Y. Guillemenet, L. Torres, and G. Sassatelli. Non-volatile run-time field-programmable gate arrays structures using thermally assisted switching magnetic random access memories. *IET Computers Digital Techniques*, 4(3):211–226, 2010.

170. D. Suzuki, M. Natsui, A. Mochizuki, S. Miura, H. Honjo, H. Sato, S. Fukami, S. Ikeda, T. Endoh, H. Ohno, and T. Hanyu. Fabrication of a 3000-6-input-LUTs embedded and block-level power-gated nonvolatile FPGA chip using p-MTJ-based logic-in-memory structure. In *2015 Symposium on VLSI Technology*, pages C172–C173, 2015.

171. W. Zhao, E. Belhaire, Q. Mistral, E. Nicolle, T. Devolder, and C. Chappert. Integration of spin-RAM technology in FPGA circuits. In *2006 8th International Conference on Solid-State and Integrated Circuit Technology Proceedings*, pages 799–802, 2006.

172. Y. Guillemenet, L. Torres, G. Sassatelli, N. Bruchon, and I. Hassoune. A non-volatile run-time FPGA using thermally assisted switching MRAMS. In *2008 International Conference on Field Programmable Logic and Applications*, pages 421–426, 2008.

173. D. Suzuki, M. Natsui, S. Ikeda, H. Hasegawa, K. Miura, J. Hayakawa, T. Endoh, H. Ohno, and T. Hanyu. Fabrication of a nonvolatile lookup-table circuit chip using magneto/semiconductor-hybrid structure for an immediate-power-up field programmable gate array. In *2009 Symposium on VLSI Circuits*, pages 80–81, 2009.

174. D. Suzuki, M. Natsui, T. Endoh, H. Ohno, and T. Hanyu. Six-input lookup table circuit with 62% fewer transistors using nonvolatile logic-in-memory architecture with series/parallel-connected magnetic tunnel junctions. *Journal of Applied Physics*, 111(7):07E318, 2012.

175. R. Marculescu, U.Y. Ogras, Li Shiuan Peh, N.E. Jerger, and Y. Hoskote. Outstanding research problems in NoC design: System, microarchitecture, and circuit perspectives. *IEEE Transactions on Computer-Aided Design of Integrated Circuits and Systems*, 28(1):3–21, 2009.

176. D. E. Nikonov and I. A. Young. Overview of beyond-CMOS devices and a uniform methodology for their benchmarking. In *Proceedings of the IEEE*, Vol. 101, pages 2498–2533, 2013.

177. W. Zhao, L. Torres, Y. Guillemenet, L. Vitório Cargnini, Y. Lakys, J.-O. Klein, D. Ravelosona, G. Sassatelli, and C. Chappert. Design of MRAM based logic circuits and its applications. *Proceedings of the ACM Great Lakes Symposium on VLSI*, pages 431–436, 2011.

178. E. Deng, Y. Zhang, W. Kang, B. Dieny, J.-O. Klein, G. Prenat, and W. Zhao. Synchronous 8-bit non-volatile full-adder based on spin transfer torque magnetic tunnel junction. *IEEE Transactions on Circuits and Systems I: Fundamental Theory and Applications*, 62(7):1757–1765, 2015.

179. E. Deng, W. Kang, Y. Zhang, J.-O. Klein, C. Chappert, and W. Zhao. Design optimization and analysis of multicontext STT-MTJ/CMOS logic circuits. *IEEE Transactions on Nanotechnology*, 14(1):169–177, 2015.

180. H. Cai, Y. Wang, W. Zhao, and L.A. de Barros Naviner. Multiplexing sense-amplifier-based magnetic flip-flop in a 28-nm FDSOI technology. *IEEE Transactions on Nanotechnology*, 14(4):761–767, 2015.

181. Crocus Technologies. http://www.crocus-technology.com/technology. Accessed: June 10, 2017.

182. S. Manipatruni, D.E. Nikonov, and I.A. Young. Energy-delay performance of giant spin Hall effect switching for dense magnetic memory. *Applied Physics Express*, 7(10):103001, 2014.

183. Y. Kim, X. Fong, K.-W. Kwon, M.-C. Chen, and K. Roy. Multilevel spin-orbit torque MRAMs. *IEEE Transactions on Electron Devices*, 62(2):561–568, 2015.

184. T. Windbacher, V. Sverdlov, S. Selberherr, and H. Mahmoudi. Spin torque magnetic integrated circuit, 2016. EP Patent 2784020.

185. T. Windbacher, H. Mahmoudi, V. Sverdlov, and S. Selberherr. Rigorous simulation study of a novel non-volatile magnetic flip flop. In *Proceedings of the International Conference on Simulation of Semiconductor Processes and Devices (SISPAD)*, pages 368–371, 2013.

186. T. Windbacher, H. Mahmoudi, V. Sverdlov, and S. Selberherr. Novel MTJ-based shift register for non-volatile logic applications. In *Proceedings of the IEEE/ACM International Symposium on Nanoscale Architectures (NANOARCH)*, pages 36–37, 2013.

187. T. Windbacher, A. Makarov, V. Sverdlov, and S. Selberherr. Influence of magnetization variations in the free layer on a non-volatile magnetic flip flop. *Solid-State Electronics*, 108:2–7, 2015.

188. T. Windbacher, J. Ghosh, A. Makarov, V. Sverdlov, and S. Selberherr. Modelling of multipurpose spintronic devices. *International Journal of Nanotechnology*, 12(3/4):313–331, 2015.

189. T. Windbacher, A. Makarov, V. Sverdlov, and S. Selberherr. Novel buffered magnetic logic gate grid. *ECS Transactions*, 66(4):295–303, 2015.

190. U. Tietze and C. Schenk. *Electronic Circuits – Handbook for Design and Applications*, 2nd ed. Springer, 2008.

191. T. Windbacher, A. Makarov, V. Sverdlov, and S. Selberherr. The exploitation of magnetization orientation encoded spin-transfer torque for an ultra dense non-volatile magnetic shift register. In *2016 46th European Solid-State Device Research Conference (ESSDERC)*, pages 311–314, 2016.

192. T. Windbacher, A. Makarov, V. Sverdlov, and S. Selberherr. Layer coupling and read disturbances in a buffered magnetic logic environment. In *Proceedings of SPIE*, Vol. 9931, page 9931, 2016.

193. T. Windbacher, A. Makarov, V. Sverdlov, and S. Selberherr. Analysis of a spin-transfer torque based copy operation of a buffered magnetic processing environment. In *Proceedings of the 21st World Multi-Conference on Systemics, Cybernetics and Informatics (WMSCI)*, pages 142–146, 2017.

194. D. E. Nikonov, G. I. Bourianoff, and T. Ghan. Proposal of a spin torque majority gate logic. *IEEE Transactions on Electron Devices*, 32:1128–1130, 2011.

195. D. E. Nikonov, G.I. Bourianoff, and T. Ghani. Nanomagnetic circuits with spin torque majority gates. In *Proceedings of the IEEE Conference on Nanotechnology (IEEE-NANO)*, pages 1384–1388, 2011.

196. D. E. Nikonov, S. Manipatruni, and I. A Young. Cascade-able spin torque logic gates with input-output isolation. *Physica Scripta*, 90(7):074047, 2015.

197. D. D. Awschalom and M. E. Flatte. Challenges for semiconductor spintronics. *Nature Physics*, 3(3):153–159, 2007.

198. B. Behin-Aein, A. Sarkar, S. Srinivasan, and S. Datta. Switching energy-delay of all spin logic devices. *Applied Physics Letters*, 98(12):1–3, 2011.

199. S. P. Dash, S. Sharma, R. S. Patel, M. P. de Jong, and R. Jansen. Electrical creation of spin polarization in silicon at room temperature. *Nature*, 462(7272):491–494, 2009.

200. S. Srinivasan, A. Sarkar, B. Behin-Aein, and S. Datta. All-spin logic device with inbuilt nonreciprocity. *IEEE Transactions on Magnetics*, 47(10):4026–4032, 2011.

201. M. C. Chen, Y. Kim, K. Yogendra, and K. Roy. Domino-style spin-orbit torque-based spin logic. *IEEE Magnetics Letters*, 6:1–4, 2015.

202. Z. Liang and S. S. Sapatnekar. Energy/delay tradeoffs in all-spin logic circuits. *IEEE Journal on Exploratory Solid-State Computational Devices and Circuits*, 2:10–19, 2016.

203. Z.-Z. Guo. Effects of the channel material parameters on the spin-torque critical current of lateral spin valves. *Superlattices and Microstructures*, 75:468–476, 2014.

204. S. C. Chang, R. M. Iraei, S. Manipatruni, D. E. Nikonov, I. A. Young, and A. Naeemi. Design and analysis of copper and aluminum interconnects for all-spin logic. *IEEE Transactions on Electron Devices*, 61(8):2905–2911, 2014.

205. R. K. Kawakami. Spin amplification by controlled symmetry breaking for spin-based logic. *2D Materials*, 2(3):034001, 2015.

206. L. Su, W. Zhao, Y. Zhang, D. Querlioz, Y. Zhang, J.-O. Klein, P. Dollfus, and A. Bournel. Proposal for a graphene-based all-spin logic gate. *Applied Physics Letters*, 106(7):072407, 2015.

207. L. Su, Y. Zhang, J.-O. Klein, Y. Zhang, A. Bournel, A. Fert, and W. Zhao. Current-limiting challenges for all-spin logic devices. *Scientific Reports*, 5:14905, 2015.

208. D. E. Nikonov and I. Young. Benchmarking of beyond-CMOS exploratory devices for logic integrated circuits. *IEEE Journal on Exploratory Solid-State Computational Devices and Circuits*, 1:3–11, 2015.

209. J. Hu, N. Haratipour, and S. J. Koester. The effect of output-input isolation on the scaling and energy consumption of all-spin logic devices. *Journal of Applied Physics*, 117(17):17B524, 2015.

210. S. Manipatruni, D. E. Nikonov, and I. A. Young. Material targets for scaling all-spin logic. *Physical Review Applied*, 5:014002, 2016.

211. Z. Zhang, Y. Zhang, Z. Zheng, G. Wang, L. Su, Y. Zhang, and W. Zhao. Energy consumption analysis of graphene based all spin logic device with voltage controlled magnetic anisotropy. *AIP Advances*, 7(5):055925, 2017.

212. K. Y. Camsari, S. Ganguly, and S. Datta. Modular approach to spintronics. *Scientific Reports*, 5:10571, 2015.
213. P. Bonhomme, S. Manipatruni, R. M. Iraei, S. Rakheja, S. C. Chang, D. E. Nikonov, I. A. Young, and A. Naeemi. Circuit simulation of magnetization dynamics and spin transport. *IEEE Transactions on Electron Devices*, 61(5):1553–1560, 2014.
214. S. C. Chang, S. Manipatruni, D. E. Nikonov, I. A. Young, and A. Naeemi. Design and analysis of Si interconnects for all-spin logic. *IEEE Transactions on Magnetics*, 50(9):1–13, 2014.
215. S. Verma, M. S. Murthy, and B. K. Kaushik. All spin logic: A micromagnetic perspective. *IEEE Transactions on Magnetics*, 51(10):1–10, 2015.
216. T. Moriyama, G. Finocchio, M. Carpentieri, B. Azzerboni, D. C. Ralph, and R. A. Buhrman. Phase locking and frequency doubling in spin-transfer-torque oscillators with two coupled free layers. *Physical Review B*, 86:060411, 2012.
217. X. Fong, Y. Kim, R. Venkatesan, S. H. Choday, A. Raghunathan, and K. Roy. Spin-transfer torque memories: Devices, circuits, and systems. In *Proceedings of the IEEE*, Vol. 104, pages 1449–1488, 2016.
218. S. Ghosh, A. Iyengar, S. Motaman, R. Govindaraj, J. W. Jang, J. Chung, J. Park, X. Li, R. Joshi, and D. Somasekhar. Overview of circuits, systems, and applications of spintronics. *IEEE Journal on Emerging and Selected Topics in Circuits and Systems*, 6(3):265–278, 2016.
219. X. Fong, Y. Kim, K. Yogendra, D. Fan, A. Sengupta, A. Raghunathan, and K. Roy. Spin-transfer torque devices for logic and memory: Prospects and perspectives. *IEEE Transactions on Computer-Aided Design of Integrated Circuits and Systems*, 35(1):1–22, 2016.
220. S. Fukami, M. Yamanouchi, K. J. Kim, T. Suzuki, N. Sakimura, D. Chiba, S. Ikeda et al. 20-nm magnetic domain wall motion memory with ultralow-power operation. In *IEEE International Electron Devices Meeting (IEDM)*, pages 3.5.1–3.5.4, 2013.
221. J. A. Currivan, Y. Jang, M. D. Mascaro, M. A. Baldo, and C. A. Ross. Low energy magnetic domain wall logic in short, narrow, ferromagnetic wires. *IEEE Magnetics Letters*, 3:3000104, 2012.
222. J. A. Currivan-Incorvia, S. Siddiqui, S. Dutta, E. R. Evarts, C. A. Ross, and M. A. Baldo. Spintronic logic circuit and device prototypes utilizing domain walls in ferromagnetic wires with tunnel junction readout. In *IEEE International Electron Devices Meeting (IEDM)*, pages 32.6.1–32.6.4, 2015.
223. K. Huang and R. Zhao. Magnetic domain-wall racetrack memory-based nonvolatile logic for low-power computing and fast run-time-reconfiguration. *IEEE Transactions on Very Large Scale Integration (VLSI) Systems*, 24(9):2861–2872, 2016.
224. J. A. Currivan-Incorvia, S. Siddiqui, S. Dutta, E. R. Evarts, J. Zhang, D. Bono, C. A. Ross, and M. A. Baldo. Logic circuit prototypes for three-terminal magnetic tunnel junctions with mobile domain walls. *Nature Communications*, 7:10275, 2016.
225. D. Morris, D. Bromberg, J. G. Zhu, and L. Pileggi. mLogic: Ultra-low voltage non-volatile logic circuits using STT-MTJ devices. In *Design Automation Conference (DAC)*, pages 486–491, 2012.
226. D. M. Bromberg, D. H. Morris, L. Pileggi, and J. G. Zhu. Novel STT-MTJ device enabling all-metallic logic circuits. *IEEE Transactions on Magnetics*, 48(11):3215–3218, 2012.
227. S. S. P. Parkin, M. Hayashi, and L. Thomas. Magnetic domain-wall racetrack memory. *Science*, 320(5873):190–194, 2008.
228. H. Mahmoudi, T. Windbacher, V. Sverdlov, and S. Selberherr. Reliability analysis and comparison of implication and reprogrammable logic gates in magnetic tunnel junction logic circuits. *IEEE Transactions on Magnetics*, 49:5620–5628, 2013.
229. A. Whitehead and B. Russell. *Principia Mathematica*. Cambridge at the University Press, 1910.
230. C. E. Shannon. A symbolic analysis of relay and switching circuits. *Electrical Engineering*, 57(12):713–723, 1938.
231. J. Borghetti, G. S. Snider, P. J. Kuekes, J. J. Yang, D. R. Stewart, and R. S. Williams. Memristive switches enable stateful logic operations via material implication. *Nature*, 464:873–876, 2010.

232. H. Mahmoudi, T. Windbacher, V. Sverdlov, and S. Selberherr. Performance analysis and comparison of two 1T/1MTJ-based logic gates. In *Proceedings of the International Conference on Simulation of Semiconductor Processes and Devices (SISPAD)*, pages 163–166, 2013.

233. H. Mahmoudi, T. Windbacher, V. Sverdlov, and S. Selberherr. High performance MRAM-based stateful logic. In *International Conference on Ultimate Integration of Silicon (ULIS)*, pages 117–120, 2014.

234. H. Mahmoudi, T. Windbacher, V. Sverdlov, and S. Selberherr. MRAM-based logic array for large-scale non-volatile logic-in-memory applications. In *Proceedings of the 2013 IEEE/ACM International Symposium on Nanoscale Architectures (NANOARCH)*, pages 26–27, 2013.

235. H. Mahmoudi, T. Windbacher, V. Sverdlov, and S. Selberherr. RRAM implication logic gates, 2014. EP Patent App. 2736044.

236. F. S. Marranghello, M. G. A. Martins, V. Callegaro, A. I. Reis, and R. P. Ribas. Exploring factored forms for sequential implication logic synthesis. In *Nanotechnology (IEEE-NANO), 2014 IEEE 14th International Conference on*, pages 268–273. IEEE, 2014.

237. T. Breuer, A. Siemon, E. Linn, S. Menzel, R. Waser, and V. Rana. A HfO2-based complementary switching crossbar adder. *Advanced Electronic Materials*, 1(10), 2015.

238. J. Lee, D. I. Suh, and W. Park. The universal magnetic tunnel junction logic gates representing 16 binary boolean logic operations. *Journal of Applied Physics*, 117(17):17D717, 2015.

239. G. C. Adam, B. D. Hoskins, M. Prezioso, and D. B. Strukov. Optimized stateful material implication logic for three-dimensional data manipulation. *Nano Research*, 9(12):3914–3923, 2016.

240. M. Hosomi, H. Yamagishi, T. Yamamoto, K. Bessho, Y. Higo, K. Yamane, H. Yamada et al. A novel nonvolatile memory with spin torque transfer magnetization switching: Spin-RAM. In *IEEE International Electron Devices Meeting (IEDM)*, pages 459–462, 2005.

241. S. I. Kiselev, J. C. Sankey, I. N. Krivorotov, N. C. Emley, R. J. Schoelkopf, R. A. Buhrman, and D. C. Ralph. Microwave oscillations of a nanomagnet driven by a spin-polarized current. *Nature*, 425(6956):380–383, 2003.

242. W. H. Rippard, M. R. Pufall, S. Kaka, T. J. Silva, and S. E. Russek. Current-driven microwave dynamics in magnetic point contacts as a function of applied field angle. *Physical Review B*, 70:100406, 2004.

243. S. Bonetti, P. K. Muduli, F. Mancoff, and J. Åkerman. Spin torque oscillator frequency versus magnetic field angle: The prospect of operation beyond 65 GHz. *Applied Physics Letters*, 94(10):102507, 2009.

244. S. Bonetti, V. Tiberkevich, G. Consolo, G. Finocchio, P. K. Muduli, F. Mancoff, A. Slavin, and J. Åkerman. Experimental evidence of self-localized and propagating spin wave modes in obliquely magnetized current-driven nanocontacts. *Physical Review Letters*, 105:217204, 2010.

245. P. Villard, U. Ebels, D. Houssameddine, J. Katine, D. Mauri, B. Delaet, P. Vincent et al. A GHz spintronic-based RF oscillator. *IEEE Journal of Solid-State Circuits*, 45(1):214–223, 2010.

246. M. R. Pufall, W. H. Rippard, S. Kaka, T. J. Silva, and S. E. Russek. Frequency modulation of spin-transfer oscillators. *Applied Physics Letters*, 86(8):082506, 2005.

247. P. K. Muduli, Ye. Pogoryelov, S. Bonetti, G. Consolo, F. Mancoff, and J. Åkerman. Nonlinear frequency and amplitude modulation of a nanocontact-based spin-torque oscillator. *Physical Review B*, 81:140408, 2010.

248. Y. Pogoryelov, P. K. Muduli, S. Bonetti, E. Iacocca, F. Mancoff, and J. Åkerman. Frequency modulation of spin torque oscillator pairs. *Applied Physics Letters*, 98(19):192501, 2011.

249. Y. Pogoryelov, P. K. Muduli, S. Bonetti, F. Mancoff, and J. Åkerman. Spin-torque oscillator linewidth narrowing under current modulation. *Applied Physics Letters*, 98(19):192506, 2011.

250. P. K. Muduli, Ye. Pogoryelov, F. Mancoff, and J. Åkerman. Modulation of individual and mutually synchronized nanocontact-based spin torque oscillators. *IEEE Transactions on Magnetics*, 47(6):1575–1579, 2011.

251. P. K. Muduli, Ye. Pogoryelov, Y. Zhou, F. Mancoff, and J. Åkerman. Spin torque oscillators and RF currents-modulation, locking, and ringing. *Integrated Ferroelectrics*, 125(1):147–154, 2011.

252. A. A. Tulapurkar, Y. Suzuki, A. Fukushima, H. Kubota, H. Maehara, K. Tsunekawa, D. D. Djayaprawira, N. Watanabe, and S. Yuasa. Spin-torque diode effect in magnetic tunnel junctions. *Nature*, 438(7066):339–342, 2005.

253. H. Maehara, H. Kubota, Y. Suzuki, T. Seki, K. Nishimura, Y. Nagamine, K. Tsunekawa, A. Fukushima, A. M. Deac, K. Ando, and S. Yuasa. Large emission power over 2 μW with high Q factor obtained from nanocontact magnetic-tunnel-junction-based spin torque oscillator. *Applied Physics Express*, 6(11):113005, 2013.

254. S. Sani, J. Persson, S.M. Mohseni, Y. Pogoryelov, P. K. Muduli, A. Eklund, G. Malm, M. Käll, A. Dmitriev, and J. Åkerman. Mutually synchronized bottom-up multi-nanocontact spin-torque oscillators. *Nature Communications*, 4:2731, 2013.

255. E. Iacocca, P. Dürrenfeld, O. Heinonen, J. Åkerman, and R. K. Dumas. Mode-coupling mechanisms in nanocontact spin-torque oscillators. *Physical Review B*, 91:104405, 2015.

256. Z. M. Zeng, P. Upadhyaya, P. K. Amiri, K. H. Cheung, J. A. Katine, J. Langer, K. L. Wang, and H. Jiang. Enhancement of microwave emission in magnetic tunnel junction oscillators through in-plane field orientation. *Applied Physics Letters*, 99(3):032503, 2011.

257. C. H. Sim, M. Moneck, T. Liew, and J.-G. Zhu. Frequency-tunable perpendicular spin torque oscillator. *Journal of Applied Physics*, 111(7):07C914, 2012.

258. Z. Zeng, G. Finocchio, B. Zhang, P. Khalili Amiri, J. A. Katine, I. N. Krivorotov, Y. Huai et al. Ultralow-current-density and bias-field-free spin-transfer nano-oscillator. *Scientific Reports*, 3:1426, 2013.

259. A. Makarov, V. Sverdlov, and S. Selberherr. Magnetic oscillation of the transverse domain wall in a penta-layer MgO-MTJ. In *Proceedings of the International Symposium Nanostructures*, pages 338–339, 2013.

260. A. Makarov. Modeling of emerging resistive switching based memory cells. PhD thesis, TU Wien, 2014.

261. A. Makarov, V. Sverdlov, and S. Selberherr. Concept of a bias-field-free spin-torque oscillator based on two MgO-MTJs. *Extended Abstracts of the International Conference on Solid State Devices and Materials (SSDM)*, pages 796–797, 2013.

262. A. Makarov, V. Sverdlov, and S. Selberherr. Geometry optimization of spin-torque oscillators composed of two MgO-MTJs with a shared free layer. In *Proceedings of the International Conference on Nanoscale Magnetism (ICNM)*, page 69, 2013.

263. A. Makarov, T. Windbacher, V. Sverdlov, and S. Selberherr. Efficient high-frequency spin-torque oscillators composed of two three-layer MgO-MTJs with a common free layer. In *Proceedings of the Iberchip Workshop*, pages 1–4, 23, 2015.

264. T. Windbacher, A. Makarov, H. Mahmoudi, V. Sverdlov, and S. Selberherr. Novel bias-field-free spin transfer oscillator. *Journal of Applied Physics*, 115(17):17C901–1–17C901–3, 2014.

265. A. V. Nazarov, H. M. Olson, H. Cho, K. Nikolaev, Z. Gao, S. Stokes, and B. B. Pant. Spin transfer stimulated microwave emission in MgO magnetic tunnel junctions. *Applied Physics Letters*, 88(16):162504, 2006.

266. A. M. Deac, A. Fukushima, H. Kubota, H. Maehara, Y. Suzuki, S. Yuasa, Y. Nagamine, K. Tsunekawa, D. D. Djayaprawira, and N. Watanabe. Bias-driven high-power microwave emission from MgO-based tunnel magnetoresistance devices. *Nature Physics*, 4(10):803–809, 2008.

267. D. Houssameddine, S. H. Florez, J. A. Katine, J.-P. Michel, U. Ebels, D. Mauri, O. Ozatay et al. Spin transfer induced coherent microwave emission with large power from nanoscale MgO tunnel junctions. *Applied Physics Letters*, 93(2):022505, 2008.

268. P. K. Muduli, O. Heinonen, and J. Åkerman. Bias dependence of perpendicular spin torque and of free- and fixed-layer eigenmodes in MgO-based nanopillars. *Physical Review B*, 83:184410, 2011.

269. F. Mancoff, N. D. Rizzo, B. N. Engel, and S. Tehrani. Phase-locking in double-point-contact spin-transfer devices. *Nature*, 437(7057):393–395, 2005.

270. S. Kaka, M. R. Pufall, W. H. Rippard, T. J. Silva, S. E. Russek, and J. A. Katine. Mutual phase-locking of microwave spin torque nano-oscillators. *Nature*, 437(7057):389–392, 2005.

271. A. Slavin and V. Tiberkevich. Theory of mutual phase locking of spin-torque nanosized oscillators. *Physical Review B*, 74:104401, 2006.
272. J. Persson, Y. Zhou, and J. Åkerman. Phase-locked spin torque oscillators: Impact of device variability and time delay. *Journal of Applied Physics*, 101(9):09A503, 2007.
273. B. Georges, J. Grollier, V. Cros, and A. Fert. Impact of the electrical connection of spin transfer nano-oscillators on their synchronization: An analytical study. *Applied Physics Letters*, 92(23):232504, 2008.
274. Xi Chen and R. H. Victora. Phase locking of spin-torque oscillators by spin-wave interactions. *Physical Review B*, 79:180402, 2009.
275. A. Ruotolo, V. Cros, B. Georges, A. Dussaux, J. Grollier, C. Deranlot, R. Guillemet, K. Bouzehouane, S. Fusil, and A. Fert. Phase-locking of magnetic vortices mediated by antivortices. *Nature Nanotechnology*, 4(8):528–532, 2009.
276. E. Iacocca and J. Åkerman. Destabilization of serially connected spin-torque oscillators via non-Adlerian dynamics. *Journal of Applied Physics*, 110(10):103910, 2011.
277. W. H. Rippard, M. R. Pufall, S. Kaka, T. J. Silva, S. E. Russek, and J. A. Katine. Injection locking and phase control of spin transfer nano-oscillators. *Physical Review Letters*, 95:067203, 2005.
278. Y. Zhou, J. Persson, and J. Åkerman. Intrinsic phase shift between a spin torque oscillator and an alternating current. *Journal of Applied Physics*, 101(9):09A510, 2007.
279. B. Georges, J. Grollier, M. Darques, V. Cros, C. Deranlot, B. Marcilhac, G. Faini, and A. Fert. Coupling efficiency for phase locking of a spin transfer nano-oscillator to a microwave current. *Physical Review Letters*, 101:017201, 2008.
280. Y. Zhou, J. Persson, S. Bonetti, and J. Åkerman. Tunable intrinsic phase of a spin torque oscillator. *Applied Physics Letters*, 92(9):092505, 2008.
281. S. Urazhdin, P. Tabor, V. Tiberkevich, and A. Slavin. Fractional synchronization of spin-torque nano-oscillators. *Physical Review Letters*, 105:104101, 2010.
282. S. Y. Martin, N. de Mestier, C. Thirion, C. Hoarau, Y. Conraux, C. Baraduc, and B. Diény. Parametric oscillator based on nonlinear vortex dynamics in low-resistance magnetic tunnel junctions. *Physical Review B*, 84:144434, 2011.
283. A. Dussaux, A. V. Khvalkovskiy, J. Grollier, V. Cros, A. Fukushima, M. Konoto, H. Kubota et al. Phase locking of vortex based spin transfer oscillators to a microwave current. *Applied Physics Letters*, 98(13):132506, 2011.
284. A. Hamadeh, N. Locatelli, V. V. Naletov, R. Lebrun, G. de Loubens, J. Grollier, O. Klein, and V. Cros. Perfect and robust phase-locking of a spin transfer vortex nano-oscillator to an external microwave source. *Applied Physics Letters*, 104(2):022408, 2014.
285. S. Urazhdin, V. Tiberkevich, and A. Slavin. Parametric excitation of a magnetic nanocontact by a microwave field. *Physical Review Letters*, 105:237204, 2010.
286. P. Bortolotti, E. Grimaldi, A. Dussaux, J. Grollier, V. Cros, C. Serpico, K. Yakushiji et al. Parametric excitation of magnetic vortex gyrations in spin-torque nano-oscillators. *Physical Review B*, 88:174417, 2013.
287. R. L. Stamps, S. Breitkreutz, J. Åkerman, A. V. Chumak, Y. Otani, G. E. W. Bauer, J.-U. Thiele et al. The 2014 magnetism roadmap. *Journal of Physics D: Applied Physics*, 47(33):333001, 2014.
288. R. H. Liu, W. L. Lim, and S. Urazhdin. Spectral characteristics of the microwave emission by the spin Hall nano-oscillator. *Physical Review Letters*, 110:147601, 2013.
289. V. E. Demidov, S. Urazhdin, H. Ulrichs, V. Tiberkevich, A. Slavin, D. Baither, G. Schmitz, and S. O. Demokritov. Magnetic nano-oscillator driven by pure spin current. *Nature Materials*, 11(12):1028–1031, 2012.
290. Z. Duan, A. Smith, L. Yang, B. Youngblood, J. Lindner, V. E. Demidov, S. O. Demokritov, and I. N. Krivorotov. Nanowire spin torque oscillator driven by spin orbit torques. *Nature Communications*, 5:5616, 2014.
291. V. E. Demidov, S. Urazhdin, A. Zholud, A. V. Sadovnikov, and S. O. Demokritov. Nanoconstriction-based spin-Hall nano-oscillator. *Applied Physics Letters*, 105(17):172410, 2014.
292. V. E. Demidov, S. Urazhdin, E. R. J. Edwards, M. D. Stiles, R. D. McMichael, and S. O. Demokritov. Control of magnetic fluctuations by spin current. *Physical Review Letters*, 107:107204, 2011.

293. A. Slavin and V. Tiberkevich. Spin wave mode excited by spin-polarized current in a magnetic nanocontact is a standing self-localized wave bullet. *Physical Review Letters*, 95:237201, 2005.

294. H. Ulrichs, V. E. Demidov, and S. O. Demokritov. Micromagnetic study of auto-oscillation modes in spin-Hall nano-oscillators. *Applied Physics Letters*, 104(4):042407, 2014.

295. A. Fukushima, K. Yakushiji, H. Kubota, and S. Yuasa. Spin dice (physical random number generator using spin torque switching) and its thermal response. *2015 IEEE Magnetics Conference (INTERMAG)*, pages 1–1, 2015.

296. A. Fukushima, T. Seki, K. Yakushiji, H. Kubota, H. Imamura, S. Yuasa, and K. Ando. Spin dice: A scalable truly random number generator based on spintronics. *Applied Physics Express*, 7(8):083001, 2014.

297. H. Lee, C. Grezes, A. Lee, F. Ebrahimi, P. Khalili Amiri, and K. L. Wang. A spintronic voltage-controlled stochastic oscillator for event-driven random sampling. *IEEE Electron Device Letters*, 38(2):281–284, 2017.

298. K. Pagiamtzis and A. Sheikholeslami. Content-addressable memory (CAM) circuits and architectures: A tutorial and survey. *IEEE Journal of Solid-State Circuits*, 41(3):712–727, 2006.

299. R. Karam, R. Puri, S. Ghosh, and S. Bhunia. Emerging trends in design and applications of memory-based computing and content-addressable memories. In *Proceedings of the IEEE*, Vol. 103, pages 1311–1330, 2015.

300. R. Govindaraj and S. Ghosh. Design and analysis of 6-T 2-MTJ ternary content addressable memory. In *2015 IEEE/ACM International Symposium on Low Power Electronics and Design (ISLPED)*, pages 309–314, 2015.

301. R. Govindaraj and S. Ghosh. Design and analysis of STTRAM-based ternary content addressable memory cell. *Journal of Emerging Technologies in Computing Systems*, 13(4):52:1–52:22, 2017.

302. S. Matsunaga, S. Miura, H. Honjou, K. Kinoshita, S. Ikeda, T. Endoh, H. Ohno, and T. Hanyu. A 3.14 um2 4T-2MTJ-cell fully parallel TCAM based on nonvolatile logic-in-memory architecture. In *2012 Symposium on VLSI Circuits*, pages 44–45, 2012.

303. Y. Zhang, W. Zhao, J. O. Klein, D. Ravelosona, and C. Chappert. Ultra-high density content addressable memory based on current induced domain wall motion in magnetic track. *IEEE Transactions on Magnetics*, 48(11):3219–3222, 2012.

304. S. Matsunaga, A. Mochizuki, T. Endoh, H. Ohno, and T. Hanyu. Design of an energy-efficient 2T-2MTJ nonvolatile TCAM based on a parallel-serial-combined search scheme. *IEICE Electronics Express*, 11(3):20131006, 2014.

305. S. Jain, A. Ranjan, K. Roy, and A. Raghunathan. Computing in memory with spin-transfer torque magnetic RAM. *Computing Research Repository*, abs/1703.02118, 2017.

5

Ferroelectric Tunnel Junctions as Ultra-Low-Power Computing Devices

Spencer Allen Pringle and Santosh K. Kurinec

CONTENTS

5.1 Introduction

Next-generation energy-efficient computing has recently seen strong interest, with positive pressures from the popularization of Internet-of-Things (IoT) devices and high-performance portable electronics, including phones, tablets, and laptops. A large bottleneck for energy-efficiency has long been the information storage units, which typically rely on charge-storage for programming and erasing. HfO$_2$-based ferroelectric tunnel junctions (FTJs) store data as a change in polarization state, which is read as a modification of resistance state, and represent a unique opportunity as a next-generation digital nonvolatile memory, while also being complementary metal-oxide-semiconductor-compatible (CMOS-compatible). These devices have higher reliability than filamentary resistive random-access memory (ReRAM) and lower power consumption compared to competing devices, including phase-change memory (PCM) and other state-of-the-art FTJ. As shown in Table 5.1, FTJs outperform all other production and next-generation devices in terms of energy efficiency by orders of magnitude. In this chapter, the quantum-mechanical origins for this impressive energy performance will be discussed, along with some of the production implementation challenges and engineering considerations. Finally, system design examples utilizing these devices will be presented.

TABLE 5.1

Performance Comparison of Alternative Technologies Compiled from Various Sources

Device	NAND/NOR	ReRAM	PCM	MTJ	FTJ
Visual (Red and blue boxes are two different states in time)					
Mechanism	Vt Shift	O_2 Vac./ Filamentary	Phase change	e^- Spin (Mag)	Polarization
Speed (s)	400u[4]	500n	50n	30n[4]	10n
Power (J)	15u[5]	250n	6p	1.4u	30f
Cell Area	$4F^2$ or $10F^2$6	$25F^2$	$16F^2$	$1.3F^2$	$22F^2$
Material	Si	Ag, a-Si, W	$Ge_2Sb_2Te_5$	MgO(mag)	$BaTiO_2$ (now HfO_2)

MTJ values from Aziz et al. and Lee et al.,[8,9] selected speed values from Takishita et al.,[5] and NAND/NOR power from Mohan et al.[6] and cell area from a Micron Technology technical report.[7] All other device values from Ebong et al. and Kim et al.[10,11]

5.2 History and Operation

The FTJ is a two-terminal device composed of a ferroelectric film sandwiched between two electrodes that has two resistance states, low resistance state (LRS) and high resistance state (HRS), which can be switched between by applying a sufficiently positive (e.g., high to low) or negative (e.g., low to high) voltage. This write voltage must be such that it exceeds the coercive electric field of the ferroelectric and causes it to switch polarization directions. Contreras et al. first demonstrated and termed the FTJ in 2002, reporting a device comprised of $BaTiO_3$ and (surprisingly) matching $SrRuO_3$ electrodes that achieves reasonable performance and a memory window (*HRS/LRS*) of 3x resistance change, though they could not explain the origins of this device's operation.[1] It was not until 2005 that Zhuravlev et al. explained the operation of this device, postulating that the memory window arises from band bending due to a dissimilarity in the total potential introduced by screening charges in electrodes that have dissimilar screening (Debye) lengths.[2] Simulated energy bands of HfO_2-based FTJs and their corresponding current in low and high resistance states are shown in Figures 5.1 and 5.2. Zhuravlev et al. further postulated that the device of Contreras et al. only exhibited its small memory window due to a modification of the screening charge density in one electrode, which is not present in the other.[2] Kholstedt et al. (with Contreras) confirm this, later that year, by reporting a Ruddelsen-Popper interfacial layer in the bottom electrode of their device, not present in the top electrode, which modifies the effective screening length and introduces the field asymmetry and resulting memory window.[26] Academic research on $BaTiO_3$-based FTJs is relatively abundant, including a 2013 paper by Wang et al. in which the

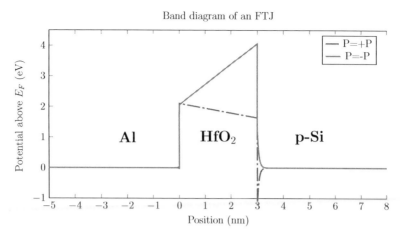

FIGURE 5.1
The energy band diagrams, at $V = 0$, of an Al/HfO$_2$/p-Si FTJ having 3 nm ferroelectric with material properties consistent with those reported by Mueller et al.[15] for positive (toward Al) and negative (toward p-Si) polarization.

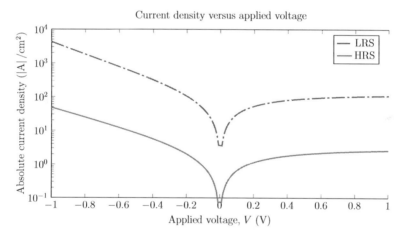

FIGURE 5.2
The Fowler-Nordheim tunneling current of an FTJ with material properties consistent with those used in Figure 5.1, showing a memory window *HRS/LRS* $\approx$ 90.

group achieves *HRS/LRS* $\approx$ 100 using Co and La$_{0.67}$Sr$_{0.33}$MnO$_3$ electrodes and a 2nm BaTiO$_3$.[3] Using rather high write voltages, $\approx 3V_c$, their device reaches write times as low as 13 ns.[3] The group even creates *Verilog-A* models for their devices.[3] Even more interestingly, Sören Boyn published a PhD thesis in 2016 evaluating FTJs for use as synaptic devices in neuromorphic computing, focusing heavily on BaTiO$_3$-based and BiFeO$_3$-based devices.[4] Some of the best-performing devices that they report (a Co/BiFeO$_3$/CCMO FTJ) have a reasonably-low coercive voltage (1.5 V) with a large memory window (*HRS/LRS* $\approx$ 1000) thanks to a large remanent polarization (P_r = 100 μC/cm^2) and achieves endurance of 4×10^6 cycles![4] Though these devices are not incredibly fast ($\approx$ 100 ns write times), their performance does correspond to a best-case surface power density (Pd = $V\,I = V^2/R_{HRS}$) of 45 W/cm^2, about an order of magnitude lower than typical BaTiO$_3$-based FTJ performance.[4]

5.3 HFO$_2$-Based FTJs

Ferroelectricity was first reported in hafnium oxide films by researchers from NaMLab gGmbH, in Dresden, Germany, in 2011.[12] The group, Böscke et al., discovered that by doping the HfO$_2$ film with 4 mol. % Si (forming SiO$_2$ bonds) and encapsulating it, before annealing, with a TiN top electrode, the TiN mechanically damps formation of the monoclinic crystal phase and instead promotes the growth of orthorhombic phase crystals, which exhibit both piezoelectric and ferroelectric response.[12] Importantly, this resulting film has a suitable remanent polarization ($P_r = 10$ μC/cm^2) and, more interestingly, a rather low coercive electric field of $\varepsilon_c = 1$ MV/cm^2.[12] This combination of material properties, combined with the prominent industry use of non-ferroelectric hafnium oxide as a high-k dielectric in most modern transistors, makes it particularly interesting for next-generation ultra-low-power computing. As shown in Table 5.2, the low coercive electric field, high dielectric constant, and good remanent polarization combine to enable the design of (simulated) FTJ with lower surface power density than all other competing FTJ.[18] Additionally, its reasonably high endurance is promising for industry-grade reliability.[15,16]

TABLE 5.2

Tables Comparing Performance of HfO$_2$-Based FTJs to Competing State of the Art FTJs

Researcher [cite]	Device	Materials	E_c (V/cm)	Area	CMOS?
Chanthbouala et al.[13]	FTJ	Co/BaTiO$_3$/LSMO	1×10^7	0.096	No
Abuwasib et al.[14]	FTJ	Co/BaTiO$_3$/LSMO	1×10^7	0.303	No
Boyn et al.[4]	FTJ	Co/BiFeO$_3$/CCMO	3.3×10^6	0.125	No
Mueller, Schroeder et al.[15,16]	MIM	Pt/TiN/Si-HfO$_2$/TiN	1×10^6	10,000	Yes
Polakowski et al.[17]	MIM	TiN/Al:HfO$_2$/TiN	1×10^6	50	Yes
Pringle (Based on Simulations)[18]	FTJ	Al/Al:HfO$_2$/p-Si	1×10^6	0.25	Yes

Research	d (nm)	ρ_{HRS} (Ω·cm^2)	HRS/LRS	P_r	Speed	Endurance	V_c	Pd (W/cm^2)
13	2	4.8×10^{-2}	300	—	10 ns	—	2.0	83.3
14	2	4.8×10^{-3}	60	—	—	—	2.0	823
4	4.6	5×10^{-2}	1000	100	100 ns	4×10^6	1.5	45
15,16	10	—	—	17	10 ns	1×10^{10}	1.0	
17	12	—	—	15	not rep.	2×10^9	1.2	
18	2	1.8×10^{-4}	6.2				0.2	222
	3	1.54×10^{-1}	90	15	10 ns	2×10^9	0.3	0.584
	5	2.3×10^5	500,000				0.5	1×10^{-6}

LSMO stands for La$_x$Sr$_{1-x}$MnO$_3$ and CCMO is Ca$_{0.96}$Ce$_{0.04}$MnO$_3$. Area units are μm^2. P_r is remanent polarization in (μC/cm^2) and Pd is surface power density.

5.4 Circuit Implementations

FTJs are particularly promising as synaptic devices for neuromorphic computing applications. Neuromorphic, or brain-inspired, computing is an emerging paradigm that mimics the operation of the human brain, which applies very well to solving complex multi-dimensional or temporal classification and regression tasks including video, image-recognition, audio-processing, and deep learning.[4] A simple neural net, shown in Figure 5.3, is comprised of i inputs connected to each of a layer of j neurons by synaptic devices $W_{i,j}$, which provide a varying connection strength between each input-neuron pair. Either directly or through further nested nets, with synaptic Band diagram of an FTJ connections between all neurons in a layer and the next, the neurons are connected to a final output layer of k neurons, which usually corresponds to a classification of up to k different groups. Classically, the neurons and synaptic devices are simulated using a field-programmable gate array (FPGA), though this is typically very energy-inefficient, because the stored synaptic weights and resulting waveforms must be calculated with floating-point digital logic, and therefore limits the maximum size and complexity of the system.[4] Interestingly, great improvements in power-efficiency can be made by instead implementing the stored synaptic weight changes using a spike-timing dependent plasticity (STDP) mimicking paradigm.[4] Analog waveforms, originally proposed by Linares-Barranco et al. in 2009, are instead generated asynchronously and passed to the neurons of the system at any time, to varying effect.[19] A synapse that receives a pre-neuron waveform shortly before a post-neuron waveform exhibits an increase in synaptic conductivity; conversely, a synapse that receives a post-neuron signal shortly before a pre-neuron waveform experiences a decrease in conductivity.[4,19] Boyn uses this mechanism to great effect in their 2016 thesis while implementing neuromorphic computing using FTJs.[4] Importantly, by using this particular STDP implementation

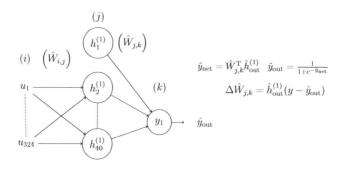

FIGURE 5.3

A simple neural net diagram and corresponding functions for the net output ($\hat{y}_{net}$), the neuron output functions ($\hat{y}_{out}$), and the update function for the second weight matrix ($\Delta\hat{W}_{j,k}$) as a function of desired net output (y), the actual output, and the internal neuron outputs ($\hat{h}_{out}^{(1)}$).

with FTJs the computation does not require extra steps in the neuron and requires no steps within the FTJ synapse.[4] The FTJ, by nature, responds desirably to the overlapping waveforms (consistent with the Linares-Barranco implementation[19]), simplifying the design and improving energy-efficiency.[4] Though Boyn doesout not discuss the power-efficiency of their proposed systems, this author calculates the base surface power density (Pd = V_c^2/ρ_{HRS}) to be 45 W/cm^2, which is rather good compared to BaTiO$_3$-based FTJs. However, HfO$_2$-based FTJ, while still maintaining reasonably fast operation, have the potential to reach surface power density 2 orders of magnitude smaller![18]

Neuromorphic computing is only one of many applications for which memristive implementations can improve energy efficiency. In 2017, Singh et al. reported exclusive inverted-OR (XNOR) using 3 memristors and 3 NMOS transistors, which they then use to design a 1-bit full adder.[20] Compared to conventional CMOS full adder, this design achieves delay 10% and power consumption of 1.3 mW, 1% of the CMOS version![20] Later that year, Lebdeh et al. reported a heterogeneous memristive (HM) XNOR, which they simulate in SPICE for ReRAM devices.[21] These XNOR are entirely composed of memristors and the group even designs and simulates 3-input exclusive OR (XOR) using HM XNOR.[21] These 3-in XOR dissipate only 52 fJ, 50% the energy of other state-of-the-art devices, with 50% the latency.[21]

Memristors can also be used in majority-logic systems for implementing arithmetic structures in supercomputing applications.[22] Specifically, any memristor can be used in programmable majority logic array (PMLA), charge-sharing threshold gate (CSTG), and current-mirror threshold gate (CMTG)-based majority logic systems.[22] Kvatinsky et al. describe the last two of these systems as it pertains to their implication logic (IMPLY) system and present an 8-bit full adder using logic built entirely with a memristor crossbar.[23]

Earlier this year, 2018, Fayyazi et al.[24] reported HSPICE simulation results of digital-to-analog converter (DAC) and analog-to-digital converter (ADC) circuits for IoT applications using 90-nm CMOS and a memristor SPICE model by Yakopcic et al.[25] Their systems, based on a low triangle neural net (LTNN), achieve average power and delay 3 orders of magnitude smaller than other state-of-the-art systems, with power as low as 9 μW and delay of $\approx$ 6 ps.[24] Initializing the circuit does take longer compared to conventional static random access memory (SRAM) implementations, since the synaptic weights must be written over a sufficient number of epochs, though the memristor circuit still only consumes ($\approx$ 1.7 pJ) 25% of the energy SRAM cache uses while writing.[24]

5.5 Conclusions

FTJs have been shown to reduce energy consumption by up to 9 orders of magnitude compared to competing state-of-the-art CMOS and outperform alternative memristor devices.[6,10] Though BaTiO$_3$- and BiFeO$_3$-based FTJs can achieve decently low power consumption, in comparison the lower coercive electric field, better film thickness scalability, and better CMOS-compatibility of ferroelectric-HfO$_2$ makes it ideal for FTJs enabling next-generation ultra-low-power computing. Memristor-based logic systems for XOR, XNOR, full-adder, DAC, and ADC outperform CMOS with as low as 50% the delay and 0.1% the power consumption!

References

1. J. R. Contreras, J. Schubert, H. Kohlstedt, and R. Waser, "Memory device based on a ferroelectric tunnel junction," in *Device Research Conference, 2002. 60th DRC. Conference Digest*, June 2002, pp. 97–98.

2. M. Zhuravlev, R. Sabirianov, S. Jaswal, and E. Tsymbal, "Giant electroresistance in ferroelectric tunnel junctions," *Physical Review Letters*, vol. 94, no. 24, 2005.

3. Z. Wang, W. Zhao, A. Bouchenak-Khelladi, Y. Zhang, W. Lin, J. O. Klein, D. Ravelosona, and C. Chappert, "Compact modelling for Co/BTO/LSMO ferroelectric tunnel junction," in *Nanotechnology (IEEE-NANO), 2013 13th IEEE Conference on*, August 2013, pp. 229–232.

4. S. Boyn, "Ferroelectric tunnel junctions: Memristors for neuromorphic computing," PhD dissertation, University of Paris-Saclay, 2016.

5. H. Takishita, T. Onagi, and K. Takeuchi, "Storage class memory based ssd performance in consideration of error correction capabilities and write/read latencies," in *2016 IEEE Silicon Nanoelectronics Workshop (SNW)*, June 2016, pp. 92–93.

6. V. Mohan, T. Bunker, L. Grupp, S. Gurumurthi, M. R. Stan, and S. Swanson, "Modeling power consumption of nand flash memories using flashpower," *IEEE Transactions on Computer-Aided Design of Integrated Circuits and Systems*, vol. 32, no. 7, pp. 1031–1044, 2013.

7. Micron Technology Inc., "Tn-29-19: Nand flash 101," Technical Report, 2006. Available: https://www.micron.com/~/media/documents/products/technical-note/nand-flash/tn2915.pdf

8. A. Aziz, N. Shukla, S. Datta, and S. K. Gupta, "Coast: Correlated material assisted STT MRAMs for optimized read operation," in *2015 IEEE/ACM International Symposium on Low Power Electronics and Design (ISLPED)*, July 2015, pp. 1–6.

9. S. Lee, K. Kim, K. Kim, U. Pi, Y. Jang, U. I. Chung, I. Yoo, and K. Kim, "Highly scalable STT-MRAM with 3-dimensional cell structure using in-plane magnetic anisotropy materials," in *2012 Symposium on VLSI Technology (VLSIT)*, June 2012, pp. 65–66.

10. I. E. Ebong and P. Mazumder, "CMOS and memristor-based neural network design for position detection," *Proceedings of the IEEE*, vol. 100, no. 6, pp. 2050–2060, 2012.

11. S. Kim, H. Wong, Y. Cui, Y. Nishi, and S. U. D. of Electrical Engineering, *Scalability and Reliability of Phase Change Memory*. Stanford University, 2010. Available: https://books.google.com/books?id=4WZq0q4wCAAC

12. T. S. Boescke, J. Mueller, D. Brauhaus, U. Schroeder, and U. Boettger, "Ferroelectricity in hafnium oxide thin films," *Applied Physics Letters*, vol. 99, no. 10, p. 102903, 2011. Available: http://ezproxy.rit.edu/login?url=http://search.ebscohost.com/login.aspx?Direct=true&db=afh&AN=65503915&site=ehost-live

13. A. Chanthbouala, A. Crassous, V. Garcia, K. Bouzehouane, S. Fusil, X. Moya, J. Allibe et al., "Solid-state memories based on ferroelectric tunnel junctions," *Nature Nanotechnology*, vol. 7, no. 2, pp. 101–104, 2012.

14. M. Abuwasib, H. Lee, P. Sharma, C. B. Eom, A. Gruverman, and U. Singisetti, "Cmos compatible integrated ferroelectric tunnel junctions (ftj)," in *2015 73rd Annual Device Research Conference (DRC)*, June 2015, pp. 45–46.

15. S. Mueller, J. Muller, U. Schroeder, and T. Mikolajick, "Reliability characteristics of ferroelectric Si:HfO$_2$ thin films for memory applications," *IEEE Transactions on Device and Materials Reliability*, vol. 13, no. 1, pp. 93–97, March 2013.

16. U. Schroeder, E. Yurchuk, S. Mueller, J. Mueller, S. Slesazeck, T. Schloesser, M. Trentzsch, and T. Mikolajick, "Non-volatile data storage in hfo2-based ferroelectric fets," in *2012 12th Annual Non-Volatile Memory Technology Symposium Proceedings*, October 2013, pp. 60–63.

17. P. Polakowski, S. Riedel, W. Weinreich, M. Rudolf, J. Sundqvist, K. Seidel, and J. Muller, "Ferroelectric deep trench capacitors based on al:hfo2 for 3d nonvolatile memory applications," in *2014 IEEE 6th International Memory Workshop (IMW)*, May 2014, pp. 1–4.

18. S. Pringle, "Modeling and Implementation of HfO2-based Ferroelectric Tunnel Junctions," Theses, December 2017. Available: http://scholarworks.rit.edu/theses/9710

19. B. Linares-Barranco and T. Serrano-Gotarredona, "Memristance can explain spike-time-dependent-plasticity in neural synapses," *Nature Preceedings*. Available: http://precedings. nature.com/documents/3010/, 2009.

20. A. Singh, "Memristor based xnor for high speed area efficient 1-bit full adder," in *2017 International Conference on Computing, Communication and Automation (ICCCA)*, May 2017, pp. 1549–1553.

21. M. A. Lebdeh, H. Abunahla, B. Mohammad, and M. Al-Qutayri, "An efficient heterogeneous memristive xnor for in-memory computing," *IEEE Transactions on Circuits and Systems I: Regular Papers*, vol. 64, no. 9, pp. 2427–2437, September 2017.

22. G. Jaberipur, B. Parhami, and D. Abedi, "Adapting computer arithmetic structures to sustainable supercomputing in low-power, majority-logic nanotechnologies," *IEEE Transactions on Sustainable Computing*, pp. 1–1, 2018.

23. S. Kvatinsky, G. Satat, N. Wald, E. G. Friedman, A. Kolodny, and U. C. Weiser, "Memristor-based material implication (imply) logic: Design principles and methodologies," *IEEE Transactions on Very Large Scale Integration (VLSI) Systems*, vol. 22, no. 10, pp. 2054–2066, 2014.

24. A. Fayyazi, M. Ansari, M. Kamal, A. Afzali-Kusha, and M. Pedram, "An ultra low-power memristive neuromorphic circuit for internet of things smart sensors," *IEEE Internet of Things Journal*, vol. 5, no. 2, pp. 1011–1022, 2018.

25. C. Yakopcic, T. M. Taha, G. Subramanyam, and R. E. Pino, "Generalized memristive device spice model and its application in circuit design," *IEEE Transactions on Computer-Aided Design of Integrated Circuits and Systems*, vol. 32, no. 8, pp. 1201–1214, 2013.

26. H. Kohlstedt, N. Pertsev, J. Contreras, and R. Waser, "Theoretical current-voltage characteristics of ferroelectric tunnel junctions," *PHYSICAL REVIEW B*, vol. 72, no. 12, SEP 2005.

Section II

Sensors, Interconnects, and Rectifiers

6

X-ray Sensors Based on Chromium Compensated Gallium Arsenide (HR GaAs:Cr)

Anton Tyazhev and Oleg Tolbanov

CONTENTS

6.1 HR GaAs:Cr Technology

Unlike the well-known Liquid Encapsulated Czochralski LEC technology of crystal growth semi-insulating gallium arsenide (LEC SI GaAs:EL2), high resistive GaAs:Cr (HR GaAs:Cr) wafers are fabricated by means of compensation of LEC n-GaAs wafers with chromium (Cr) impurity. The high-temperature annealing is used for the compensation [1,2]. The results of Hall measurement of HR GaAs:Cr and SI GaAs:EL2 samples are presented in Table 6.1 [3].

An analysis of the data from Table 6.1 shows the HR GaAs:Cr (**H**igh **R**esistance - HR GaAs, *compensated with chromium*) resistivity to be more than an order of magnitude higher than that of SI GaAs. Another important feature of HR GaAs:Cr is that the concentration of holes (p) exceeds the concentration of electrons (n). One can say that Cr acting as a deep acceptor becomes dominant level and defines electrophysical characteristics of HR GaAs:Cr. Ionized Cr impurity forms a hole trapping level (0.78 eV above valence band) in comparison with ionized EL2 centers in SI GaAs, which is a trap for electrons and has a strong dependence on a cross-capture section of electric field strength. Exceeding the hole concentration over electron one suppresses the formation of moving domains of strong

TABLE 6.1

The Results of Hall Measurement

Sample	Resistivity, Gohm×cm	R×σ, (cm²/V×s)	n, (10^5 cm⁻³)	p, (10^5 cm⁻³)	Electron Mobility mn, (cm²/V×s)	Hole Mobility mp, (cm²/V×s)
SI GaAs:EL2	0.1–0.2	−6900	70–100	4–6	5100–5800	340–390
HR GaAs:Cr	0.9–1.7	−2500	2–3	120–200	3200–4700	210–320

electric field in HR GaAs:Cr sensors. The drift of the domains from cathode to anode leads to a generation of stochastic current pulses in SI GaAs:EL2, which makes it difficult to use SI GaAs:EL2 sensors for X-ray imaging [2].

Calculation of resistivity and R × σ dependence on n/n_i ratio (n_i - is the intrinsic concentration of GaAs) shows that maximal resistivity corresponds to n value of 5×10^5 cm⁻³ (Figure 6.1a). At the same time, the R × σ value is about minus 2600 cm²/Vxs. The calculation was carried out taking into account the electron to hole mobilities ratio (b) of 15. Points A, B, and C (Figure 6.1b) correspond to electron concentration at which R × σ becomes positive (A), resistivity reaches the maximal value (B), and condition of intrinsic GaAs ($n = p = n_i$) realizes (C).

One can see that the calculated (Figure 6.1) and experimental (Table 6.1) values are in good agreement.

Thereby, HR GaAs technology allows for reaching a resistivity that is close to the theoretical limit. Producing the high resistive HR GaAs:Cr wafers requires keeping of n/n_i ratio within narrow range of 0.2–0.4. Measurement of R × σ value can be used to control whether the requirement is satisfied.

To measure the resistivity distribution through the test wafer thickness, a rapid method based on measuring the current-voltage characteristics of a point contact along the angle lap made of HR GaAs:Cr wafers was used. The measurement results are shown in Figure 6.2 [3].

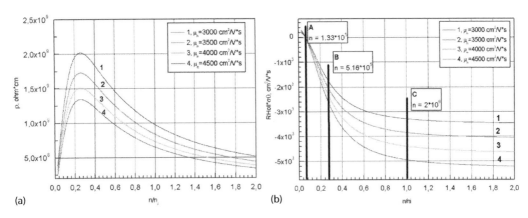

FIGURE 6.1

(a,b) Calculation of resistivity and R × σ dependences on n/n_i (n_i - is the intrinsic concentration of GaAs) ratio.

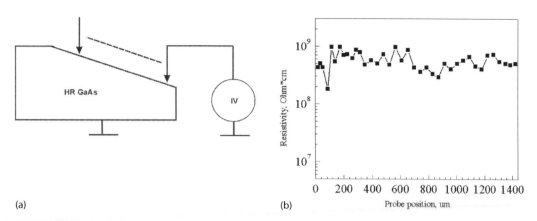

(a)   (b)

FIGURE 6.2

Circuit for measuring of IV curves of point contact on HR GaAs:Cr angle lap (a), resistivity distribution through the thickness of HR GaAs:Cr sample (b).

Since the HR GaAs:Cr technology is based on compensation of shallow donor impurity in n-GaAs, the distribution of resistivity and mobility-lifetime ($\mu\,\tau$) product in HR GaAs:Cr wafers will be determined by both technological conditions of compensation with Cr atoms and distribution profile of the carrier concentration (CC) in n-GaAs wafers. Thus, the CC concentration in n-GaAs wafers should be measured prior to the high-temperature annealing. The CC concentration was monitored by measuring the IR absorption on free-charge carriers (electrons). It was found that non-uniformity of CC distribution exists not only across the n-GaAs crystal length, but also across the diameter of the crystal (Figures 6.3 and 6.4). It should be noted that this non-uniformity is growing with the increase of the charge carrier concentration from the beginning (Figures 6.3a and 6.4a) of the ingot to its end (Figures 6.3b and 6.4b) [3,4].

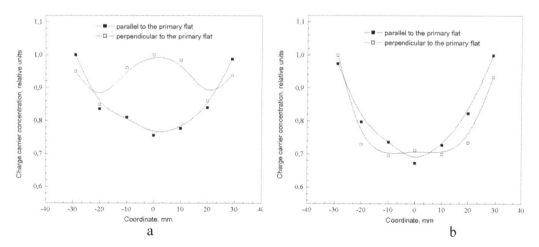

FIGURE 6.3

Charge carrier mapping in 3 inches n-GaAs wafers from the beginning (a) and end (b) of the ingot.

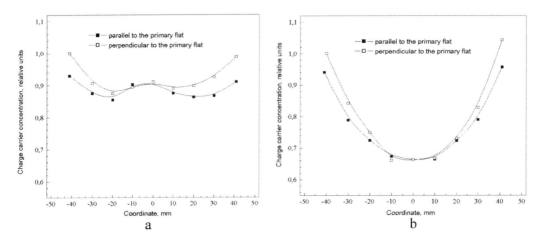

FIGURE 6.4
Charge carrier mapping in 4 inches n-GaAs wafers from the beginning (a) and end (b) of the ingot.

System of contactless resistivity measurement COREMA WT has been used for resistivity mapping of 3-inch and 4-inch HR GaAs:Cr wafers. Measurement results are shown in Figures 6.5 and 6.6 [3,4].

It can be seen that the resistivity mapping correlates with the distribution profile of the CC in n-GaAs wafers. Nevertheless HR GaAs:Cr technology allows producing of HR GaAs:Cr with average resistivity value exceeding of 1 GOhm×cm and non-uniformity no more than 15%.

It should be emphasized that the results of resistivity measurement by means of a contactless method (Figures 6.5 and 6.6) are consistent with Hall measurements (Table 6.1) and with calculated ones (Figure 6.1a). This allows to use the contactless method for non-destructive testing of the resistivity distribution in HR GaAs:Cr wafers.

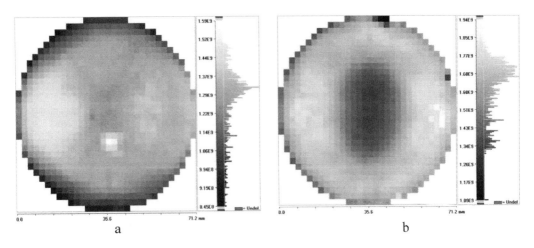

FIGURE 6.5
Resistivity mapping in 3 inches HR GaAs:Cr wafers from the beginning (a) and end (b) of the ingot.

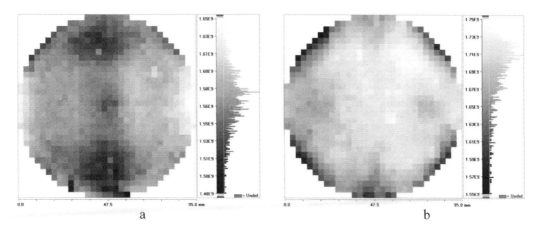

a   b

FIGURE 6.6
Resistivity mapping in 4 inches HR GaAs:Cr wafers from the beginning (a) and end (b) of the ingot.

6.2 Electrical Characteristics of Me-HR GaAs:Cr-Me Structures

As an object of investigation, HR GaAs:Cr pad sensors with an area of 0.1–0.25 cm^2 and a sensitive layer thickness in the range of 250–1000 μm were used. Metal contacts were fabricated by means of chemical deposition of nickel (Ni) of 0.5–1 μm thickness or by electron-beam evaporation of Cr/Ni films of 0.1–0.2 μm total thickness on both sides of a pad sensor. Thus, the sensors had a symmetric Ni - HR GaAs:Cr - Ni or Ni/Cr - HR GaAs:Cr – Cr/Ni structure. Prior to measurements, the sensor lateral faces were treated in an etchant based on sulphuric acid solution followed by rinsing in deionized water and drying in the atmosphere at a temperature of 150°C for 10 minutes. Measurement of the current-voltage characteristics was performed at room temperature using a Keithley 2410 source meter. Figure 6.7 shows the current-voltage characteristics of the samples under study. It should be noted that, as applied to the investigated symmetric structures, the terms "forward" and "reverse" bias denote positive or negative potential on one of the sensor contacts [4,5].

The analysis of the I–V curves (Figure 6.7) suggests that in the abovementioned range the sensor's current-voltage characteristics are symmetric and can be described by the function

$$J \propto U^B, \tag{6.1}$$

where J is the current density, U is the voltage on the sensor, and B is the index of a power of the current density as a function of the voltage.

The current-voltage characteristics of all sensors have three typical sections: section 1 (from 0.02 to 0.5 V) demonstrates linear dependence of the current density on the bias (B ≈ 1); section 2 (from 1 to 15 V) is the sublinear one (B ≈ 0.7 – 0.9); and section 3 (>20 V) is the superlinear dependence of the current density on the bias (B ≈ 1.1 – 1.3). The above sections are most distinctly seen when the experimental I–V characteristics are processed using the expression:

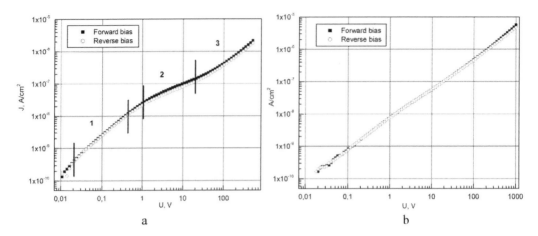

FIGURE 6.7
I-V curves of HR GaAs:Cr pad sensors with Ni (a) and Cr/Ni (b) contacts having thickness of 500 μm and 980 μm, correspondently.

$$\rho_{\text{diff}} = \frac{1}{d} \cdot \left[\frac{dJ}{dU} \right]^{-1}, \qquad (6.2)$$

where ρ_{diff} is differential resistivity, d - sensor thickness.

The differential resistivity of the HR GaAs:Cr pad sensors as a function of bias is shown in Figure 6.8.

The results presented in Figure 6.8 show that, regardless of the type of metal contact, the ρ_{diff} dependencies on the voltage consist of three sections: section 1 (from 0.02 to 0.5 V)—ρ_{diff} is independent of voltage; Section 2 (from 1 to 15 V)—an increase of the ρ_{diff} value; Section 3 (from 20 V and more)—attainment of saturation and a decrease of the ρ_{diff} value with increasing voltage. It should be noted that the ρ_{diff} value in section 1 is in the range from 0.7 to 1.5 GΩ×cm, which is in agreement with the HR GaAs:Cr resistivity

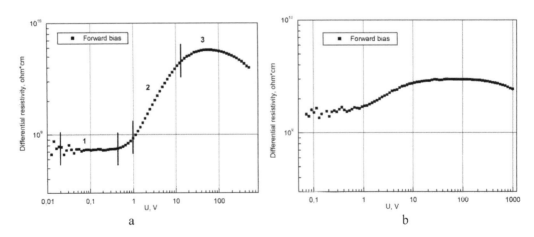

FIGURE 6.8
Dependences of differential resistivity of the HR GaAs:Crpad sensors with Ni (a) and Cr/Ni (b) contacts on bias.

values obtained with Hall Effect measurements (Table 6.1, Figures 6.2b, 6.5, and 6.6), as well as with the calculated values (Figure 6.1a).

It was shown [5] that a band diagram (Figure 6.9a) and an equivalent circuit of a HR GaAs:Cr pad sensor (Figure 6.9b) consist of series-connected Schottky diodes (SD 1, 2) and the resistor R. The Schottky diodes are -Ni-HR GaAs:Cr- interfaces, whereas R is the bulk material resistance. It was demonstrated that calculated (Figure 6.10a) and experimental I-V curves (Figure 6.10b) of Ni-HR GaAs:Cr-Ni structures have satisfactory agreement in the bias range of 0.02–15 V. The calculation was carried out within the framework of thermionic emission theory, where the potential lowering of the barrier height and bulk material resistance R were taken into account.

Experimental results (Figure 6.7) and simulations (Figure 6.10) allow for the conclusion that section 1 (from 0.02 to 0.5 V) of I–V curve is defined by bulk resistivity of HR GaAs. Thus, bulk resistivity of HR GaAs:Cr can be derived in accordance with equation (6.2) from experimental I–V curves.

The temperature stability of the Ni, Ni/Cr contacts, as well as HR GaAs:Cr itself were studied by annealing of Me-HR GaAs:Cr-Me pad sensors at temperatures of 250°C and 400°C in hydrogen for 10 minutes, followed by treatment in the sulphuric acid solution,

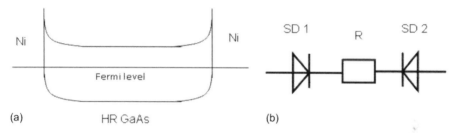

FIGURE 6.9

A band diagram of a HR GaAs:Cr pad sensor in the equilibrium state (a) and an equivalent circuit design (b).

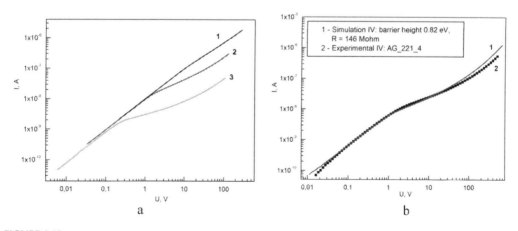

FIGURE 6.10

Calculated I–V characteristics of the HR GaAs:Cr sensor taking into account the potential barrier height lowering at different potential barrier height φ_0: 0.75 eV (1), 0.80 eV (2), and 0.85 eV(3) and serial resistor R = 100 MOhm (a). Calculated (1) and experimental (2) current-voltage characteristics the HR GaAs:Cr pad sensor (b).

rinsing in deionized water, and drying in the atmosphere at a temperature of 150°C for 10 minutes [5,6]. Figure 6.11 shows the current-voltage characteristics of the samples under study. The equation (6.2) was used to calculate bulk resistivity of HR GaAs:Cr. The dependence of HR GaAs:Cr resistivity on temperature of the annealing is demonstrated in Figure 6.12.

The analysis of the study of the Me-HR GaAs:Cr-Me structures thermal stability allows us to draw the following conclusions:

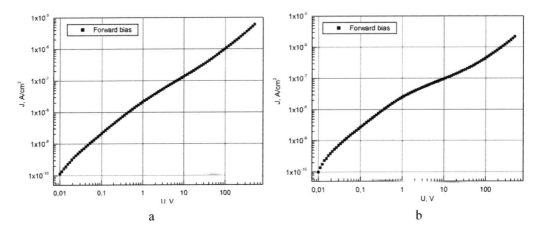

FIGURE 6.11
The current-voltage characteristics of Ni – HR GaAs:Cr - Ni pad sensors with a thickness 500 μm after low temperature annealing at 250°C (a) and 400°C (b) in hydrogen for 10 minutes. (From Tolbanov, O. et al., Investigation of the current-voltage characteristics, the electric field distribution and the charge collection efficiency in X-ray sensors based on chromium compensated gallium arsenide, *Proceedings of the SPIE 9213, Hard X-Ray, Gamma-Ray, and Neutron Detector Physics*, XVI, 92130G, September 5, 2014.)

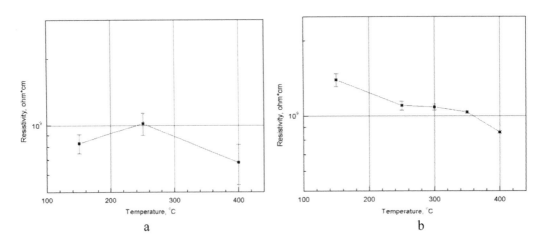

FIGURE 6.12
Dependency of bulk resistivity of Ni – HR GaAs:Cr - Ni(a) and Ni/Cr – HR GaAs:Cr – Cr/Ni(b) pad sensors on annealing temperature. (From Novikov, V. et al., Impact of thermal treatment on electrical characteristics and charge collection efficiency of Me-GaAs:Cr-Me x-ray sensors, *Proceedings of the SPIE 9593, Hard X-Ray, Gamma-Ray, and Neutron Detector Physics* XVII, 95930F, August 26, 2015.)

- Annealing in the temperature range 150°C–400°C has little effect on the pad sensor current-voltage characteristics of both types of contacts -Ni-HR GaAs:Cr-Ni- and -Ni / Cr-HR GaAs:Cr-Cr / Ni-, which is explained by the thermal stability of the Me-HR GaAs:Cr interface.

- The resistivity of HR GaAs:Cr decreases by 18%–30% from the initial value after annealing at 400°C.

It is known that when assembling HR GaAs:Cr matrix sensors with ASIC via flip-chip method, various low-melting metals and alloys and, in particular, indium [7] are used as bumps. Interaction of molten indium with metallization of a pixel that forms an electrical contact can lead to an uncontrolled change in the electrical and mechanical characteristics of the contact. In this connection, the -Ni-HR GaAs:Cr -Ni / In- characteristics of pad sensors were studied, which were created by alloying In into one of the pad contacts Ni – HR GaAs:Cr - Ni at temperatures of 250°C and 400°C. The alloying was carried out in an hydrogen atmosphere for 10 minutes. The current-voltage characteristics and the dependence of the differential resistance on the voltage of pad sensors based on -Ni-HR GaAs:Cr-Ni / In- structures are shown in Figures 6.13 and 6.14, respectively.

The main difference between the current-voltage characteristics of the -Ni-HR GaAs:Cr-Ni / In- structures is a sharp increase of the current density at a negative potential at the -Ni / In- contact, which is observed at a bias of about 70 V (Figure 6.13). Dependences of the differential resistance on the voltage applied to the structure (Figure 6.14) show that a sharp decrease in the value of the differential resistivity begins at 35 V when a negative potential is applied at the Ni / In contact. It is important to note that, with positive polarity on the -Ni / In- contact, the current density and differential resistance versus the voltage of the -Ni-HR GaAs:Cr-Ni / In- structures are similar to those of all other types of structures (Figures 6.7 through 6.11).

Investigation of the photovoltaic effect in the contacts of different metals with HR GaAs:Cr confirms the results presented in Figures 6.7, 6.9, 6.10, and 6.13 and show that contacts based on In, in contrast to Ni and Cr contacts, do not form barrier contacts for electrons [8].

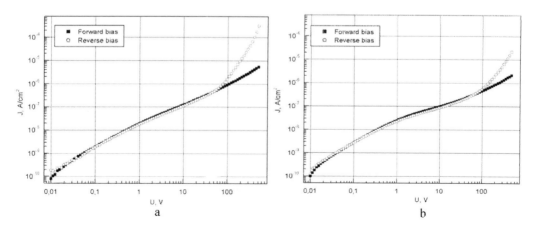

FIGURE 6.13
Current-voltage characteristics of 500 μm thick Ni - HR GaAs:Cr – Ni/In pad sensors after low temperature annealing at 250°C (a) and 400°C (b) in hydrogen for 10 minutes.

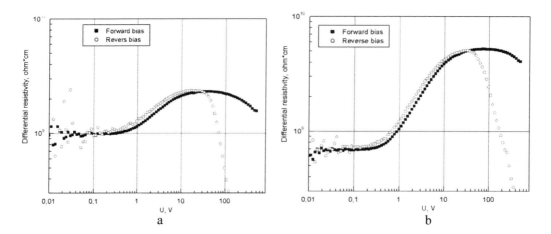

FIGURE 6.14
Dependences of differential resistivity of the Ni - HR GaAs:Cr – Ni/In pad sensors on bias after low temperature annealing at 250°C (a) and 400°C (b).

It is well known that to ensure the efficient operation of the X-ray sensor, it is necessary to provide high values of the electric field strength in the entire sensitive volume while maintaining low dark current [9]. The most well-known method for solving this problem is to create sensors based on such diode structures as Schottky barrier, p-n junction, or p-i-n diode. In this case, the thickness of the sensing region of the sensor is determined by the thickness of the space charge region (SCR), which ultimately depends on the doping level of the semiconductor and the bias voltage on the sensor [10]:

$$d_{SCR} \propto \sqrt[a]{\frac{U}{N}},\tag{6.3}$$

где d_{SCR} – SCR thickness, U – applied voltage; N – shallow donor concentration, a – power coefficient $1/2 - 1/3$.

The analysis of expression (6.3) shows that, to achieve large values of the SCR thickness, it is necessary to use a semiconductor material with the lowest possible level of doping.

At present, the minimum doping level of semi-insulating gallium arsenide crystals grown by the Czochralski method (LEC SI GaAs: EL2) is 10^{14} cm^{-3} [11]. This leads to fundamental limitations in achieving large thicknesses of SCR in sensors based on LEC SI GaAs:EL2. Nevertheless, the analysis of open-access publications shows that most developers and researchers of gallium arsenide X-ray sensors use a standard approach—fabricating diode structures on a material with the lowest possible level of doping. As a consequence, there are limitations both on the maximum achievable thickness of the SCR and on the stability of the SI GaAs: EL2 sensors [12–17].

In X-ray sensors based on the metal-HR GaAs:Cr-metal structures, another concept is realized: the thickness of the sensitive layer is determined not by the thickness of the SCR, but by the thickness of the layer that determines the volume resistance (R) of the sensor (Figure 6.9a). This is possible if the condition is:

$$R_{eff} \leq R, \tag{6.4}$$

where R_{eff} - effective SCR resistance, R - volume resistance of the sensor

Practical fulfillment of condition (6.4) was possible because of the high resistivity of HR GaAs:Cr and the relatively low value of R_{eff}. In this case, most of the bias voltage will fall on the sensor volume. Since the resistivity is distributed uniformly over the entire thickness of the metal-HR GaAs:Cr-metal sensor, the electric field strength will also have a corresponding uniform distribution over the entire thickness of the HR GaAs:Cr layer.

Methods of investigating the electric field strength distribution in HR GaAs:Cr pad sensors are based on the use of the Pockels effect [5]. Measurements were performed using infrared radiation with 920 nm wavelength. The distribution of the intensity of the infrared radiation depending on the bias voltage at the HR GaAs:Cr and SI LEC GaAs:EL2 pad sensors are demonstrated in Figures 6.15 and 6.16, respectively.

The analysis of the results shown in Figures 6.15 and 6.16 allows one to assert that in the sensors based on HR GaAs:Cr the electric field is stable in time and homogeneously distributed through the whole sensor thickness in the voltage range under study. In sensors based on SI LEC GaAs:EL2 and having the structure of -"Schottky barrier – SI LEC

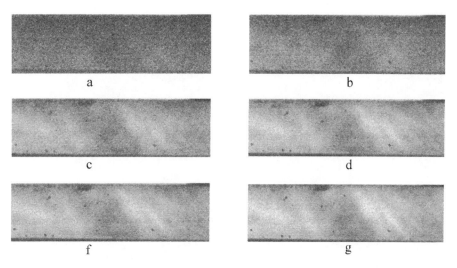

FIGURE 6.15
IR transmission images of 500 μm thick Ni/Cr – HR GaAs:Cr – Cr/Npad sensor at different bias voltages: a – 0 V; b – 100 V; c – 200 V; d – 300 V; f – 400 V; g – 500 V. (From Tolbanov, O. et al., Investigation of the current-voltage characteristics, the electric field distribution and the charge collection efficiency in X-ray sensors based on chromium compensated gallium arsenide, *Proceedings of the SPIE 9213, Hard X-Ray, Gamma-Ray, and Neutron Detector Physics*, XVI, 92130G, September 5, 2014.)

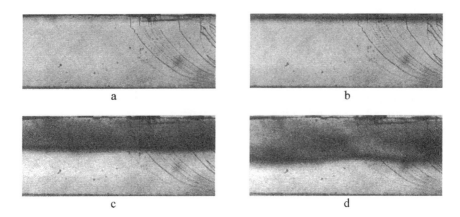

FIGURE 6.16

IR transmission images of 630 μm thick Schottky barrier – SI LEC GaAs:EL2 – ohmic contact pad sensor at different bias voltages: a – 0 V; b – 100 V; c – 300 V; d – 400 V. (From Tolbanov, O. et al., Investigation of the current-voltage characteristics, the electric field distribution and the charge collection efficiency in X-ray sensors based on chromium compensated gallium arsenide, *Proceedings of the SPIE 9213, Hard X-Ray, Gamma-Ray, and Neutron Detector Physics*, XVI, 92130G, September 5, 2014.)

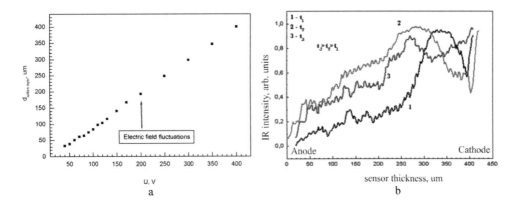

FIGURE 6.17

Electric field into Schottky barrier – SI LEC GaAs:EL2 – ohmic contact pad sensor. (a) Dependence of active layer thickness on bias voltage. (From Tolbanov, O. et al., Investigation of the current-voltage characteristics, the electric field distribution and the charge collection efficiency in X-ray sensors based on chromium compensated gallium arsenide, *Proceedings of the SPIE 9213, Hard X-Ray, Gamma-Ray, and Neutron Detector Physics*, XVI, 92130G, September 5, 2014.) and (b) temporal dependence of electric field profile at 300 V. (From Budnitsky, D.L. et al., Electric field distribution in chromium compensated GaAs, *12th International Conference on Semiconducting & Insulating Materials*, Smolenice Castle, Slovakia, June 30 to July 5, pp. 23–26, Institute of Electrical and Electronics Engineers, 2002.)

GaAs:EL2 – ohmic contact-," the active region thickness depends on the bias voltage with a coefficient of approximately 1 μm/V (Figure 6.17).

It should be noted that, unlike the results shown in Figure 6.15, the strong field region in the sensors -"Schottky barrier – SI LEC GaAs:EL2 – ohmic contact-" is illustrated by the dark section in the infrared image of the lateral face (Figure 6.16). This suggests an effective absorption of the infrared radiation by the EL2 centers in the region of the strong electric field. Besides, at a voltage higher than 200 V, generation of moving domains of strong electric field is observed into SI LEC GaAs: EL2 sensors (Figure 6.17). The movement of the domains leads to stochastic current oscillations in the external circuit. This is one of the main disadvantages of SI LEC GaAs: EL2 X-ray imaging sensors.

6.3 Sensitivity of Me-HR GaAs:Cr-Me Pad Sensors to Ionizing Radiations

The experimental investigation and simulation of pulse response of HR GaAs:Cr pad sensors irradiated with 5.5 MeV alpha particles of ^{241}Am source were carried out [18–20]. Some results are demonstrated in Figure 6.18.

One can see that the duration of the experimental pulses (Figure 6.18a) exceeds the simulated values of 2–3 ns (Figure 6.18b). Also, there is no saturation of maximum current pulse amplitude (Figure 6.18a) comparing to the calculated values (Figure 6.18b). In addition, experimental values of output current pulse are lower than the calculated ones under the equal bias voltages. It was mentioned previously [5], that such differences may be related to the plasma effect caused by heavily ionizing alpha particles. The electron-hole plasma generated in the alpha-particle track screens off the external electric field, which results in decreasing the spacing between electrons and holes and causes their intensive trapping and recombination. According to the simulation [21], the time of existence of the electron-hole plasma, i.e., the plasma time, accounts for up to 30% of the lifetime of nonequilibrium charge carriers. As a result, there occurs a considerable deficiency of the charge collected by a sensor. It reduces charge collection efficiency (CCE) and leads to a significant underestimation of the ($\mu \times \tau$) values. The use of weakly ionizing 1 MeV beta-particles (a minimum ionizing particle, MIP), or gamma-quanta, avoids the plasma effects and obtains more correct $\mu \times \tau$ values. Table 6.1 shows the results of calculating the $\mu \times \tau$ values based on the CCE dependences on bias voltage for alpha (CCE_α), beta (CCE_β), and gamma (CCE_γ) radiations [5]. Besides, the using of MIP allows for estimating the $\mu \times \tau$ values, both for electrons and holes in the HR GaAs:Cr material [22] (Table 6.2).

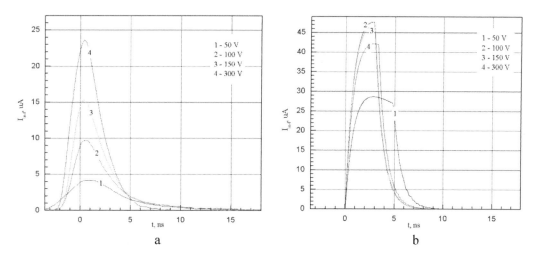

a   b

FIGURE 6.18

Experimental (a) and calculated (b) dependencies of output current pulse on bias voltage. Cathode of 250 um thick HR GaAs:Cr pad sensor was irradiated with 5.5 MeV alpha-particles. (From Dragone, A. et al., Feasibility study of a "4H" X-ray camera based on GaAs:Cr sensor, *J. Instrum.*, 2016.)

TABLE 6.2

The Results of $\mu \times \tau$ Measurements

	Based on CCE_α	Based on CCE_β	Based on CCE_γ
$(\mu \times \tau)_n$, $cm^2/V \times s$	6.8×10^{-6}	6.0×10^{-5}	7.6×10^{-5}
$(\mu \times \tau)_p$, $cm^2/V \times s$	—	3.0×10^{-7}	—

The results of $(\mu \times \tau)_{n,p}$ measurements show that in HR GaAs:Cr structures mainly electrons are collected. Optimal conditions for compensation of n-GaAs by Cr atoms make it possible to increase the value of $(\mu \times \tau)_n$ up to $2.0 \times 10^{-4}\,cm^2/Vs$ [4,25], with $(\mu \times \tau)_p$ values not exceeding $1.0 \times 10^{-6}\,cm^2/Vs$. It is important to emphasize that $(\mu \times \tau)_n$ don't change after annealing of Ni/Cr - HR GaAs:Cr – Cr/Ni sensors in the temperature range of 250°C–400°C [6].

Investigations of X-ray sensitivity of Ni/Cr - HR GaAs:Cr – Cr/Ni pad sensors were carried out [5,23]. The experimental dependences of the photocurrent (Iph) on the exposure dose rate (EDR) are shown in Figure 6.19.

Analysis of the obtained results shows that the sensor sensitivity is higher in the case of the cathode irradiation (Figure 6.19b) than when the anode is irradiated (Figure 6.19a) at other equal parameters. The I_{ph} saturation with the EDR increase is probably contingent on the sensor polarization due to trapping of holes. It results in the appearance of uncompensated charge in the sensor bulk and, consequently, in the decrease of the electric field in the sensitive region of sensor with the corresponding decrease of the CCE value. To restore linear dependence of I_{ph} on EDR in the range of 150 – 400 mR/s the bias on the sensors should be increased having electric field strength about 4 kV/cm.

Contacts characteristics play a significant role in I_{ph} dependency on EDR. The investigation of X-ray sensitivity of In/Ni - HR GaAs:Cr – Cr/Ni pad sensors has shown that type of metal contacts have great influence on I_{ph} dependence on EDR. When In contact serves

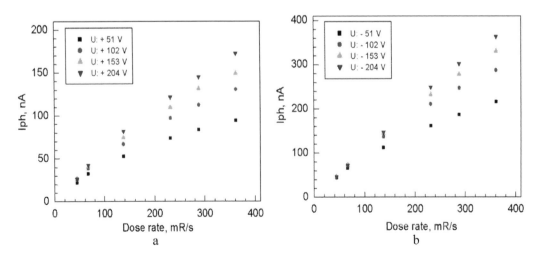

FIGURE 6.19

Photocurrent (Iph) dependence on bias voltage on 510 μm thick HR GaAs:Cr pad sensor for anode (a) and cathode (b) irradiation. X-ray tube bias 80 kV, sensor sizes of $0.3 \times 0.3\,cm^2$.

as cathode the sufficient increase of X-ray sensitivity occurs. The I_{ph} dependence on EDR becomes superlinear. On the other hand, the alloying of In doesn't influence the sensitivity, when In/Ni contact is positive biased [24].

6.4 HR GaAs:Cr X-ray Imaging Sensors

The collection of only the electronic component of the charge leads to a dependence of the charge collection efficiency and sensitivity on the generation point of the electron-hole pair in the sensitive volume of the HR GaAs:Cr pad sensor. These factors lead to a significant deterioration in energy resolution. An example of the amplitude spectrum of HR GaAs:Cr pad sensors is shown in Figure 6.20. Obviously, the peak corresponding to 60 keV is absent. [18].

It is known that a common method of improving the spectrometric characteristics of unipolar charge collection of sensors is the use of the "small pixel" effect [25–27]. The main point of the "small pixel" effect is that in a sensor, having a characteristic dimension of the contact much smaller than the thickness of the sensitive layer, the coordinate dependence of the "weight" field compensates the incomplete collection of one of the charge components. A similar type of "weight" field is realized in multi-element pixel (2D) and microstrip sensors (1D). The efficiency of the method is determined by the ratio of the diameter of the pixel (the width of the microstrip) to the thickness of the sensor's sensitive layer: the larger the ratio, the greater the efficiency. The minimum diameter of a pixel providing the best spectrometry is limited by such factors as: charge sharing, leakage of the fluorescence irradiation of sensor material and Compton scattering. Photoelectric absorption and Compton scattering become equal in GaAs around 150 keV [28]. Thus, besides charge trapping, the charge sharing and leakage of the fluorescence irradiation are the

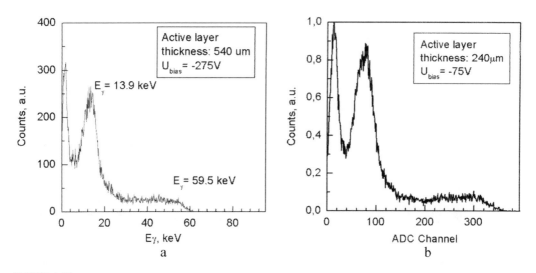

FIGURE 6.20
Pulse height distribution of HR GaAs:Cr pad sensors with thickness of 540 µm (a) and 240 µm (b). Cathode irradiation with gamma-quanta of [241]Am source.

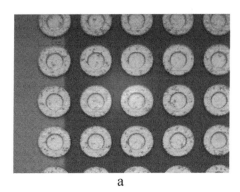
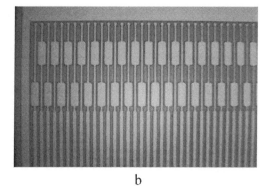

a   b

FIGURE 6.21
Images of X-ray HR GaAs:Cr pixel (a) and microstrip (b) sensors. The sensors have 55 μm (a) and 50 μm (b) pitches.

main factors that decrease energy resolution of HR GaAs:Cr pixel sensors operating with gamma-quanta having energy up to 100 keV.

Different HR GaAs:Cr pixel sensors were produced (Figure 6.21) and investigated in combination with counting [25,29,30–37] and integrating [38,39] ASICs. The issues under investigation were spatial and energy resolutions, temperature stability of CCE, high flux capability [29,33], and radiation hardness [29,40] of HR GaAs:Cr sensors.

It was found that flood images obtained with HR GaAs:Cr pixel sensors have "cellular cell" structure (Figures 6.22 and 6.23). The dark current of pixels located within wall of the "cellular cell" is higher than pixels from the middle of cells (Figure 6.22a). The count rate behavior on flood images has complex character (Figure 6.23). In both cases, "cellular cell" inhomogeneity blurs X-ray images, which leads to a decrease of spatial resolution.

Investigation of "cellular cell" network origin was done by means of white beam X-ray topography [34]. The results are depicted on Figure 6.24. One can see that "cellular cell" structure has the same shape for n-GaAs (Figure 6.24a) and HR GaAs:Cr (Figure 6.24b)

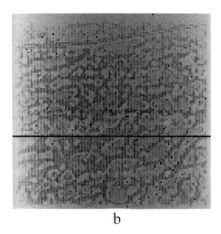

a   b

FIGURE 6.22
Dark current (a) and photocurrent (b) mapping of 50 μm pixel pitch HR GaAs:Cr pixel sensor assembled with integrating ASIC. The ASIC size is 128 × 128 pixels.

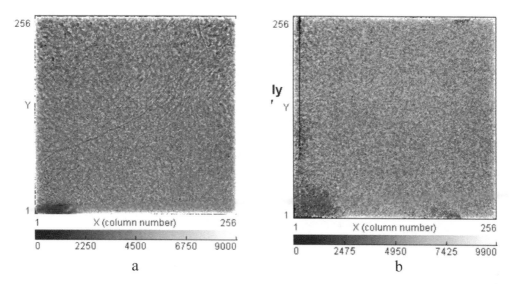

FIGURE 6.23
Count rate mapping of 55 μm pitch HR GaAs:Cr pixel sensors assembled with counting Timepix (a) and MedipixR (b) ASICs. The threshold energies were 20 keV for both assemblies. (From Hamann, E., Characterization of high resistivity GaAs as sensor material for photon counting semiconductor pixel detectors, PhD Thesis, University of Freiburg.)

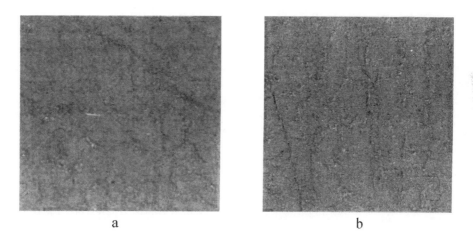

FIGURE 6.24
White beam X-ray topography of n-GaAs (a) and HR GaAs:Cr (b) wafers having thickness of 500 μm. The size of images is 4×4 mm². (From Hamann, E., Characterization of high resistivity GaAs as sensor material for photon counting semiconductor pixel detectors, PhD Thesis, University of Freiburg.)

wafers. It was confirmed that origin of the "cellular cell" network is connected with growth method of n-GaAs crystal. Similar features are well-known for crystals of other complex semiconductors [41].

One of the main advantages of HR GaAs:Cr sensors in comparison with SI GaAs:EL2 is absence of the electric field moving domains. It means that the X-ray images obtained with HR GaAs:Cr pixel sensors are stable in time. A simple flat field correction of raw image (Figure 6.25a) sufficiently improves the image quality (Figure 6.25b). It is evident that the

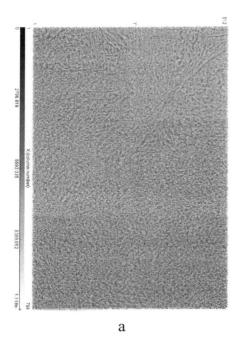 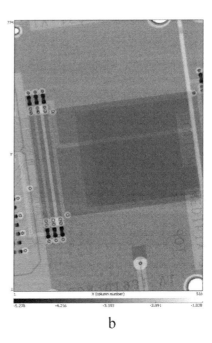

a   b

FIGURE 6.25

Images obtained with a 512 × 76 pixel GaAs sensor brmp-bonded to Timepix ASICs: a flood image (a) and flat-field corrected image of test object (b). (From Budnitsky, D. et al., *J. Instrum.*, 9, 2014.)

sensors provide good spatial resolution and enable high picture quality to be obtained. Assembly of Si pixel sensor and Timepix ASIC on PCB was used as test object (Figure 6.25b).

The current results of spatial resolution investigation are summarized in Table 6.3. The sensors are shown to provide spatial resolution corresponding to the pixel pitch.

Energy resolution of HR GaAs:Cr pixel sensors were tested by means of HEXITEC [25] and Medipix3RX [30,34] ASICs. HEXITEC ASIC has 250 pA limit on maximum leakage current per pixel, which required cooling of HR GaAs:Cr sensor down to 7°C. HR GaAs:Cr HEXITEC sensor anodes had 250 μm pitch with 50 μm spacing and 500 μm thickness. Providing HR GaAs:Cr HEXITEC sensor cathode irradiation with [241]Am source the pulse height distributions were obtained (Figure 6.26). The raw distribution was improved with

TABLE 6.3

Dependence of Spatial Resolution of HR GaAs:Cr Pixel Sensors on Photon Energy

Photon energy:	10 keV	15 keV	15 keV	25 keV	25 keV	25 keV
Threshold:	8.5 keV	8.5 keV	12 keV	8.5 keV	12 keV	15 keV
70% MTF	9 mm^{-1}	6.8 mm^{-1}	9.5 mm^{-1}	6.2 mm^{-1}	7.5 mm^{-1}	8.7 mm^{-1}
30% MTF	16.5 mm^{-1}	12.5 mm^{-1}	17.5 mm^{-1}	11.4 mm^{-1}	13.8 mm^{-1}	15.9 mm^{-1}
MTF at $f_{Nyquist}$	69%	53%	72%	46%	59%	67%

Source: Hamann, E., Characterization of high resistivity GaAs as sensor material for photon counting semiconductor pixel detectors, PhD Thesis, University of Freiburg.

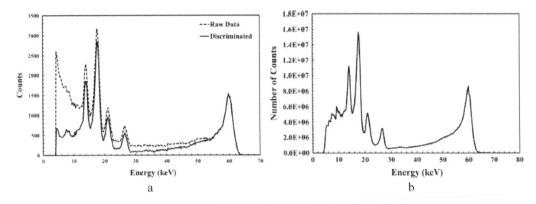

FIGURE 6.26
Effect of charge sharing discrimination on the spectroscopic performance of a typical single pixel spectrum (a) and average corrected spectroscopic response of all 6400 pixels (b). (From Veale, M. et al., *Nucl. Instr. Meth. A*, 752, 6, 2014.)

charge sharing discrimination correction (Figure 6.26a). The average corrected spectroscopic response of all 6400 pixels is shown on Figure 6.26b.

The removal of shared events leads to a 30% reduction in the total number of counts. But resulting energy resolution (FWHM) was found to be about 3 keV at 60 keV photon energy.

Medipix3RX ASIC has embedded circuitry to compensate the charge sharing events [30,34]. The ASIC has four operation modes. Two of them are modes of single photon counting (SPM) and charge summing mode (CSM). Corresponding HR GaAs:Cr sensors had pixel anodes of 55 μm pitch and 10 μm spacing. The sensor thicknesses were about 500 μm. The spectroscopic response of HR GaAs:Cr Medipix3RX assemblies operated in SPM and CSM are demonstrated in Figure 6.27a and b, correspondently. The experiments were carried out at room temperature. The results of FWHM investigation of HR GaAs:Cr Medipix3RX assemblies are summarized in Table 6.4 [30,34]. It is evident that FWHM values obtained with Medipix3RX and HEXITEC ASICs are in a good agreement.

Investigation of HR GaAs:Cr pixel sensor active layer thickness (Figure 6.28a) as well as CCE stability within "minus" 20°C to "plus" 20°C temperature range (Figure 6.28b) were performed [34]. The results show that active layer thickness is equal to geometrical

TABLE 6.4

Dependence of FWHM of HR GaAs:Cr Medipix3RX Assembly on Photon Energy

Photon energy, keV	8	15	22.5	25	40
FWHM, keV	2.8	3.4	4.34	4.16	4.5
FWHM, keV	34.6	22.8	19.3	16.7	11.2

Source: Hamann, E., Characterization of high resistivity GaAs as sensor material for photon counting semiconductor pixel detectors, PhD Thesis, University of Freiburg.

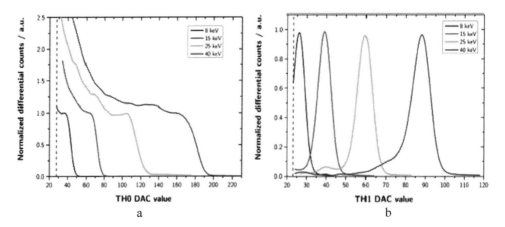

FIGURE 6.27
Spectroscopic performance of HR GaAs:Cr Medipix3RX assembly operated in SPM (a) and CSM (b) modes. (From Hamann, E. et al., *IEEE Trans. Med. Imaging*, 34, 707–715, 2015.)

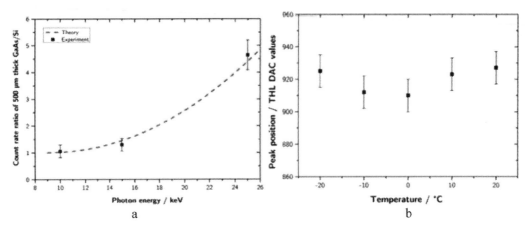

FIGURE 6.28
Spectroscopic performance of HR GaAs:Cr Medipix3RX assembly operated in SPM (a) and CSM (b) modes. (From Hamann, E., Characterization of high resistivity GaAs as sensor material for photon counting semiconductor pixel detectors, PhD Thesis, University of Freiburg.)

thickness of the sensors, which confirm results of IR transmission experiments obtained earlier (Figure 6.17). In comparison with Si pixel sensors, the HR GaAs:Cr ones demonstrate higher count rate starting from 18 keV (Figure 6.28a). The 60 keV photon peak positions are stable and consistent within their errors (Figure 6.28ba).

High flux capability was investigated using HR GaAs:Cr pixel sensors in combination with Timepix ASICs [23,30,34] as well as in combination with integrating LPD ASIC [38].

Figure 6.29 demonstrates the test results obtained from 500 μm thick HR GaAs:Cr pixel sensor bonded with Timepix ASIC [23]. A tungsten anode X-ray tube served for an X-ray source. It is evident that for a bias voltage of −50 V and at dose rate (DR) more than 140 mR/s, the count rate decreases (Figure 6.29a). This effect can be compensated by

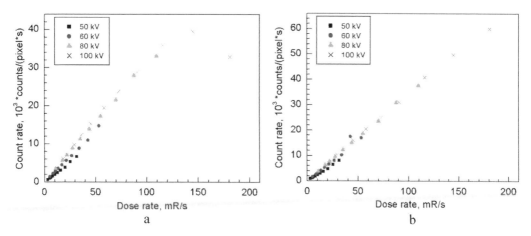

FIGURE 6.29
The dependence of 500 μm thick HR GaAs:Cr pixel sensor count rate on the dose rate at different bias voltage:minus 50 V (a) and minus 200 V (b) (From Budnitsky, D. et al., *J. Instrum.*, 9, 2014.)

increasing the bias voltage up to −200 V (Figure 6.29b). The results presented in Figure 6.29 are in agreement with those shown in Figure 6.19.

Results of investigations of count rate dependence on bias voltage are shown in Figure 6.30a. It was demonstrated that maximum count rate of HR GaAs:Cr pixel sensor in combination with Timepix ASIC can reach 800 kHz/pixel at bias voltage of minus 800 V and for X-ray tube voltage of 80 kV (Figure 6.30b).

Analysis shows that for an X-ray flux intensity exceeding 3×10^6 photons/pixel×s, the increase in the maximum count rate is saturated. A conceivable reason for the saturation is a limited Timepix chip speed of response (0.5–1 MHz/pixel) [23]. It was shown that increase of Ikrum DAC value allows reaching of the count rate up to 3×10^6 photons/pixel×s (10^9 photons/(mm²×s)). It was noted that with the 10 hours exposure of the sensor there were no effects connected to polarization or radiation damage [33,34].

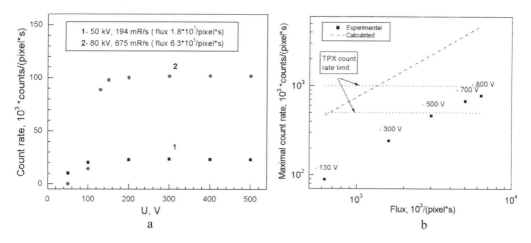

FIGURE 6.30
The dependence of the count rate of the 500 μm thick HR GaAs:Cr pixel sensor on the bias voltage (a) and on X-ray flux intensity at different bias voltages applied on the sensor (b).

Investigation of 500 μm thick HR GaAs:Cr pixel sensor in combination with integrating LPD ASIC was carried out on Diamond Light Source Synchrotron. It was demonstrated that for 20 keV X-ray flux intensity of 10^7–10^{11} photons/(mm^2×s) polarization of HR GaAs:Cr pixel sensor is observed. The possible reason of the polarization is the build-up of positive charge due to hole trapping. Increase of sensor bias voltage up to 1000 V reduce the polarization and allows X-ray imaging at a frame rate of 3.7 MHz [38].

The HR GaAs:Cr pixel sensor based LAMBDA system for fast X-ray imaging was tested. The system allows X-ray imaging with 2 kHz frame rate. The 500 μm thick HR GaAs:Cr pixel sensor has 768 × 512 pixels with 55 μm pitch. The system was tested at PETRA-III beamline P02.2. The X-ray photon energy was 42 keV; at this energy, the absorption efficiency of 500 μm thick GaAs is 74% in comparison with 4.6% for the same thickness of silicon. Results of the X-ray diffraction on CeO_2 powder are presented in Figure 6.31 [32].

Studies of radiation hardness of HR GaAs:Cr sensors were made using 12 keV photons [29] as well as using of 8.5–10 MeV electrons [40]. It was found that the radiation hardness of the sensors are at least 1 MGy for 12 keV X-rays (Figure 6.32) and about 1.5 MGy for high energy beta-particles (Figure 6.33).

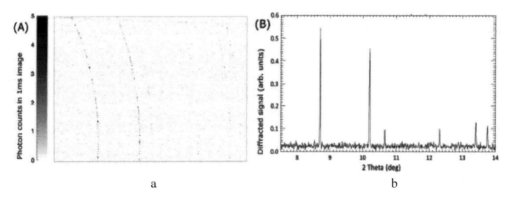

a   b

FIGURE 6.31
Diffraction pattern obtained in 1 ms with the HR GaAs:Cr LAMBDA detector (a) and diffracted X-ray intensity dependence on scattering angle (b). (From Pennicard, D. et al., *J. Instrum.*, 9, Article number C12026, 2014.)

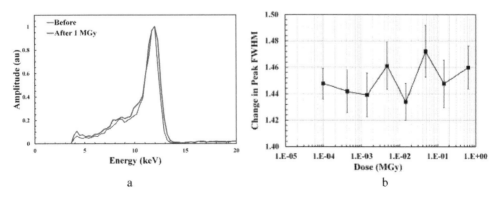

a   b

FIGURE 6.32
The spectrum of the 12 keV measured before and after a 1 MGy exposure (a) and the variation in the FWHM of the 12 keV peak as a function of dose (b). (From Veale, M. et al., *J. Instrum.*, 2014.)

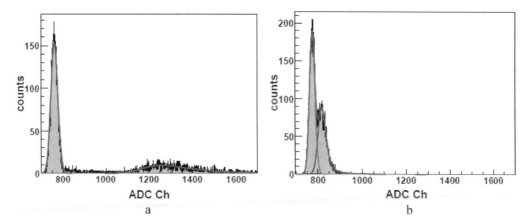

FIGURE 6.33
The spectrum of the MIP measured before (a) and after (b) a 1.5 MGy exposure with 8.5–10 MeV electrons. (From Afanaciev, K. et al., *J. Instrum.*, 2012.)

6.5 Summary

6.5.1 Resistivity

It is experimentally shown that a material with resistivity of about 10^9 Ohm × cm, which is close to the theoretical limit, and can be produced using the technology of n-type gallium arsenide compensation with chromium. Optimization of compensation conditions allows for the achieving of good uniformity of the resistivity distribution, both across the area and thickness of the wafers.

It was found that the current-voltage characteristics of HR GaAs:Cr sensors measured at room temperature are symmetrical and nearly linear in a wide range of bias voltages. Importantly, unlike the SI GaAs:EL2 sensors, there are no current oscillations throughout the whole range of bias voltages in HR GaAs:Cr sensors.

6.5.2 Active Layer Thickness

As was previously reported, the technology of n-type gallium arsenide compensation with chromium allows obtaining of high-resistive layers of up to 1 mm thickness. As a result of experiments, it was shown that the electric field in HR GaAs:Cr sensors is distributed uniformly throughout the thickness of high-resistive layer.

6.5.3 Spectrometry and μ×τ

The investigations of the charge collection efficiency (CCE) have shown that HR GaAs:Cr is characterized mainly by the collection of electron component of the charge. The value of $(\mu \times \tau)_n$ for electrons reaches 10^{-4} cm^2/V, and for holes $(\mu \times \tau)_p$ is about 10^{-6} cm^2/V.

It was shown that by using the "small pixel effect" it is possible to obtain an energy resolution of about 2–3 keV at 60 keV line in the temperature range from 7°C to ambient. The study of HR GaAs pixel sensor in combination with Medipix3RX ASIC showed that employment of charge summing mode (CSM) leads to a significant suppression of the charge sharing effect and allows achieving of energy resolution that is acceptable to spectroscopic CT application.

The temperature dependence showed that the CCE is stable in the temperature range from −20°C to +20°C.

6.5.4 X-ray Imaging and Spatial Resolution

The study of the spatial resolution of HR GaAs:Cr pixel sensors with 55 μm pitch was conducted using Timepix and Medipix3RX ASICs. It was shown that the spatial resolution of the sensor is about 8 lp/mm, which is close to the theoretical limit determined by the pixel pitch. In the original (raw X-ray images) images, a cellular structure is observed due to the dislocation network emerging during crystal growth of n-GaAs. However, the temporal stability of the electric field distribution allows to use flat field correction, which greatly improves the image quality.

6.5.5 Radiation Hardness and High Flux Operation

Studies of radiation hardness of HR GaAs:Cr sensors were made using X-ray at DLS synchrotron. It was found that the radiation hardness of the sensor is at least 1 MGy when it's irradiated with 12 keV X-rays.

Studies of high flux X-ray imaging using HR GaAs:Cr Timepix assemblies have shown that the count rate of 16 keV photons reaches values of $1 \times 10^9 \, s^{-1} mm^{-2}$ [4,10]. It is noted that with the 10-hour exposure of the sensor there were no effects connected to polarization or radiation damage.

6.5.6 HR GaAs:Cr Pixel Sensor Technology and Future Forecast

Currently, HR GaAsCr technology enables the production of pixel and microstrip sensors on 2-, 3-, 4-inch HR GaAs:Cr wafers with a thickness from 300 to 1000 microns (Figure 6.34).

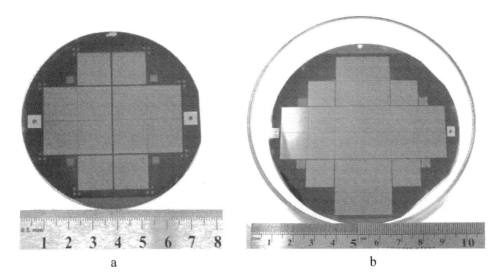

a   b

FIGURE 6.34
The HR GaAs:Cr pixel sensors compatible with Medipix ASICs located on 3 inches (a) and 4 inches (b) HR GaAs wafers.

The minimum pitch is of 35 microns and 50 microns for pixel and microstrip sensors, respectively.

The plan for the near future is developing technology of 6-inch HR GaAs:Cr wafers and sensors keeping the resistivity and $(\mu\times\tau)_n$ value within the same range as for 3, 4 inches wafers. Additional efforts will be done to increase capability of HR GaAs sensors:Cr work for high flux X-ray imaging.

Figure 6.34. The HR GaAs:Cr pixel sensors compatible with Medipix ASICs located on 3 inches (a) and 4 inches (b) HR GaAs:Cr wafers

Acknowledgments

The authors would like to thank the teams from Functional Electronics Laboratory of Tomsk State University (Russia), DESY (Germany), ESRF (France), RAL STFC (UK), KIT (Germany), FMF of Freiburg University (Germany), CHESS of Cornell University (USA), and PSI (Switzerland) for their valuable contribution to the study of HR GaAs:Cr X-ray imaging sensors.

References

1. G.I. Ayzenshtat et al. 2001. GaAs structures for X-ray imaging detectors, *Nuclear Instruments and Methods in Physics Research*, 466, 25–32. doi:10.1016/S0168-9002(02)00951-8

2. D.L. Budnitsky, O.P. Tolbanov, A.V. Tyazhev, 2002. Electric field distribution in chromium compensated GaAs. *12th International Conference on Semiconducting & Insulating Materials*, Smolenice Castle, Slovakia, June 30 to July 5, pp. 23–26, Institute of Electrical and Electronics Engineers.

3. D. Budnitsky et al. "GaAs Pixel Detectors". MRS Online Proceedings Library, 1576, (2013) mrss13-1576-ww08-01, doi:10.1557/opl.2013.1144

4. D. Budnitsky, V. Novikov, A. Lozinskaya, I. Kolesnikova, A. Zarubin, A. Shemeryankina, T. Mikhailov, M. Skakunov, O. Tolbanov, A. Tyazhev, 2017. Characterization of 4 inch GaAs:Cr wafers, *Journal of Instrumentation*. doi:10.1088/1748-0221/12/01/C01063

5. O. Tolbanov et al. 2014. Investigation of the current-voltage characteristics, the electric field distribution and the charge collection efficiency in X-ray sensors based on chromium compensated gallium arsenide, *Proceedings of the SPIE 9213, Hard X-Ray, Gamma-Ray, and Neutron Detector Physics*, XVI, 92130G, September 5. doi:10.1117/12.2061302

6. V. Novikov, A. Zarubin, O. Tolbanov, A. Tyazhev, A. Lozinskaya, D. Budnitsky, M. Skakunov, 2015. Impact of thermal treatment on electrical characteristics and charge collection efficiency of Me-GaAs:Cr-Me x-ray sensors, *Proceedings of the SPIE 9593, Hard X-Ray, Gamma-Ray, and Neutron Detector Physics*, XVII, 95930F, August 26. doi:10.1117/12.2187181

7. M. Juergen Wolf, G. Engelmann, L. Dietrich, H. Reichl,. 2006. Flip chip bumping technology—Status and update, *Nuclear Instruments and Methods in Physics Research A* 565, 290–295. doi:10.1016/j.nima.2006.05.046

8. D.L. Budnitskii et al. 2012. A photovoltaic effect in the metal-high-resistive GaAs:Cr contact, *Russian Physics Journal*, 55(7), 748–751. doi:10.1007/s11182-012-9876-4

9. H. Spieler,. 2005. Radiation detectors and signal processing, Lecture Notes - XV. Heidelberger Graduate Lectures in Physics, University of Heidelberg, October 10–14, http://www-physics.lbl.gov/~spieler/Heidelberg_Notes_2005/index.html

10. S.M. Sze, 1981. *Physics of Semiconductor Devices*, 2nd ed. Jon Wiley & Sons.

11. A. Cola, L. Reggiani, L. Vasanelli, 1997. Field-assisted capture of electrons in semi-insulating GaAs, *Journal of Applied Physics*, 81 (2). doi:10.1063/1.364190

12. B. Zat'ko et al., On the detection performance of semi-insulating GaAs detectors coupled to multichannel ASIC DX64 for X-ray imaging applications, *Nuclear Instruments and Methods in Physics Research A*. doi:10.1016/j.nima.2008.03.031

13. F. Dubecký, V. Nečas, 2007. Study of radiation detectors based on semi-insulating GaAs and InP: Aspects of material and electrode technology, *Proceedings of the SPIE 6706, Hard X-Ray and Gamma-Ray Detector Physics* IX, 67061D. doi:10.1117/12.734173

14. B. Zat'koa et al., 2007. Imaging performance study of the quantum X-ray scanner based on GaAs detectors, *Nuclear Instruments and Methods in Physics Research A*, 576 66–69, doi:10.1016/j.nima.2007.01.122

15. A. Perd'ochová-Šagátová et al., 2006. Experimental analysis of the electric field distribution in GaAs radiation detectors, *Nuclear Instruments and Methods in Physics Research A*, 563 187–191. doi:10.1016/j.nima.2006.01.092

16. B. Zat'ko et al., 2006. Performance of a Schottky surface barrier radiation detector based on bulk undoped semi-insulating GaAs at reduced temperature, *IEEE Transactions on Nuclear Science*, 53(2). doi:10.1109/TNS.2006.870446

17. J.H.H.Y.K. Kim et al., 2005. Development of bulk semi-insulating GaAs semiconductor radiation detector at room temperature, *Nuclear Science Symposium Conference Record*, IEEE. doi:10.1109/NSSMIC.2005.1596578

18. A.V. Tyazhev et al., 2005. GaAs radiation imaging detectors for nondestructive testing, medical, and biological applications, *Proceedings of the SPIE 5922, Hard X-Ray and Gamma-Ray Detector Physics* VII, 59220Q, September 17. doi:10.1117/12.613410

19. A.V. Tyazhev et al., Impulse response of GaAs radiation imaging detectors, *2005 Siberian Conference on Control and Communications*, doi:10.1109/SIBCON.2005.1611205

20. A. Dragone et al., 2016. Feasibility study of a "4H" X-ray camera based on GaAs:Cr sensor, *Journal of Instrumentation*, doi:10.1088/1748-0221/11/11/C11042

21. G.I. Ayzenshtat et al., 2001. Modeling of processes of charge division and collection in GaAs detectors taking into account trapping effects, *Nuclear Instruments and Methods in Physics Research A*, 466, 1–8. doi:10.1016/S0168-9002(01)00817-8

22. A.N. Zarubin et al., 2006. Non-equilibrium charge carriers life times in semi-insulating GaAs, compensated with chromium, *International Workshops and Tutorials on Electron Devices and Materials*, July 1–5. doi:10.1109/SIBEDM.2006.231988

23. D. Budnitsky et al., 2014. Chromium-compensated GaAs detector material and sensors, *Journal of Instrumentation*, doi:10.1088/1748-0221/9/07/C07011

24. A. Lozinskaya et al., 2015. Dependence of X-ray sensitivity of GaAs:Cr sensors on material of contacts, *Journal of Instrumentation*, doi:10.1088/1748-0221/10/01/C01020

25. M. Veale et al., 2014. Chromium compensated gallium arsenide detectors for X-ray and γ-ray spectroscopic imaging, *Nuclear Instruments and Methods A*, 752, 6. doi:10.1016/j.nima.2014.03.033

26. W.C. Barber, J.C. Wessel, E. Nygard, J.S. Iwanczyk, 2015. Energy dispersive CdTe and CdZnTe detectors for spectral clinical CT and NDT applications. *Nuclear Instruments and Methods in Physics Research A*, 784, 531–537. doi:10.1016/j.nima.2014.10.079

27. J.S. Iwanczyk et al., 2009. Photon counting energy dispersive detector arrays for X-ray imaging. *IEEE Transactions on Nuclear Science*, 56(3), 535–542. doi:10.1109/TNS.2009.2013709

28. D. Pennicard, H. Graafsma, 2011. Simulated performance of high-Z detectors with Medipix3 readout. *Journal of Instrumentation*, doi:10.1088/1748-0221/6/06/P06007

29. M. Veale et al., 2014. Investigating the suitability of GaAs:Cr material for high flux X-ray imaging, *Journal of Instrumentation*, doi:10.1088/1748-0221/9/12/C12047

30. E. Hamann et al., 2015. Performance of a Medipix3RX spectroscopic pixel detector with a high resistivity gallium arsenide sensor, *IEEE Transactions on Medical Imaging*, 34(3), 707–715. doi:10.1109/TMI.2014.2317314

31. E. Hamann, A. Cecilia, A. Zwerger, T. Baumbach, M. Fiederle, 2013. Characterization of photon counting pixel detectors based on semi-insulating GaAs sensor material, /*Journal of Physics: Conference Series*, 425(6), publ. 062015. doi:10.1088/1742-6596/425/6/062015
32. D. Pennicard et al., 2014. The LAMBDA photon-counting pixel detector and high-Z sensor development, *Journal of Instrumentation*, 9(12), Article number C12026.
33. E. Hamann et al., Investigation of GaAs:Cr Timepix assemblies under high flux irradiation, *Journal of Instrumentation*, doi:10.1088/1748-0221/10/01/C01047(2015)
34. E. Hamann, Characterization of high resistivity GaAs as sensor material for photon counting semiconductor pixel detectors, PhD Thesis, University of Freiburg.
35. C. Ponchut et al., 2017. Characterisation of GaAs:Cr pixel sensors coupled to Timepix chips in view of synchrotron applications, *Journal of Instrumentation*, doi:10.1088/1748-0221/12/12/C12023
36. P. Smolyanskiy et al., 2017. Study of a GaAs:Cr-based Timepix detector using synchrotron facility, *Journal of Instrumentation*, doi:10.1088/1748-0221/12/11/P11009
37. M. Ruat et al., 2018. Photon counting microstrip X-ray detectors with GaAs sensors, *Journal of Instrumentation*, https://doi.org/10.1088/1748-0221/13/01/C01046
38. M.C. Veale et al., 2017. MHz rate X-Ray imaging with GaAs:Cr sensors using the LPD detector system, *Journal of Instrumentation*, doi:10.1088/1748-0221/12/02/P02015
39. J. Becker et al., 2018. Characterization of chromium compensated GaAs as an X-ray sensor material for charge-integrating pixel array detectors, *Journal of Instrumentation*, https://doi.org/10.1088/1748-0221/13/01/P01007
40. K. Afanaciev et al., 2012. Investigation of the radiation hardness of GaAs sensors in an electron beam, *Journal of Instrumentation*, doi:10.1088/1748-0221/7/11/P11022
41. P. Rudolph, 2005. Dislocation cell structures in melt-grown semiconductor compound crystals, *Crystal Research and Technology*, 40(1/2), 7–20. doi:10.1002/crat.200410302

7

Vertical-Cavity Surface-Emitting Lasers for Interconnects

Werner H. E. Hofmann

CONTENTS

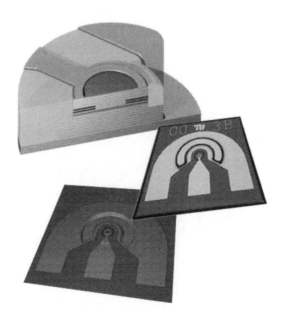

Chapter Outline

This chapter deals with the application of Vertical-Cavity Surface-Emitting Lasers (VCSELs) as directly-modulated light-sources in optical interconnects. Devices from 850 nm to 1550 nm are discussed. This includes recent developments in the highest modulation speeds, highest operation temperatures, and highest energy-efficiency. The chapter is divided as follows:

1. The application of VCSELs in Interconnects

   This section clarifies that VCSEL-based optical interconnects are going to replace the conventional copper-based technology. VCSELs can deliver higher bandwidths below the cost of copper and enable highly scalable solutions with their small footprint. Future supercomputers can only be realized by the broad application of VCSELs.

2. Speed limitations in VCSELs

   As the system costs scale with the number of interconnects, high serial bandwidths are desired to master the growing data-traffic at low cost. Directly modulated VCSELs have bandwidth limitations, which must be conquered to enable the highest bandwidths at ultra-low cost.

3. High-Speed, Temperature-Stable 980-nm VCSELs

   The recent development of ultra-high-speed temperature-stable 980-nm VCSEL devices is shown in this chapter. Data-rates over 40 Gb/s and data-transmission at temperatures as high as 155°C are demonstrated.

4. Energy-Efficient 850-nm VCSELs

   850-nm is the standard-wavelength for links in the range of some 100 m via multimode fiber. Recently, devices with energy-efficiencies below 100 fJ/bit could be demonstrated.

5. Advantages of Long-Wavelength VCSELs for Interconnects

   Having seamless solutions and silicon-photonics in mind, long-wavelength VCSELs emitting around 1.3–1.6 µm are very attractive light-sources. Meanwhile, VCSELs at these wavelengths are also commercially available.

7.1 The Application of VCSELs in Interconnects

This section clarifies that VCSEL-based optical interconnects are going to replace the conventional copper-based technology. VCSELs can deliver higher bandwidths below the cost of copper and enable highly scalable solutions with their small footprint. Future supercomputers can only be realized by the broad application of VCSELs.

The maximum data-rate an interconnect can carry is limited by Shannon's law. According to that, bandwidth and/or the signal-to-noise ratio must be increased for ultimate data-rates. As copper suffers from higher damping and crosstalk at higher frequencies, there are physical restrictions in interconnect speeds based on copper. Of course, one can push out the border by higher signal levels and multiple parallel lines. This has happened in the past with CPUs growing more and more pins and carrying a bigger and bigger heat-sink on top.

Microelectronics can deliver much more compact circuits than integrated optics due to the De Broglie wavelength of electrons being much shorter. Taking the unimaginable amounts that have gone into the development of silicon microelectronics into account, it is unrealistic that microelectronics will be abolished soon. On the other hand, assuming a certain bandwidth-length product, optics is much more energy-efficient for interconnects.

Even though optical interconnects are the way to go, the world has tried to stay with copper for interconnects, postponing the inevitable transition to optics for many times. One reason for this might have been that optical communications technologies were almost exclusively driven by the long-haul market. Even though the technological achievements were remarkable, scaling down optical long-haul equipment in price and energy-consumption is not straightforward, if not impossible. Consequently, novel technologies had to be developed to meet the requirements of short optical interconnects.

Vertical-cavity surface-emitting lasers (VCSELs) are an answer to the question: Which optical technology could be a workhorse of future interconnects? VCSELs are capable of delivering the highest modulation speeds beyond 40 Gbps at high operation temperatures [1]. At the same time, they consume orders of magnitude less power and can be mass-fabricated at a very low cost.

7.1.1 The End of Copper-Based Interconnects

With optical interconnects developed for long-haul applications not cost-competitive to copper, electrical interconnects were squeezed to their limits.

Today's silicon chips are limited by their thermal budget. Surprisingly, most of the generated heat comes from the signal and clock lines [2]. A transition from CMOS-compatible self-passivating aluminum lines to encapsulated copper has already been realized some years ago. This great technological effort was motivated only by a quite small difference in conductivity. This example shows how uncompromising roadmaps and scaling rules are.

Each new generation of supercomputers, to give another example, is required to deliver a vast boost in computational speed but has to keep cost and energy-consumption to a moderate level. This is no longer possible with copper-based interconnects. Therefore, the 2008 IBM's PetaFlop Supercomputer incorporated 48,000 optical links. Following the roadmap dictated by the market, 2020s ExaFlop mainframes will need 320 million optical interconnects running at 5x the speed at 1/50 of the energy consumption and 1/400 of the price, compared to the optical links used in 2008 [3]. One should also note, that most of the energy consumed by these computers is needed for the interconnects.

Supercomputers used optical interconnects at a quite early stage for two reasons. Firstly, the bandwidth requirements of these high-performance systems could only be accommodated by optics, and secondly the tolerance of higher cost. Other systems would already have greatly benefited of optical interconnects as well, but the cost constraints were too strong inhibiting the widespread use of optical interconnects in the past.

7.1.2 VCSELs for Optical Interconnects

The easiest high-speed optical source is a directly modulated laser. If lasers have to provide excellent beam quality at low cost and energy-consumption, VCSELs are the current technology of choice. For example, this is the reason why each laser-mouse for the personal computer is a VCSEL-mouse [4].

In order to be applied in short optical interconnects, VCSELs have to deliver high serial bandwidths, with small footprints allowing dense packaging and uncooled operation. The serial bandwidth is given by system design rules, like the Amdahl's law, stating that any data processing installation needs to provide certain interconnect bandwidth and memory capacity matching its computational power to avoid bottlenecks.

On the other hand, the amount of optical links ramps up complexity and cost. Moreover, the scalability of a certain technology is also limited by the amount of links that can be connected. In 2011, system designers from Google stated that 40 Gbps would be the bandwidth desired for their next generation of data centers [5].

For system scalability, very dense packaging of the optical chips is a necessity. The VCSEL footprint, being one order of magnitude smaller than edge-emitters, is also in favor of this technology. Furthermore, to ensure a compact hybrid package, bottom-emitting devices with the electrical fanout on the one side and the optical on the other have been proposed by IBM's TERABUS project [6]. This requires the substrate to be transparent requiring longer wavelengths like 980 nm and cannot be accommodated by the 850-nm VCSEL standard designed for multi-mode fiber links targeting at some 100 m link-lengths. Additionally, 980-nm VCSEL devices allow the use of binary distributed Bragg mirrors with much better thermal conductivity than their ternary counterparts. This enables these devices to operate at much higher ambient temperatures, which occur in uncooled dense arrays or when integrated on top of a high-performance silicon CPU or memory. Last, but not least, VCSELs have a much smaller energy consumption than other types of laser diodes and would just for that reason be the light source of choice in future application having limited natural resources in mind [7].

As the optical interconnect market requires a volume of billions of devices to be available, the mature GaAs-based technology seems to be the workhorse for the years to come. On the other hand, for integrated optoelectronics on Silicon, longer wavelengths like 1.3 or 1.55 μm are very attractive for seamless solutions. Data could be transmitted throughout fibers directly into the silicon waveguides of integrated chips. So, longer wavelength VCSELs might be the next step of evolution [8].

Further shrinking of footprint and energy consumption, together with higher bandwidth are future requirements. Novel technologies, like nano-lasers, might replace the newly evolving VCSEL technology one day [9].

7.2 Speed Limitations in VCSELs

As the system costs scale with the number of interconnects, high serial bandwidths are desired to master the growing data-traffic at low cost. Directly modulated VCSELs have bandwidth limitations that must be conquered to enable the highest bandwidths at ultra-low cost.

The continuously rising bandwidth demand requires faster lasers. Nowadays, novel communication standards, like 100G Ethernet, must be fixed without having devices ready for mastering such kind of modulation speed. Therefore, it is of the utmost importance to understand the mechanisms in a laser under modulation, in order to identify the limiting factors and ultimately minimize them.

One of the main advantages of semiconductor lasers is that they can be directly modulated at high speeds. This means the direct conversion of an electrical input signal into an optical signal. Directly modulated lasers share the same circuit for biasing and modulation, and no external modulator is needed. Therefore, more cost-effective optical communications solutions, especially interconnects are possible. In contrast, DFB lasers with external Mach-Zehnder modulators are state-of-the-art for long-haul transmission. However, metro-range links, fiber to the home (FTTH) solutions, optical interconnects, and directly modulated VCSELs are favored as they can provide broadband data links at a much lower system cost [10]. Furthermore, VCSELs have a much lower energy consumption and footprint compared to edge-emitters, enabling small form factor communication modules with low power budgets. Especially this point becomes more and more important, as power consumption and heat dissipation is now becoming the limiting factor for data centers and central offices [11,12]. "Green IT" is an upcoming issue, demanding novel technologies, like VCSEL-based optical interconnects.

For these applications in digital communication, the maximum speed at which information can be transmitted is dependent on the laser's modulation bandwidth. Additionally, the transfer function of the laser should be free of peaking or resonances within the used frequency range. VCSELs, with their high intrinsic damping, are superior in this aspect compared to edge-emitters. Therefore, VCSELs usually allow quite high data-rates compared to their bandwidth.

7.2.1 High-Speed VCSELs

To identify the limits of modulation performance, a rate-equation analysis is a useful instrument [13,14]. For directly modulated lasers, the easiest model assumes a photon and a carrier reservoir that interact by stimulated emission.

The rates of change in carrier and photon densities follow the following expressions:

$$\frac{dN}{dt} = J_{inj} - J_{th} - R_{stim}S \tag{7.1}$$

$$\frac{dS}{dt} = S\left(\Gamma R_{stim} - v_g(\alpha_i + \alpha_m)\right) + \Gamma J_{sp} \tag{7.2}$$

with N as carrier density and S as photon density in the active region and the optical cavity, respectively. As one has to look at the particles flowing per unit time, the particle numbers have to be scaled by the volumes of the active region and the photon reservoir. Here, this is taken into account by the confinement factor Γ. The first equation states that the carrier density change equals to the injected carrier density J_{inj} minus the carriers recombining due to spontaneous emission or losses (J_{th}) and stimulated emission ($R_{stim}S$), whereby R_{stim} is the stimulated emission rate.

For investigations of the dynamic response of the laser, we must analyze these rate equations with the time derivates included. Unfortunately, an exact analytical solution of the full rate equations cannot be obtained. Linearizing the system matrix for small signals, the determinant yields a well-known 2-pole filter function.

The system matrix becomes

$$\begin{pmatrix} j\omega + \mu_{11} & \mu_{12} \\ -\mu_{21} & j\omega + \mu_{22} \end{pmatrix} \begin{pmatrix} dN \\ dS \end{pmatrix} = \begin{pmatrix} dJ_{inj} \\ 0 \end{pmatrix} \equiv \mathbf{Mx} = \mathbf{i} \tag{7.3}$$

with

$$\mu_{11} = \frac{\delta}{\delta N} J_{th} + v_g aS \cong v_g aS \tag{7.4}$$

$$\mu_{12} = v_g g - v_g a_p S \cong v_g g_{th} \tag{7.5}$$

$$\mu_{21} = \Gamma \frac{\delta}{\delta N} J_{sp} + \Gamma v_g aS \cong \Gamma v_g aS \tag{7.6}$$

$$\mu_{22} = -\Gamma v_g g + v_g (\alpha_i + \alpha_m) + \Gamma v_g a_p S = J_{sp} \Gamma / S + \Gamma v_g a_p S \cong 0 \tag{7.7}$$

The assumptions in Eqs. 2.4 through 2.7 can be made above threshold neglecting gain compression.

Solving the equation system 2.3 we yield

$$\mathbf{x} = \mathbf{M}^{-1}\mathbf{i} = \frac{dJ_{inj}}{\det\mathbf{M}} \begin{pmatrix} j\omega + \mu_{22} \\ \mu_{21} \end{pmatrix} = \begin{pmatrix} dN \\ dS \end{pmatrix} \tag{7.8}$$

and

$$\det\mathbf{M} = \omega_R^2 + j\omega\gamma - \omega^2 \tag{7.9}$$

with

$$\omega_R^2 = \mu_{11}\mu_{22} + \mu_{12}\mu_{21} \text{ and } \gamma = \mu_{11} + \mu_{22} \tag{7.10}$$

to be identified as relaxation oscillation frequency ω_R and damping factor γ.

As these equations include many approximations and assumptions, we should state that numerical methods would give better results. On the other hand, the nonlinearity and uncertainties in real devices are so large, that understandable and intuitive formulas are still very helpful.

Damping and resonance are coupled by the *K*-factor giving the limits for over-damping. For easy modeling, parasitics can be added by first-order low-pass filter. In various textbooks, we can find further calculations finding limits caused by parasitic, device-heating, damping, and so on. Unfortunately, the assumptions made to keep these formulas easy might be valid for edge-emitting lasers operating around 2.5 Gb/s; however, they do not hold for VCSELs beyond 10 Gb/s. The real limits for high-speed VCSELs are:

- Parasitics that are directly limiting the over-damped responses
- Damping that gives a real physical limit and increases the effect of parasitics
- Limitations of the relaxation oscillation frequency f_R by device heating, the active medium, and the laser cavity
- Other nonlinear effects like transport, spatial hole burning, current crowding, modal properties, and so on.

To ide ntify intrinsic limitations, we have to look at the original equations gained by the small-signal rate-equation analysis:

$$4\pi^2 \cdot f_R^2 = v_g^2 \cdot g_{th} \cdot \Gamma \cdot a \cdot S \tag{7.11}$$

$$K \equiv \frac{\gamma - \gamma_0}{f_R^2} \approx \gamma/f_R^2 = \frac{4\pi^2}{v_g} \cdot \frac{1}{g_{th}} \left(\frac{1}{\Gamma} + \frac{a_p}{a} \right) \tag{7.12}$$

The resonance-frequency f_R, as given in Eq. 2.11, is limited by group velocity v_g, threshold gain g_{th}, the confinement Γ of the optical wave, the differential gain a, and the photon density S. For high modulation speed, a high resonance-frequency is preferential. By raising the resonance-frequency, the damping also rises causing a limitation of the performance. The figure-of merit to judge how fast damping rises with higher resonance-frequencies is the K-factor, which is also connected to device parameters. The interrelationship is given in Eq. 2.12 with a_p as the gain compression factor. Even though the damping flattens out the response, which can be beneficial in some special cases [15], small K-factors are preferential for ultimate modulation speeds. Consequently, threshold gain, optical confinement, and differential gain have to be maximized. Additionally, high photon densities at low gain-compression are needed. Last, but not least, a very good thermal design is crucial, as otherwise a high photon density will never be achieved, and all other figures of merit will degrade! Generally speaking, high-speed performance always requires very good static laser performance as a pre-condition.

Even though these considerations are generally independent of the laser wavelength, the desired emission wavelength also has an impact on the potential device speed. This is caused by two reasons. First, longer-wavelength devices require lower-bandgap materials, which provide inferior carrier confinement. Second, non-radiative effects, like Auger, are more pronounced, and free-carrier absorption rises by the power of two with wavelength [16].

On the other hand, there are only a couple of device technologies available grown on GaAs or InP substrates. Each of these technologies has some intrinsic limitations based on the desired emission wavelength. These facts are schematically depicted in Figure 7.1, giving an overview of certain laser wavelengths favored by nature assuming state-of-the-art device technology. It turns out that emission wavelengths around 1 µm seem to be a very promising candidate for directly-modulated laser-sources operating at ultimate speeds [17].

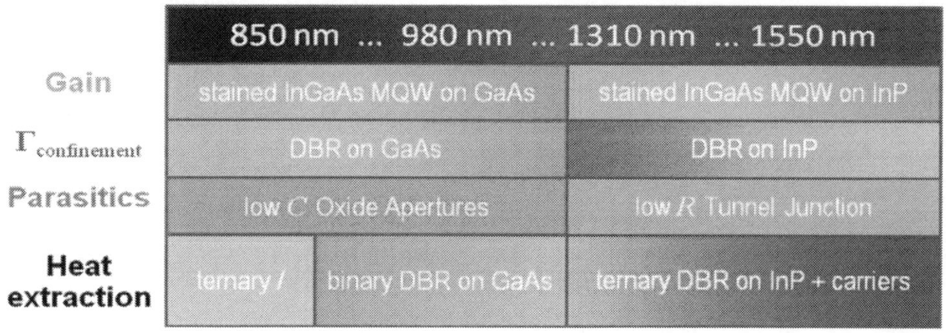

	850 nm ... 980 nm ... 1310 nm ... 1550 nm		
Gain	stained InGaAs MQW on GaAs		stained InGaAs MQW on InP
$\Gamma_{\text{confinement}}$	DBR on GaAs		DBR on InP
Parasitics	low C Oxide Apertures		low R Tunnel Junction
Heat extraction	ternary /	binary DBR on GaAs	ternary DBR on InP + carriers

FIGURE 7.1
The impact of emission-wavelength dependent device technology on potential modulation speed.

7.3 High-Speed, Temperature-Stable 980-nm VCSELs

The recent development of ultra-high-speed temperature-stable 980-nm VCSEL devices is shown in this chapter. Data-rates over 40 Gb/s and data-transmission at temperatures as high as 155°C are demonstrated.

Compared to 850 nm, the wavelength of 980 nm has several advantages for use in short distance optical interconnects. The transparency of gallium arsenide at this wavelength allows for the realization of bottom emitting devices, which enables high packaging density and simple integration to silicon photonics. Additionally, binary GaAs/AlAs distributed Bragg reflectors (DBR) can be used. By replacing the ternary alloys typically used by binary material, the thermal conductivity of the mirrors can be enhanced significantly leading to effective heat extraction from the VCSELs. This is essential for highly temperature stable high-speed performance. Temperature stability can contribute to low power consumption of optical interconnects, because high-speed operation at constant current and voltage driving parameters provides the opportunity to dispose of cooling systems and to use simpler driver feedback circuits [21,22].

The High-Speed, Temperature-Stable 980-nm VCSELs made at the Center of Nanophotonics at the Technical University of Berlin [1] were processed in a coplanar low-parasitic ground-signal-ground contact pad layout incorporating thick BCB passivation layers as depicted in Figure 7.2a. Moreover, we widely varied the device layout to confirm the best device geometries as presented in Figure 7.2b. To identify the best design, we used our home-built fully automated wafer prober to do the mapping of various device characteristics of a whole tree inch wafer. This allowed us to identify the best device design for various purposes. Consequently, we fabricated VCSELs with a wide range of oxide-aperture diameters from 1 to 10 µm. These VCSELs were here built upon the

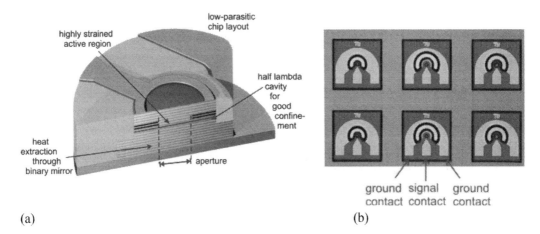

(a)   (b)

FIGURE 7.2

980-nm high speed VCSEL (a) Schematic of VCSEL chip in high-speed layout. (b) Photographic picture of VCSEL-chips on wafer showing systematic design variation.

previously reported oxide-confined 980-nm devices [18]. To achieve a higher speed, we reduced the cavity length to $\lambda/2$. GaAs/AlAs mirror pairs are used in the n-doped bottom DBR not only to reduce penetration and effective cavity length, but also to increase heat extraction from the active region. The mirror-reflectivity of the out-coupling mirror was reduced to lower the photon-lifetime respectively threshold gain and achieve highest modulation speed. Multiple oxide apertures were used to reduce the parasitic device capacitance. Furthermore, the apertures were moved very close to the active region to avoid damping of the resonance-frequency response by carrier transport. The apertures were formed by an in-situ controlled wet oxidation process utilizing our home-built oxidation furnace with precise vapour pressure, temperature, and flux control. To avoid spatial hole burning, we replaced the phosphorous containing barriers by pure GaAs with higher electron mobility and better thermal conductivity. The graded interface around the active region was replaced by an abrupt interface for better carrier confinement at high operation temperatures. The mode-gain offset was 15 nm for optimal room-temperature performance. Furthermore, the highly strained InGaAs multiple quantum wells (MQW) were tailored to achieve maximum photoluminescence without gain broadening. From this measure, better gain properties are expected, being advantageous for the overall device performance. To avoid carrier reservoirs and storage for hot carriers, any tapered border layers around the active region were eliminated. Sometimes these tapers are named "separate confinement heterostructure"-layers (SCH), as the epi-design looks very similar to the successful SCH-confinement of edge-emitting lasers.

To sum up, the High-Speed, Temperature-Stable 980-nm VCSELs made at the Center of Nanophotonics at the Technical University of Berlin feature:

- Low parasitic chip design with
  - Multiple oxide apertures (low capacitance)
  - Sophisticated modulation doping (low resistance, low optical loss)
  - Optimized geometrical chip and contact pad layout
- High-speed active region
  - Highly strained InGaAs MQWs for a large differential gain
  - GaAs barriers for high electrical and thermal homogeneity to avoid gain compression
- Excellent optical confinement
  - Half-lambda cavity
  - Short DBR penetration due to the high index-contrast binary bottom mirror
- Minimized transport effects
  - No "separate confinement heterostructure"-like gradings next to the active region
  - Oxide aperture very next to the active region avoiding current crowding
- Highly optimized thermal design with
  - Heat extraction via binary GaAs/AlAs mirror supported by a double-mesa chip
  - Active region placed very close to heat-sink
  - Strong carrier confinement by $Al_{0.90}Ga_{0.10}As$ and the elimination of carrier reservoirs

7.3.1 High-Speed VCSEL Modulation

In order to evaluate the high-speed performance of a fiber-based optical system, both sender and receiver must be able to provide enough bandwidth. As the 980-nm waveband is rather new for optical communication systems, and ultra-high-speed lasers were not available until recently, there are no proper receiver modules in the market yet. Therefore, an optical receiver module within a collaborative research project capable to receive data-rates in excess of 40 Gb/s also had to be developed. The experiments at data-rates beyond 25 Gb/s were carried out using a demonstrator from u²t with ~30 GHz bandwidth, multi-mode fiber input, a responsivity of ~0.26 A/W at 980 nm and a matched trans-impedance limiting amplifier. The demonstrator-module was an adaption of a u²t photo-receiver. The setup is schematically depicted in Figure 7.3.

Judging the device performance, highest speed, best energy-efficiency, or uncooled operation can be required by the application. To achieve the highest ratings in any of these categories, one has to compromise the others. On the other hand, better devices are usually better in all categories. Consequently, the devices were characterized by top-speed versus temperature [19]. The measurements were performed using butt-coupling and a short 3-m multimode fiber link. Typical 6-μm oxide-aperture diameter devices showed threshold currents of 0.9 mA at 20°C with a differential resistance of 75 Ω and a maximum optical output power exceeding 8 mW at a rollover current of 22 mA. The static characteristics are given in Figure 7.4a and b.

The devices operate continuous wave up to 200°C ambient temperature [20]. In Figure 7.4a, the *L–I* (Light-vs.-Current) characteristic for a representative VCSEL with a 6-μm oxide aperture at temperatures between 20°C and 200°C is depicted. The operation voltage is typically 2–3 V. The emission wavelength is between 980 and 985 nm within the temperature range. For 6-μm aperture VCSELs, the spectrum is quasi-single-mode and the shape does not change with temperature, which is important for stable fiber coupling. The spectrum is given in Figure 7.4b. Fiber coupled powers at a constant 6 mA drive current were measured [21] to be 2.3 mW, 1.2 mW, and 200 μW for temperatures of 85°C, 155°C, and 200°C, respectively. The threshold current is around 1 mA, slightly increasing with temperature, indicating a too small mode gain offset and room for improvement looking at maximum ambient temperature ratings.

Data-transmission experiments were carried out under ambient temperatures from −14 to +155°C allowing us to identify error-free operation at bit rates from 12.5 to 49 Gb/s. We achieved error-free operation of directly-modulated VCSELs with non-return-to-zero (NRZ) coding by a 2^7-1 bits long pseudo-random-bit-sequence (PRBS) at record-high bit

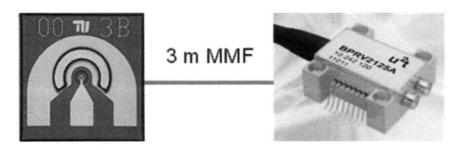

FIGURE 7.3
Setup for characterization. The "quasi back-to-back" configuration should emulate the application within a short optical interconnect.

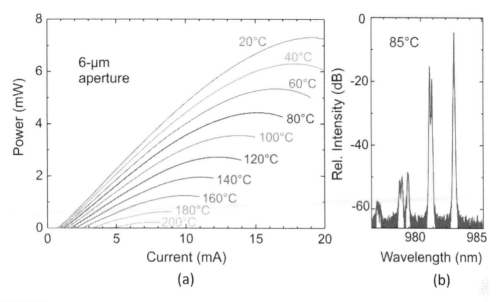

FIGURE 7.4
Static output characteristics (a) *L–I* characteristics from 20°C to 200°C. (b) Corresponding spectrum at 85°C.

rates of 12.5 Gb/s at 155°C, 17 Gb/s at 145°C, 25 Gb/s at 120°C, 38 Gb/s at 85°C, 40 Gb/s at 75°C, 44 Gb/s at 25°C, 47 Gb/s at 0°C, and 49 Gb/s at −14°C [1,19,20,22]. Due to the very high temperature stability, longer bit-patterns up to PRBS sequences of 2^{31}-1 showed no or only minor degradation. The results of the large-signal transmission experiments are depicted in Figure 7.5. In (a), we show a Bit-Error-Rate plot at room temperature demonstrating error-free performance beyond 40 Gb/s. In (b), all large signal experiments are summed up in a temperature versus error-free bit rate plot and compared with the state-of-the-art of 2011. The result presented in Figure 7.5a translates to the 44 Gb/s and 25°C point. In Figure 7.5b, we see 44 Gb/s with a BER below 10^{-12} in Figure 7.5a [19].

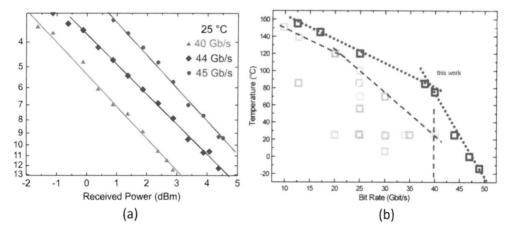

FIGURE 7.5
High-speed modulation of 980-nm VCSELs (a) Bit-error-rate plots of data transmission experiments at room temperature. (b) Error-free NRZ-bitrate versus ambient temperature including the State-of-the-art in 2010 and the results achieved at the TU Berlin in 2011.

This experimental result is in agreement with our theoretical considerations given in Chapter 2. Especially for high-speed VCSELs, we believe that a good thermal design is crucial. Consequently, we were able to move the entire border with a single device-run. The drop in performance towards higher bit rates could be due to limitations (receiver bandwidth below 30 GHz, not intended to be used up to 40 Gb/s) on the receiver side.

Even though these lasers are the most energy-efficient devices beyond 30 Gb/s [22], they were clearly optimized for the highest bandwidths. Devices optimized for lowest energy consumption per bit are discussed in the next section.

7.4 Energy-Efficient 850-nm VCSELs

850-nm is the standard-wavelength for links in the range of some 100 m via multi-mode fiber. Recently, devices with energy-efficiencies below 100 fJ/bit could be demonstrated.

7.4.1 Energy Efficient Data-Transmission

According the International Technology Roadmap for Semiconductors (ITRS), lasers for future optical interconnects should highly energy efficient. In 2015, energy efficient high-speed lasers operating at 100 mW/Tbps (100 fJ/bit) will be required [11,23]. These numbers refer to the *dissipated electrical energy* per bit to fit the cooling budget of the data center. This figure of merit can be defined as heat-to-bit rate ratio (HBR) [32] (mW/Tbps)

$$HBR = P_{diss}/BR \qquad (7.13)$$

where P_{diss} is the dissipated heat ($P_{diss} = P_{el} - P_{optical}$) of the laser and BR is the bit rate. As "green photonics" means that the *total energy* consumed per transmitted amount of data is of equal importance [32]. Consequently, the electrical energy-to-data ratio (EDR) (fJ/bit) can be defined as

$$EDR = P_{el}/BR \qquad (7.14)$$

where $P_{el} = V \cdot I$ is the total consumed electrical power with V and I as the laser's operating bias point [7]. Additionally, the modulation power absorbed by the laser should be taken into account [22]. Even though, depending on the used electronics, it might be the case that the power consumed in the VCSEL is smaller than the power needed by the driving electronics, we believe that it is the most crucial one. This is due to its multiplier effect of the energy consumed by the light-source on the power consumption of the whole system. Please note, that efficiency per bit is not the same as the wall-plug efficiency (WPE), which can be expressed in terms of HBR and EDR:

$$WPE = 1 - HBR/EDR \tag{7.15}$$

Furthermore, this means that the most power-efficient lasers in terms of data-transmission are not necessarily the lasers with the highest WPE, nor is the driving condition for best HBR or EDR identical with the point of highest WPE.

To identify the figures of merit for designing an efficient laser looking at the energy per bit, we have to recap the rate-equation analysis in Chapter 2. From Eq. 7.11, we find

$$f_R^2 \propto S \propto I - I_{th} \tag{7.16}$$

with the latter proportionality only being valid for small driving currents. Please note, that this is especially not the case for strongly biased ultra-high speed VCSELs. Assuming linear behaviour, we can define a D-factor and a modified factor D_{BR} based on the bit rate:

$$f_R \equiv D\sqrt{I - I_{th}} \propto BR \equiv D_{BR}\sqrt{I - I_{th}} \tag{7.17}$$

To model the electrical power consumed by the VCSEL, we use an ideal diode with series resistance and define

$$U \equiv U_{th} + R_d I \tag{7.18}$$

with U_{th} as threshold voltage and R_d as the differential series resistance.

Like that we can write the EDR as defined in 7.14 as follows:

$$EDR = \frac{U \cdot I}{BR} = \frac{1}{BR}\left[\left(\left(\frac{BR}{D_{BR}}\right)^2 + I_{th} \right) U_{th} + \left(\left(\frac{BR}{D_{BR}}\right)^2 + I_{th} \right)^2 R_d \right] \tag{7.19}$$

$$= U_{th}\left(\frac{BR}{D_{BR}^2} + \frac{I_{th}}{BR} \right) + R_d\left(\left(\frac{BR^{3/4}}{D_{BR}}\right)^2 + \frac{I_{th}}{\sqrt{BR}} \right)^2 \tag{7.20}$$

To achieve a small value of EDR and an efficient laser, R_D and U_{th} have to be as small as possible. This means we need to optimize for a VCSEL with very low electrical losses. A large bit rate would be required to compensate high threshold currents. On the other hand, for a given D-factor, a higher bit rate translates into an inferior efficiency. This is why we suggest the use of lasers with very small threshold currents. If we neglect the threshold currents in Eq. 7.20, we yield:

$$EDR \mid_{I_{th}=0} = U_{th}\frac{BR}{D_{BR}^2} + R_d\frac{BR^3}{D_{BR}^4} \tag{7.21}$$

From Eq. 7.21, we can learn that low EDR-values become more and more difficult for higher bit rates. A high D-factor, on the other hand, is crucial for efficient lasers. The D-factor can be written as:

$$D = \frac{1}{2\upsilon} \sqrt{\frac{v_g}{e} \cdot \frac{\eta_i a}{V_{res}}} \tag{7.22}$$

with η_i as differential quantum efficiency and V_{res} as the volume of the optical resonator. This makes clear that small-aperture VCSELs are beneficial in many ways for energy-efficient links. Firstly, we achieve threshold at low currents and secondly benefit from a larger D-factor.

Efficient high-speed VCSELs have been developed in the wavebands from 850 to 1550 nm [24] and can be potentially optimized for highest data-transmission efficiencies. In 2011, the topic of Green Photonics became a focus of public and scientific interest. Conferences focusing on that topic and scientific awards acknowledging achievements in this field were initiated [25]. VCSELs operating at 10 Gb/s at 1060 nm with 140 mW/Tbps HBR have been reported lately [26]. Longer wavelengths use less energy per photon and there have an intrinsic advantage. Furthermore, active materials with better gain properties can be used [21]. Nevertheless, 850 nm remains the current standard wavelength for fiber-based links. On the other hand, for the application in short optical interconnects also proprietary solutions with other wavelengths can address the market. Vast efforts have been made in boosting both bit rate [1,17,21,22,27–31] and energy efficiency [22,25–27,30,32–36] significantly to meet the requirements of future data centers and supercomputers. Researchers in Taiwan could demonstrate single-mode devices with high wall-plug efficiencies and a remarkable HBR of 109 mW/Tbps [34]. Please note, that higher bit rates require a quadratic increase in current-densities at a given device-technology. This makes energy-efficient devices operating at higher bit rates more challenging. An EDR of 500 fJ/bit or more is usually needed to achieve bit rates of 30 Gb/s or more [27]. Consequently, at bit rates as high as 35-Gb/s HBR and EDR values in the order of 200–300 fJ/bit are also outstanding results [22,33].

7.4.2 Energy-Efficient VCSELs for Interconnects

In a first order approximation, for a given directly modulated VCSEL device, the resonance frequency rises with the square-root of the VCSEL power. Therefore, it is trivial to understand that high-speed VCSELs typically consume more energy per bit when they are operated at higher bit rates. However, VCSELs designed to work at ultra-high bit rates don't automatically get more energy efficient just by simply reducing the pumping current and the bit rate. In order to realize energy efficient high-speed performance, large resonance frequencies must be achieved at a low drive current [7,35].

Recently, achievements on the field of energy-efficient VCSELs were made at the Center of Nanophotonics at the Technical University of Berlin. These devices were optimized for the highest energy efficiency. The results for devices in the 850-nm waveband are given in Figure 7.6. The HBR is 69 mW/Tbps at 17 Gb/s and 99 mW/Tbps at 25 Gb/s [32]. A 100-m fiber link shows negligible power penalties. By heating up the device to 55°C, we yielded a record-low EDR of 81 fJ/bit and an HBR of 70 mW/Tbps at 17 Gb/s. Data transmission over 1 km of multi-mode fiber is has also been accomplished [35].

Record-high data-rates and efficiencies are not achieved at the same driving conditions. On the other hand, as the efficiency of the device is not affected by ambient temperature in a wide range, uncooled systems saving large amounts energy are feasible.

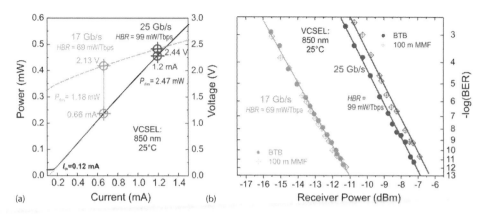

FIGURE 7.6
Energy-efficient 850-nm VCSEL, room temperature: (a) LIV-characteristics; the biasing point of the data-transmission experiment is indicated. (b) Bit-error-rate measurement of that device at 17 and 25 Gb/s. Energy-Efficiency: 17 Gb/s: HBR = 69 mW/Tbps, EDR = 83 fJ/bit, Modulation energy = 10 fJ/bit; 25 Gb/s: HBR = 99 mW/Tbps, EDR = 117 fJ/bit, Modulation energy = 6 fJ/bit.

7.5 Advantages of Long-Wavelength VCSELs for Interconnects

Having seamless solutions and silicon-photonics in mind, long-wavelength VCSELs emitting around 1.3–1.6 μm are very attractive light-sources. Meanwhile, VCSELs at these wavelengths are also commercially available.

Longer wavelengths have the advantage of needing less energy per photon. This translates into lower driving voltages of the active devices. Photon energies below one electron-volt make it easier to create energy-efficient CMOS driver chips. Furthermore, silicon becomes transparent at these longer wavelengths. Therefore, this kind of laser-source seems to be an ideal candidate for future integrated optics and silicon photonics. Also, seamless solutions from a metro-range fiber link via printed circuit boards into the core of a silicon photonics chip could be realized at these wavelengths. On the other hand, the lower energy per photon makes these lasers more vulnerable to device heating. Auger processes become more and more dominant at higher wavelengths. Free-carrier absorption scales with wavelength at the power of two [16]. Active region and VCSEL chip prefer different base substrates. This makes it more challenging to realize such kind of devices.

7.5.1 Challenges of Long-Wavelength VCSELs

With VCSELs, there are two principal issues to conquer that are mostly material related. Firstly, a low-loss, high-q laser cavity has to be realized, and secondly, the laser current has to be confined to the active area while concurrently avoiding excessive heating. To reach lasing threshold, mirror and cavity losses have to be compensated by the gain of the active laser section. As the overlap of the active region with the optical mode only happens in a thin vertical area, a high-q cavity with high-reflectivity mirrors is needed. GaAs-based devices operating continuous wave at room temperature were already reported in 1988 by Fumio Koyama et al. [37], 11 years after the laser-concept had been suggested in 1977 by

Kenichi Iga [38]. Realizing VCSELs in the GaAs-material system is relatively easy, as both binary mirrors and stable wet oxidation techniques are available.

Unfortunately, there is no easy solution for realizing long-wavelength VCSELs emitting around the desirable waveband around 1.3–1.6 μm:

- Optical losses in *p*-conducting materials scale with wavelength squared
- On InP heat extraction through quaternary mirror stacks is not efficient
- On GaAs there is no classic material for the active region
- No oxidation of the current aperture on InP
- Very thick epitaxial layers have to be grown precisely and homogeneously

Nevertheless, there has been vast effort in realizing such kind of devices in different concepts. Devices in the InGaAlAs material system and grown on InP substrate were presented in 1999 [39] (pulsed) and 2000 in CW operation [40,41]. Another approach in realizing long-wavelength (LW) VCSEL devices is growing an active region of GaInNAs material (diluted nitrides) on GaAs substrate [42]. This allows using mature GaAs-based VCSEL technology with its mirrors with good thermal conductivity and oxidized apertures. However, nitrogen containing materials are not completely understood yet and tend to decompose under certain conditions, like extreme temperatures or current densities. VCSELs, on the other hand, are typically driven at rather high current densities, and the active region suffers from self-heating. Especially targeting at high modulation speeds, high carrier and photon densities are needed to boost the relaxation oscillation frequency; that is, the intrinsic bandwidth of the laser. Therefore, top performance and reliability are somehow contradictory in this approach and will depend on the quality of the nitrogen containing layers. Active regions based on quantum dots could be the better alternative. Here, the main challenge is the growth of these novel active materials [43].

Another approach is growing the distributed Bragg reflectors (DBRs) on GaAs wafers and the active region on InP. The final layer structure is generated by two wafer-bonding steps [44]. The bonded interfaces, however, show quite bad electrical properties making lateral intra-cavity contacts necessary [45]. This necessity makes device-processing more sophisticated and leads to higher electrical parasitic limiting modulation speeds.

Buried-tunnel-junction long-wavelength VCSELs, which were reported by Ortsiefer et al. [40], enabled room temperature CW operation with superior performance.

To sum up, quite different design concepts based on GaAs or InP substrates have matured and are commercially available. Schematics of the competing design concepts are depicted in Figure 7.7.

VCSELs might be the answer to the question how to quench the never-ending thirst for more and more bandwidth at lower costs and energy consumption. Due to the challenges discussed before, several different designs of long-wavelength VCSELs have been developed and matured to commercial availability. Industry, however, usually prefers to buy standard products with exceptional performance from a variety of suppliers rather than choosing from proprietary solutions with its pros and cons. Furthermore, recent progress in VCSEL research might have been underestimated. On the other hand, novel applications like silicon photonics requiring these wavebands might change minds.

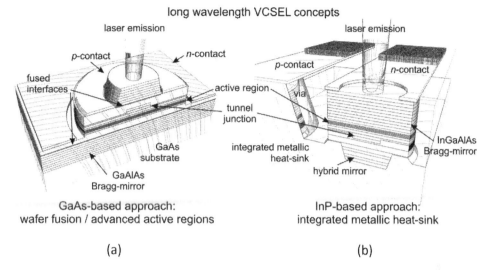

FIGURE 7.7
Long-wavelength VCSELs: (a) GaAs-based devices with active regions based on diluted nitrides or quantum dots or InP-based active regions and wafer-fused GaAs-based mirrors; (b) Monolithic concept on InP with hybrid mirror and integrated metallic heat sink.

References

1. W. Hofmann, P. Moser, P. Wolf, A. Mutig, M. Kroh, D. Bimberg, "44 Gb/s VCSEL for optical interconnects," *OFC/NFOEC*, PDPC5, pp. 1–3, 2011.
2. T. Mudge, "Power: A first-class architectural design constraint," *Computer*, 34, pp. 52–58, 2001.
3. F. Doany, C. Schow, C. Baks, D. Kuchta, P. Pepeljugoski, L. Schares, R. Budd et al., "160 Gb/s Bidirectional Polymer-Waveguide Board-Level Optical Interconnects Using CMOS-Based Transceivers," *IEEE Trans. Adv. Packag.*, 32, pp. 345–359, 2009.
4. D. Wiedenmann, M. Grabherr, R. Jäger, R. King, "High Volume Production of Single-Mode VCSELs," *Proc. SPIE*, 6132-2, 2006.
5. A. Vahdat, H. Liu, X. Zhao, C. Johnson, "The Emerging Optical Data Center," *OFC/NFOEC*, OTuH2, pp. 1–3, 2011.
6. L. Schares, J. Kash, F. Doany, C. Schow, C. Schuster, D. Kuchta, P. Pepeljugoski et al., "Terabus: Terabit/Second-Class Card-Level Optical Interconnect Technologies," *IEEE J. Sel. Top. Quantum Electron.*, 12, pp. 1032–1044, 2006.
7. W. Hofmann, P. Moser, D. Bimberg, "Energy Efficient Interconnects," in Breakthroughs in Photonics 2011, *IEEE Photonics Journal*, 2012.
8. W. Hofmann, "Evolution of high-speed long-wavelength vertical-cavity surface-emitting lasers," *Semicond. Sci. Technol.*, 26, pp. 014011, 2011.
9. S. Chuang, D. Bimberg, "Metal-Cavity Nanolasers," in Breakthroughs in Photonics 2010, *IEEE Photonics Journal*, pp. 288, 2011.
10. E. Wong, M. Mueller, P. Dias, C. Chan, M.-C. Amann, "Energy saving strategies for VCSEL ONUs," *OFC/NFOEC*, OTu1H5, pp. 1–3, 2012.
11. D. Miller, "Device Requirements for Optical Interconnects to Silicon Chips," *Proc. IEEE*, 97, pp. 1166–1185, 2009.
12. M. Taubenblatt, "Optical Interconnects for High-Performance Computing," *J. Lightwave Technol.*, 30, p. 448, 2012.

13. L. Coldren, S. Corzine, Chapter 5, "Dynamic Effects," in *Diode Lasers and Photonic Integrated Circuits*, pp. 184–212, Wiley, New York, 1995.

14. W. Hofmann, Chapter 2.3, "Laser Dynamics," in *InP-based Long-Wavelength VCSELs and VCSEL Arrays for High-Speed Optical Communication*, pp. 25–40, Selected Topics of Semiconductor Physics and Technology, Vol. 99, Munich, Germany, 2009.

15. L. Graham, H. Chen, D. Gazula, T. Gray, J. Guenter, B. Hawkins, R. Johnson et al., "The next generation of high speed VCSELs at Finisar," *Proc. SPIE*, 8276, 827602, 2012.

16. J. Buus, M.-C. Amann, *Tunable Laser Diodes and Related Optical Sources*, Wiley-VCH, Weinheim, Germany, 2005.

17. Y. Chang, C. Wang, L. Coldren, "High-Efficiency, High-Speed VCSELs with 35 Gb/s Error-Free Operation," *Electron. Lett.*, 43, pp. 1022–1023, 2007.

18. A. Mutig, *High Speed VCSELs for Optical Interconnects* (Springer Theses), XIV168, Springer, Berlin, Germany, 2011.

19. P. Wolf, P. Moser, G. Larisch, M. Kroh, A. Mutig, W. Unrau, W. Hofmann, D. Bimberg, "High-performance 980 nm VCSELs for 12.5 Gbit/s data transmission at 155°C and 49 Gbit/s at -14°C," *Electron. Let.*, 48, pp. 389–390, 2012.

20. W. Hofmann, P. Moser, A. Mutig, P. Wolf, W. Unrau, D. Bimberg, "980-nm VCSELs for Optical Interconnects at 25 Gb/s up to 120°C and 12.5 Gb/s up to 155°C," *Proc. CLEO/QELS*, pp. 1–2, 2011.

21. W. Hofmann, P. Moser, P. Wolf, G. Larisch, W. Unrau, D. Bimberg, "980-nm VCSELs for optical interconnects at bandwidths beyond 40 Gb/s," *Proc. SPIE*, 8276, 827605, 2012.

22. P. Moser, P. Wolf, A. Mutig, G. Larisch, W. Unrau, W. Hofmann, D. Bimberg, "85°C error-free operation at 38 Gb/s of oxide-confined 980-nm vertical-cavity surface-emitting lasers," *Appl. Phys. Lett.*, 100, pp. 081103, 2012.

23. International Technology Roadmap for Semiconductors, 2007 Edition, accessed January 2012, http://www.itrs.net/Links/2007ITRS/ExecSum2007.pdf

24. A. Larsson, "Advances in VCSELs for Communication and Sensing," *IEEE J. Sel. Top. Quantum Electron.*, 17, pp. 1552–1567, 2011.

25. P. Moser, J. Lott, P. Wolf, G. Larisch, A. Payusova, G. Fiol, N. Ledentsov, W. Hofmann, D. Bimberg, "Energy-efficient Vertical-Cavity Surface-Emitting Lasers (VCSELs) for "Green" Data and Computer Communication," *Proceedings of SPIE*, Photonics West, 8276-18, Green Photonics Award in Communications, San Francisco, CA, 2012.

26. S. Imai, K. Takaki, S. Kamiya, H. Shimizu, J. Yoshida, Y. Kawakita, T. Takagi et al., "Recorded Low Power Dissipation in Highly Reliable 1060-nm VCSELs for "Green" Optical Interconnection," *IEEE J. Sel. Top. Quantum Electron.*, 17, pp. 1614–1620, 2011.

27. P. Westbergh, J. Gustavsson, A. Haglund, A. Larsson, F. Hopfer, G. Fiol, D. Bimberg, A. Joel, "32 Gbit/s multimode fibre transmission using high-speed, low current density 850 nm VCSEL," *Electron. Lett.*, 45, pp. 366–368, 2009.

28. P. Westbergh, J. Gustavsson, B. Kögel, A. Haglund, A. Larsson, A. Mutig, A. Nadtochiy, D. Bimberg, A. Joel, "40 Gbit/s error-free operation of oxide-confined 850 nm VCSEL," *Electron. Lett.*, 46, pp. 1014–1016, 2010.

29. S. Blokhin, J. Lott, A. Mutig, G. Fiol, N. Ledentsov, M. Maximov, A. Nadtochiy, V. Shchukin, and D. Bimberg, "Oxide confined 850 nm VCSELs operating at bit rates up to 40 Gbit/s," *Electron. Lett.*, 45, 2009.

30. T. Anan, N. Suzuki, K. Yashiki, K. Fukatsu, H. Hatakeyama, T. Akagawa, K. Tokutome, M. Tsuji, "High-Speed 1.1 μm-Range InGaAs VCSELs," *OFC/NFOEC* 2008, San Diego, CA, OThS5, pp. 1–3, 2008.

31. W. Hofmann, M. Müller, P. Wolf, A. Mutig, T. Gründl, G. Böhm, D. Bimberg, M.-C. Amann, "40 Gbit/s modulation of 1550 nm VCSEL," *Electron. Lett.*, 47, pp. 270–271, 2011.

32. P. Moser, W. Hofmann, W., P. Wolf, J. Lott, G. Larisch, A. Payusov, N. Ledentsov, D. Bimberg, "81 fJ/bit energy-to-data ratio of 850 nm vertical-cavity surface-emitting lasers for optical interconnects," *Appl. Phys. Lett.*, 98, pp. 231106, 2011.

33. Y. Chang, L. Coldren, "Efficient, High-Data-Rate, Tapered Oxide-Aperture Vertical-Cavity Surface-Emitting Lasers," *IEEE J. Sel. Top. Quantum Electron.*, 15, pp. 704–715, 2009.

34. J. Shi, W. Weng, F. Kuo, J. Chyi, "Oxide-Relief Vertical-Cavity Surface-Emitting Lasers with Extremely High Data-Rate/Power-Dissipation Ratios," *OFC/NFOEC, OThG2,* pp. 1–3, Los Angeles, CA, 2011.

35. P. Moser, J. Lott, P. Wolf, G. Larisch, A. Payusov, N. Ledentsov, W. Hofmann, D. Bimberg, "99 fJ/(bit·km) Energy to Data-Distance Ratio at 17 Gb/s Across 1 km of Multimode Optical Fiber With 850-nm Single-Mode VCSELs," *IEEE Photon. Technol. Lett.,* 24, pp. 19–21, 2012.

36. M.-C. Amann, M. Müller, E. Wong, "Energy-Efficient High-Speed Short-Cavity VCSELs," *Proc. OFC/NFOEC,* 2012, OTh4F.1, 2012.

37. F. Koyama, S. Kinoshita, K. Iga, "Room-temperature continuous wave lasing characteristics of GaAs vertical cavity surface-emitting laser," *Appl. Phys. Lett.,* 55, pp. 221–222, 1989.

38. K. Iga, "Surface-emitting laser—Its birth and generation of new optoelectronics field," *IEEE J. Sel. Top. Quantum Electron.,* 6, pp. 1201–1215, 2000.

39. C. Kazmierski, J. Debray, R. Madani, J. Sagnes, A. Ougazzaden, N. Bouadma, J. Etrillard, F. Alexandre, M. Quillec, "+55°C pulse lasing at 1.56 µm of all-monolithic InGaAlAs-InP vertical cavity lasers," *Electron. Lett.,* 35, pp. 811–812, 1999.

40. M. Ortsiefer, R. Shau, G. Böhm, F. Köhler, M.-C. Amann, "Room-temperature operation of index-guided 1.55 µm InP-based vertical-cavity surface-emitting laser," *Electron. Lett.,* 36, pp. 437–438, 2000.

41. W. Yuen, G. Li, R. Nabiev, J. Boucart, P. Kner, R. Stone, D. Zhang et al., "High-Performance 1.6 µm single-epitaxy top-emitting VCSEL," *Electron. Lett.,* 36, pp. 1121–1123, 2000.

42. H. Riechert, A. Ramakrishnan, G. Steinle, "Development of InGaAsN-based 1.3 µm VCSELs," *Semicond. Sci. Technol.,* 17, pp. 892–897, 2002.

43. N. Ledentsov, F. Hopfer, D. Bimberg, "High-Speed Quantum-Dot Vertical-Cavity Surface-Emitting Lasers," *Proc. IEEE,* 95, pp. 1741–1756, 2007.

44. A. Syrbu, A. Mircea, A. Mereuta, A. Caliman, C. Berseth, G. Suruceanu, V. Iakovlev, M. Achtenhagen, A. Rudra, E. Kapon, "1.5mW single-mode operation of wafer-fused 1550 nm VCSELs," *IEEE Photon. Technol. Lett.,* 16, pp. 1230–1232, 2004.

45. A. Mereuta, G. Suruceanu, A. Caliman, V. Iakovlev, A. Sirbu, E. Kapon, "10-Gb/s and 10-km error-free transmission up to 100°C with 1.3-µm wavelength wafer-fused VCSELs," *Opt. Express,* 17, pp. 12981–12986, 2009.

8

Low-Power Optoelectronic Interconnects on Two-Dimensional Semiconductors

D. Keith Roper

CONTENTS

8.1 Introduction

Mobile computing, autonomous systems, live-streaming video, and the creation of smart cities in which 5G wireless and delivery at more than 100 Gigabits per second support cyberphysical connectivity is driving unprecedented growth in Internet and data center traffic. The concomitant demand for dynamically allocable, secure, and resilient optoelectronic data infrastructure is straining data centers and networks.

Energy-efficient scalability of data processing is limited in part by hardware bottlenecks, such as aggregation points where tiers of grooming, switching, and routing nodes electronically process optical signals from low speed access links or host connections to high-capacity backbone networks.[1] Without low-power, high-speed connectivity, peak household access rates have stalled at 0.01–0.1 Gigabits per second (Gbps), while commercial fiber to the home and industry standard Universal Serial Bus 3.1 have risen to 10 Gbps. Universities often connect to the Internet at a campus-wide aggregate with a capacity less than 10 Gbps.

Unique optical, electronic, and transport features of two-dimensional semiconductors could contribute to overcoming hardware bottlenecks, such as those that limit the expansion of big data networks. At 100 Gbps, envisioned in the recent 40/100 Gigabit Ethernet standard,[2] conventional electronics appear less efficient than optical technologies. Hybrid optical/electronic and all-optical signal grooming and processing are anticipated to deliver innovative, reduced-energy, scalable aggregation, particularly at high baud rates, for example, in spatially multiplexed waveguides. However, components with higher switching speeds would be required. High values of carrier mobility (100–700 cm^2 $V^{-1}s^{-1}$) reported for monolayer molybdenum disulfide, MoS_2, for example, could provide switching speeds of 10^{-7} to $10^{-9}s^{-1}$. These estimates rival switching speeds of top metal oxide semiconductor field effect transistors (MOSFETs), while avoiding their prohibitive thermal profile.[3] Besides higher switching speeds, exceptional optical activity of monolayer MoS_2 could support giant laser pulse formation (Q-switching), mode-locking, and optical limiting functionalities.[4] But development, implementation, and use of MoS_2 or other monolayer transition metal dichalcogenides for such applications is constrained by limited characterization and tunability of their nonlinear and optoelectronic properties.[5]

Plasmonic nanoparticles show promise for modulating optoelectronic as well as nonlinear optical activity of semiconductor nanocrystals like two-dimensional transition metal dichalcogenides (TMDs). Broken inversion symmetry and strong light-matter interactions in TMDs supports superior nonlinear optical activity.[6,7] Nonlinear susceptibilities reported for monolayer MoS_2, for example, exceed those of conventional materials by $>10^2$ pm/V with related increases in second harmonic generation (SHG) intensity[8–13] and two-photon absorption probability. Nanoparticles could enhance the nonlinear activity of monolayer MoS_2 by up to 10^3-fold to improve its photostability, wavelength multiplexing, and switching capabilities.[14,15] But, a framework for developing structure-function relations for nanoparticle-decorated TMDs remains largely undeveloped.

This chapter examines direct measures of optoelectronic interactions at interconnecting interfaces between nanoparticles and two-dimensional semiconductor nanocrystals. The increased spatiotemporal resolution provided by emerging methods advances understanding and use of heterostructures consisting of nanoparticles and two-dimensional semiconductor crystals. Corresponding development of comprehensive theoretical frameworks to provide accurate, validated structure-function relations in order to guide rational design and development of envisioned devices is considered. Advances in economic and reliable methods to fabricate, analyze, and evaluate heterostructures of nanoparticles and two-dimensional semiconductors for consideration in technologies are summarized. New understanding that emerges from these instrumental, computational, and fabrication approaches advances design and development of atom-thin flexible integrated circuits, field effect transistors, and closed-loop resonators enabled by low-power optoelectronic nanoparticle interconnects on two-dimensional semiconductors.

8.2 Two-Dimensional Semiconductors, Plasmonic Nanoparticles, and Their Interactions

8.2.1 Two-Dimensional Semiconductors

Two-dimensional transition metal dichalcogenides are nanocrystalline semiconductor analogs of graphene with unique optoelectronic functionality. Transition metal dichalcogenides, or TMDs, consist of a transition metal such as molybdenum, Mo, or tungsten, W, stoichiometrically bonded to two chalcogen atoms, such as sulfur, S, selenium, Se or tellurium, Te. This results in a general chemical formula of MX_2, where M represents the transition metal and X corresponds to the chalcogen atom. Monolayer TMDs exhibit two primary phases—a more stable semiconducting phase in which a transition metal (Mo or tungsten, W) occupies trigonal prismatic sites between two hexagonal layers of chalcogen (sulfur, S; selenium, Se or tellurium, Te), and a less stable octahedral metallic phase.[16]

Bulk TMDs are earth abundant. Using mechanical or chemical exfoliation, it is possible to break the weak van der Waals bonds that hold adjacent TMD layers and economically obtain the monolayer crystal. The nanoscale dimensions of monolayer TMDs enhance their optoelectronic properties relative to bulk material. The distance for carrier transport is decreased, for example. In contrast to graphene, monolayer TMDs are efficient direct bandgap light absorbers. TMDs show high catalytic activity attributable to edge sulfur groups with unoccupied electronic orbitals. Like graphene, TMDs exhibit high electron mobility as well as optical-, electronic-, and strain-driven phase transitions related to novel topological states that exhibit, for example, superconducting or quantum spin Hall insulating behavior. Such properties have broad applicability for two-dimensional electronic and photonic devices. In particular, MoS_2 exhibits a thickness-dependent ca. 1.9 eV direct bandgap, strong light emission[6,8], and observed transistor switching performance.[6,17]

Broken inversion symmetry and strong light-matter interactions in monolayer TMD support near-ideal nonlinear optics.[6,7] As a result, two-dimensional TMDs exhibit remarkable nonlinear activity. As examples, nonlinear susceptibilities reported for MoS_2 and tungsten disulfide, WS_2, exceed those of conventional materials by $>10^2$ pm/V with concomitant increases in SHG intensity[8–13] and two-photon absorption probability. Values of second-order nonlinear susceptibility, $\chi^{(2)}$, calculated for MoS_2 and WS_2 are 10^2 to ca. 10^3 pm/V, respectively. These values are up to 10^3 larger than common nonlinear crystals.[18] As a result, atomically-thin TMD offer higher measurability due to reduced autofluorescence background, and photostability to support wavelength multiplexing.[14,15] Beyond applications envisioned in hybrid optical/electronic and all-optical signal grooming and processing, these features offer benefits for nonlinear-enhanced biosensing,[19] molecular spectroscopy, biological microscopy,[20,21] high-speed imaging,[22] and photoinduction[23] (e.g., drug delivery, ablation, optogenetics, and tissue regeneration).

However, the utility of two-dimensional semiconducting TMDs in applications such as photodetectors for optical communications and sensing[24,25] has been limited to date by the difficulty in tuning their intrinsic optoelectronic properties.[5] As an example, WS_2 absorbs less than two percent of UV-vis radiation, largely at wavelengths less than 620 nm, and exhibits a layer-number-dependent photoluminescent response.[6,26] In largely empirical studies, interfacing TMDs with plasmonic nanoparticles have been shown to enhance optical absorption, photoluminescent response (more than measured 10-fold[27] values), charge carrier injection at sub-bandgap energies, and charge separation. These features offer a lowered detection threshold and could enable mid-infrared communication.

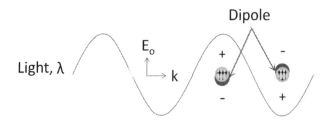

FIGURE 8.1
Light at resonant frequency (λ) induces plasmon oscillation (alternating negative sign, −) and corresponding local enhanced electromagnetic field (purple), on two subwavelength nanoparticle dipoles spaced at λ/2.

8.2.2 Plasmonic Nanoparticles

Subwavelength metal particles adjacent to a frequency-dependent dielectric can support collective oscillation of unbound electrons driven a by a resonant electromagnetic field and opposed by a Coulombic restoring force. A quantum of this oscillating mode that is concentrated at the metal-dielectric interface is known as a localized surface plasmon, or "plasmon" for short. Figure 8.1 illustrates a plasmon, represented by a negative sign (−), and its associated local electromagnetic field (purple) supported on two subwavelength nanoparticles. The particles are separated by a distance of one-half the wavelength of incident photon energy that induces plasmon oscillation at the localized surface plasmon resonance frequency. The value of resonant plasmon frequency changes with the size,[28–32] shape,[33] and composition of the nanoparticle, as well as the configuration of a nanoparticle relative to neighboring particles or dielectric. The electric field distribution on excited subwavelength particles is commonly dipolar, consisting of negatively charged electrons oscillating relative to positively charged stationary nuclei. Increases in particle size perturb the electric field distribution, which results in higher order (e.g., quadrupolar, electric modes). Dipolar, quadruplor, and higher order bright modes have a net non-zero dipole moment. Bright modes are inducible, or damped, via interaction with photons. Dark modes have a net zero dipole moment, and thus cannot couple with light and are neither induced by incident photons nor scattered by reradiated photons.

Plasmon resonance of noble metal particles occurs at visible frequencies corresponding to wavelengths about 400–700 nm resulting in a variety of colorimetric phenomena. Gold and silver dust of about 70 nm was dispersed by ancient Roman artisans into glass to produce the dichroic Lycurgus Cup whose walls both reflect and absorb incident green wavelength: light appears green when reflected back from the front of the artifact but red when it is transmitted from behind it. Dichroic glass in medieval church windows has similar optical properties. Current applications of plasmons make use of the fact that plasmon fields, local to subwavelength elements, are up to a million-fold larger than incident electromagnetic energy. This photon localization typically takes place within tens of nm from the metal-dielectric interface and completely decays within about one micron. It results in, for example, significantly enhanced Raman scattering and thermal dissipation.

8.2.3 Optoelectronic Interactions between Nanoparticles and Two-Dimensional Semiconductors

Electromagnetic interactions between a nanoparticle and a two-dimensional semiconductor, the transition metal dichalcogenide MoS_2, were first reported in 2013. Gold-nanoparticle decoration was observed to increase photocatalytic hydrogen evolution from a TMD.[34] Photoluminescence was enhanced more than ten-fold in monolayer WS_2 by nanoparticles.[35] Plasmonic nanoparticles were observed to enhance photocurrent in a MoS_2 field effect transistor.[36] Heterostructures of nanoparticles and two-dimensional semiconductors have been observed to enhance charge transport and separation photovoltaics[37,38] and photocatalysts[34,39] used in water treatment and hydrogen evolution.[34,39–41] In general, such empirical observations of nanoparticle enhancement have outpaced development of theoretical and computational frameworks based on direct measures of optoelectronic interactions between nanoparticles and two-dimensional semiconductors.

Nanoparticle enhancement of carrier dynamics and energetics in two-dimensional semiconductors has been attributed to the intense local electric fields induced by plasmons. The local field enhancement is hypothesized to reduce carrier recombination. Figure 8.2 illustrates an energy diagram of the heterointerface between a metal nanoparticle (left) and a two-dimensional nanocrystal (right). The respective metal work function, $q\theta$, and electron affinity, $q\chi$ relative to vacuum are shown. The Fermi energy level E_F of the metal and the valence, E_v, and conduction, E_c, band energies of the nanocrystal are identified. The Schottky barrier, $q\theta_{SB}$, which corresponds to the difference between the metal work function and the electron affinity, is identified. This barrier is about 1 eV for gold and WS_2. Interfacing a two-dimensional semiconductor with a metal has been observed to p-dope the semiconductor.

Resonant excitation of a metal electron above the Fermi level can overcome the Schottky barrier and result in direct electron transfer to the adjacent semiconductor. Such 'hot electron transfer' occurs on the scale of femtoseconds, before thermalization of carriers into phonon modes. Hot electron transfer represents a potential source of carriers at frequencies below the bandgap when $q\theta_{SB}$ is less than $E_c - E_v$. Direct electron transfer could provide

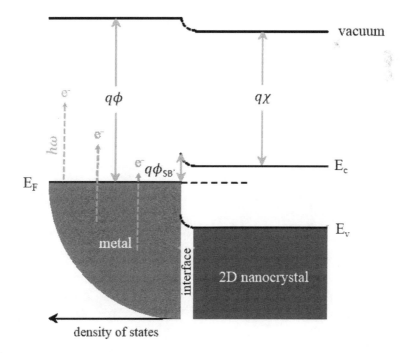

FIGURE 8.2
Energy diagram of a heterointerface of a metal nanoparticle (left) and a two-dimensional nanocrystal (right) showing their respective metal work function, $q\theta$, and electron affinity, $q\chi$ relative to vacuum; the Fermi energy level E_F; the valence, E_v, and conduction, E_c, band energies of the nanocrystal; and the Schottky barrier, $q\theta_{SB}$.

an intrinsic overpotential in cases where electron energies are much higher than the redox potential, thereby increasing reaction rates. Hot electron transfer is a possible source of electrons for physisorbed chemical species, which would facilitate particular reactions. This enhancement is adjunct to thermalization of plasmons into phonon vibration modes, which increases reaction kinetics via an Arrhenius effect.

8.3 Fabricating Nanoparticles on Two-Dimensional Semiconductors

8.3.1 Preparing Monolayer Two-Dimensional Semiconductors

Monolayer transition metal dichalcogenides (TMDs) nanocrystalline sheets are primarily prepared either by chemical vapor deposition or by mechanical, chemical or liquid-phase exfoliation of bulk TMD. Use and commercial implementation of monolayer TMDs has been limited to date in part by inaccessibility of these materials due to unscalable and economic fabrication methods.[5] Gaseous chalcogen is reacted with oxidized transition metal at high temperature and vacuum pressure to form TMD monolayer by chemical vapor deposition. Resulting triangular monolayers typically range from 1 to 100 microns, but can reach centimeter scales depending on purity of precursor and substrate and processing conditions.[42] Mechanical exfoliation using adhesive tape yields 1–10 micron TMD nanocrystals but at low throughput and with little size control. Chemical exfoliation using intercalation with n-Butyllithium reagent, for example, returns the metastable metallic 1T-polytype[43,44] Liquid-phase exfoliation, in contrast, preserves the semiconducting 2H-polytype useful for optoelectronic applications. It employs ultrasonication of bulk TMD in a selected solvent containing an aggregation-limiting surfactant like sodium cholate to produce a colloidally stable dispersion of TMD.[45,46] Thickness polydispersity of liquid-exfoliated TMD is reduced compared to mechanically exfoliated TMD. Liquid-phase exfoliation offers an economic, method capable of preparing large quantities of solution-processable nanocrystalline TMD monolayer sheets.[47] Figure 8.3 shows electron micrographs MoS_2 and WS_2 few-to-monolayer nanocrystals prepared by liquid phase exfoliation.[48]

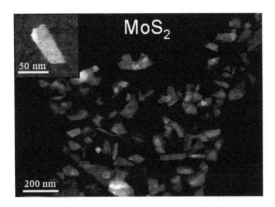 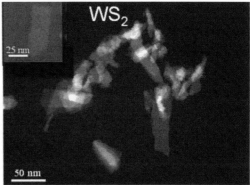

FIGURE 8.3
Electron micrographs of transition metal dichalcogenides molybdenum disulfide (a) and tungsten disulfide (b) few-to-monolayer nanocrystals prepared by liquid phase exfoliation. (Adapted from Dunklin, J.R., Plasmon-enhanced interfacial energy transfer in nanocomposite media, 2017.)

The method of preparing TMD nanocrystal influences it optoelectronic activity. Mid-bandgap states increase the probability of two-photon absorption and increase intensity of monolayer TMDs. Liquid-exfoliated TMD is anticipated to exhibit more mid-bandgap edge states per unit area than CVD crystals.[13] Liquid exfoliation of TMD avoids atomic-scale defects such as ripples or lattice strains resulting from transfer of chemical vapor deposited TMD to transmission electron microscope membranes.

8.3.2 Depositing Nanoparticles on Two-Dimensional Semiconductors

Decoration of two-dimensional semiconductors with plasmonic nanoparticles is possible using top-down and/or bottom-up approaches to lithography and nanoparticle deposition, resulting in regular or irregular configurations. The lithography, or patterning, step creates or applies a template in order to isolate a nanoparticle from its surroundings at a particular location on an underlying substrate. Deposition consists of introducing the material that composes the nanoparticle, typically gold or silver metal, onto the substrate either randomly, or as guided by the patterned template.

Regularly configured nanoparticles on two-dimensional semiconductors may be obtained by top-down approaches to patterning and deposition. These approaches manipulate a fine electromagnetic or ion beam or probe to alter a substrate or template in one dimension using electronic or mechanical forces. Electronic approaches include lithography using focused beams of electrons (electron beam lithography[49-55]), ions (focused ion beam[56,57]), reactive ions (reactive ion etching[58]), or light (laser interference lithography[59-64]). Mechanical approaches to templating include nanoscale manipulation of a nanoscale element (i.e., atomic force microscopy[65] and dip pen nanolithography[66,67]) or impression of soft matter with a previously-templated harder material "negative" (nanoimprint lithography[68,69]). Capital expense, severe operating conditions, material requirement, limited throughput, and availability of trained personnel typically limit implementation of top-down approaches.[70,71]

Bottom-up approaches to regularly configure nanoparticles rely on manipulating local surface forces using thermodynamic state and/or electromagnetism to guide self-assembly of particle components. The principal technique is nanosphere lithography:[72,73] monodisperse beads suspended in a solvent coalesce into ordered colloidal crystals as the solvent evaporates.[74-76] Different bead sizes are used to tune the size and of the resulting hexagonally close packed lattice onto which nanoparticles may be deposited. Bottom-up methods can provide a simpler, more cost-effective approach to order nanoparticles in two-[77] or three-dimensions,[78] but it is harder to create arbitrary template shapes, defects in resulting arrays are difficult to eliminate, and maintenance of order across long scales remains challenging.

Hybrid approaches offer promising, scale-able means to arrange nanoparticles on semiconductors. These consist of variations of nanoimprint lithography[68,69] in which a master pattern fabricated by top-down is a used to form a "negative" polymer image or template from which multiple copies of the original master are created in an imprinted resist. The negative and the copies form from curing of a polymer that flow freely into nanoscale voids in the master or template. Thermal nanoimprint lithography,[79,80] substrate conformal imprint lithography,[81] step and flash imprint lithography,[82] and microcontact printing[83] are variations to nanoimprint lithography. The time required, achievable line thickness, and aspect ratio of patterns formed by such nanoimprint lithographies are limited by the interplay of forces due to surface interactions and chemical equilibria with transport phenomena (i.e., momentum, thermal and mass diffusion).

Nanoparticles may be deposited regularly onto templated semiconductor via evaporation, sputtering, or electroplating of metal ion precursors. Evaporation requires 10^{-8} torr and high temperature. The directional nature of evaporation and sputtering limits the range of NP shapes that can be deposited. Good adhesion of gold to silica that overcomes the inherent lattice mismatch typically necessitates an adhesive layer of chromium, which changes the local dielectric and affects optical and electronic response characteristics.[84] A solution containing dispersed nanoparticles can be drop-cast onto a two-dimensional semiconductor substrate heated to the boiling point of the solvent. This results in irregular deposition. Coatings like polyvinylpyrrolidone used to synthesize and/or prevent aggregation of solution-dispersed nanoparticles insulate the nanoparticle from adjacent semiconductor.

8.3.3 Electroless Deposition of Nanoparticles on Transition Metal Dichalcogenides

Electroless metal plating of nanoparticles provides a "bottom-up" alternative to evaporation, sputtering, electroplating, or drop casting for depositing metal nanoparticles. Electroless plating relies on differences in redox potential between an aqueous dissolved metal ion (gold), a substrate-associated coordinating atom like sulfur on MoS_2 or WS_2, and the solvent.[85] In the seventeenth century, Andreas Cassius published a method for coating ceramics via a reduction of Au salt by Sn (II) in aqua regia to form *Cassius Purpur,*[86] which was later shown by Nobel laureate Zsigmondy to be Au(0) formed concurrent with tin oxide (SnO_2).[87] Deposition of metal nanoparticles by electroless plating allows rapid decoration of fragile, three dimensional or internal surfaces at ambient conditions without requiring conductive substrates or applied fields. The deposition rate, bath stability, and resulting nanoparticle structure and optical characteristics are determined by solution composition, pH and redox potential, temperature, and processing conditions.[88–93] Sensitization of a cleaned inorganic or organic substrate by divalent tin, its oxidation to reduce monovalent silver ions, and subsequent galvanic displacement of silver by monovalent gold is a common method.[94–98] It has been shown that trivalent gold may be reduced in preference to monovalent gold despite the higher reduction potential of the latter.[85] The rate of reducing aqueous trivalent gold ions to gold nanoparticles has been increased to more than 6 hours at concentrations up to 2 milligrams per liter using 3,3′,5,5′-tetramethylbenzidine dihydrochloride as the reducing agent at balanced redox conditions.

Direct reduction of aqueous trivalent gold ions onto uncoordinated sulfur atoms at edge sites of the semiconducting (2H) form of liquid phase exfoliated WS_2 using an electroless method was recently demonstrated.[99] Covalent attachment of gold nanoparticles to sulfur groups of $2H$-WS_2 was confirmed by high resolution transmission electron microscopy (Figure 8.4[47]) and X-ray photoelectron spectroscopy. Covalent attachment was in contrast to previous reduction of aqueous gold to elemental nanoparticles on chemically exfoliated metallic (1T) form of TMDs resulting in gold nanoparticles physisorbed on the nanosheets.[100–103] Resulting gold nanoparticle-decorated WS_2 nanosheets exhibited Tafel plots with slopes close to 100 mV/decade and electrode reaction rates over 10 fold higher than comparable TMD-based nanomaterial network electrodes. UV-vi, and Raman photoluminescence spectroscopy confirmed that optical properties of the $2H$-WS_2 were retained after gold nanoparticle reduction.

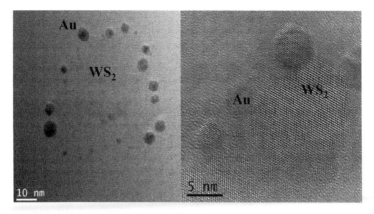

FIGURE 8.4

High resolution transmission electron microscope images of gold nanoparticles reduced onto sulfur atoms at edges of tungsten disulfide monolayer nanocrystals. Left hand image indicates a size range of ca. 1–8 nm. Right hand image illustrates crystal atoms organized in a faceted spheroidal nanomorphology. (Adapted from Bonaccorso, F. et al., *Adv. Mater.*, 28, 6136–6166, 2016.)

8.4 Modeling Nanoparticles on Two-Dimensional Semiconductors

The first exact solution of Maxwell's equations for scattering of incident electromagnetic plane waves on an isolated dielectric sphere was published by Gustav Mie.[104] A corresponding Fortran code developed by Bohren and Huffman[105] is publicly available in a convenient graphical user interface (MiePlot, v. 4.2.03, philiplaven.com). Exact solutions to Maxwell's equations for complicated nanoparticle scenarios may be obtained using finite difference time domain or T-matrix methods to characterize local plasmon electromagnetic fields. Finite difference time domain and T-matrix solutions are capable of describing structures illuminated by resonant light that has a wavelength on the same order of magnitude as the characteristic dimensions of the structure. Finite difference time domain employs the Yee algorithm[106], which uses the curl form of Maxwell's equations to compute the electric and magnetic fields for each unit cell of a grid representing a physical structure (e.g., its permittivity, permeability, and conductivity). The illumination source is modeled by a Gaussian pulse. A staggered, or "leapfrog," stepping scheme is used to streamline calculation of future field values from the past values and avoid computationally intensive solution of a large system of equations. However, the grid must be truncated to store electric, and magnetic fields computed at each unit cell. Absorbing boundary conditions are adequate for simple simulations, but not when a dielectric is introduced because of their dependence on a velocity of wave propagation of unity. So perfectly matched layers are widely used to truncate the grid. These consist of a lossy material used at the boundaries of the simulation domain to absorb outgoing radiation and prevent reflection back onto the scatterer.

Despite its versatility, finite different time domain has drawbacks that diminish its utility for certain problem geometries. For example, its conventional formulation uses a rectangular grid that can introduce "staircasing" errors when representing curved geometries depending on the grid coarseness. A total simulation domain much larger than the characteristic dimension of smallest structure of interest introduces additional difficulties. The T-matrix solution, in contrast, provides an exact multipole solution.[29] A T-matrix solution for electromagnetic scattering can be constructed using the null-field or extended boundary condition method. This method generalizes Mie principles to arbitrary particle geometries. Multilayer particles and clusters in both the near and far field regime have been solved by T-matrix. However, numerical implementation of T-matrix incurs large computation costs due to the number of integrations of spherical Bessel and spherical harmonic functions over the surface of the particle. Round-off errors tend to increase with particle size leading to an ill-conditioned problem. Both FDTD and T-matrix incur large computational costs and additional complexities that limit the number and geometry of elements that are possible to describe.

An alternative method for solving Maxwell's equations uses the discrete dipole approximation (DDA). The DDA approximates arbitrary nanoparticle shapes and configurations by representing them using a three dimensional array of electromagnetic dipoles. Each dipole i has a polarization, P_i, that is induced incident, electrostatic, and radiative electromagnetism in proportion to the its polarizability, α. The coupled dipole approximation (CDA) solves Maxwell's equations for situations in which isolated nanospheroids may be approximated by single dipoles in the array. Localized surface plasmon resonance is modeled effectively as a single dipole on a nanoparticle in the quasistatic limit. Both DDA and CDA provide good agreement between experiment and data.[33,107] Dipole polarizability has been described using a quasistatic approximation,[108] its modified long wavelength approximation (MLWA), and a dynamic polarizability introduced by Doyle.[109] The dynamic polarizability analyzes phase variations due to structural elements (e.g., lattice constant, d) whose characteristic length is on the order of the incident wavelength with an expression obtained from Mie theory. Depolarization and radiative dipole effects are modeled in this polarizability using *dynamic* terms[109,110]

$$\alpha = -4\pi \frac{3iR^3}{2x^3} a_1 \qquad (8.1)$$

where R is the particle radius, $x = \eta_0 kR$ is the size parameter with η_0 as the medium refractive index, and k as the vacuum wavenumber. The 4π coefficient is included for conversion to S.I. units. The term a_1 is the first Mie coefficient corresponding to the electric dipole. The polarizability is used to calculate the extinction cross section, C_{ext}, for an array of scattering nodes. Experimental spectra validate accuracy of DDA in representing complex spectral features heterostructures containing metal nanoparticle(s) and semiconducting and/or dielectric substrate and superstrate.[111–113]

Figure 8.5 compares extinction spectra simulated by DDA (dashed lines) for a gold nanosphere interacting with monolayer MoS_2 with measured UV-vis spectra (solid lines).[114] Results are shown for monolayer MoS_2 (blue), 76 nm diameter gold nanosphere (green), and 76 nm gold nanosphere atop monolayer MoS_2 (red). Each measured and simulated sample was performed atop a silica glass substrate. Dielectric data for the simulations was obtained from the literature for bulk gold (Johnson and Christy[115]), chemical vapor deposited monolayer MoS_2 (Mukherjee et al.[116]), and silica glass support (Rubin[117]). A micrograph of the gold nanosphere – MoS_2 – silica glass sample from which measurements were obtained is inset in Figure 8.5a (5 μm scalebar). Enhancement in local electric near field at (b) 566 nm and (c) 655 nm due to nanoparticle plasmon and monolayer MoS_2 A exciton, respectively, were simulated using DDSCAT with target and parameter files generated using a custom MATLAB tool available on nanoHUB.[118] The electric field within the AuNS was set equal to 1.

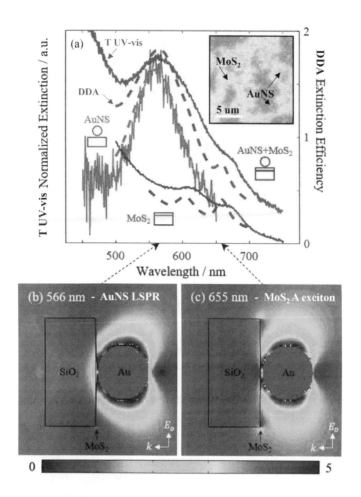

FIGURE 8.5

Optoelectronic interaction of gold nanosphere on monolayer MoS_2. (a) Extinction spectra simulated by DDA (dashed lines) and measured by transmission UV-vis spectroscopy (solid lines) for monolayer MoS_2 (blue), 76 nm diameter gold nanosphere (green), and 76 nm gold nanosphere atop monolayer MoS_2 (red). A silica glass substrate was included in each DDA simulation and experiment. A micrograph of the gold-nanosphere—MoS_2—silica glass sample is inset (5 μm scalebar). DDA was used to simulate enhancements in electric near-field at (b) 566 nm and (c) 655 nm due to nanoparticle plasmon and monolayer MoS_2 A exciton, respectively. Electric field within the AuNS was set equal to 1. (Adapted from Forcherio, G.T. and Roper, D.K., *Adv. Opt. Mater.*, 4, 2016.)

8.5 Measuring Optoelectronics of Nanoparticles on Two-Dimensional Semiconductors

Both optical and electron-based methods have been used to physically characterize optoelectronic interactions between nanoparticles and two-dimensional semiconductors. Optical excitation of bright plasmon modes on gold nanoparticles deposited on monolayer MoS_2 has been used to infer direct electron transfer from nanoparticle to semiconductor.[34–36,38,39,119] Optical photoluminescence was used to decompose radiative and non-radiative contributions to plasmon damping by measuring changes in plasmon photoluminescence bandwidth.[120] But optical characterization is constrained by limits imposed by photon diffraction. Resolution of optical measures is diffraction limited. Signal from optical measurements may be confounded due to concurrent optical excitation of both nanoparticle (plasmon) and semiconductor (exciton) modes from direct electron-hole pair generation, for example. Acquired optical data typically accrues signal from multiple nanoparticles with various geometries and local environments.

Electron energy loss spectroscopy (EELS), on the other hand, offers high-resolution, direct, unambiguous measures of optoelectronic interactions between nanoparticles and two-dimensional semiconductors that avoid the confounding effects intrinsic to optical measures. EELS is performed in a scanning transmission electron microscope. Nanometer-scale resolution is provided by the de Broglie wavelength of the electron source. Electron optics of the EELS monochromation system are used to distinguish the amount of electron energy lost at a particular frequency. This energy loss corresponds to the probability that an electromagnetic event such as a plasmon or exciton occurs. Irradiation by a fast-moving electron beam is the electromagnetic equivalent of interrogation by a broadband electromagnetic source. It may result in induction of bright, dark, and or hybrid plasmon modes on metal nanoantennas. The appearance, energy, and bandwidth of these modes varies as a result of the impact point of the sub-nanometer electron probe[121,122] as well as the composition, geometry and environment of the nanoparticle and/or two-dimensional semiconductor.[123–131] EELS was used as early as 2014 to measure direct electron transfer from gold ellipses to graphene at up to 20% quantum efficiency.[130]

Table 8.1 lists physicochemical and electromagnetic characteristics of a range of shapes and sizes of gold and silver nanoparticles deposited on monolayer nanocrystals that appear relevant to their optoelectronic interactions with the underlying nanocrystals. Gold ellipses were evaporated onto monolayer graphene prepared by chemical vapor deposition that was patterned by electron beam lithography. Two intervening nanometers of evaporated chromium were used to compensate for the gold-graphene lattice mismatch. Polyvinylpyrrolidone (PVP)-coated gold nanospheres of 78 nanometer diameter were drop-cast onto monolayer MoS_2 prepared by liquid phase exfoliation. PVP-coated silver nanoprisms with an average edge length of 53 nanometers were similarly drop-cast onto monolayer MoS_2 prepared by liquid phase exfoliation. Gold nanoparticles of approximately 20 nm diameter were reduced directly onto edge sulfur atoms of monolayer WS_2 prepared by liquid phase exfoliation.

Across the four samples in Table 8.1, characteristics of the nanoparticle, semiconductor, heterostructure and heterointerface relevant to their respective optoelectronic interactions varied significantly. As examples, the metal-TMD contact at the heterointerface ranged from minimal, i.e., a single point of contact between gold nanosphere and PVP-coated MoS_2 to significant, i.e., gold nanoparticle reduced directly onto uncoordinated sulfur

TABLE 8.1

Characteristics of Nanoparticle—Nanocrystalline Monolayer Heterostructures

	AuNP-Graphene[130]	AuNS-MoS$_2$[132]	AgNPr-MoS$_2$[114]	AuNP-WS$_2$[133]
Metal shape	Ellipse	Nanosphere	Nanoprism	Nanoparticle
Dimension (nm)	225 × 540	78 ± 2	53 ± 19 (edge)	ca. 20
2DNC decorated by	Evaporated	Drop-cast	Drop-cast	Direct reduction
Interface: thickness (nm)	Cr: 2	PVP: ca. 10	PVP: ca. 10	Au-S bond: ~0
2DNC: electron affinity, qχ (eV)	Gr: 4.5	MoS$_2$: 4.2	MoS$_2$: 4.2	WS$_2$: 4.5
Metal: work function, qϕ (eV)	gold: 5.1	gold: 5.1	silver: 4.3	gold: 5.1
Schottky barrier, qϕ_{SB} (eV)	0.6	0.9	0.1	0.6

TABLE 8.2

Direct Electron Transfer between Nanoparticle and Nanocrystalline Monolayer Measured by Electron Energy Loss Spectroscopy

	AuNP-graphene[130]	AuNS-MoS$_2$[132]	AgNPr-MoS$_2$[114]	AuNP-WS$_2$[133]
LSPR Γ_c bandwidth (eV)	0.15	0.08 ± 0.02	10 (nm)	0.19 ± 0.08
Plasmon dephasing time, $T_{2,i}$ (fs)	9.2	16 ± 4	—	7 ± 3
Quantum efficiency, η (%)	20	6 ± 1	16	11 ± 5

atoms at edges of liquid-phase exfoliated WS$_2$. The Schottky barrier ranged from value of ca. 0.1 for PVP-coated silver nanoprisms drop-cast onto liquid-exfoliated MoS$_2$ to a value of ca. 0.9 for PVP-coated gold nanospheres similarly drop-cast onto liquid-exfoliated MoS$_2$.

Table 8.2 lists important measured spectral features corresponding to direct electron transfer from the metal nanoparticle to the monolayer nanocrystal for each heterostructure pair listed in Table 8.1. Broadening of the linewidth of the plasmon resonance measured by EELS, or UV-vis spectroscopy (for AgNPr atop MoS$_2$), after accounting for damping due to photon and photon scattering, screening, and interfacial interactions, as well as the effects of instrumentation and methodology is attributable to direct electron transfer. Increases in plasmon bandwidth ranged from 0.08 for PVP-coated gold nanospheres drop-cast onto liquid-exfoliated MoS$_2$ to 0.19 for gold nanoparticle reduced directly onto uncoordinated sulfur atoms at edges of liquid-phase exfoliated WS$_2$. Corresponding plasmon decay times ranged from 16 femtoseconds in the former case to 7 femtoseconds in the latter. Measured plasmon decay times were within a range of values reported previously for gold nanorods using EELS[134] (2.5–18 femtoseconds) and optical dark field scattering[120] (160 femtoseconds). Quantum efficiency for direct electron transfer relative compared with damping due to all mechanisms ranged from six percent for PVP-coated gold nanospheres drop-cast onto liquid-exfoliated MoS$_2$ to up to 20% for gold nanoparticles evaporated onto graphene synthesized by chemical vapor deposition. Direct reduction of gold onto TMD edge sulfurs yielded higher damping bandwidths and shorter dephasing times than drop-cast or evaporated metal nanoparticles on TMD or graphene, whereas quantum efficiency appeared highest for nanoparticles like evaporated ellipses or nanoprisms that had relatively high hetero-interfacial metal-to-substrate contact area. Direct electron transfer from plasmons to TMDs has been reported to induce a 2H to 1T crystal phase transition that increases conductivity.

8.6 Nonlinear Activity of Two-Dimensional Semiconductors

Large values of second order susceptibility, $\chi^{(2)}$, at optical frequencies and significant electro-optical coefficients in the low-frequency limit make two-dimensional TMDs[13,135–137] potentially valuable for nonlinear optical devices and electric optical switches.[136,138–140] Values of $\chi^{(2)}$ calculated for TMDs using density functional theory with and without scissors correction,[141,142] density matrix theory,[143] and Bethe-Salpeter equation[144] increase from 10^2 pm V^{-1} off-resonance to ca. 10^3 pm V^{-1} on-resonance. Density functional theory determines $\chi^{(2)}$ frequency dependence via the WS$_2$ density of states at various excitonic transitions in its electronic band structure.[144] Second order susceptibility values of trilayers are ca. 1/3 that of monolayers.[142] Second harmonic generation (SHG) decreases 50% from monolayer to trilayer TMD and 80% from trilayer to pentalayer.[136] Second harmonic generation is precluded in even-layer TMDs for which crystalline inversion symmetry is restored[11,136] due to spin-valley-layer locking, absent an electrical bias.[145]

Reported measurements of $\chi^{(2)}$ for TMD monolayers[10-12,18] vary from 10^2 to 10^4 pm V^{-1} based on indirect classical calculations subject to electrodynamic consideration[141] as "bulk" or "sheet" material. Two-dimensional TMD possesses D_{3h} point-group symmetry that yields a single non-zero $\chi^{(2)}$ tensor element $\chi^{(2)}_{xxx} = -\chi^{(2)}_{xyy} = -\chi^{(2)}_{yyx} = -\chi^{(2)}_{yxy}$, where x and y are crystal axes. Analysis of $\chi^{(2)}_{xxx}$ for MoS$_2$ was performed by scaling its second harmonic generation intensity to an α-quartz reference.[11,141] D_{3h} point-group symmetry permits facile estimation of $\chi^{(2)}_{xxx}$ from orientation-averaged bulk-like $\langle\chi^{(2)}\rangle$ via $\chi^{(2)}_{xxx} = (21/8)^{1/2}\langle\chi^{(2)}\rangle$.[146] Bulk-like $\langle\chi^{(2)}\rangle$ may in turn be obtained from orientation-averaged first hyperpolarizability, $\langle\beta\rangle$ when scattered second harmonic originates from the underlying 3-atom plane thick TMD crystalline volume, V, using $\langle\chi^{(2)}\rangle = 2\langle\beta\rangle/V$,[18,147,148] where V is estimable from mean length and layer quantity extracted from absorbance spectra by published protocols.[149]

Orientation-averaged hyperpolarizability, $\langle\beta\rangle$, may be determined from concentration-dependent second harmonic (SH) scattering from a subwavelength colloidal nanoparticle dispersion by Hyper Rayleigh Scattering (HRS),[147,150,151] which was originally used to measure orientation-averaged nonlinear efficiency of dissolved molecules.[152-154] The intensity of HRS, I_{HRS}, corresponds to the incoherent sum of second harmonic frequencies scattered by a colloidal media, *viz.*

$$I_{HRS} = I_\omega^2 G \sum_i N_i F_i \langle \beta_i^2 \rangle \tag{8.2}$$

where I_ω is the pump intensity, $\langle\beta\rangle$ is the orientation-averaged first hyperpolarizability of molecule i with concentration N, F is a local field factor, and G is an experimental constant. There is relatively small contribution of exfoliated TMD edges, which include a stochastic combination of thiols and disulfides to overall $\chi^{(2)}$ at ca. 1550 nm.[13] This HRS method translates to a per-nanoparticle measure of the magnitude of a second-order efficiency $\langle\beta\rangle$ to $\chi^{(2)}$, an intrinsic material property that may be compared directly to second harmonic signal measured from single chemical vapor deposited or cleaved TMD.

Figure 8.6 shows a value of $\langle\chi^{(2)}\rangle = 250 \pm 12$ pm V^{-1} for liquid exfoliated two-dimensional WS$_2$ obtained from SH measured by HRS at 1064 nm irradiation.[155] Broad background photoluminescence present in SHG from WSe$_2$ reported by Fryett et al.[156] was absent from the HRS signal. The HRS-measured value was within 21% of a value of 315 pm V^{-1} calculated by density functional theory with[143] scissors correction, Δ_{SCI}. The density functional calculation without Δ_{SCI}[18] was 115 pm V^{-1}. Shown for comparison are reported values of $\chi^{(2)}$ for WS$_2$ of 8900-9000 pm V^{-1} measured using microscopy at 832 nm irradiation for suspended WS$_2$ and WS$_2$ on SiO$_2$/Si based on a Green's function calculation.[18]

The value of $\chi^{(2)}$ for liquid exfoliated WS$_2$ obtained using HRS in Figure 8.6 is the first reported use of concentration-dependent HRS measures to characterize the bulk-like $\chi^{(2)}$ of two-dimensional material. HRS appears usable to measure $\chi^{(2)}_{xxx}$ of TMD and/or other two-dimensional materials across important spectral regions. Deriving $\chi^{(3)}$[157] to compare with available z-scan measures[135] could benchmark use of HRS in analysis of frequency mixing and harmonic generation. HRS could be used to characterize harmonic generation of the less stable octahedral metallic 1T phase, other TMD polytypes, or nanoparticle-decorated plasmon-enhancement.[133]

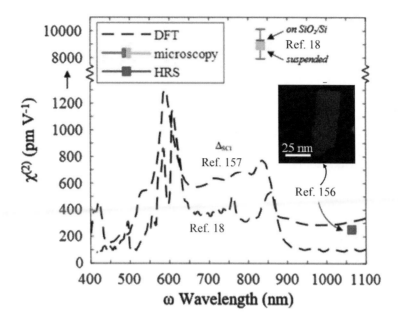

FIGURE 8.6

Second order susceptibility, $\chi^{(2)}$ for WS$_2$ crystal (example in inset) calculated by density functional theory (black dashed lines) with and without scissors correction (Δ_{sci}) and measured by HRS[156] (blue square) and microscopy[18] based on a Green's function calculation (red and green). Error of HRS value is within the data point. (Adapted from Forcherio, G.T. et al., *Opt. Lett.*, 42, 2017.)

To date, SHG of chemical vapor deposited[10,13] and mechanically exfoliated[9-12] TMD has been studied in the absence of nanoparticle decoration. Fryett et al. was first to report 200x enhancement of second harmonic generation (SHG) from a sub-region of mechanically-exfoliated monolayer tungsten diselenide, WSe$_2$, following deconvolution of SHG from broad background SHG and photoluminescence.[156] A linearly-polarized Si photonic crystal cavity was reported to augment local electromagnetic field strength in the subregion.

Wang et al. subsequently showed excitation of lateral gap plasmons within 20 nm trenches of an Au film that supported a WSe$_2$ monolayer. Gap plasmons were reported to enhance SHG 47-fold at locations within the trench and 7000-fold relative to WSe$_2$ on sapphire.[158] Malard et al. reported a blue-shifted SHG associated with increased density of states in MoS$_2$ at the direct C exciton transition near the point.[11] Plasmon-induced increases in second and third harmonic generation from 60- to 1700-fold, including increases in surface SHG,[159] are reported for a variety of nanomaterials.[160-166] But, evaluation of plasmonic enhancement of nonlinear activity on nanoparticle-decorated two-dimensional semiconductor crystals remains largely unexplored.

8.7 Summary

Low-power optoelectronics on two-dimensional semiconductors supported by plasmonic nanoparticles has been explored by fabricating covalent metal-semiconductor interconnects using novel electrochemical reduction and liquid phase exfoliation methods.

Characterization of optoelectronic activity using discrete dipole approximation of Maxwell's equations and high spatiotemporal resolution electron energy loss spectroscopy distinguished effects of physicochemical, geometric, and electromagnetic parameters on direct electron transfer from metal to nanocrystal. Direct translation of orientation averaged first hyperpolarizability to intrinsic second order susceptibility of monolayer semiconductors using Hyper Rayleigh Scattering was validated by density functional theory. These complementary approaches revealed possibilities for augmentation and tunability of unique optical, electronic, and transport properties in two-dimensional semiconductors by enhanced plasmonic near fields and electronic transitions. Increased understanding provided by these emerging approaches to fabrication, theoretical description, and instrumental analysis advance design and development of atom-thin flexible integrated circuits, field effect transistors, and closed-loop resonators using tunable low-power optoelectronic interconnects on two-dimensional semiconductors.

Funding

Preparation of this work was made possible by generous support from the Charles W. Oxford Chair of Emerging Technologies in the Ralph E. Marin Department of Chemical Engineering at the University of Arkansas.

Acknowledgment

The author gratefully acknowledges collaborative interactions with W. Ahn, M. Benamara, P. Blake, L. Bonacina, D. DeJarnette, J.R. Dunklin, G.T. Forcherio, R. Le Dantec, Y. Mugnier, and J. Riporto who contributed to the development of ideas, figures, and information discussed in the manuscript. Any opinions, findings, conclusions, or recommendations expressed are those of the author.

References

1. Kilper, D. C., Rastegarfar, H., Fang, Q., Li, C., Tu, X., Duan, N., Chen, K., Tern, R.-C. & Liow, T.-Y. Energy challenges in optical access and aggregation networks. *Philos. Trans. R. Soc. A Math. Phys. Eng. Sci.* **374**, 20140435 (2016).
2. Finisar Corporation – Finisar First to Demonstrate 40 Gigabit Ethernet LR4 CFP Transceiver Over 10 km of Optical Fiber at ECOC. Available at: http://investor.finisar.com/releasedetail.cfm?ReleaseID=410286
3. Nathan, A. & Al., E. Flexible Electronics: The Next Ubiquitous Platform. *Proc. IEEE* **100**, 1486–1517 (2012).
4. Bikorimana, S., Lama, P., Walser, A., Dorsinville, R., Anghel, S., Mitioglu, A., Micu, A. et al. Nonlinear optical responses in two-dimensional transition metal dichalcogenide multilayer: WS 2, WSe 2, MoS 2 and Mo 0.5 W 0.5 S 2. *Opt. Express* **24**, 699–712 (2012).

5. Ferrari, A. C., Bonaccorso, F., Falko, V., Novoselov, K. S., Roche, S., Bøggild, P., Borini et al. Science and technology roadmap for graphene, related two-dimensional crystals, and hybrid systems. *Nanoscale* **7**, 4598–4810 (2014).

6. Wang, Q. H., Kalantar-Zadeh, K., Kis, A., Coleman, J. N. & Strano, M. S. Electronics and optoelectronics of two-dimensional transition metal dichalcogenides. *Nat. Nanotechnol.* **7**, 699–712 (2012).

7. Boyd, R. W. *Nonlinear Optics*. (Academic Press, 2008).

8. Neshev, D. & Kivshar, Y. Nonlinear optics pushed to the edge. *Science* **344**, 483–484 (2014).

9. Wagoner, G. A., Persans, P. D., Van Wagenen, E. A. & Korenowski, G. M. Second-harmonic generation in molybdenum disulfide. *J. Opt. Soc. Am. B* **15**, 1017–1021 (1998).

10. Kumar, N., Najmaei, S., Cui, Q., Ceballos, F., Ajayan, P., Lou, J. & Zhao, H. Second harmonic microscopy of monolayer MoS2. *Phys. Rev. B* **87**, 161403 (2013).

11. Malard, L. M., Alencar, T. V., Barboza, A. P. M., Mak, K. F. & de Paula, A. M. Observation of intense second harmonic generation from MoS2 atomic crystals. *Phys. Rev. B* **87**, 201401 (2013).

12. Li, Y., Rao, Y., Mak, K. F., You, Y., Wang, S., Dean, C. R. & Heinz, T. F. Probing symmetry properties of few-layer MoS2 and h-BN by optical second-harmonic generation. *Nano Lett.* **13**, 3329–3333 (2013).

13. Yin, X., Ye, Z., Chenet, D. A, Ye, Y., O'Brien, K., Hone, J. C. & Zhang, X. Edge nonlinear optics on a MoS2 atomic monolayer. *Science* **344**, 488–490 (2014).

14. Schell, A. W., Tran, T. T., Takashima, H., Takeuchi, S. & Aharonovich, I. Non-linear excitation of quantum emitters in two-dimensional hexagonal boron nitride. *ArXiv* **1606.09364**, 1–11 (2016).

15. Chimene, D., Alge, D. L. & Gaharwar, A. K. Two-dimensional nanomaterials for biomedical applications: Emerging trends and future prospects. *Adv. Mater.* **27**, 7261–7284 (2015).

16. Mak, K. F., Lee, C., Hone, J., Shan, J. & Heinz, T. F. Atomically thin MoS2: A new direct-gap semiconductor. *Phys. Rev. Lett.* **105**, 136805 (2010).

17. Radisavljevic, B., Radenovic, A., Brivio, J., Giacometti, V. & Kis, A. Single-layer MoS2 transistors. *Nat. Nanotechnol.* **6**, 147–150 (2011).

18. Janisch, C., Wang, Y., Ma, D., Mehta, N., Elías, A. L., Perea-López, N., Terrones, M., Crespi, V. & Liu, Z. Extraordinary second harmonic generation in tungsten disulfide monolayers. *Sci. Rep.* **4**, 5530 (2014).

19. Bonacina, L. Nonlinear nanomedecine: Harmonic nanoparticles toward targeted diagnosis and therapy. *Mol. Pharm.* **10**, 783–792 (2013).

20. Extermann, J., Bonacina, L., Cuña, E., Kasparian, C., Mugnier, Y., Feurer, T. & Wolf, J-P. Nanodoublers as deep imaging markers for multi-photon microscopy. *Opt. Express* **17**, 15342–15349 (2009).

21. Bonacina, L., Mugnier, Y., Courvoisier, F., Le Dantec, R., Extermann, J., Lambert, Y., Boutou, V., Galez, C. & Wolf, J.-P. Polar Fe(IO3)3 nanocrystals as local probes for nonlinear microscopy. *Appl. Phys. B* **87**, 399–403 (2007).

22. Magouroux, T., Extermann, J., Hoffmann, P., Mugnier, Y., Le Dantec, R., Jaconi, M. E., Kasparian, C., Ciepielewski, D., Bonacina, L. & Wolf, J.-P. High-speed tracking of murine cardiac stem cells by harmonic nanodoublers. *Small* **8**, 2752–2756 (2012).

23. Staedler, D., Magouroux, T., Passemard, S., Schwung, S., Dubled, M., Schneiter, G. S., Rytz, D., Gerber-Lemaire, S., Bonacina, L. & Wolf, J.-P. Deep UV generation and direct DNA photo-interaction by harmonic nanoparticles in labelled samples. *Nanoscale* **6**, 2929–2936 (2014).

24. Mueller, T., Xia, F. & Avouris, P. Graphene photodetectors for high-speed optical communications. *Nat. Photonics* **4**, 20 (2010).

25. Zhai, T., Fang, X., Liao, M., Xu, X., Zeng, H., Yoshio, B. & Golberg, D. A comprehensive review of one-dimensional metal-oxide nanostructure photodetectors. *Sensors* **9**, 6504–6529 (2009).

26. Backes, C., Smith, R. J., McEvoy, N., Berner, N. C., McCloskey, D., Nerl, H. C., O'Neill et al. Edge and confinement effects allow in situ measurement of size and thickness of liquid-exfoliated nanosheets. *Nat. Commun.* **5**, 4576 (2014).

27. Kern, J., Trügler, A., Niehues, I., Ewering, J., Schmidt, R., Schneider, R., Najmaei, S. et al. Nanoantenna-enhanced light-matter interaction in atomically thin WS_2. *ACS Photonics* **2**, 1260–1265 (2015).

28. Klar, T., Perner, M., Grosse, S., von Plessen, G., Spirkl, W. & Feldmann, J. Surface-plasmon resonances in single metallic nanoparticles. *Phys. Rev. Lett.* **80**, 4249–4252 (1998).

29. Zhao, L., Kelly, K. L. & Schatz, G. C. The extinction spectra of silver nanoparticle arrays: Influence of array structure on plasmon resonance wavelength and width. *J. Phys. Chem. B* **107**, 7343–7350 (2003).

30. Jain, P. K., Lee, K. S., El-Sayed, I. H. & El-Sayed, M. A. Calculated absorption and scattering properties of gold nanoparticles of different size, shape, and composition: Applications in biological imaging and biomedicine. *J. Phys. Chem. B* **110**, 7238–7248 (2006).

31. Kinnan, M. K. & Chumanov, G. Plasmon coupling in two-dimensional arrays of silver nanoparticles: II. Effect of the particle size and interparticle distance. *J. Phys. Chem. C* **114**, 7496–7501 (2010).

32. Link, S. & El-Sayed, M. A. Size and temperature dependence of the plasmon absorption of colloidal gold nanoparticles. *J. Phys. Chem. B* **103**, 4212–4217 (1999).

33. Haynes, C. L., McFarland, A. D., Zhao, L., Van Duyne, R. P., Schatz, G. C., Gunnarsson, L., Prikulis, J., Kasemo, B. & Käll, M. Nanoparticle optics: The importance of radiative dipole coupling in two-dimensional nanoparticle arrays. *J. Phys. Chem. B* **107**, 7337–7342 (2003).

34. Kang, Y., Gong, Y., Hu, Z., Li, Z., Qiu, Z., Zhu, X., Ajayan, P. M. & Fang, Z. Plasmonic hot electron enhanced MoS_2 photocatalysis in hydrogen evolution. *Nanoscale* **7**, 4482–4488 (2015).

35. Sobhani, A., Lauchner, A., Najmaei, S., Ayala-Orozco, C., Wen, F., Lou, J. & Halas, N. J. Enhancing the photocurrent and photoluminescence of single crystal monolayer MoS2 with resonant plasmonic nanoshells. *Appl. Phys. Lett.* **104**, 31112 (2014).

36. Lin, J., Li, H., Zhang, H. & Chen, W. Plasmonic enhancement of photocurrent in MoS2 field-effect-transistor. *Appl. Phys. Lett.* **102**, 203109 (2013).

37. Britnell, L., Ribeiro, R. M., Eckmann, A., Jalil, R., Belle, B. D., Mishchenko, A., Kim, Y.-J. et al. Strong light-matter interactions in heterostructures of atomically thin films. *Science* **340**, 1311–1314 (2013).

38. Yang, X., Liu, W., Xiong, M., Zhang, Y., Liang, T., Yang, J., Xu, M., Ye, J. & Chen, H. Au nanoparticles on ultrathin MoS2 sheets for plasmonic organic solar cells. *J. Mater. Chem. A* **2**, 14798 (2014).

39. Yin, Z., Chen, B., Bosman, M., Cao, X., Chen, J., Zheng, B. & Zhang, H. Au nanoparticle-modified MoS2 nanosheet-based photoelectrochemical cells for water splitting. *Small* **10**, 3537–3543 (2014).

40. Thurston, T. R. & Wilcoxon, J. P. Photooxidation of organic chemicals catalyzed by nanoscale MoS2. *J. Phys. Chem. B* **103**, 11–17 (1999).

41. Yang, J. & Shin, H. S. Recent advances in layered transition metal dichalcogenides for hydrogen evolution reaction. *J. Mater. Chem. A* **2**, 5979–5985 (2014).

42. Forcherio, G. T. Infrared energy conversion in plasmonic fields at two-dimensional semiconductors (2017). Theses and Dissertations. 1896. https://scholarworks.uark.edu/etd/1896

43. Chhowalla, M., Shin, H. S., Eda, G., Li, L.-J., Loh, K. P. & Zhang, H. The chemistry of two-dimensional layered transition metal dichalcogenide nanosheets. *Nat. Chem.* **5**, 263–275 (2013).

44. Benavente, E., Santa Ana, M., Mendizábal, F. & González, G. Intercalation chemistry of molybdenum disulfide. *Coord. Chem. Rev.* **224**, 87–109 (2002).

45. Nicolosi, V., Chhowalla, M., Kanatzidis, M. G., Strano, M. S. & Coleman, J. N. Liquid Exfoliation of Layered Materials. *Science* **340**, 1226419 (2013).

46. Tao, H., Zhang, Y., Gao, Y., Sun, Z., Yan, C. & Texter, J. Scalable exfoliation and dispersion of two-dimensional materials – An update. *Phys. Chem. Chem. Phys.* **19**, 921–960 (2017).

47. Bonaccorso, F., Bartolotta, A., Coleman, J. N. & Backes, C. 2D-crystal-based functional inks. *Adv. Mater.* **28**, 6136–6166 (2016).

48. Dunklin, J. R. Plasmon-enhanced interfacial energy transfer in nanocomposite media (2017). Theses and Dissertations. 1986. https://scholarworks.uark.edu/etd/1986

49. Félidj, N., Grand, J., Laurent, G., Aubard, J., Lévi, G., Hohenau, A., Galler, N., Aussenegg, F. R. & Krenn, J. R. Multipolar surface plasmon peaks on gold nanotriangles. *J. Chem. Phys.* **128**, 94702 (2008).

50. Bakker, R. M., Yuan, H.-K., Liu, Z., Drachev, V. P., Kildishev, A. V., Shalaev, V. M., Pedersen, R. H., Gresillon, S. & Boltasseva, A. Enhanced localized fluorescence in plasmonic nanoantennae. *Appl. Phys. Lett.* **92**, 43101 (2008).

51. Smythe, E. J., Dickey, M. D., Bao, J., Whitesides, G. M. & Capasso, F. Optical antenna arrays on a fiber facet for in situ surface-enhanced Raman scattering detection. *Nano Lett.* **9**, 1132–1138 (2009).

52. Rechberger, W., Hohenau, A., Leitner, A., Krenn, J. R., Lamprecht, B. & Aussenegg, F. R. Optical properties of two interacting gold nanoparticles. *Opt. Commun.* **220**, 137–141 (2003).

53. Pompa, P. P., Martiradonna, L., Della Torre, A., Della Sala, F., Manna, L., De Vittorio, M., Calabi, F., Cingolani, R. & Rinaldi, R. Metal-enhanced fluorescence of colloidal nanocrystals with nanoscale control. *Nat. Nanotechnol.* **1**, 126–30 (2006).

54. Bakker, R. M., Drachev, V. P., Yuan, H.-K. & Shalaev, V. M. Enhanced transmission in near-field imaging of layered plasmonic structures. *Opt. Express* **12**, 3701–3706 (2004).

55. Salerno, M., Krenn, J. R., Lamprecht, B., Schider, G., Ditlbacher, H., Félidj, N., Leitner, A. & Aussenegg, F. R. Plasmon polaritons in metal nanostructures: The optoelectronic route to nanotechnology. *Opto-Electronics Rev.* **10**, 217–224 (2002).

56. Dhawan, A., Gerhold, M., Madison, A., Fowlkes, J., Russell, P. E., Vo-Dinh, T. & Leonard, D. N. Fabrication of nanodot plasmonic waveguide structures using FIB milling and electron beam-induced deposition. *Scanning* **31**, 139–146 (2009).

57. Lin, C. H., Jiang, L., Chai, Y. H., Xiao, H., Chen, S. J. & Tsai, H. L. A method to fabricate 2D nanoparticle arrays. *Appl. Phys. A* **98**, 855–860 (2010).

58. Zheng, Y. B., Hsiao, V. K. S. & Huang, T. J. *Active plasmonic devices based on ordered Au nanodisk arrays. MEMS 2008. IEEE 21st International Conference* (IEEE, 2008). doi:10.1109/MEMSYS.2008.4443754

59. Ellman, M., Rodríguez, A., Pérez, N., Echeverria, M., Verevkin, Y. K., Peng, C. S., Berthou, T., Wang, Z., Olaizola, S. M. & Ayerdi, I. High-power laser interference lithography process on photoresist: Effect of laser fluence and polarisation. *Appl. Surf. Sci.* **255**, 5537–5541 (2009).

60. Lai, N. D., Liang, W. P., Lin, J. H., Hsu, C. C. & Lin, C. H. Fabrication of two- and three-dimensional periodic structures by multi-exposure of two-beam interference technique. *Opt. Express* **13**, 9605–11 (2005).

61. Geyer, U., Hetterich, J., Diez, C., Hu, D. Z., Schaadt, D. M. & Lemmer, U. Nano-structured metallic electrodes for plasmonic optimized light-emitting diodes. in *Proceedings of SPIE* **7032**, 70320B–1–10 (SPIE, 2008).

62. Lasagni, A. F., Yuan, D., Shao, P. & Das, S. Periodic micropatterning of polyethylene glycol diacrylate hydrogel by laser interference lithography using nano- and femtosecond pulsed lasers. *Adv. Eng. Mater.* **11**, B20–B24 (2009).

63. Lai, N. D., Huang, Y. Di, Lin, J. H., Do, D. B. & Hsu, C. C. Fabrication of periodic nanovein structures by holography lithography technique. *Opt. Express* **17**, 3362–9 (2009).

64. Ma, F., Hong, M. H. & Tan, L. S. Laser nano-fabrication of large-area plasmonic structures and surface plasmon resonance tuning by thermal effect. *Appl. Phys. A* **93**, 907–910 (2008).

65. Maier, S. A., Brongersma, M. L., Kik, P. G., Meltzer, S., Requicha, A. A. G. & Atwater, H. A. Plasmonics-A route to nanoscale optical devices. *Adv. Mater.* **13**, 1501–1505 (2001).

66. Zhang, H., Chung, S.-W. & Mirkin, C. A. Fabrication of sub-50-nm solid-state nanostructures on the basis of dip-pen nanolithography. *Nano Lett.* **3**, 43–45 (2003).

67. Zhang, H., Amro, N. A., Disawal, S., Elghanian, R., Shile, R. & Fragala, J. High-throughput dip-pen-nanolithography-based fabrication of Si nanostructures. *Small* **3**, 81–5 (2007).

68. Chou, S. Y., Krauss, P. R. & Renstrom, P. J. Nanoimprint lithography. *J. Vac. Sci. Technol. B Microelectron. Nanom. Struct.* **14**, 4129–4133 (1996).

69. Jung, J.-M., Yoo, H., Stellacci, F. & Jung, H. Two-photon excited fluorescence enhancement for ultrasensitive DNA detection on large-area gold nanopatterns. *Adv. Mater.* **22**, 2542–2546 (2010).

70. Ahn, W., Blake, P., Shultz, J., Ware, M. E. & Roper, D. K. Fabrication of regular arrays of gold nanospheres by thermal transformation of electroless-plated films. *J. Vac. Sci. Technol. B Nanotechnol. Microelectron.* **28**, 638–642 (2010).

71. Blake, P., Ahn, W. & Roper, D. K. Enhanced uniformity in arrays of electroless plated spherical gold nanoparticles using tin presensitization. *Langmuir* **26**, 1533–1538 (2010).

72. Sepúlveda, B., Angelomé, P. C., Lechuga, L. M. & Liz-Marzán, L. M. LSPR-based nanobiosensors. *Nano Today* **4**, 244–251 (2009).

73. Yonzon, C. R., Stuart, D. A., Zhang, X., McFarland, A. D., Haynes, C. L. & Van Duyne, R. P. Towards advanced chemical and biological nanosensors-An overview. *Talanta* **67**, 438–48 (2005).

74. Jiang, P. & McFarland, M. J. Large-scale fabrication of wafer-size colloidal crystals, macroporous polymers and nanocomposites by spin-coating. *J. Am. Chem. Soc.* **126**, 13778–13786 (2004).

75. Li, H., Low, J., Brown, K. S. & Wu, N. Large-area well-ordered nanodot array pattern fabricated with self-assembled nanosphere template. *IEEE Sensors* **8**, 880–884 (2008).

76. Pan, F., Zhang, J., Cai, C. & Wang, T. Rapid fabrication of large-area colloidal crystal monolayers by a vortical surface method. *Langmuir* **22**, 7101–7104 (2006).

77. Denkov, N. D., Velev, O. D., Kralchevsky, P. A., Ivanov, I. B., Yoshimura, H. & Nagayama, K. Two-dimensional crystallization. *Nature* **361**, 26 (1993).

78. Jiang, P., Bertone, J. F., Hwang, K. S. & Colvin, V. L. Single-crystal colloidal multilayers of controlled thickness. *Chem. Mater.* **11**, 2132–2140 (1999).

79. Perez Toralla, K., De Girolamo, J., Truffier-Boutry, D., Gourgon, C. & Zelsmann, M. High flowability monomer resists for thermal nanoimprint lithography. *Microelectron. Eng.* **86**, 779–782 (2009).

80. Zelsmann, M., Perez Toralla, K., De Girolamo, J., Boutry, D. & Gourgon, C. Comparison of monomer and polymer resists in thermal nanoimprint lithography. *J. Vac. Sci. Technol. B* **26**, 2430–2434 (2008).

81. Ji, R., Hornung, M., Verschuuren, M. A., van de Laar, R., van Eekelen, J., Plachetka, U., Moeller, M. & Moormann, C. UV enhanced substrate conformal imprint lithography (UV-SCIL) technique for photonic crystals patterning in LED manufacturing. *Microelectron. Eng.* **87**, 963–967 (2010).

82. Chauhan, S., Palmieri, F., Bonnecaze, R. T. & Willson, C. G. Feature filling modeling for step and flash imprint lithography. *J. Vac. Sci. Technol. B* **27**, 1926–1932 (2009).

83. Bergmair, I., Mühlberger, M., Lausecker, E., Hingerl, K. & Schöftner, R. Diffusion of thiols during microcontact printing with rigid stamps. *Microelectron. Eng.* **87**, 848–850 (2010).

84. Simsek, E. On the surface plasmon resonance modes of metal nanoparticle chains and arrays. *Plasmonics* **4**, 223–230 (2009).

85. Jang, G.-G. & Roper, D. K. Balancing redox activity allowing spectrophotometric detection of Au(I) using tetramethylbenzidine dihydrochloride. *Anal. Chem.* **83**, 1836–1842 (2011).

86. Ali, H. O. & Christie, I. R. A. A review of electroless gold deposition processes. *Gold Bull.* **17**, 118–127 (1984).

87. Mallory, G. O., Hajdu, J. B. & American electroplaters and surface finishers society. *Electroless plating: fundamentals and applications.* (AESF, 1990). Available at: https://searchworks.stanford.edu/view/10269645

88. Yasseri, A. A., Sharma, S., Young Jung, G. & Kamins, T. I. Electroless deposition of Au nanocrystals on Si(111) surfaces as catalysts for epitaxial growth of Si nanowires. *Electrochem. Solid-State Lett.* **9**, C185 (2006).

89. Zhao, L., Siu, A. C.-L., Petrus, J. A., He, Z. & Leung, K. T. Interfacial bonding of gold nanoparticles on a H-terminated Si(100) substrate obtained by electro- and electroless deposition. *J. Am. Chem. Soc.* **129**, 5730–5734 (2007).

90. Brown, K. R. & Natan, M. J. Hydroxylamine seeding of colloidal Au nanoparticles in solution and on surfaces. *Langmuir* **14**, 726–728 (1998).

91. Menzel, H., Mowery, M. D., Cai, M. & Evans, C. E. Surface-confined nanoparticles as substrates for photopolymerizable self-assembled monolayers. *Adv. Mater.* **11**, 131–134 (1999).

92. Jin, Y., Kang, X., Song, Y., Zhang, B., Cheng, G. and & Dong, S. Controlled nucleation and growth of surface confined gold nanoparticles on a (3-aminopropyl)trimethoxysilane-modified glass slide: A strategy for SPR substrates. *Anal. Chem* **73**, 2843–2849 (2001).

93. Hrapovic, S., Liu, Y., Enright, G., Bensebaa, F. & Luong, J. H. T. New strategy for preparing thin gold films on modified glass surfaces by electroless deposition. *Langmuir* **19**, 3958–3965 (2003).

94. Roper, D. K., Ahn, W., Taylor, B. J. & Dall'Asen, A. G. U.S. Patent 8,097,295. Method of making nanoparticles by electroless plating (2012).

95. Ahn, W., Taylor, B., Dall'Asén, A. G. & Roper, D. K. Electroless gold island thin films: Photoluminescence and thermal transformation to nanoparticle ensembles. *Langmuir* **24**, (2008).

96. Ahn, W. & Roper, D. K. Transformed gold island film improves light-to-heat transduction of nanoparticles on silica capillaries. *J. Phys. Chem. C* **112**, 12214–12218 (2008).

97. Jang, G.-G. & Roper, D. K. Continuous flow electroless plating enhances optical features of Au films and nanoparticles. *J. Phys. Chem. C* **113**, 19228–19236 (2009).

98. Wei, X. & Roper, D. K. Tin sensitization for electroless plating review. *J. Electrochem. Soc.* **161**, (2014).

99. Dunklin, J. R., Lafargue, P., Higgins, T. M., Forcherio, G. T., Benamara, M., McEvoy, N., Roper, D. K., Coleman, J. N., Vaynzof, Y. & Backes, C. Production of monolayer-rich gold-decorated 2H–WS2 nanosheets by defect engineering. *npj 2D Mater. Appl.* **1**, 43 (2018).

100. Kim, J., Byun, S., Smith, A. J., Yu, J. & Huang, J. Enhanced electrocatalytic properties of transition-metal dichalcogenides sheets by spontaneous gold nanoparticle decoration. *J. Phys. Chem. Lett.* **4**, 1227–1232 (2013).

101. Shi, Y., Huang, J.-K., Jin, L., Hsu, Y.-T., Yu, S. F., Li, L.-J. & Yang, H. Y. Selective decoration of Au nanoparticles on monolayer MoS2 single crystals. *Sci. Rep.* **3**, 1839 (2013).

102. Zhang, P., Lu, X., Huang, Y., Deng, J., Zhang, L., Ding, F., Su, Z., Wei, G. & Schmidt, O. G. MoS2 nanosheets decorated with gold nanoparticles for rechargeable Li-O2 batteries. *J. Mater. Chem. A* **3** (2015).

103. Topolovsek, P., Cmok, L., Gadermaier, C., Borovsak, M., Kovac, J. & Mrzel, A. Thiol click chemistry on gold-decorated MoS$_2$: Elastomer composites and structural phase transitions. *Nanoscale* **8**, 10016–10020 (2016).

104. Mie, G. Contributions to the optics of turbid media, particularly of colloidal metal solutions. *Ann. Phys.* **25**, 377–445 (1908).

105. Bohren, C. F. & Huffman, D. R. *Absorption and Scattering of Light By Small Particles.* (Wiley, 1998).

106. Yee, K. S. Numerical solution of initial boundary value problems involving maxwell's equations in isotropic media. *IEEE Trans. Antennas Propag.* **14**, 302–307 (1966).

107. Zou, S. & Schatz, G. C. Narrow plasmonic/photonic extinction and scattering line shapes for one and two dimensional silver nanoparticle arrays. *J. Chem. Phys.* **121**, 12606–12 (2004).

108. Kelly, K. L., Coronado, E., Zhao, L. L. & Schatz, G. C. The optical properties of metal nanoparticles: The influence of size, shape, and dielectric environment. *J. Phys. Chem. B* **107**, 668–677 (2003).

109. Doyle, W. T. Optical properties of a suspension of metal spheres. *Phys. Rev. B* **39**, 9852–9858 (1989).

110. Roper, D. K., Ahn, W., Taylor, B. & Dall'Asen, A. G. Enhanced spectral sensing by electromagnetic coupling with localized surface plasmons on subwavelength structures. *IEEE Sens. J.* **10**, 531–540 (2010).

111. Lisunova, M., Norman, J., Blake, P., Forcherio, G. T., Dejarnette, D. F. & Roper, D. K. Modulation of plasmonic Fano resonance by the shape of the nanoparticles in ordered arrays. *J. Phys. D. Appl. Phys.* **46**, (2013).

112. Forcherio, G. T., Blake, P., Seeram, M., DeJarnette, D., Roper, D. K. Coupled dipole plasmonics of nanoantennas in discontinuous, complex dielectric environments. *J. Quant. Spectrosc. Radiat. Transf.* **166**, 93–101 (2015).

113. DeJarnette, D., Blake, P., Forcherio, G. T. & Roper, D. K. Far-field Fano resonance in nanoring arrays modeled from extracted, point dipole polarizability. *J. Appl. Phys.* **115**, 24306 (2014).

114. Forcherio, G. T. & Roper, D. K. Spectral characteristics of noble metal nanoparticle–Molybdenum disulfide heterostructures. *Adv. Opt. Mater.* **4**, (2016).

115. Johnson, P. B. & Christy, R. W. Optical constants of the noble metals. *Phys. Rev. B* **6**, 4370–4379 (1972).

116. Mukherjee, B., Tseng, F., Gunlycke, D., Kumar, K., Eda, G. & Simsek, E. Complex electrical permittivity of the monolayer molybdenum disulfide (MoS2) in near UV and visible. *Opt. Mater. Express* **5**, 447–455 (2015).

117. Rubin, M. Optical properties of soda lime silica glasses. *Sol. Energy Mater.* **12**, 275–288 (1985).

118. Seeram, M., Forcherio, G. T., Blake, P. & Roper, D. K. Shape generator for the discrete dipole approximation. *nanoHUB* (2015).

119. Fang, Z., Wang, Y., Liu, Z., Schlather, A., Ajayan, P. M., Koppens, F. H. L., Nordlander, P. & Halas, N. J. Plasmon-induced doping of graphene. *ACS Nano* **6**, 10222–10228 (2012).

120. Hoggard, A., Wang, L. Y., Ma, L., Fang, Y., You, G., Olson, J., Liu, Z., Chang, W. S., Ajayan, P. M. & Link, S. Using the plasmon linewidth to calculate the time and efficiency of electron transfer between gold nanorods and graphene. *ACS Nano* **7**, 11209–11217 (2013).

121. Egerton, R. F. Electron energy-loss spectroscopy in the TEM. *Reports Prog. Phys.* **72**, 16502 (2009).

122. Kociak, M. & Stéphan, O. Mapping plasmons at the nanometer scale in an electron microscope. *Chem. Soc. Rev.* **43**, 3865–83 (2014).

123. Scholl, J. A., Koh, A. L. & Dionne, J. A. Quantum plasmon resonances of individual metallic nanoparticles. *Nature* **483**, 421–427 (2012).

124. Schmidt, F.-P., Ditlbacher, H., Hohenester, U., Hohenau, A., Hofer, F. & Krenn, J. R. Dark plasmonic breathing modes in silver nanodisks. *Nano Lett.* **12**, 5780–5783 (2012).

125. Nelayah, J., Kociak, M., Stéphan, O., García de Abajo, F. J., Tencé, M., Henrard, L., Taverna, D., Pastoriza-Santos, I., Liz-Marzán, L. M. & Colliex, C. Mapping surface plasmons on a single metallic nanoparticle. *Nat. Phys.* **3**, 348–353 (2007).

126. Bosman, M., Keast, V. J., Watanabe, M., Maaroof, A. I. & Cortie, M. B. Mapping surface plasmons at the nanometre scale with an electron beam. *Nanotechnology* **18**, 165505 (2007).

127. Schaffer, B., Grogger, W., Kothleitner, G. & Hofer, F. Comparison of EFTEM and STEM EELS plasmon imaging of gold nanoparticles in a monochromated TEM. *Ultramicroscopy* **110**, 1087–1093 (2010).

128. Diaz-Egea, C., Sigle, W., van Aken, P. A. & Molina, S. I. High spatial resolution mapping of surface plasmon resonance modes in single and aggregated gold nanoparticles assembled on DNA strands. *Nanoscale Res. Lett.* **8**, 337 (2013).

129. Koh, A. L., Bao, K., Khan, I., Smith, W. E., Kothleitner, G., Nordlander, P., Maier, S. A. & Mccomb, D. W. Electron energy-loss spectroscopy (EELS) of surface plasmons in single silver nanoparticles and dimers: Influence of beam damage and mapping of dark modes. *ACS Nano* **3**, 3015–3022 (2009).

130. DeJarnette, D. & Roper, D. K. Electron energy loss spectroscopy of gold nanoparticles on graphene. *J. Appl. Phys.* **116**, 54313 (2014).

131. Forcherio, G. T., DeJarnette, D. F., Benamara, M. & Roper, D. K. Electron energy loss spectroscopy of surface plasmon resonances on aberrant gold nanostructures. *J. Phys. Chem. C* **120**, 24950 (2016).

132. Forcherio, G. T., Benamara, M. & Roper, D. K. Electron energy loss spectroscopy of hot electron transport between gold nanoantennas and molybdenum disulfide by plasmon excitation. *Adv. Opt. Mater.* **5**, 1600572 (2017).

133. Forcherio, G. T., Dunklin, J. R., Backes, C., Vaynzof, Y., Benamara, M. & Roper, D. K. Gold nanoparticles physicochemically bonded onto tungsten disulfide nanosheet edges exhibit augmented plasmon damping. *AIP Adv.* **7**, 75103 (2017).

134. Bosman, M., Ye, E., Tan, S. F., Nijhuis, C. A., Yang, J. K. W., Marty, R., Mlayah, A., Arbouet, A., Girard, C. & Han, M.-Y. Surface plasmon damping quantified with an electron nanoprobe. *Sci. Rep.* **3**, 1312 (2013).

135. Dong, N., Li, Y., Zhang, S., McEvoy, N., Zhang, X., Cui, Y., Zhang, L., Duesberg, G. S. & Wang, J. Dispersion of nonlinear refractive index in layered WS2 and WSe2 semiconductor films induced by two-photon absorption. *Opt. Lett.* **41**, 3936–3939 (2016).

136. Li, D., Xiong, W., Jiang, L., Xiao, Z., Rabiee Golgir, H., Wang, M., Huang, X. et al. Multimodal nonlinear optical imaging of MoS2 and MoS2-based van der Waals heterostructures. *ACS Nano* **10**, 3766–3775 (2016).

137. Mak, K. F. & Shan, J. Photonics and optoelectronics of 2D semiconductor transition metal dichalcogenides. *Nat. Photonics* **10**, 216–226 (2016).

138. Seyler, K. L., Schaibley, J. R., Gong, P., Rivera, P., Jones, A. M., Wu, S., Yan, J., Mandrus, D. G., Yao, W. & Xu, X. Electrical control of second-harmonic generation in a WSe2 monolayer transistor. *Nat. Nanotechnol.* **10**, 407–411 (2015).

139. Janisch, C., Mehta, N., Ma, D., Elías, A. L., Perea-López, N., Terrones, M. & Liu, Z. Ultrashort optical pulse characterization using WS_2 monolayers. *Opt. Lett.* **39**, 383–385 (2014).

140. Hsu, W., Zhao, Z., Li, L., Chen, C., Chiu, M. & Chang, P. Second harmonic generation from artificially stacked transition metal dichalcogenide twisted bilayers. *ACS Nano* **8**, 2951–2958 (2014).

141. Clark, D. J., Senthilkumar, V., Le, C. T., Weerawarne, D. L., Shim, B., Jang, J. I., Shim, J. H. et al. Strong optical nonlinearity of CVD-grown MoS2 monolayer as probed by wavelength-dependent second-harmonic generation. *Phys. Rev. B* **90**, 121409 (2014).

142. Wang, C. Y. & Guo, G. Y. Nonlinear optical properties of transition-metal dichalcogenide MX2 (M = Mo, W; X = S, Se) monolayers and trilayers from first-principles calculations. *J. Phys. Chem. C* **119**, 13268–13276 (2015).

143. Rhim, S. H., Kim, Y. S. & Freeman, A. J. Strain-induced giant second-harmonic generation in monolayered 2H-MoX2 (X = S, Se, Te). *Appl. Phys. Lett.* **107**, 241908 (2015).

144. Trolle, M. L., Seifert, G. & Pedersen, T. G. Theory of excitonic second-harmonic generation in monolayer MoS2. *Phys. Rev. B* **89**, 235410 (2014).

145. Yu, H., Talukdar, D., Xu, W., Khurgin, J. B. & Xiong, Q. Charge-induced second-harmonic generation in bilayer WSe2. *Nano Lett.* **15**, 5653–5657 (2015).

146. Cyvin, S. J., Rauch, J. E. & Decius, J. C. Theory of hyper-raman effects (nonlinear inelastic light scattering): Selection rules and depolarization ratios for the second-order polarizability. *J. Chem. Phys.* **43**, 4083 (1965).

147. Joulaud, C., Mugnier, Y., Djanta, G., Dubled, M., Marty, J.-C., Galez, C., Wolf, J.-P., Bonacina, L. & Le Dantec, R. Characterization of the nonlinear optical properties of nanocrystals by Hyper Rayleigh Scattering. *J. Nanobiotechnology* **11**, S1–S8 (2013).

148. Karvonen, L., Säynätjoki, A., Mehravar, S., Rodriguez, R. D., Hartmann, S., Zahn, D. R. T., Honkanen, S. et al. Investigation of second- and third-harmonic generation in few-layer gallium selenide by multiphoton microscopy. *Sci. Rep.* **5**, 10334 (2015).

149. Backes, C., Szydłowska, B. M., Harvey, A., Yuan, S., Vega-Mayoral, V., Davies, B. R., Zhao, P. et al. Production of highly monolayer enriched dispersions of liquid-exfoliated nanosheets by liquid cascade centrifugation. *ACS Nano* **10**, 1589–1601 (2016).

150. Le Dantec, R., Mugnier, Y., Djanta, G., Bonacina, L., Extermann, J., Badie, L., Joulaud, C. et al. Ensemble and individual characterization of the nonlinear optical properties of ZnO and BaTiO3 nanocrystals. *J. Phys. Chem. C* **115**, 15140–15146 (2011).

151. Staedler, D., Magouroux, T., Hadji, R., Joulaud, C., Extermann, J., Schwung, S., Passemard, S. et al. Harmonic nanocrystals for biolabeling: A survey of optical properties and biocompatibility. *ACS Nano* **6**, 2542–2549 (2012).

152. Hubbard, S. F., Petschek, R. G. & Singer, K. D. Spectral content and dispersion of hyper-Rayleigh scattering. *Opt. Lett.* **21**, 1774–1776 (1996).

153. Stadler, S., Dietrich, R., Bourhill, G. & Brauchle, C. Long-wavelength first hyperpolarizability measurements by hyper-Rayleigh scattering. *Opt. Lett.* **21**, 251–253 (1996).

154. Hendrickx, E., Clays, K. & Persoons, A. Hyper-Rayleigh scattering in isotropic solution. *Acc. Chem. Res.* **31**, 675–683 (1998).

155. Forcherio, G. T., Riporto, J., Dunklin, J. R., Mugnier, Y., Le Dantec, R., Bonacina, L. & Roper, D. K. Nonlinear optical susceptibility of two-dimensional WS2 measured by hyper Rayleigh scattering. *Opt. Lett.* **42**, (2017).

156. Fryett, T. K., Seyler, K. L., Zheng, J., Liu, C., Xu, X. & Majumdar, A. Silicon photonic crystal cavity enhanced second-harmonic generation from monolayer WSe2. *2D Mater.* **4**, 15031 (2016).

157. Schmidt, C., Riporto, J., Uldry, A., Rogov, A., Mugnier, Y., Dantec, R. Le, Wolf, J. & Bonacina, L. Multi-order investigation of the nonlinear susceptibility tensors of individual nanoparticles. *Sci. Rep.* **6**, 25415 (2016).

158. Wang, Z., Dong, Z., Zhu, H., Jin, L., Chiu, M.-H., Li, L.-J., Xu, Q.-H. et al. Selectively plasmon-enhanced second-harmonic generation from monolayer tungsten diselenide on flexible substrates. *ACS Nano* **12**, 1859–1867 (2018).

159. McMahon, M. D., Ferrara, D., Bowie, C. T., Lopez, R. & Haglund Jr., R. F. Second harmonic generation from resonantly excited arrays of gold nanoparticles. *Appl. Phys. B* **87**, 259–265 (2007).

160. Chen, K., Durak, C., Heflin, J. R. & Robinson, H. D. Plasmon-enhanced second-harmonic generation from ionic self-assembled multilayer films. *Nano Lett.* **7**, 254–258 (2007).

161. Ishifuji, M., Mitsuishi, M. & Miyashita, T. Bottom-up design of hybrid polymer nanoassemblies elucidates plasmon-enhanced second harmonic generation from nonlinear optical dyes. *J. Am. Chem. Soc.* 4418–4424 (2009). doi:10.1021/ja808749h

162. Pu, Y., Grange, R., Hsieh, C. L. & Psaltis, D. Nonlinear optical properties of core-shell nanocavities for enhanced second-harmonic generation. *Phys. Rev. Lett.* **104**, 207402 (2010).

163. Zakharko, Y., Nychyporuk, T., Bonacina, L., Lemiti, M. & Lysenko, V. Plasmon-enhanced nonlinear optical properties of SiC nanoparticles. *Nanotechnology* **24**, 55703 (2013).

164. Grinblat, G., Rahmani, M., Cortes, E., Caldarola, M., Comedi, D., Maier, S. A. & Bragas, A. V. High-efficiency second harmonic generation from a single hybrid ZnO nanowire/Au plasmonic nano-oligomer. *Nano Lett.* **14**, 6660–6665 (2014).

165. Sánchez-García, L., Tserkezis, C., Ramírez, M. O., Molina, P., Carvajal, J. J., Aguiló, M., Díaz, F., Aizpurua, J. & Bausá, L. E. Plasmonic enhancement of second harmonic generation from nonlinear RbTiOPO4 crystals by aggregates of silver nanostructures. *Opt. Express* **24**, 8491 (2016).

166. Metzger, B., Hentschel, M., Schumacher, T., Lippitz, M., Ye, X., Murray, C. B., Knabe, B., Buse, K. & Giessen, H. Doubling the efficiency of third harmonic generation by positioning ITO nanocrystals into the hot-spot of plasmonic gap-antennas. *Nano Lett.* **14**, 2867–2872 (2014).

9

GaN-Based Schottky Barriers for Low Turn-On Voltage Rectifiers

Nishant Darvekar and Santosh K. Kurinec

CONTENTS

9.1 Introduction

The power semiconductor industry was earlier focused on saving costs and increasing productivity, but the present scenario is entirely different. Manufacturers and their customers demand energy efficient components at a competitive price. As a result, the need for these energy efficient electrical systems in the industry is increasing as the trends toward miniaturization and portability emerge.

Power semiconductor devices are vital in the generation and distribution of power throughout the world. The semiconductor diode is one of the most widely used, discrete components in electronics. Rectification, light emission, photodetectors, photovoltaics, and voltage regulation are the most popular applications of the diode along with other RF and high-frequency signal processing applications. Rectification is the ability of a semiconductor diode to convert alternating current (AC) to direct current (DC) as it allows the electrons to flow only in one direction.

It was in 1904 that the first diode was used in the conversion of radio waves to current. This diode was called the Fleming valve. However, it was a German physicist by the name of Ferdinand Braun who discovered that electrons flow freely only in one direction when lead sulfide crystals were probed with the point of a thin metal wire. Interestingly, Thomas Edison also made a device that allowed flow of current in one direction, but he later realized there were not any applications for this. In 1938, Walter Schottky created the theory that explained the rectifying behavior of a metal-semiconductor contact as dependent on a barrier layer at the surface of contact between the two materials. The metal semiconductor diodes, which were later built based on this theory, are called Schottky barrier diodes (SBD). Since then, the industry has come a long way with many types of diodes (Figure 9.1) for a variety of applications ranging from very low power signals to high power signals.

The Schottky barrier diodes discussed in this chapter are primarily for use in the AC voltage to DC voltage inverters in power supplies. The efficiency of conversion of AC to

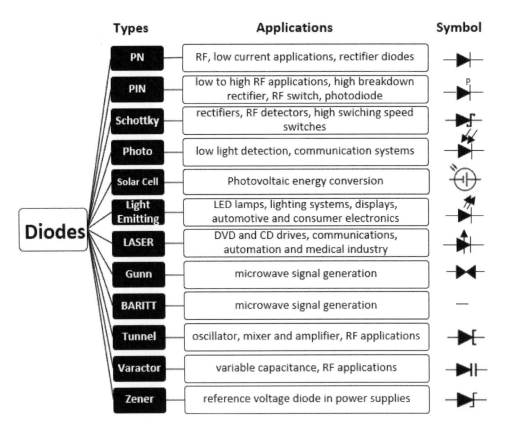

FIGURE 9.1
Types of diodes, their applications, and symbols.

DC depends on the voltage drop across every device in the power system. A lower voltage drop (V_f) would mean a more efficient conversion of AC to DC, thus reducing power dissipation.

9.2 Motivation

The mains supplies in most countries (excluding Japan and North America) is a 230 V three phase AC. This voltage needs to be converted to the operation range of 3 V DC to 20 V DC for computer power supplies and other electronic devices. Conversion occurs in three stages. First, the 230 V AC is rectified to a high DC voltage (Figure 9.2). Second, this is converted to a low AC voltage using an inverter. Finally, the low AC voltage is rectified to a low DC voltage in the specified range mentioned above. Various devices and techniques can be used to increase the efficiency of the last two stages. The efficiency of the system is greatly affected by the first stage, which has a full wave bridge rectifier shown below.

In the positive AC cycle, diodes D1 and D2 conduct, whereas in the negative cycle, diodes D3 and D4 conduct. The smoothening capacitor "C" charges and discharges to form a constant DC voltage. The ripple in the DC voltage can be further reduced using other techniques.

Schottky diodes have the advantage of a higher switching speed, fast reverse recovery time (t_{rr}) and low on-resistance (R_{on}) over conventional PN diodes, which is useful in high-frequency applications, but has a relatively low breakdown voltage (V_{br}). Silicon PN junction diodes typically have a forward turn-on voltage, V_f of ~ 0.7 V, while Schottky diodes can go much lower than that. If the supply voltage is 230 V AC, with a rectifying system of high turn-on voltage diodes, the power loss could be significant. A Schottky diode with a lesser V_f of ~ 0.4 V or lower could reduce the power loss and increase the efficiency of rectification. Some techniques used to reduce V_f and increase V_{br} are discussed in this chapter.

Figure 9.3 illustrates the diodes available in the industry for rectification applications. It is evident that the combination of low V_f and high V_{br} is difficult to achieve in silicon process technology. The challenge is to manufacture a reliable and inexpensive Schottky

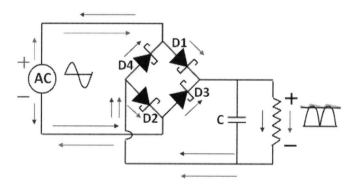

FIGURE 9.2
AC to DC bridge rectifier circuit with Schottky diodes and smoothening capacitor.

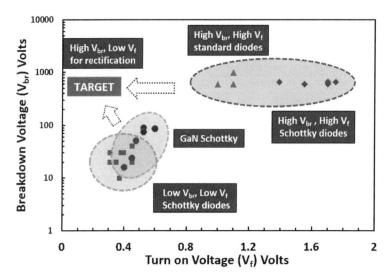

FIGURE 9.3
Application map of commercial Schottky diodes. Target shows the range of operation needed for GaN/AlGaN SBD for power supplies. (From Digi-Key Corporation; Schmitz, A.C. et al., *J. Electron. Mater.*, 27, 255–260, 1998.)

diode with V_f less than 0.4 V and V_{br} greater than 400 V. The wide and direct bandgap AlGaN/GaN is one of the material systems in which these requirements could be met.

9.3 Basic Semiconductor Theory

To understand the band structure and flow of electrons in a metal-semiconductor (MS) junction, it is essential to understand a few terms related to band physics. The conduction band, valence band, vacuum level, and Fermi level are denoted by E_C, E_V, E_0, and E_F, respectively. The electron affinity ($q\chi$) of a semiconductor is defined as the energy needed for an electron to jump from the conduction band to the vacuum level. Its value is 4.1 eV for GaN. The energy required for an electron to escape from the metal is called the work function of the metal ($q\phi_m$). It is the difference between Fermi level and vacuum level. Similarly, the work function of the semiconductor ($q\phi_s$) is defined as the difference between its Fermi level and vacuum level. It therefore depends on the level of doping in the semiconductor. For a degenerately doped n-type semiconductor, the Fermi level is close to or even above the conduction band. For a p-type semiconductor, it is close to the valence band.

$$q\chi = E_0 - E_C \tag{9.1}$$

$$q\phi_s = q\chi + \frac{Eg}{2} + (E_i - E_F) \tag{9.2}$$

E_i is the intrinsic Fermi-level.

It must be noted that χ and ϕ_B are expressed in volts and are multiplied by q to express in eV.

9.3.1 Schottky Diode Electrostatics

When the metal work function (ϕ_m) is larger than the semiconductor work function (ϕ_s) in a metal – n-type semiconductor (with doping N_D) junction, the electrons in the semiconductor near the junction move into the metal to maintain equilibrium. In such a case, a depletion region is created in the semiconductor, which contains positive donor ions near the junction. This leads to a barrier for the electrons flowing from the semiconductor to the metal and is denoted by ϕ_B, known as the Schottky barrier height [2]. The following equations explain Figure 9.4 in detail.

$$\phi_B = \phi_m - \chi. \tag{9.3}$$

$$qV_{bi} = q\phi_B - q\phi_n \tag{9.4}$$

The built in voltage of the junction (V_{bi}) depends on $q\phi_m$ and $q\phi_n$, which is the energy difference between the E_C and E_{FS} that depends on the doping in the semiconductor. It increases as ϕ_m increases for different metals and decreases if the semiconductor is highly n-type doped.

The depletion region is only in the semiconductor side in a MS junction. The depletion width (W) is a function of doping, applied voltage, and the permittivity of the semiconductor (ε_s). It reduces in width with a higher doped semiconductor and increases with increase in reverse voltage V_r.

$$W = \sqrt{\frac{2\varepsilon_s\left(V_{bi} + V_r\right)}{qN_D}} \tag{9.5}$$

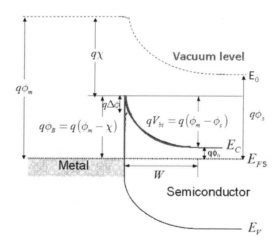

FIGURE 9.4

Metal and n-type semiconductor contact under thermal equilibrium for $\phi_m > \phi_s$ showing the Schottky barrier height (SBH) $q\phi_B$ created for electrons on the metal side and built-in potential, qV_{bi} for electrons on the semiconductor side. Barrier height lowering due to image force $\Delta\phi$ is also shown. (From Neamen, D.A., *Semiconductor Physics and Devices: Basic Physics*, 4th ed., McGraw-Hill, New York, 2012)

The difference between E_C and E_F can be found out by Joyce-Dixon approximation [3] given by

$$-\frac{E_C - E_F}{kT} \approx ln\frac{n}{N_c} + \sum_{m=1}^{4} A_m \left(\frac{n}{N_c}\right)^m \tag{9.6}$$

Equation (9.6) can be approximated to two terms to get the following equation.

$$E_C - E_F = kT\left[ln\left(\frac{n}{N_c}\right) + \frac{1}{\sqrt{8}}\frac{n}{N_c} \right] \tag{9.7}$$

or Nilsson approximation [4] that takes into account degenerate doping levels, as well, expressed in Equation (9.8).

$$E_C - E_F = q\phi_n = kT\left[\frac{lnr}{1-r} + \left(\frac{3r\sqrt{\pi}}{4}\right)^{\frac{2}{3}} + \frac{8r\sqrt{\pi}}{3\left(4 + r\sqrt{\pi}\right)^2} \right] \tag{9.8}$$

where r is n/N_C, n is the carrier concentration, and N_C is the effective density of states given by the following equation

$$N_c = 2\left(\frac{m^*T}{2\pi\,\hbar^2}\right)^{3/2} \tag{9.9}$$

where m^* is the effective mass of the electron.

9.3.2 Schottky Diode Current-Voltage Characteristics

The forward current density in a Schottky diode may consist of a number of current conduction mechanisms; thermionic emission (TE) current J_{TE}, tunneling currents (J_{TFE} and J_{FE}), generation-recombination current J_{GR}, and shunt leakage current J_{SH} [5].

$$J_{Total} = J_{TE} + J_{TFE} + J_{FE} + J_{GR} + J_{SH} \tag{9.10}$$

The I-V characteristic of an ideal Schottky diode is explained in terms of thermionic emission mechanism. It is analogous to the PN junction equation but has a different saturation current density (J_s). It is a function of a material specific Richardson constant (A^*) and Schottky barrier height. An ideality factor (n) is also included, the value for which is $1 < n < 2$ for PN junction diodes and close to 1 for an ideal Schottky diode. The current density is J is given by

$$J = J_s\left(e^{qV/nkT} - 1 \right) \tag{9.11}$$

$$J_s = A^* T^2 e^{\left(-q\phi_B / nkT \right)} \tag{9.12}$$

where, J_s is the saturation current density of emission (A/cm^2), $q\Phi_B$ is barrier height (eV), k is Botlzmann's constant $8.62 \cdot 10^{-5}$ eV/K, and A^* is Richardson constant (A/cm^{2}K^2) given by

$$A^* = \frac{4\pi q m^* k^2}{\hbar^3} \tag{9.13}$$

where, m^* is the effective mass of an electron, q is electron charge, and h is Planck's constant. A lower Richardson constant implies lower effective electron mass, which translates to better on resistance, R_{on} and lower leakage currents.

A correction factor ($\Delta\phi$) needs to be incorporated into the barrier height to account for barrier lowering due to image force, given by

$$\Delta\phi = \sqrt{\frac{qE}{4\pi\varepsilon_s}} \tag{9.14}$$

where E is the electric field at the semiconductor surface. The effect of image force is more pronounced for low barrier height contacts. The saturation current density with image correction factor is given by

$$J_s = A^* T^2 e^{-q(\Phi_B - \Delta\Phi) / nkT} \tag{9.15}$$

Thermionic emission theory assumes that the electrons will cross the barrier, provided they have energy greater than the barrier, and if they are near it with velocity in the direction of current flow. The shape of the barrier is hereby ignored.

In the case of this n-type semiconductor, the minority carriers are holes. Diffusion (generation-recombination) current is also part of the total current but is generally considered negligible as it is orders of magnitude lower than the thermionic emission current. The presence of only one type of carrier (electrons in this case) enables Schottky diodes to operate at frequencies up to 1000 times higher than conventional PN junction diodes. As the depletion width approaches electron de Broglie wavelength due to the increase in semiconductor doping, tunneling starts to contribute to the MS current. Tunneling currents are discussed in Section 9.4.3.

9.3.3 Structure of Power Schottky Diodes

Figure 9.5 describes the schematic structure of a basic Schottky diode. For power diodes, only n-doped semiconductors are used in practice because of the higher mobility of electrons [6]. Epitaxy is generally used for the low-doped n$^-$ layer, which is grown on a substrate with high n+ doping. Real Schottky diodes need additional junction termination structures. The low-doped layer, x_B, called the drift region must sustain the reverse voltage. Since a Schottky diode is a unipolar device, the ohmic voltage drop in the forward direction is directly proportional to the width of the drift region.

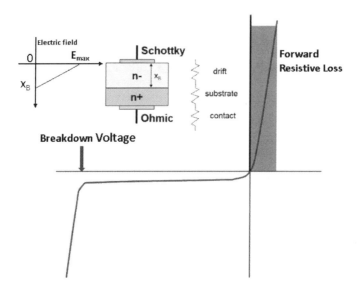

FIGURE 9.5
Structrfure of a basic Schottky diode with its forward and reverse I–V characteristics.

The on resistance, R_{on}, is dominated by the drift region and the contact resistance. Breakdown happens at a reverse voltage V_{br}, when E_{max} reaches the critical electric field E_{crit}. As both the drift region resistance and breakdown voltage depend on doping, R_{on} can be expressed as

$$R_{on} = \frac{4V_{br}^2}{\varepsilon_s \mu_e E_{crit}^3} \tag{9.16}$$

where μ_e is electron mobility in the drift region and E_{crit} is the critical electric field. The ratio $(V_{br})^2/R_{on}$ is known as the Baliga Figure of Merit (BFOM). Thus, a high breakdown voltage and low on-resistance is needed to get the optimum BFOM.

9.3.4 Gallium Nitride

Compound wide bandgap material systems, like GaN and SiC, have several advantages over silicon in high power and high-frequency applications [7]. The ability of III–V semiconductor devices to operate successfully at high voltages and high temperatures make them the optimum choice for power semiconductor devices. A comparison is shown in Table 9.1 [7,8]. GaN has a higher electron mobility, wider bandgap, and higher breakdown field. BFOM for GaN is ~187 times the value of silicon (Table 9.1). There are multiple FOMs to compare the performance and capabilities of various devices and material systems. Johnson's figure of merit (JFOM), is a measure of the ultimate high-frequency capability of the material. A combined figure of merit (CFOM), accounts for high frequency, high power, and high temperature performance (Figure 9.6).

The critical electric field E_{crit}, is defined as the maximum field in a one-sided junction at the onset of avalanche breakdown. In such a junction, the electric field profile is a linear function of position, and the maximum field occurs at the junction itself [9]. The critical electric field, in turn, scales approximately as the square of the semiconductor bandgap [10].

TABLE 9.1

Properties of Si, SiC-4H, and GaN

Properties	Si	SiC-4H	GaN
Bandgap (eV), E_g	1.12 Indirect	3.26 Indirect	3.45 Direct
Electron mobility (cm²/V-sec), μ_e	1500	1000	1250
Critical electric field (MV/cm), E_{crit}	0.25	3	4
Effective electron mass (kg), m[a]	$1.18m_0$	$0.77m_0$	$0.22m_0$
Richardson's constant (Acm⁻²K⁻²) A[a]	131	93	26
Dielectric permittivity, ε_s	$11.9\varepsilon_0$	$9.66\varepsilon_0$	$9.5\varepsilon_0$
Thermal conductivity (W/cmK), κ	1.5	4.9	1.3
BFOM[a], $\varepsilon_s\mu_e E_{crit}^3/4$	1.0	223.1	186.7
JFOM[a], $E_{crit}v_{sat}/2\pi$	1.0	215	215
CFOM[a], $\kappa\varepsilon_s\mu_e v_{sat}E_{crit}^2$	1.0	458	489

[a] Relative to values in silicon.

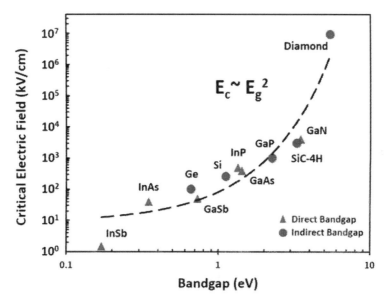

FIGURE 9.6
Critical electric field (kV/cm) as a function of semiconductor bandgap (eV). (From Ozpineci, B. and Tolbert, L.M., *Comparison of Wide-Bandgap Semiconductors for Power Electronics Applications*, Oak Ridge National Laboratory, U.S. Department of Energy, Washington, DC, Report 4/5, 2003; Tsao, J.Y., *Adv. Electron. Mater.*, 4, 1600501, 2018.)

Although SiC is comparable to GaN when it comes to switching properties, the performance of GaN is better with increasing temperatures [11]. The wider bandgap of GaN implies that the forward voltage of diodes is more than SiC, and the breakdown voltage is lower than Si and SiC, but a lot of engineering can be done to improve these characteristics. Even if GaN devices are made on suitable substrates with effective heat dissipation, SiC is much better at higher temeperatures in this regard. GaN is found in zinc-blende (FCC) and wurtzite (HCP) crystal structures, but the latter is the most common form.

9.4 Advanced Models for AlGaN/GaN

The GaN/AlGaN interface has excellent electron transport properties due to the formation of a two-dimensional electron gas (2DEG) [12] explained in Section 9.5.2. GaN Schottky diodes have a problem of high leakage current under reverse bias. This was earlier explained by thermionic emission (TE) model based on the Schottky barrier height and metal work function (ϕ_m) explained in the earlier section. It was later found out that a more accurate model was required to explain the current mechanisms in a GaN Schottky junction. This section provides insight into the different models used to explain the current mechanisms in III-V nitride semiconductors, with a focus on the AlGaN/GaN system.

9.4.1 Gap States

The insensitivity of ϕ_m to ϕ_B gave rise to various models, which then led to the discovery of gap states present at the MS interface. Fermi-level pinning (explained in detail later) is caused by these states at the interface due to which the metal work function has no effect on the barrier height. These gap states can either be metal induced or defect induced. This leads to the assumption of a thin interfacial layer at the surface of the interface to account for this pinning due to the interaction between metal and the semiconductor surface.

The metal-induced gap states (MIGS) model proposes the overlap of the conduction band of metal and semiconductor due to the tailing of the metal electron wave function into the semiconductor. The Fermi level pinning can also be due to defect induced gap states (DIGS) generated from the metal deposition process. Both these models have similar pinning mechanisms; however, the reason for gap states is different: DIGS have an extrinsic pinning mechanism while MIGS is intrinsic [13].

The barrier height is highly influenced not only by the metal work function, but also by the interface states. Two assumptions can be made while evaluating the barrier height: (a) a thin interficial layer (shown by δ in Figure 9.7) between the metal and semiconductor, which is transparent to electrons although it can withstand a potential across it, and (b) the interface states are just a property of the semiconductor and are independent of the metal. Figure 9.7 shows a metal-semiconductor contact with an interfacial layer [14].

where the various quantities depicted in Figure 9.7 are

ϕ_{BO} = Barrier height without image force lowering

ϕ_B = Barrier height with image force lowering

E_{CNL} = Charge neutrality level

ϕ_0 = E_{CNL} measured from vacuum level

N_{IS} = Density of interface states per unit energy (cm^{-2}eV)

ϕ_N = Neutrality level (above E_v)

Δ = Potential across interfacial layer

δ = Thickness of interfacial layer

Q_{sc} = Space charge density in semiconductor

Q_{ss} = Interface trap charge density

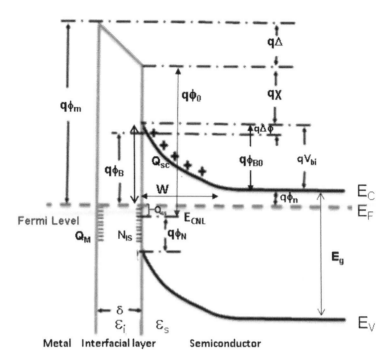

FIGURE 9.7
Band structure of MS contact with interfacial layer and surface states. (From Szatkowski, J. and Sierański, K., *Solid-State Electron.*, 35, 1013–1015, 1992.)

Q_M = Surface charge density on metal

ε_i = Permittivity of interfacial layer

ε_s = Permittivity of semiconductor

The charge neutrality level (CNL) is the boundary energy below which the charge states are donor-like and above which the charge states are acceptor-like. CNL is the weighted average of the density of states. It is repelled by a large density of states in the valence or conduction band. It is an important reference point for interfaces.

The Schottky barrier height (ϕ_B) to the n-type semiconductor can be given by the following expression:

$$\phi_B = S(\phi_m - \chi) + (1-S)(\phi_o - \chi) = (\phi_o - \chi) + S(\phi_m - \phi_o) \tag{9.17}$$

where "S" is the slope of ϕ_B versus ϕ_m graph shown in Figure 9.8, ϕ_o is the location of the CNL from the vacuum level, which is universally accepted to be at 4.5 eV [6]. The slope describes the degree of fermi-level pinning.

$$S = \frac{\varepsilon_i}{\varepsilon_i + q^2 \delta N_{is}} \tag{9.18}$$

N_{is} is the density of states at the CNL due to the gap states (MIGS/DIGS), given by:

$$N_{is} = \frac{\varepsilon_i (1-S)}{Sq^2\delta} \tag{9.19}$$

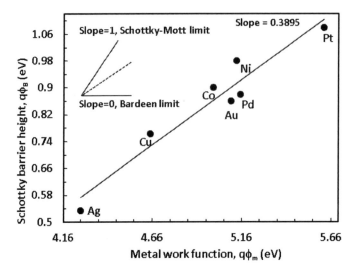

FIGURE 9.8
Schottky barrier height (SBH) as a function of metal work function. (From Morkoç, H., *Nitride Semiconductor Devices: Fundamentals and Applications*, Wiley-VCH, Weinheim, Germany, 2013.)

In the case of an ideal MS interface, the N_{is} is "0," which gives the value of S as "1," implying that the contact is at the ideal Schottky-Mott limit. Consequently, if N_{is} is "∞," the slope S becomes "0" corresponding to strong pinning called the Bardeen limit. The graph in Figure 9.8 shows the dependence of metal work function on barrier heights to various Schottky contacts to GaN. The slope is on the higher side, which suggests that surface states play a very influential role in GaN.

9.4.2 Thin Surface Barrier Model (TSB)

Hasegawa et al. [13] developed a model for III-V nitride semiconductors, which involves a Thin Surface Barrier (TSB) near the semiconductor surface, shown in Figure 9.9. This model considers a high density of defect donors near the semiconductor surface that act as dopants. Oxygen impurities, nitrogen vacancies, or impurities during the processing of GaN/AlGaN could be some of the reasons for these defects. Ionization of these donors reduces the width of the Schottky barrier, increasing tunneling probability of electrons by various tunneling mechanisms like thermionic field emission (TFE) and field emission (FE).

9.4.3 Tunneling Currents

The electrons with energy lower than $q\phi_B$ can tunnel through the Schottky barrier by thermionic field emission (TFE) or field emission (FE). When tunneling occurs at the Fermi level (E_f) of the metal it is FE, whereas when electrons with higher energy tunnel it is TFE. Various other mechanisms like trap-assisted tunneling (TAT), dislocation assisted tunneling, hopping (HOP), Fowler-Nordheim tunneling and Frankel-Poole emission (PFE) are also responsible for tunneling through the barrier shown in Figure 9.9. Current flow in the forward bias regime was traditionally explained by TE only. Now, it is universally accepted that other mechanisms also play a role in the flow of electrons. Trapping states caused by defects like threading dislocations propagated to the surface and other contamination

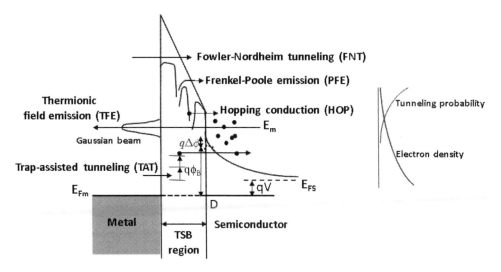

FIGURE 9.9

TSB Model with various current transport mechanisms. (From Eller, B.S. et al., *J. Vac. Sci. Technol. A*, 31, 050807, 2013.)

issues could be the reasons behind this. Generally, the flow of carriers is due to emission (TE, TFE) and trap-assisted tunneling (TAT, HOP), which are functions of temperature. TAT usually happens at lower temperatures, whereas TE at higher temperatures.

Tunneling of electrons through a barrier is purely a quantum mechanical phenomenon. The tunneling current or FE current is expressed as

$$J = \frac{A^*T}{k}\left(\exp\left(\frac{E_{FS}}{kT}\right) - \exp\left(\frac{E_{FM}}{kT}\right)\right)\int_0^\infty \exp\left(-\frac{E_x}{kT}\right)T(E_x)dE_x \qquad (9.20)$$

where $T(E_x)$ is the tunneling transmission probability. Using the Wentzel-Kramers-Brillouin (WKB) approximation, and assuming that the barrier is triangular, the expression for tunneling is given as

$$T = \exp\left(\frac{-4q\phi_B}{3E_{00}}\right) \qquad (9.21)$$

$$E_{00} = \frac{q\hbar}{2}\sqrt{\frac{n}{\varepsilon_s m^*}} \qquad (9.22)$$

where, n is electron concentration in cm^{-3}, m^* is effective electron mass and ε_s is relative permittivity of material. The ratio kT/E_{00} determines if tunneling or thermionic emission is the dominating phenomenon. If $E_{00} \gg kT$ then field emission dominates, else if $E_{00} \ll kT$ thermionic emission dominates (Figure 9.10).

The electron concentration for Si can be approximated to the value of doping concentration (N_D) assuming complete ionization but not for wide bandgap semiconductors like GaN [16]. The electron concentration in wide bandgap semiconductors can be given by Equation (9.23) assuming that $n = N_D - N_A$.

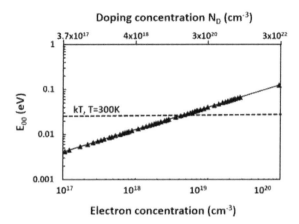

FIGURE 9.10

E_{00} versus electron and doping concentration in GaN.

$$n = \frac{2(N_D - N_A)}{1 + \alpha N_A + \sqrt{(1 + \alpha N_A)^2 + 4\alpha(N_D - N_A)}} \qquad (9.23)$$

where:

$$\alpha = \frac{2}{N_C} \exp\left(\frac{E_d}{kT}\right) \qquad (9.24)$$

N_C is the effective density of states in the conduction band. The values of N_C at 300 K for Si, GaN, and SiC are 2.86×10^{19} cm^{-3}, 2.24×10^{18} cm^{-3}, and 1.7×10^{19} cm^{-3}, respectively [16]. E_d is the electron energy level of the donors. At lower values of E_d, the electron concentration is high, while at higher values of $E_d > 100$ meV, the electron concentration decreases considerably. The graph in Figure 9.11 is plotted using Equations (9.23) and (9.24) and shows the electron concentration versus doping for different E_d. Acceptor concentration N_A is assumed to be 2×10^{17} cm^{-3}. For silicon, it is safe to assume $n = N_D$ at $T = 300$ K but not for GaN. The electron concentration is lower than the doping concentration in GaN.

In the case of pure thermionic emission, the transmission probability over the barrier is explained by a step function. Current density for an applied voltage "V" is shown by Equation 9.25.

$$J_{TE} = A^{**}T^2 \exp\left(\frac{-q\phi_B}{kT}\right)\left[\exp\left(\frac{qV}{kT}\right) - 1\right] \qquad (9.25)$$

The modified reduced Richardson constant (A^{**}) is used to consider additional effects, such as reflection probability and diffusion. Its value is generally lower than that of A^*. Current densities for TFE and FE depend on ϕ_B and doping given by the following expressions [17]

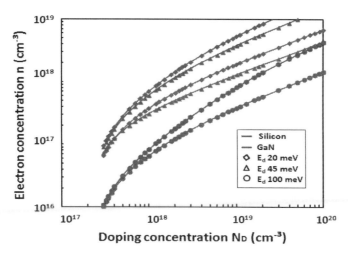

FIGURE 9.11
Electron concentration versus doping concentration for Si and GaN for different E_d at T = 300 K.

$$J_{TFE} = \frac{A^{**}T\sqrt{\pi E_{00}q(\phi_B+\phi_n-V)}}{k\cdot\cosh\left(\dfrac{E_{00}}{kT}\right)}\exp\left[\frac{q\phi_B}{kT}-\frac{q(\phi_B-\phi_n)}{E_0}\right]\exp\left(\frac{qV}{E_0}\right) \tag{9.26}$$

$$J_{FE} = \frac{A^{**}T\pi\exp\left[\dfrac{-q(\phi_B-V)}{E_{00}}\right]}{c_1 k\sin(\pi c_1 kT)}\cdot\left[1-\exp(-c_1 qV)\right] \tag{9.27}$$

where the variables c_1 and E_0 are defined as

$$c_1 = \frac{1}{2E_{00}}\ln\left(\frac{4\phi_B}{\phi_n}\right) \tag{9.28}$$

$$E_0 = E_{00}\coth\left(\frac{E_{00}}{kT}\right) \tag{9.29}$$

In the forward current voltage characteristics, FE dominates at lower voltages and TFE at higher voltages. The thin barrier provides more tunneling and thus an increase in current is observed. While at lower voltages, the electrons available for conduction are more, as depicted in the right inset in Figure 9.9.

In the presence of series resistance R_S and leakage path giving a shunt resistance, R_{SH} in parallel, the device I–V characteristics can be written as [18]

$$I_{TE} = I_{0,TE}\left(\exp\frac{qV-IR_s}{kT}-1\right) \tag{9.30}$$

$$I_{TFE} = I_{0,TFE}\left(exp\frac{qV - IR_s}{E_0} - 1 \right) \tag{9.31}$$

$$I_{GR} = I_{0,GR}\left(exp\frac{qV - IR_s}{2kT} - 1 \right) \tag{9.32}$$

$$I_{SH} = \frac{V - IR_s}{R_{SH}} \tag{9.33}$$

Figure 9.12 presents the measured and modeled I–V characteristics of AlGaN/GaN Schottky diodes showing the various current mechanism contributions. The results show higher tunneling currents indicating that the dislocation-governed tunneling is dominant in AlGaN/GaN Schottky diodes [18].

9.4.4 Specific Contact Resistivity

The specific contact resistivity ρ_c (Ohm-cm^2) of the junction is defined as

$$\rho_c = \left(\frac{\partial J}{\partial V} \right)^{-1}\Bigg|_{V=0} \tag{9.34}$$

The specific resistivity for TE, TFE, and FE can be found by the following analytical expressions.

$$\rho_{c,TE} = \frac{k}{qA^*T}\exp\left(\frac{q\phi_B}{kT} \right) = 3.3 \times 10^{-6}\exp\left(\frac{q\phi_B}{kT} \right) \tag{9.35}$$

$$\rho_{c,TFE} = \frac{k^2}{qA^*} \cdot \frac{\sqrt{E_0}}{E_{00}\sqrt{\pi q\left(\phi_B + \phi_n \right)}}\cosh\left(\frac{E_{00}}{kT} \right)\exp\left(\frac{q\phi_B}{E_0} - \frac{q\phi_n}{kT} \right) \tag{9.36}$$

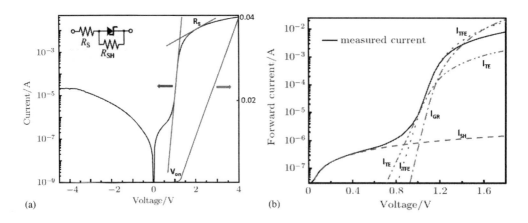

FIGURE 9.12
(a) Measured I–V curves for AlGaN/GaN Schottky diodes; (b) The fitting of different conduction components to the data in (a). (Reproduced from Yuan-Jie, L. et al., *Chin. Phys. B*, 23, 027101, 2014. With permission.)

$$\rho_{c,FE} = \frac{k^2}{qA^*} \cdot \left[\frac{\pi kT}{\sin(\pi c_1 kT)} \cdot \exp\left(-\frac{q\phi_B}{E_{00}}\right) - \frac{1}{c_1} \exp\left(-\frac{q\phi_B}{E_{00}} - c_1 q\phi_n\right) \right]^{-1} \quad (9.37)$$

These expressions can be simplified for GaN by evaluating all the constants at room temperature. The resistivities are in ohm-cm^2.

$$\rho_{c,TE} = 3.3 \times 10^{-6} \exp\left(\frac{q\phi_B}{kT}\right) \quad (9.38)$$

$$\rho_{c,TFE} = 2.845 \times 10^{-10} \cdot \left[\frac{\sqrt{E_0}}{E_{00}\sqrt{\pi q(\phi_B+\phi_n)}} \cosh\left(\frac{E_{00}}{kT}\right) \exp\left(\frac{q\phi_B}{E_0} - \frac{q\phi_n}{kT}\right) \right] \quad (9.39)$$

$$\rho_{c,FE} = 2.845 \times 10^{-10} \cdot \left[\frac{\pi kT}{\sin(\pi c_1 kT)} \cdot \exp\left(-\frac{q\phi_B}{E_{00}}\right) - \frac{1}{c_1} \exp\left(-\frac{q\phi_B}{E_{00}} - c_1 q\phi_n\right) \right]^{-1} \quad (9.40)$$

The Richardson constant A^* is taken to be 26 Acm^{-2}K^{-2}, and Area of the contact is assumed 1 cm^2. As stated earlier, E_{00} is a function of electron concentration for wide bandgap semiconductors. The Nilsson approximation is used to calculate ϕ_n.

The values of specific contact resistivity are calculated using Equations (9.38), (9.39), and (9.40) and are plotted in Figure 9.13. Field emission is the transport mechanism at high doping, whereas thermionic field emission dominates at lower doping values. A kink is observed when one emission mechanism takes over the other. Lower contact resistivities can be obtained at higher electron concentration (i.e., at high doping concentration), whereas contact resistivities increase at lower doping.

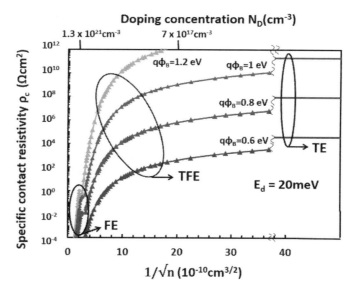

FIGURE 9.13
Specific contact resistivity as a function of electron and doping concentration for n-GaN with donor level of 20 meV.

Specific contact resistivity in thermionic emission is independent of electron concentration, hence the straight line. The contact resistivity is proportional to barrier height. Ohmic contacts require lower resitivities, hence lower barrier heights with high electron concentration are chosen. Schottky contacts require higher barrier heights. At lower barrier heights and high electron concentrations, very less field emission is observed.

9.5 GaN Based Schottky Diode Structures

Due to the band gap difference between AlGaN and GaN, a quantum well is formed at the junction that results in quantum confinement in the direction perpendicular to the junction, creating a two-dimensional electron concentration known as two-dimensional electron gas (2DEG). The 2DEG formed in AlGaN/GaN heterostructure has unique features, such as (1) a high density of two-dimensional electron gas with high electron mobility and saturation velocity, (2) a large breakdown field enabled by the wide bandgap, and (3) the high temperature stability and the chemical inertness providing the promise for an excellent reliability. These structures are created on GaN substrates.

9.5.1 Growth of GaN Substrates

Much of the earlier research into GaN-based devices was hindered due to a lack of suitable substrates, causing increased dislocation densities. Hexagonal GaN has a lattice constant of 3.19 Å. Generally, sapphire is the preferred substrate for GaN due to its hexagonal structure, availability, ease of handling, and stability at high temperatures, although the lattice mismatch is about 16%. A thicker layer of GaN must be grown to obtain a good quality structure. Many other substrates were also used to grow III-V nitrides, SiC, MgO, and ZnO, to name a few [19] (Table 9.2).

The growth of high-quality crystal of GaN has previously been done by Metal Organic Chemical Vapor Deposition (MOCVD), Molecular Beam Epitaxy (MBE), and Hydride Vapor Phase Epitaxy (HVPE). GaN substrates are made using HVPE, while the other two methods are usually used to deposit high-quality GaN films.

TABLE 9.2

Substrate Used for GaN/AlGaN Devices with Lattice Mismatch to GaN

Material	Symmetry	Plane	a Lattice Constant (Å)	% Mismatch
AlN	Hexagonal	(0001)	3.112	2.47
InN	Hexagonal	(0001)	3.544	−10.03
4H-SiC	Hexagonal	(0001)	3.073	3.77
6H-SiC	Hexagonal	(0001)	3.081	3.51
Sapphire	Trigonal	(0001) rotated 30°	2.747	16.1
ZnO	Hexagonal	(0001)	3.253	−1.97
MgO	Cubic	(111)	2.981	4.93
Si	Cubic	(111)	3.840	−16.96
GaAs	Cubic	(111)	3.997	−20.22

MOCVD relies on vapor transport. The precursors for the GaN/AlGaN growth are TriMethyl-Gallium (TMGa) and TriMethyl-Aluminum (TMAl). They flow over a heated substrate and react to form a thin film. It is a relatively faster process than MBE. On the other hand, MBE produces better films due to an ultra-high vacuum and *in situ* monitoring capabilities. The adsorption and diffusion of Ga and N atoms along the surface take place on the heated substrate to form a high-quality film. HPVE provides thick layers along with a high growth rate. The process of growth involves the reaction of liquid Gallium with HCl to form GaCl, which is later transported to the chamber along with NH_3, nitrogen, and hydrogen to form GaN [18].

GaN is inherently an n-type material due to the nitrogen vacancies in the material typically of the order of 10^{17} cm^{-3}. In the late 1990s, a lot of effort was put into increasing the incorporation of nitrogen. Silicon is used as a common n-type dopant for GaN while various p-type dopants like Mg, Zn, Li, Be were used to make p-GaN. The GaN system has very strong bond strength, hence most of the wet chemical etchants do not work well, though it has been stated that the use of hot alkali solutions etches GaN at slow rates. Reactive ion etch for GaN, which was not explored until the mid-1990s, is now widely used to etch III-V nitrides using Chlorine and Argon gas.

9.5.2 Two-Dimensional Electron Gas (2DEG)

The formation of two-dimensional electron gas (2DEG) does not need any intentional doping or other processes, unlike in the GaAs system, but it is polarization induced due to the thin (~20 nm) AlGaN layer (Figure 9.14).

The electrons for transport are provided by the surface states and are confined in a region. The carrier density is of the order of 10^{12} cm^{-2}. Since the carriers are restricted to this region only, a high breakdown voltage and higher currents are obtained. The density and mobility of carriers in 2DEG is determined by the mole fraction of Al (x) in $Al_xGa_{(1-x)}N$, the roughness at the interface, and dislocation densities. The density is directly proportional to the mole fraction of Al in AlGaN, but is also linearly related to leakage currents [20]. The current conduction mechanisms are illustrated in Figure 9.15.

9.5.3 Recessed Anode Structure

This structure is different from a conventional diode in the following ways. Firstly, the Schottky anode is recessed into the AlGaN layer with a thin AlGaN layer between the GaN and anode. Secondly, the anode is laterally extended for better reverse bias

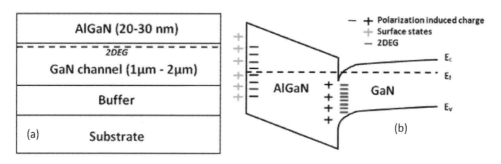

FIGURE 9.14
(a) AlGaN/GaN heterostructure; (b) Formation of 2DEG in GaN.

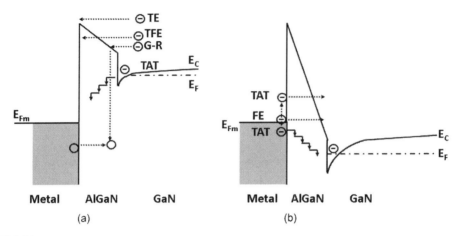

FIGURE 9.15

Metal/n-AlGaN/GaN Schottky diode under forward bias (a) and reverse bias (b) showing possible current conduction mechanisms.

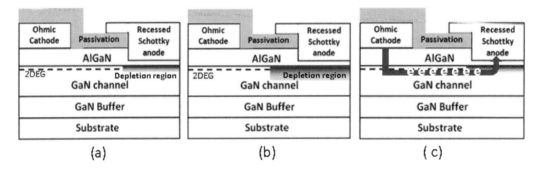

FIGURE 9.16

(a) AlGaN/GaN rectifier with recessed Schottky anode at zero bias applied at anode; (b) Depletion region extends under high reverse bias; (c) Electron flow under forward bias.

characteristics. Lastly, the presence of a passivation nitride dielectric. An applied reverse bias at the anode depletes the carriers in the 2DEG channel and pinches it off. This 2DEG facilitates the flow of electrons in the GaN channel from cathode to anode in the forward bias regime. The 2DEG under the gate region is determined by a gate (anode) voltage. The gate is not in contact with GaN, but has a small gap. The gap must be small enough for the zero biased gate to fully deplete the 2DEG (Personal communication with Dr. Ho-Young Cha, Hongik University, Korea, 2014). The operation of this device at difference bias voltage is shown below (Figure 9.16).

9.6 Metal Contacts

Generally, a metal with a higher work function (ϕ_m) than the semiconductor electron affinity (χ) is used for a Schottky contact, and a metal with lower work function is used for an ohmic contact. As stated earlier, due to the large amount of nitrogen vacancies in the GaN structure, it is inherently n-type. To make reliable high-temperature operation devices, excellent metal contacts to the semiconductor surface are needed. The surface of GaN is unique from other III-V semiconductors. Its highly ionic nature enables monotonous increase in the barrier height (ϕ_B) with corresponding increase in ϕ_m, although it does not scale proportionally [21]. The ohmic contacts to AlGaN require a rapid thermal anneal, whereas the Schottky contacts are as-deposited. The importance of a clean surface of the semiconductor prior to metal deposition cannot be understated. Various wet chemistries and techniques are used to make the surface free of oxides and other impurities, although the oxides formed on GaN are not always detrimental. The metals are deposited by thermal or e-beam evaporation.

9.6.1 Schottky Contacts

Metals with a higher ϕ_m like Au, Pt, Pd, Ni, Re, Co, and Cu are commonly used in research as Schottky materials to (Al)GaN, although Ni, Au, Pt, and Pd are the most favored. Schottky barriers at large voltage bias and high temperature have the problem of leakage current, due to the high electric field around the metal contacts. With an increase in reverse bias voltage, the high defect density in GaN causes defect-assisted tunneling. This is one of the major concerns addressed by the research groups working on GaN (Figure 9.17).

As stated in the earlier sections, the current transport mechanism in the Schottky contacts to AlGaN/GaN is not purely thermionic [25–27]. In the forward bias regime, tunneling is observed at low operating voltages and thermionic emission at higher voltages.

TE explained in the earlier section assumes that ϕ_B and ideality (n) do not depend much on temperature. If this is considered, then to achieve a barrier height of around 0.9 eV and n close to 1 is not possible. Hence, TE and TFE also take place in addition to TE. TE is only

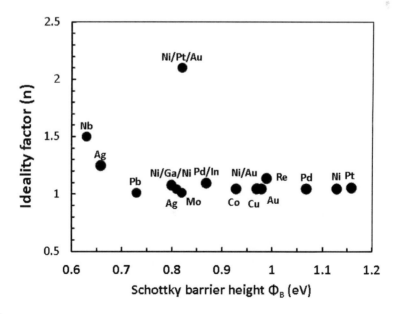

FIGURE 9.17
Ideality factor (n) versus Schottky barrier height (ϕ_B) in eV for different metal stacks. (From Quay, R. et al., *Gallium Nitride Electronics*, Springer, Berlin, Germany, 2008; Venugopalan, H.S. et al., *Semicond. Sci. Technol.*, 14, 757–761, 1999; Kampen, T.U. and Mönch, W., *Appl. Surf. Sci.*, 117, 388–393, 1997.)

valid in a semiconductor with moderate doping $<10^{18}$ cm^{-3} and contains few defects and surface charges. On the other hand, thermionic field emission (TFE), which includes tunneling through the barrier along with TE, is effective in higher doped ($>10^{19}$ cm^{-3}) semiconductors. GaN is unique in the sense that, although n-GaN is moderately doped, it exhibits TFE.

At a high reverse bias, the thermionic emission is reduced while other tunneling processes dominate. The high temperature excites the electrons to a trap state in the interface and then tunnel through or the electrons hop through the barrier. When tunneling through the semiconductor, some of the thermally excited electrons from the metal are nullified by the polarization charges, while others move toward the quantum well. As a result, the donor atoms in the AlGaN are ionized, and these electrons also move toward the quantum well, thus increasing the depletion region [28].

With the decrease in ϕ_B, the tunneling of carriers is greatly reduced and so is the forward voltage drop. Thus, these mechanisms are needed to lower the forward drop along with maintaining a high breakdown voltage and maintaining the quality of the functioning of diode (i.e., ideality factor). The ideality factor increases with temperature and also if TE is the only conduction mechanism.

9.6.2 Ohmic Contacts

The formation of reliable low-resistance ohmic contacts to the AlGaN surface is a key factor in the successful operation of the device. Good ohmic contacts are needed to reduce the knee voltage and heating of the contact, increase frequency of operation and current, improve thermal stability for high-temperature performance, and flat morphology for good edge acquity [29]. Lowering the barrier height or creating a thin barrier for tunneling by high doped AlGaN surface are the two ways by which a low resistance contact can be made to AlGaN. One of the concerns in contact technology is the formation of planar and stable contacts at high temperatures. Some of the commonly used ohmic metal stacks are stated in this chapter. The process factors affecting the contact resistance (R_c) are anneal temperatures, diffusion of metals, and the ratio of metals in the stack. The RTA temperatures, for most of the stacks, are between 800°C and 900°C in a N$_2$ or Ar ambient. The table below shows some of the metal stacks used in the structures investigated. Linear Transfer Length Method (LTLM) is widely used to measure the contact resistances [30]. Ti/Al-based metal stack is most commonly used with research groups focusing on optimizing the metal ratios and adding other metal barriers to improve the performance (Table 9.3).

The ohmic contact formation between the metal stack and AlGaN/GaN is primarily due to reactions with Ti/Al. It is usually explained by the following theories: (a) Ti reacts with the Nitrogen vacancies to form a highly ohmic TiN, or AlTi$_2$N, compound when annealed [35], (b) the N vacancies in AlGaN itself conduct [36], (c) depletion of N from AlGaN bulk to form a highly doped nitride surface at the MS interface, which allows bending of the bands for tunneling of electrons [35], (d) AlTi alloy penetrates into the AlGaN layer and touches the 2DEG at the AlGaN/GaN interface [36], (e) electrons are transported from the metal to the 2DEG through the TiN formed around the threading dislocations [40], and (f) The passivation layer acts as a source of N vacancies for Ti [36]. The authors proposed that, as the annealing temperature of the Ti/Al contacts reaches 520 K, the metal stack reacts with AlGaN increasing the sheet electron density (n_s) and the mobility (μ) by reducing the effective electron mass (m_e) in the 2DEG. This phenomenon is due to the tensile strain induced by the larger Al atom from the metal stack diffusing into the AlGaN layer resulting in an increase in the polarization charge at the AlGaN/GaN interface. As Al has a low melting

TABLE 9.3

Comparison of Substrates, Metal Stacks, Composition (nm), and Contact Resistance (ohm-mm)

Substrate	Stack	Composition (nm)	Contact Resistance (ohm-mm)
Silicon [31]	Ti/Al	—	0.35
Sapphire [32]	Ti/Al/Cr/Mo/Au	(25/125/50/45/55)	0.21
Unknown (not GaN) [33]	Ti/Al/Mo/Au	(15/60/35/50)	0.28
	Ti/Al/Ni/Au	(15/60/35/50)	0.44
Sapphire [34]	Ti/Al/Mo/Au	(15/90/40/-)	0.54
	Ti/Al/Ti/Au	(15/80/45/80)	0.48
Silicon Carbide [35]	Ti/Al/Pt/Au	(50/25/50/10)	1.73
	Ti/Al/Pd/Au	(-/-/-/-)	0.18
Sapphire [30]	Ti/Al/Ni/Au	(15/220/40/50)	0.15
Sapphire [36]	Ti/Al/Mo/Au	(15/60/35/50)	0.3
Sapphire [37]	Ti/Al/Ni/Au	(30/180/40/150)	0.2
Sapphire [38]	Ti/Al/Ti/Au	(20/120/40/50)	0.25
Silicon [39]	Ti/Al/Ni/Au	(15/60/35/50)	0.4
	Ti/Al/Pt/Au	(15/60/35/50)	0.4
	Ti/Al/Mo/Au	(15/60/35/50)	0.33

point of 660°C, the surface and edge exhibit a slight roughness, which affects the reliability of the stack due to poor topology and morphology.

Gold is used to reduce the oxidation of the stack and sheet resistance during the rapid thermal treatment. In one of the metal stacks, Au diffusion into AlGaN could have created a conduction path for electrons. This diffusion into the metal stack could also result in the formation of an alloy between Au and Al, termed as "purple plague alloy," hence the need of a diffusion barrier arises. Nickel is known to avoid the formation of this alloy. Cr, Mo, Pt, and Pd are some other diffusion barriers used in these devices. Chromium is a good oxide reducing agent and makes a good ohmic contact (Cr-N complex) to AlGaN. The incorporation of chromium in the Ti/Al/Cr/Mo/Au stack helps reduce the effect of anneal temperatures on the surface roughness caused due to Al. It is claimed that Ni/Al (1:1) stacks are as good as Ti/Al stacks, if not better [41]. Nickel, too, has the advantage of resisting oxidation, hence eliminating the need for Au and is not affected by the increasing mole fraction in $Al_xGa_{(1-x)}N$, which is responsible for leakage currents as stated in the earlier section. Ni, when incorporated as a barrier for Au in the Au/Ni/Al/Ti stack, swaps its location with Au during anneal and forms a NiAl top layer. In reference [29], the authors achieved the lowest contact resistance of 0.15 Ω-mm of all the work described in this paper at the lowest anneal temperature of 700°C in Argon ambient. Replacing Ni with Mo yielded better morphology and lower resistance (0.28 Ω-mm vs. 0.44 Ω-mm) at a lower temperature (850°C vs. 900°C) [33]. Similarly, Mo is also considered superior to Ti in terms of metal spilling, roughness and resistance [34].

Generally, for all the work reported in this section, anneal temperatures higher than 900°C were detrimental to the device in terms of increased resistance, possibly due to increased oxidation. An optimum temperature is required for every stack, which is a function of the metals, their ratio, and surface topology. The substrates also play an important role in the formation of a low resistance ohmic contact. The contact resistance on silicon was more than that on SiC and sapphire. A trade-off between the substrate cost and the current, frequency of operation, and delay of the device must be made.

9.7 Engineering V_f and V_{br}

To reduce the power losses in a device, it is necessary to reduce the forward drop, on-resistance, leakage currents, and improve the reverse breakdown characteristics. Various geometries such as recessed dual gate anode structures, nitride passivation, fluorine treatments, and so on are used to improve the performance of the device.

The Schottky anode can either be in contact with the AlGaN or directly to the 2DEG in GaN. When recessed into the AlGaN, the thin AlGaN layer between the 2DEG and the anode acts as a barrier for tunneling of electrons. On the other hand, the barrier height can be significantly reduced if the Schottky metal is directly in contact with the 2DEG in GaN [42]. The recessed gates with a dual gate structure usually have a lower turn-on voltage than conventional Schottky diodes, but a lower slope [43]. This is due to the fact that in recessed dual anode, the current at small forward voltages below the Schottky turn voltage $V_{on,Sh}$, ($V < V_{on,Sh}$) easily flows to the ohmic anode. When a forward bias becomes large enough to turn on the recessed Schottky contact and, thereafter, the unrecessed Schottky contact, the forward current is now composed of both ohmic and Schottky current components, as illustrated in Figure 9.18.

In conventional SBDs the electrons flow only through the Schottky gate due to absence of the ohmic anode. Thus, the current in the recessed gates is more due to an additional ohmic path. This predominantly affects the characteristics only at low forward bias and disappears at high forward voltages. It has been proved that the recessed gates have no effect on breakdown voltages of the devices. Reduction in the turn-on voltage has also been done by a fluorine plasma treatment [44]. The fluorine ions in the contact deplete the 2DEG channel below at reverse bias enhancing the "pinch-off" mechanism. In a conventional SBD, the region under the contacts can also be doped with silicon to enhance contact resistance (R_c) along with V_f, V_{br}, and R_{on} [45]. The only drawback of recessed contacts is the precision of etch chemistry needed. Since GaN has a strong bond strength, the development of chemistry for an anisotropic etch is always a challenge.

High breakdown voltage is achieved by increasing the distance between the anode and cathode and growing thick layers of semiconductors with low concentration of carriers. On the other hand, these parameters increase R_{on}. Thus, there always has to be a trade-off between these two terms to meet the required Figure of Merit (FOM). Impact ionisation is the reason for avalanche breakdown in power electronic devices. Methods for edge

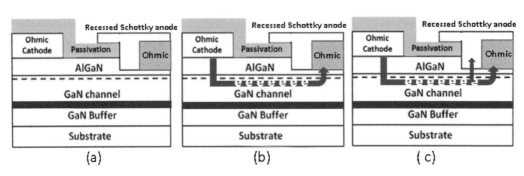

FIGURE 9.18
(a) AlGaN/GaN rectifier with dual Schottky ohmic anode and buried doped layer; (b) at forward bias $< V_{on,Sh}$; (c) at forward bias $> V_{on,Sh}$.

termination, like ring gaurds and field plates, are used in devices. Field plate engineering is commonly performed to enhance the performance of the device in reverse bias mode by extending the depletion region even further. The incorporation of a field plate has known to increase the breakdown voltage up to 5 times for similar device geometries [46]. The length of the field plate along with the nitride passivation play an important role in dictating breakdown voltages. The nitride film should be thick enough to sustain a high electric field without breaking down and thin enough for the electric field to influence the 2DEG. Thus, optimum nitride passivation thickness and field plate length needs to be chosen [47]. A magnesium-doped buried layer or a carbon doped barrier in the GaN channel also improves the breakdown characteristics [42,46]. The breakdown is also directly related to the diameter of the Schottky contact. Circular contacts help weaken the electric field at the periphery.

9.8 Conclusions

To conclude, GaN Schottky diodes have a huge potential in the power semiconductor industry due to their superior high voltage and temperature performance. On the other hand, suitable substrates, interface states, and defects continue to hinder the high-volume manufacturing of these devices. Various process techniques and geometries, such as recessed anode, field plate, and guard rings, can be used to achieve sub-0.4 V turn-on without compromising the breakdown voltage. The combination of AlGaN/GaN hetero-structures, 2DEG, choice of metal for ohmic, and Schottky contacts along with desired substrate can be used to realize a Schottky barrier diode for rectification applications.

References

1. Schmitz AC, Ping AT, Khan MA, Chen Q, Yang JW, Adesida I. Metal contacts to n-type GaN. *J Electron Mater.* 1998;27(4):255–260.
2. Neamen DA. *Semiconductor Physics and Devices: Basic Physics*, 4th ed. New York: McGraw-Hill, 2012.
3. Joyce WW, Dixon RW. Analytical approximations for the Fermi energy of an ideal Fermi gas. *Appl Phys Lett*, 1977;31:354.
4. Nilsson NG. An accurate approximation of the generalized Einstein relation for degenerate semiconductors. *Phys Status Solidi* 1973;A19:K75.
5. Sze SM, Ng KK. *Physics of Semiconductor Devices*, 3rd ed. New York: John Wiley & Sons, 2007.
6. Baliga BJ. *Advanced Power Rectifier Concepts*. Dordrecht, the Netherlands: Springer Science + Business Media, LLC, 2009.
7. Morkoç H. *Nitride Semiconductor Devices: Fundamentals and Applications*. Weinheim, Germany: Wiley-VCH, 2013.
8. Ozpineci B, Tolbert LM. *Comparison of Wide-Bandgap Semiconductors for Power Electronics Applications*. Washington, DC: Oak Ridge National Laboratory, U.S. Department of Energy, Report 4/5, 2003.
9. Hudgins HJ, Simin GS, Santi E, Khan MA. An assessment of wide bandgap semiconductors for power devices, *IEEE Trans. Power Elec.* 2003;18(3):907.
10. Tsao JY. Ultra-wide bandgap semiconductors: Research opportunities and challenges. *Adv Electron Mater.* 2018;4(1):1600501. doi:10.1002/aelm.201600501.

11. Amano H et al. The 2018 GaN power electronics roadmap. *J Phys D: Appl Phys*. 2018;51:163001 (48pp). doi:10.1088/1361-6463/aaaf9d.

12. Mishra UK et al. GaN Based RF power devices and amplifiers. *Proc IEEE*. 2008;96(2):287–305.

13. Hasegawa H, Akazawa M. Current transport, Fermi level pinning, and transient behavior of group-III nitride Schottky barriers. *J Korean Phys Soc*. 2009;55(3):1167–1179.

14. Szatkowski J, Sierański K. Simple interface-layer model for the nonideal characteristics of the Schottky-barrier diode. *Solid-State Electron*. 1992;35(7):1013–1015.

15. Eller BS, Yang J, Nemanich RJ. Electronic surface and dielectric interface states on GaN and AlGaN. *J Vac Sci Technol A*. 2013;31(5):050807.

16. Perez-Tomas A, Fontsere A, Placidi M, Jennings M, Gammon P. Modelling the metal-semiconductor band structure in implanted ohmic contacts to GaN and SiC. *Modell Simul Mater Sci Eng*. 2013;21(3).

17. Tung RT. Recent advances in Schottky barrier concepts. *Mater Sci Eng R*. 2001;35(1):1–138.

18. Yuan-Jie L, Zhi-Hong F, Zhao-Jun L, Guo-Dong G, Shao-Bo D, Jia-Yun Y, Ting-Ting H, Shu-Jun C. Comparison of electrical characteristic between AlN/GaN and AlGaN/GaN heterostructure Schottky diodes. *Chin Phys B*. 2014;23(2):027101.

19. Kung P, Razeghi M. III-nitride wide bandgap semiconductors: A survey of the current status and future trends of the material and device technology. *Opto-Electronics Rev*. 2000;8(3):201–239.

20. Terano A, Tsuchiya T, Mochizuki K. Investigation of relationship between 2DEG density and reverse leakage current in AlGaN/GaN Schottky barrier diodes. *Electron Lett*. 2012;48(5):274.

21. Schmitz AC, Ping AT, Khan MA, Chen Q, Yang JW, Adesida I. Schottky barrier properties of various metals on n-type GaN. *Semicond Sci Technol*. 1996;11(10):1464–1467.

22. Quay R, *Gallium Nitride Electronics*. Berlin, Germany: Springer, 2008.

23. Venugopalan HS, Mohney SE, DeLucca JM, Molnar RJ. Approaches to designing thermally stable Schottky contacts to n-GaN. *Semicond Sci Technol*. 1999;14(9):757–761.

24. Kampen TU, Mönch W. Barrier heights of GaN Schottky contacts. *Appl Surf Sci*. 1997;117:388–393.

25. Liu QZ, Lau SS. A review of the metal–GaN contact technology. *Solid State Electron*. 1998;42(5):677–691.

26. Yan D, Jiao J, Ren J, Yang G, Gu X. Forward current transport mechanisms in Ni/Au-AlGaN/GaN Schottky diodes. *J Appl Phys*. 2013;114(14).

27. Nam T, Jang J, Seong T. Carrier transport mechanism of strained AlGaN/GaN Schottky contacts. *Curr Appl Phys*. 2012;12(4):1081–1083.

28. Xiaoling Z, Fei L, Changzhi L, Xuesong X, Ying L, Mohammad N. High-temperature characteristics of AlxGa1−xN/GaN Schottky diodes. *J Semicond*. 2009;30:034001.

29. Bright AN, Thomas PJ, Weyland M, Tricker DM, Humphreys CJ, Davies R. Correlation of contact resistance with microstructure for Au/Ni/Al/Ti/AlGaN/GaN ohmic contacts using transmission electron microscopy. *J Appl Phys*. 2001;104(6):3143–3150.

30. Grover S, Effect of Transmission Line Measurement (TLM) Geometry on Specific Contact Resistivity Determination (2016). Thesis. Rochester Institute of Technology. Accessed from http://scholarworks.rit.edu/theses/9343.

31. Hajlasz M, Donkers JJTM, Sque SJ, Heil SBS, Gravesteijn DJ, Rietveld FJR et al. Sheet resistance under ohmic contacts to AlGaN/GaN heterostructures. *Appl Phys Lett*. 2014;104(24):242109.

32. Lan Y, Lin H, Liu H, Lee G, Ren F, Pearton SJ et al. Improved surface morphology and edge definition for ohmic contacts to AlGaN/GaN heterostructures, 2009.

33. Maeta R, Tokuda H, Kuzuhara M. Formation of low ohmic contacts to AlGaN/GaN heterostructures using Ti/Al-based metal stack. *IEEE International Meeting for Future of Electron Devices*, Kansai 2013.

34. Vertiatchikh A, Kaminsky E, Teetsov J, Robinson K. Structural properties of alloyed Ti/Al/Ti/Au and Ti/Al/Mo/Au ohmic contacts to AlGaN/GaN. *Solid State Electron*. 2006;50(7):1425–1429.

35. Hilton K, Han Y, Harrison I, Uren M, Balmer R, Martin T et al. Structural and electrical characterization of AuPtAlTi ohmic contacts to al ga N/ga N with varying annealing temperature and al content. *J Appl Phys*. 2008;103(7):074501.

36. Tokuda H, Kojima T, Kuzuhara M. Role of al and ti for ohmic contact formation in AlGaN/GaN heterostructures. *Appl Phys Lett.* 2012;101(26):262104.
37. Jacobs B, Kramer MCJCM, Geluk EJ, Karouta F. Optimisation of the Ti/Al/Ni/Au ohmic contact on AlGaN/GaN FET structures. *J Cryst Growth.* 2002;241(1):15–18.
38. Zhou XJ, Qiu K, Ji CJ, Zhong F, Li XH, Wang YQ. Low-resistance ohmic contact on undoped AlGaN/GaN heterostructure with surface treatment using CCl2F2 reactive ion etching. *Appl Phys Lett.* 2007;75(10):103511.
39. Mohammed FM, Wang L, Adesida I, Piner E. The role of barrier layer on ohmic performance of Ti/Al-based contact metallizations on AlGaN/GaN heterostructures. *J Appl Phys.* 2006;104(2):023708.
40. Tsuneishi K. A novel Ohmic contact for AlGaN/GaN using silicide electrodes. Master thesis, Tokyo Institute of Technology, Tokyo, Japan, 2013.
41. Ingerly DB, Chen Y, William RS, Takeuchi T, Chang YA. Low resistance ohmic contacts to n-GaN and n-AlGaN using NiAl. *Appl Phys Lett.* 2000;75(3):382–384.
42. Bahat-Treidel E, Hilt O, Zhytnytska R, Wentzel A, Meliani C, Wurfl J et al. Fast-switching GaN-based lateral power Schottky barrier diodes with low onset voltage and strong reverse blocking. *IEEE Electron Device Lett.* 2012;33(3):357–359.
43. Lee J, Park B, Cho C, Seo K, Cha H. Low turn-on voltage AlGaN/GaN-on-si rectifier with gated ohmic anode. *IEEE Electron Device Lett.* 2013;34(2):214–216.
44. Wong K, Chen K, Chen W, Huang W. High-performance al ga N/ga N lateral field-effect rectifiers compatible with high electron mobility transistors. *Appl Phys Lett.* 2008;92(25):253501.
45. Lian Y, Lin Y, Yang J, Cheng C, Hsu S. AlGaN/GaN Schottky barrier diodes on silicon substrates with selective si diffusion for low onset voltage and high reverse blocking. *IEEE Electron Device Lett.* 2013;34(8):981–983.
46. Tang C, Xie G, Zhang L, Guo Q, Wang T, Sheng K. Electric field modulation technique for high-voltage AlGaN/GaN Schottky barrier diodes. *Chinese Phys B.* 2013;22(10).
47. Remashan K, Huang W, Chyi J. Simulation and fabrication of high voltage AlGaN/GaN based Schottky diodes with field plate edge termination. *Microelectron Eng.* 2007;84(12):2907–2915.

10

Compound Semiconductor Oscillation Device Fabricated by Stoichiometry Controlled-Epitaxial Growth and Its Application to Terahertz and Infrared Imaging and Spectroscopy

Takeo Ohno, Arata Yasuda, Tadao Tanabe, and Yutaka Oyama

CONTENTS

10.1 Introduction

In the field of compound semiconductors, evaluation of the deviation from stoichiometry and its control are important because it seriously affects the electrical and optical characteristics [1]. In the case of gallium arsenide (GaAs), controlled vapor pressure of group V elements (e.g., arsenic in GaAs) has shown significant effects on the properties of a wide variety of compound semiconductor materials, while liquid phase epitaxy (LPE) have been performed on various III–V compounds [2–4]. The stoichiometry control is also quite important to impurity doping characteristics in the research field of vapor phase epitaxy (VPE). Indeed, it is reported that surface stoichiometry seriously affects the surface reaction of some organic metal compounds in GaAs growth [5–7].

Recently, semiconductor devices require ultra-thin layer and multi-layered structures. In other words, high-quality and heavily impurity-doped layers grown by low temperature process are desired. The crystal growth technology to satisfy those demands is essential for the fabrication of high-performance devices, such as terahertz (10^{12} Hz: THz) transistors [8,9] and oscillators [10,11]. To achieve precise film thickness control with atomic accuracy and stoichiometric composition, molecular layer epitaxy (MLE) is one of the most promising vapor phase epitaxial growth technologies that has the potential to realize a desired distribution of impurity concentration in the growth direction with an atomic order resolution. LPE growth with the temperature difference method under controlled vapor pressure (TDM-CVP) is also a useful method for compound semiconductor optical devices, such as light-emitting diodes (LED), laser diodes (LD), and photo diodes (PD). For example, in the case of mid- to far-infrared (IR) optical devices, narrow direct-gap compound semiconductors, such as lead telluride (PbTe), are grown by LPE technique, and the stoichiometry-controlled TDM-CVP growth provides the high performance required for IR devices.

The stoichiometric control effectively reduced dislocations and point defects in semiconductor crystals. By using these stoichiometry-controlled crystals, electronic and photonic devices have been realized in mid-IR and THz regions. These devices are listed in Table 10.1. Gallium phosphide (GaP) and gallium selenide (GaSe) also generate frequency-tunable THz waves based on different frequency mixing of non-linear optical effect. The stoichiometric control enables higher power generation from mid-IR to THz regions due to decrement of carrier density [12–17].

THz wave has high permeability for non-polar materials, which is expected to be used as non-destructive evaluation. Besides, THz wave has low quantum photon energy, so that it is safe for human tissues. Our group has created a database of THz permeability characteristics for industrial materials and successfully constructed non-destructive THz diagnosis of building blocks, insulated copper cable, hot-dip galvanized steel sheet, and banknotes [18–20]. Furthermore, the energy of a THz wave corresponds to molecular interactions, such as hydrogen bonding, van der Waals interactions, and lattice interactions. The lattice interaction is affected by a mechanical deformation of polymer. THz spectroscopy can be used for non-destructive diagnosis of mechanical deformation in polymers [21].

This chapter is organized as follows. In Section 10.2, an introduction of MLE, beryllium (Be) doping characteristics in MLE GaAs growth, and a tunnel-injection transit time effect diode (TUNNETT) THz oscillator are described. The introduction of TDM-CVP LPE, PbTe and PbSnTe growth, and IR LD are mentioned in Section 10.3. In Section 10.4, THz application is described. A summary is provided in Section 10.5.

TABLE 10.1

Electronic and Photonic Devices Based on Stoichiometry-Controlled Semiconductor Crystals

Approach	Device	Frequency	THz-wave Form (Operating Temp.)	Characters	Applications
Micro-electronic	TUNNETT	0.05–1 THz	CW (RT)	Low noise narrow linewidth G-bit modulation	THz communication imaging high-speed operation
Laser photonic	PbSnTe LD	20–50 THz	CW (LT)	High CW power	Spectroscopy imaging
	GaP	0.2–7.5 THz	ns pulse (RT) CW (RT)	High resolution sweepability	Spectroscopy imaging
	GaSe	0.1–100 THz	ns pulse (RT)	High resolution wide frequency sweepability	Spectroscopy imaging

10.2 GaAs Oscillator Fabricated by Molecular Layer Epitaxy (MLE)

10.2.1 MLE Growth

After the idea of atomic layer epitaxy (ALE) by Suntola and co-workers for the preparation of II–VI compound polycrystalline films on glass substrates [22,23], Nishizawa successfully applied the intermittent injection of trimethylgallium [$Ga(CH_3)_3$: TMG] and arsine (AsH_3) as precursors and achieved a mono-molecular epitaxial layer of GaAs for the first time [24]. This method of crystal growth is based on the chemical reactions of the adsorbates on semiconductor surfaces. The most important feature in MLE is self-limiting growth, which is based on self-limiting monolayer adsorption of the compound materials. In the case of GaAs MLE, it is possible to realize the self-limiting adsorption of stable monomethylgallium in a wide range of growth temperatures under a wide range of supply conditions of TMG and AsH_3. Later, the MLE method was extensively applied to silicon (Si) [25], indium phosphide (InP) [26], and antimony-related materials (GaAsSb, GaSb) [27].

Figure 10.1 shows the schematic drawing of an MLE GaAs growth system. The GaAs substrate is heated by an IR lamp. Typical growth temperatures vary from 250°C to 500°C and controlled by the IR pyrometer. The growth chamber was evacuated by a turbo molecular pump, and the background pressure was greater than 10^{-7} Pa. AsH_3 and TMG, or triethylgallium [$Ga(C_2H_5)_3$: TEG], are used as precursors for the epitaxial growth of GaAs. To grow a matrix GaAs layer, an intermittent supply of AsH_3 and TMG, or TEG, in an ultra-high vacuum is carried out. When a doping layer is formed, impurity dopant gas is separately supplied to the vacuum chamber. The evacuation duration of all gas sources is a period of several seconds in the MLE growth.

In conventional impurity doping of VPE, the matrix source gases and the dopant gas are simultaneously introduced into the chamber. Thus, the impurity concentration can be controlled only by changing the ratio of the flow rate, or gas pressure, of the dopant gas to that of the source gas. On the other hand, in the case of MLE growth, all precursors are separately introduced, and the monomolecular layer of the compound semiconductor is formed by chemical reactions on the crystal surface. Therefore, it is expected that the doping concentration is affected by not only the flow rate and pressure of gas, but also by the gas injection sequence and supply time. The MLE method has typically two sorts of

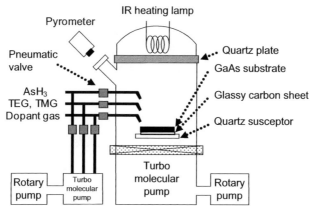

FIGURE 10.1
Experimental setup for doping MLE.

impurity doping modes [28,29]. First, the impurity gas was exposed after the GaAs surface was terminated by the exposure of Ga source gas, followed by evacuation. This is denoted as the after gallium (AG) mode. In this doping mode, the impurity gas is injected on the surface after TMG, or TEG, exposure, and then the surface is expected to be Ga-stabilized (Ga-terminated) just before the impurity gas injection. Second, the impurity gas was introduced after exposure to AsH_3, and this is denoted as the after AsH_3 (AA) mode. In the AA mode, the dopant gas is exposed on the As-stabilized surface.

In order to investigate the Be doping characteristics of GaAs from the viewpoint of the impurity doping mode related to surface stoichiometry, the sandwiched structure of undoped GaAs/Be-doped GaAs in the AG mode/undoped GaAs/Be-doped GaAs in AA mode/undoped GaAs was grown by MLE [30]. Figure 10.2 shows the impurity depth profile of a Be-doped GaAs multi-layered structure grown at 290°C on a (100) GaAs substrate, in which each doping layer was grown only by changing the impurity doping mode in one epitaxial process. Here, secondary ion mass spectroscopy (SIMS) analysis was applied to determine the Be concentration and its profile in the epitaxial layers. It is shown that the incorporation of Be is extremely enhanced when $Be(MeCp)_2$ is introduced after TEG exposure (AG mode), compared with that in AA mode.

By using the AG mode in Be doping, high carrier concentrations of p^+-GaAs can be obtained. Be doped p^+-GaAs grown by MLE at 290°C has shown carrier concentrations of up to $8 \times 10^{19}\,cm^{-3}$ and a Hall mobility of $42\,cm^2\,V^{-1}\,s^{-1}$ at 300 K, and the electrical activation ratio was almost 1 [30]. This result indicates that Be-doped GaAs MLE layers grown at low temperature have shown high carrier concentrations and good crystal quality. By using the Be-doped p^+-GaAs with high carrier concentration, sidewall p^+n^+ tunnel junctions with junction areas of around $10^{-8}\,cm^2$ were fabricated by area-selective regrowth MLE [31]. Figure 10.3a shows an SEM image of a fabricated sidewall tunnel junction. For an n^+-GaAs layer, Te and Sulfur (S) co-doping is effective in reducing the lattice strain in a heavily donor-doped GaAs epitaxial layer with a carrier concentration of $2 \times 10^{19}\,cm^{-3}$ [32]. The thickness of this layer was 50 nm, which determines the junction depth of this device. The area-selective regrowth of Be-doped p^+-GaAs was carried out to achieve a high carrier concentration of $8 \times 10^{19}\,cm^{-3}$. Before p^+-GaAs regrowth, the surface treatment, just prior to epitaxial regrowth, was carried out in a growth chamber at 350°C in AsH_3 atmosphere [33]. Figure 10.3b shows the typical current density-voltage characteristics of sidewall tunnel junctions. The tunnel junctions have shown record peak current densities of up to

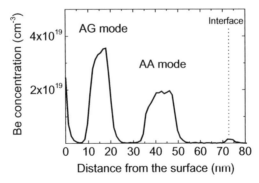

FIGURE 10.2
SIMS depth profile in multi-layered Be-doped GaAs.

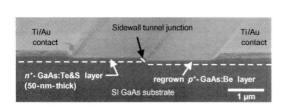

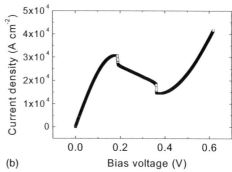

(a)   (b)

FIGURE 10.3

(a) Sidewall GaAs tunnel junction and (b) typical current density–voltage characteristic at RT. (Adapted from Oyama, Y. et al., Effects of AsH$_3$ surface treatment for the improvement of ultrashallow area-selective regrown GaAs sidewall tunnel junction, *J. Electrochem. Soc.* 151, G131–G135, Copyright 2004, with permission from The Electrochemical Society.)

31,000 A·cm^{-2} at RT [31,33]. In general, the performance of tunnel junctions (Esaki diodes) is affected by the quality of the semiconductor layer. Therefore, the results of present tunnel junctions suggest that Be-doped GaAs prepared by stoichiometry-controlled impurity doping is of good crystal quality.

10.2.2 GaAs TUNNETT THz Oscillator

TUNNETT diodes, invented by Nishizawa, are THz semiconductor oscillators based on transit-time delay and tunneling injection of electrons [11]. Figure 10.4a shows the TUNNETT device, which consists of a GaAs $p^+n^+in^+$ structure. Under a reverse bias, electrons in a TUNNETT diode are injected by tunneling from a p^+-GaAs anode to an i (or n^-)-GaAs drift layer through a depleted n^+ layer in the p^+n^+ tunnel junction. At a proper phase angle between the tunneling electron current and the displacement current in the tunneling part of the structure, the delay experienced by the electrons in the i-GaAs region results in the differential negative resistance, resulting in the diode becoming a generator. Figure 10.4b shows the radio frequency (RF) output–oscillation frequency trend for GaAs TUNNETTs fabricated by MLE growth. This oscillator in the fundamental mode reaches about 10 dBm (10 mW) at 0.1 THz, 0 dBm (1 mW) at 0.2 THz, and −44 dBm at 0.6 THz [18]. It is expected that optimization of the epitaxial structure can improve the TUNNETT oscillator performance. One of the important points is to provide a high quality p^+n^+ tunnel junction in the TUNNETT structure. The optimized cavity design should also enable oscillation at 1 THz and above, as well as increasing the generated output power.

The fabrication of a cavity-less TUNNETT oscillator has been performed [34]. Figure 10.4c illustrates a patch antenna-coupled GaAs TUNNETT oscillator for a THz-wave generation system. The fabricated patch antenna with dimensions of 390 μm × 440 μm with the bias pad of 500 μm × 500 μm was bonded to the Cu stem. Figure 10.4d shows the oscillation frequency and RF output power as a function of the bias current. The detected continuous wave (CW) frequency was 177–197 GHz with a bias current of 300–550 mA, and the output power was −35.3 dBm at 197 GHz.

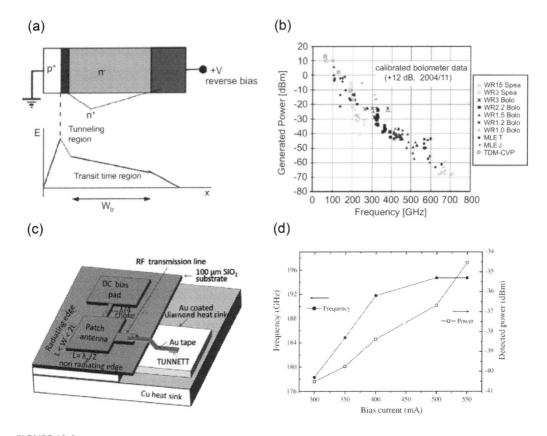

FIGURE 10.4
(a) TUNNETT oscillator and (b) RF power vs. frequency trend line for CW, fundamental-mode GaAs TUNNETTs. (Reprinted from *NDT&E International*, 42, Oyama, Y. et al., Sub-terahertz imaging of defects in building blocks, 28–33, Copyright 2009, with permission from Elsevier.) (c) Patch antenna coupled TUNNETT oscillator and (d) frequency and generated RF power against bias current. (Reprinted from *Solid-State Electron.*, 54, Balasekaran, S. et al., Patch antenna coupled 0.2 THz TUNNETT oscillators, 1578–1581, Copyright 2010, with permission from Elsevier.)

10.3 PbTe Laser Diode Fabricated by Temperature Difference Method under Controlled Vapor Pressure (TDM-CVP)

10.3.1 TDM-CVP Growth

$Pb_{1-x}Sn_xTe$ systems, group IV–VI compound semiconductor materials, are a known example with a bandgap (E_g) ranging from 0.1 to 0.3 eV; this E_g varies with varying composition and temperature (T). The E_g of $Pb_{1-x}Sn_xTe$ is estimated as [35]

$$E_g(x,T) = 171.5 - 535x + \sqrt{(12.8)^2 + 0.19(T+20)^2} \ [\text{meV}].$$

These materials are used in mid- to far-IR LED, LD, and PD. For achieving high performance, such as the performance required for IR devices, developing a robust and sophisticated crystal-growth technique is very important. One such technique, LPE, can be used to grow semiconductor crystals under a meta-stable phase; that is, from liquid to solid phase, throughout the crystal-growth process. In LPE, saturated source materials in a melt solution are grown epitaxially on substrate surfaces. LPE systems are not so expensive to produce such devices. Therefore, LPE is a suitable method to fabricate near-perfect crystals for semiconductors. Conventionally, PbTe or PbSnTe homo- or hetero-LD structures have been fabricated using the slow-cooling LPE method. Several previous studies have reported continuous lasing operation over long periods of time with PbTe/PbSnTe double hetero (DH) LDs grown using normal slow-cooling LPE or molecular beam epitaxy (MBE) [36–39]. However, adjustment of stoichiometric control during crystal growth is not easy in this approach. Particularly, in ternary system compounds, like $Pb_{1-x}Sn_xTe$, the ratio of component elements varies drastically depending on the temperature of the solution. Stoichiometric control is important because crystal defects introduced by nonstoichiometry can degrade the performance of the resulting semiconductor crystals. In addition, varying composition also causes variations in the energy band gap, which cause fluctuations in the wavelength of light emission or detection; that is, reducing the yield of the device's production process.

To address these problems, first, the TDM is proposed for epitaxial growth of group IV–VI compounds. This technique employs a stable temperature difference between the top and bottom of the melt solution (surface of substrates) using a subsidiary heater affixed on the melt cell with a cooling gas or water flow placed under the substrates. This temperature difference generates a constant density and thermal diffusion of source materials from the melts to the epitaxial crystal. Diffused source materials grow into an epitaxial crystal on the surface of the substrate without varying the component composition ratio and without causing either crystalline defects or misfit strain. In addition, growth can be achieved at a lower temperature using the TDM than conventional slow-cooling methods. Low-temperature conditions are better for achieving stoichiometry-controlled crystal growth. Moreover, the TDM is suitable for use in mass production because it does not require temperature control like the conventional slow-cooling method.

Second, to control the stoichiometry of the epitaxial crystal, vapor pressure is applied on top of the melt solution in a process called CVP. Usually, using a vapor pressing element is selected by higher evaporating pressure than lower one; for example, As is selected for GaAs crystal growth [1]. The vapor pressure is controlled by another heater, and the vapor pressure source room and the melt cells are connected by a quartz tube and are enclosed completely. With this configuration, the vapor pressure prevents the formation of nonstoichiometric defects in the epitaxial crystal. The vapor pressure is then optimized for fabricating devices, as pressures that are too high or low cause interstitial and vacancy point defects in the epitaxial crystals, respectively.

These unique techniques have been applied for III–V and II–VI compound semiconductors [1,2,40,41] for fabricating high-brightness red AlGaAs LEDs, pure green GaP LEDs, and InP epitaxial crystals. In this study, we introduce the application of these techniques for IV–VI semiconductors. PbTe and PbSnTe TDM-CVP LPE is performed using a horizontal slider graphite boat. The apparatus is equipped with a subsidiary heater and cooling N_2-gas flow under the boat to attain the temperature difference. In addition, the growth solutions are connected by a quartz tube with a Te reservoir that controls the vapor pressure applied on the growth solution. Since the vapor pressure of Te is higher than that of Pb or Sn, and a Pb-rich melt is used as usual, the Te vapor pressure could be used to control the stoichiometry. All of the systems are also conventional LPE, in quartz tube and Pd-purified H_2 flow atmosphere. The growth temperature is controlled by the main heater, and the Te vapor pressure is controlled by a second heater.

10.3.2 PbTe IR Laser Diode

In this section, a high-performance PbSnTe/PbTe DH mid-IR LD, which is fabricated using TDM-CVP LPE, is introduced [42]. The preparation and method of crystal growth were proposed in Section 10.3.1. The first growth layer is a Tl-doped (Tl concentration in the Pb-rich melts of 0.03 at%) PbTe p-type buffer with a cladding layer. The second growth layer is an active layer and is composed of Tl-doped p-type $Pb_{1-x}Sn_xTe$. Using the LD light emission wavelength, the Sn concentration, x, is changed from 0.05 to 0.11. Finally, a Bi-doped ($x_{Bi} = 2 \times 10^{19}\,cm^{-3}$, $n = 1 \times 10^{17}\,cm^{-3}$) n-type PbTe cladding layer is grown. T_g is set at 470°C, and a Te vapor pressure near the optimum value is applied to the melts to control the stoichiometry during crystal growth. Ohmic contacts are formed on both sides of the p- and n-type electrodes with Pt and Au electric plating. These samples are scribed around 500 µm × 500 µm with a diamond cutter to form a Fabry–Perot cavity. LDs are then mounted on laser stems with an indium solvent. The LD samples are placed in a cryogenic system, in which the temperature can be controlled between 15 K and RT. The duty ratio of the pulse current is set between 1/164 and 1/1840, and the pulse width is varied between 10 and 50 µs. The emission spectra are measured using a liquid-nitrogen-cooled HgCdTe detector.

Figure 10.5 shows a comparison of the lasing and spontaneous emission spectra in a $Pb_{0.95}Sn_{0.05}Te/PbTe$ DH LD. This figure demonstrates a very sharp lasing spectrum compared with the spectrum of spontaneous emissions. A longer peak shift in the lasing spectrum is also observed. Figure 10.6a shows the laser spectra of the LD, where the Sn concentration in the active layer is varied. Very sharp peaks in the lasing spectra are observed between wavelengths of 6.5 and 9.4 µm (x = 0–0.11) at 14 K without any

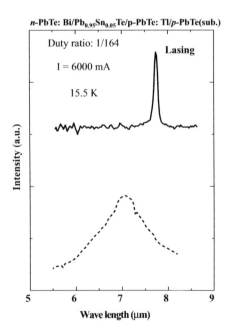

FIGURE 10.5
Comparison of the laser emission spectrum and spontaneous emission spectrum of a $Pb_{0.95}Sn_{0.05}Te/PbTe$ DH LD. (Reprinted from *Mat. Sci. Semicon. Proc.*, 27, Yasuda, A. et al., Lasing properties of PbSnTe/PbTe double hetero mid-infrared laser diodes grown by temperature difference method under controlled vapor pressure liquid-phase epitaxy, 159–162, Copyright 2014, with permission from Elsevier.)

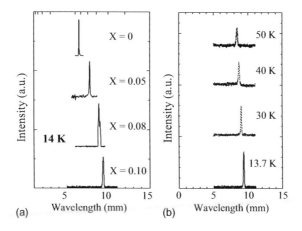

(a)   (b)

FIGURE 10.6

(a) Laser emission spectra of LDs with varying Sn concentrations in the active layer. (b) Temperature dependence of the laser emission spectrum of a $Pb_{0.89}Sn_{0.11}Te/PbTe$ DH LD. (Reprinted from *Mat. Sci. Semicon. Proc.*, 27, Yasuda, A. et al., Lasing properties of PbSnTe/PbTe double hetero mid-infrared laser diodes grown by temperature difference method under controlled vapor pressure liquid-phase epitaxy, 159–162, Copyright 2014, with permission from Elsevier.)

buffer layer. Figure 10.6b shows the dependence of the laser spectrum ($x = 0.11$) on temperature. Sharp lasing spectra are observed at temperatures of at least 50 K. The emission wavelengths are in accordance with the calculated value of PbSnTe band gap energy. These results demonstrate that DH LDs fabricated using TDM-CVP have very good quality and performance. Figure 10.7 shows the threshold current (I_{th}) as a function of temperature for DH LDs with a Bi-doped cladding layer. In these results, the I_{th} values of these LDs are lower than the I_{th} values of a LD grown by slow-cooling LPE [37]. We found that I_{th} values of DH-structure LDs are also lower than those of heavily Bi-doped homojunction LDs fabricated by TDM-CVP [43]. We can assume that

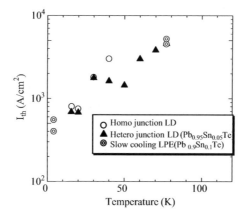

FIGURE 10.7

Threshold current densities for DH and homojunction LDs fabricated via TDM-CVP and DH LDs fabricated via slow-cooling LPE. (Reprinted from *Mat. Sci. Semicon. Proc.*, 27, Yasuda, A. et al., Lasing properties of PbSnTe/ PbTe double hetero mid-infrared laser diodes grown by temperature difference method under controlled vapor pressure liquid-phase epitaxy, 159–162, Copyright 2014, with permission from Elsevier.)

this phenomenon originates from the cladding effects of the carrier and the photons. Moreover, the $Pb_{0.95}Sn_{0.05}Te/PbTe$ DH LD operates at 77 K (liquid nitrogen temperature), at which temperature the threshold current density was approximately 4.7 kA/cm^2.

10.4 THz and IR Applications

10.4.1 THz Diagnosis of Building Blocks

A compact THz imaging system with the room-temperature operation 0.2 THz-band GaAs TUNNETT diode oscillator and its application for non-destructive and harmless inspections of timbers and concrete building blocks are shown in Figures 10.8 and 10.9, respectively [18]. Wood and concrete show a high transmittance in this frequency range, and the measured absorption coefficients correlate well with the densities. Then, the invisible grains, knots, and diffused water inside the timbers were detected. For concrete, diffused water are also detectable, as shown in Figure 10.9a. THz wave has shown a high sensitivity in detecting these defects in building blocks. Non-destructive THz diagnosis is available for invisible cracks in concrete. Figure 10.9b shows transmittance/reflection THz imaging of concrete, which includes artificial crack, where THz intensities are dependent on position and size.

10.4.2 THz Diagnosis of Insulated Copper Cable

At present, one of the main inspection methods of electric cables is visual inspection. The development of a novel non-destructive inspection technology is required because of various problems, such as water invasion by the removal of insulators. Since THz waves have high transparency to nonpolar substances, such as coatings of conductive cable, electric conductive cables are extremely suitable for THz non-destructive inspection.

Figure 10.10a shows THz transmission-reflection imaging of insulated copper cables (inset) that had been artificially damaged using a 0.14 THz IMPATT oscillator and a Schottky

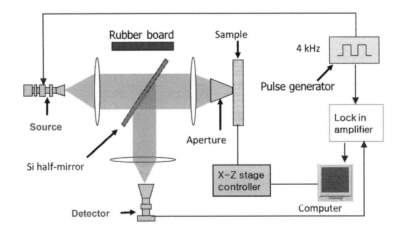

FIGURE 10.8
Reflection THz imaging system with the room-temperature operation 0.2 THz-band GaAs TUNNETT diode oscillator and Schottky barrier diode detector.

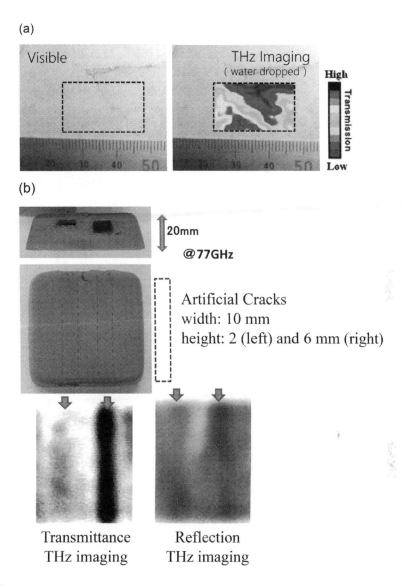

FIGURE 10.9
THz imaging of (a) diffused water and (b) artificial cracks inside concrete.

barrier detector. The internal insulated wires embedded in the opaque polyethylene can be clearly visualized by THz imaging. In defect regions, as the defect became larger, the reflection intensity decreased [19]. Besides, the reflection THz intensity is changed by the degree on corrosion. The reflectance is reduced due to surface oxide layer, as shown in Figure 10.10b, even though the oxide layer thickness is very thin compared with the wavelength of THz wave [44,45].

The quantitative detectivity for the disconnection gap is confirmed to be 0.5 mm in case of using 90 GHz GUNN diode [46]. The spatial resolution of 0.5 mm is beyond the diffraction limit, where the confocal optical configuration may be attributed to the effect of overcoming the THz wave length limit and excluding back-scattered light due to a circular aperture hole with 3 mm diameter.

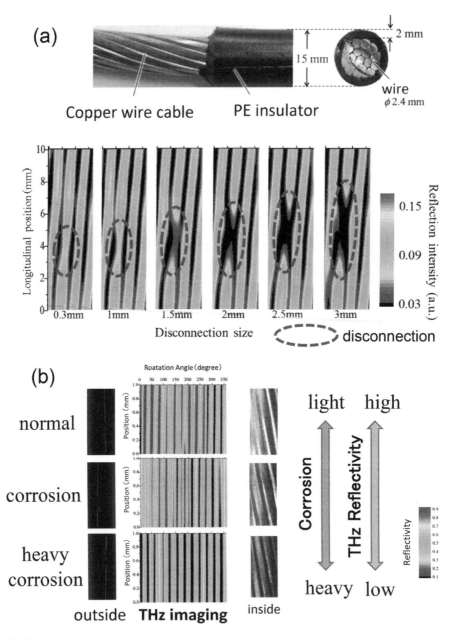

FIGURE 10.10

(a) Photographs of copper cable sample surrounded by the opaque insulating polyethylene (inset) and THz reflection imaging of the copper wires with different disconnection size and (b) that of the copper wires with different degree of corrosion.

10.4.3 THz Diagnosis of Hot-Dip Galvanized Steel Sheet

THz wave can be reflected from a metal surface covered by invisible paint, because THz wave has both the characteristics of a radio wave and light. Even the paint layer is not transparent in visible light, the reflection of THz wave detects the corrosion products under

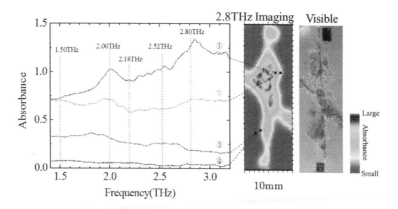

FIGURE 10.11
THz spectra of corrosion on coated steel surface (①② absorption bubble area, ③ bubble area, ④ red-colored corroded area)

the paint layer. Figure 10.11 shows 2.8 THz transmission reflection imaging of corrosion products on hot-dip galvanizing steel sheets covered by epoxy resin [20]. The sample is hot-dip galvanizing steel sheet with scratch damage on the surface, and this is corroded by 5 wt% NaCl solution spraying with 50°C for 5 days. Large absorption was observed around the scratch area. From the comparison of the THz imaging with XRF mapping, zinc hydroxychloride ($ZnCl_2 \cdot 4Zn(OH)_2$) and zinc chloride ($ZnCl_2$) can be estimated as the corrosion products. This result indicates that the THz wave technique is available for non-destructive novel corrosion diagnosis of a metal surface covered with invisible insulators or painted films [20,47].

10.4.4 THz Diagnosis of Banknotes

THz wave has high permeability for non-polar materials. THz spectroscopic imaging of banknote is expected for a non-destructive THz inspection. Ultrathin resin tape is attached to the forged banknotes of many currencies for the purpose of detecting forgeries and damage. The tape is approximately half to a quarter width of a human hair, and the most common method currently available for detecting this tape is with stylus profiling. Because physical contact is required with stylus profiling, the speed with which the tape can be detected is limited. And the physical contact also comes with the risk of damage. To prevent the forgery of banknotes, magnetic inks and security thread are used.

Figure 10.12a shows THz transmission imaging of banknotes at 2 THz [48]. The tape attached on each banknote is detected in each image, respectively. In the EURO banknote, security thread and magnetic ink are visualized. This detection is due to absorption of THz wave. The wide-frequency THz source is available to obtain spectra of banknotes. For imaging, as monochromatic THz source, semiconductor devices sources, such as TUNNETT, GUNN, and IMPATT oscillators, are expected for practical application because they are compact and operate with low electric consumption. THz imaging of banknotes with various tapes is shown in Figure 10.12b [48]. The imaging is measured by using 90 GHz wave from GUNN diode. Security thread and tapes appear in THz imaging.

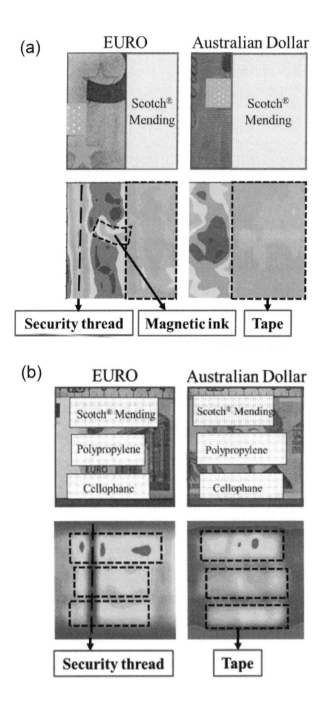

FIGURE 10.12
(a) Visible and THz images of banknotes at 2 THz and (b) visible and THz images of banknotes at 90 GHz.

10.5 Summary

In this chapter, we have discussed the stoichiometry-controlled crystal growth technique and its application to compound semiconductor oscillation devices.

In Section 10.2, we have explained the MLE method and its application for fabrication of the GaAs TUNNETT THz oscillator. By changing the impurity doping mode, stoichiometry-controlled MLE at low temperature can serve a heavily Be-doped p^+-GaAs layer with a carrier concentration up to 8×10^{19} cm^{-3} with good crystal quality. Using the Be-doped GaAs, we have achieved the record peak current densities of up to 31,000 A·cm^{-2} in the sidewall GaAs p^+n^+ tunnel junction. In addition, we have mentioned a GaAs TUNNNETT device with the p^+n^+ tunnel junction as an example of a THz oscillator.

In Section 10.3, we introduced the TDM-CVP LPE method and its application for fabrication of high-performance mid-IR Pb$_{1-x}$Sn$_x$Te LDs. In the fabricated LDs, the lasing peak was shifted to longer wavelengths than the spontaneous emission peak. Very sharp lasing spectra, between 6.5 and 9.4 µm (when x = 0–0.11), were obtained at 14 K. These results demonstrate the high quality and robust performance of the DH LDs fabricated by TDM-CVP. At different temperatures, DH LDs grown by TDM-CVP and LDs and homojunction LDs grown by the slow-cooling method exhibit similar threshold currents. We attribute this phenomenon to the efficiency of cladding. In addition, in Pb$_{0.95}$Sn$_{0.05}$Te/PbTe DH LDs operated at liquid nitrogen temperature, the threshold current density is approximately 4.7 kA/cm^2.

The expansion of IR and millimeter waves to the THz region by controlling semiconductor crystals is highlighted. The THz wave generators can be used in applications of non-destructive evaluation, as shown in Section 10.4.

Acknowledgments

We acknowledged fruitful discussions with Prof. Jun-ichi Nishizawa on the basis of his invention of molecular layer epitaxy and temperature difference method under controlled vapor pressure growth. We also want to thank Prof. Ken Suto for useful discussions.

References

1. J. Nishizawa and Y. Oyama. Stoichiometry of III–V compounds. *Mater. Sci. Eng. R* **12**, 273–426 (1994).
2. J. Nishizawa and Y. Okuno. Liquid phase epitaxy of GaP by a temperature difference method under controlled vapor pressure. *IEEE Trans. Electron Devices* **ED-22**, 716–721 (1975).
3. J. Nishizawa, Y. Oyama and K. Dezaki. Stoichiometry-dependent deep levels in *n*-type GaAs. *J. Appl. Phys.* **67**, 1884–1896 (1990).
4. Y. Oyama and J. Nishizawa. Excitation photocapacitance study of EL2 in *n*-GaAs prepared by annealing under different arsenic vapor pressures. *J. Appl. Phys.* **97**, 033705 (2005).

5. M. L. Yu. *Proc. 2nd Int. Symp. Atomic Layer Epitaxy*, Raleigh, NC (1993).

6. J. Nishizawa, H. Sakuraba and T. Kurabayashi. Surface reaction of trimethylgallium on GaAs. *J. Vac. Sci. Technol. B* **14**, 136–146 (1996).

7. M. Sasaki and S. Yoshida. Scattering of pulsed trimethylgallium beam from GaAs(100), -(110), and -(111)B surfaces. *Surf. Sci.* **356**, 233–246 (1996).

8. J. Nishizawa, Recent progress and potential of SIT. *Proc. 11th Conf. (1979 Int.) on Solid State Devices*, Tokyo, Japan, 3–11 (1979).

9. J. Nishizawa, P. Płotka and T. Kurabayashi. Ballistic and tunneling GaAs static induction transistors: Nano-devices for THz electronics. *IEEE Trans. Electron Devices* **49**, 1102–1111 (2002).

10. J. Nishizawa and Y. Watanabe. High frequency properties of the avalanching negative resistance diode. *Sci. Rep. Res. Inst. Tohoku Univ. B (Commun.)* **10**, 91 (1958).

11. J. Nishizawa, P. Płotka, T. Kurabayashi and H. Makabe. 706-GHz GaAs CW fundamental-mode TUNNETT diodes fabricated with molecular layer epitaxy. *Phys. Stat. Sol. C* **5**, 2802–2804 (2008).

12. Y. Oyama, T. Tanabe, F. Sato, A. Kenmochi, J. Nishizawa, T. Sasaki and K. Suto. Liquid-phase epitaxy of GaSe and potential application for wide frequency-tunable coherent terahertz-wave generation. *J. Cryst. Growth* **310**, 1923–1928 (2008).

13. K. Saito, T. Tanabe, Y. Oyama, K. Suto, T. Kimura and J. Nishizawa. Terahertz-wave absorption in GaP crystals with different carrier densities. *J. Phys. Chem. Solids* **69**, 597–600 (2008).

14. A. Kenmochi, T. Tanabe, Y. Oyama, K. Suto and J. Nishizawa. Terahertz wave generation from GaSe crystals and effects of crystallinity. *J. Phys. Chem. Solids* **69**, 605–607 (2008).

15. T. Onai, Y. Nagai, H. Dezaki and Y. Oyama. Liquid phase growth of bulk GaSe crystal implemented with the temperature difference method under controlled vapor pressure. *J. Cryst. Growth* **380**, 18–22 (2013).

16. K. Saito, Y. Nagai, K. Yamamoto, K. Maeda, T. Tanabe and Y. Oyama. Terahertz wave generation via nonlinear parametric process from ε-GaSe single crystal grown by liquid phase solution method. *Optics Photonics J.* **4**, 213–218 (2014).

17. Y. Nagai, K. Maeda, K. Suzuki and Y. Oyama. Comparative study of shallow acceptor levels in unintentionally doped p-type GaSe crystals prepared by the Bridgman and liquid phase solution growth methods. *J. Elec. Mat.* **43**, 3117–3120 (2014).

18. Y. Oyama, L. Zhen, T. Tanabe and M. Kagaya. Sub-terahertz imaging of defects in building blocks. *NDT&E International* **42**, 28–33 (2009).

19. S. Takahashi, T. Hamano, K. Nakajima, T. Tanabe and Y. Oyama. Observation of damage in insulated copper cables by THz imaging. *NDT&E International* **61**, 75–79 (2014).

20. Y. Nakamura, H. Kariya, A. Sato, T. Tanabe, K. Nishihara, A. Taniyama, K. Nakajima, K. Maeda and Y. Oyama. Nondestructive corrosion diagnosis of painted hot-dip galvanizing steel sheets by using THz spectral imaging. *Corrosion Eng.* **63**, 411–416 (2014).

21. T. Tanabe, K. Watanabe, Y. Oyama and K. Seo. Polarization sensitive THz absorption spectroscopy for the evaluation of uniaxially deformed ultra-high molecular weight polyethylene. *NDT&E International* **43**, 329–333 (2010).

22. M. Ahonen, M. Pessa and T. Suntola. A study of ZnTe films grown on glass substrates using an atomic layer evaporation method. *Thin Solid Films* **65**, 301–307 (1980).

23. T. Suntola. Atomic layer epitaxy. *Thin Solid Films* **216**, 84–89 (1992).

24. J. Nishizawa and Y. Kokubun. Recent progress in low temperature photochemical process. *Extended Abstract of 16th Int. Conf. on Solid State Devices and Materials*, Kobe, Japan, 1–5 (1984).

25. J. Nishizawa, A. Murai, T. Oizumi, T. Kurabayashi, K. Kanamoto and T. Yoshida. Surface morphology investigation of Si thin film grown by temperature modulation Si molecular-layer epitaxy. *J. Cryst. Growth* **226**, 39–46 (2001).

26. N. Otsuka, J. Nishizawa, H. Kikuchi and Y. Oyama. Self-limiting growth of InP by alternate trimethylindium and tertiarybutylphosphine supply in ultrahigh vacuum. *J. Cryst. Growth* **205**, 253–263 (1999).

27. T. Ohno, Y. Oyama and S. Sato. The substrate orientation dependence of GaAsSb thin layer and GaSb dots grown by molecular layer epitaxy. *Phys. Stat. Sol. C* **7**, 2510–2513 (2010).

28. J. Nishizawa, H. Abe and T. Kurabayashi. Doping in molecular layer epitaxy. *J. Electrochem. Soc.* **136**, 478–484 (1989).

29. T. Kurabayashi, H. Kikuchi, T. Hamano and J. Nishizawa. Doping method for GaAs molecular-layer epitaxy by adsorption control of impurity precursor. *J. Cryst. Growth* **229**, 147–151 (2001).

30. Y. Oyama, T. Ohno, K. Suto and J. Nishizawa. Be doping in GaAs by intermittent AsH_3/TEG supply in an ultra-high vacuum. *J. Cryst. Growth* **259**, 61–68 (2003).

31. Y. Oyama, T. Ohno, K. Tezuka, K. Suto and J. Nishizawa. Application of low-temperature area-selective regrowth for ultrashallow sidewall GaAs tunnel junctions. *Appl. Phys. Lett.* **81**, 2563–2565 (2002).

32. Y. Oyama, K. Tezuka, K. Suto and J. Nishizawa. Enhanced impurity incorporation by alternate Te and S doping in GaAs prepared by intermittent injection of triethylgallium and arsine in ultra high vacuum. *J. Cryst. Growth* **246**, 15–20 (2002).

33. Y. Oyama, T. Ohno, K. Suto and J. Nishizawa. Effects of AsH_3 surface treatment for the improvement of ultrashallow area-selective regrown GaAs sidewall tunnel junction. *J. Electrochem. Soc.* **151**, G131–G135 (2004).

34. S. Balasekaran, K. Endo, T. Tanabe and Y. Oyama. Patch antenna coupled 0.2 THz TUNNETT oscillators. *Solid-State Electron.* **54**, 1578–1581 (2010).

35. H. Preier. Recent advances in lead-chalcogenide diode lasers. *Appl. Phys.* **20**, 189–206 (1979).

36. J. F. Butler and T. C. Harman. Long-wavelength infrared $Pb_{1-x}Sn_x$Te diode lasers. *Appl. Phys. Lett.* **12**, 347–348 (1968).

37. L. R. Tomasetta and C. G. Fonstad. Liquid phase epitaxial growth of laser heterostructures in $Pb_{1-x}Sn_x$Te. *Appl. Phys. Lett.* **24**, 567–569 (1974).

38. Y. Horikoshi. Semiconductor Lasers with Wavelengths Exceeding 2 μm. Pages 93-151. In W. T. Tsang (Ed.). *Semiconductors and Semimetals*, vol. **22c**, 19. Academic Press, New York (1985).

39. J. N. Walpole, A. R. Calawa, T. C. Harman and S. H. Groves. Double-heterostructure PbSnTe lasers grown by molecular-beam epitaxy with cw operation up to 114 K. *Appl. Phys. Lett.* **28**, 552–553 (1976).

40. J. Nishizawa, Y. Okuno and K. Suto. Nearly perfect crystal growth in III-V and II-VI compounds. In J. Nishizawa (Ed.). *Japan Annual Reports in Electronics, Computers & Telecommunications (JARECT).* vol. **19**, pp. 17–80. Semiconductor Technologies, Ohmsha Ltd. and North-Holland, Tokyo (1986).

41. Y. Oyama, T. Suzuki, K. Suto and J. Nishizawa. Observation of anisotropic initial growth nucleation in liquid phase epitaxy of InP. *J. Cryst. Growth* **222**, 64–73 (2001).

42. A. Yasuda, K. Suto, J. Nishizawa, Lasing properties of PbSnTe/PbTe double hetero mid-infrared laser diodes grown by temperature difference method under controlled vapor pressure liquid-phase epitaxy. *Mat. Sci. Semicon. Proc.* **27**, 159–162, (2014).

43. W. Tamura, A. Yasuda, K. Suto, O. Itoh and J. Nishizawa. Electroluminescence and lasing properties of highly Bi-doped PbTe epitaxial layers grown by temperature difference method under controlled vapor pressure. *J. Electron. Mater.* **32**, 39–42 (2003).

44. T. Tanabe and Y. Oyama. Frequency-tunable coherent THz-wave pulse generation using two Cr:Forsterite lasers with one Nd:YAG laser pumping and applications for non-destructive THz inspection. In K. Jakubczak (Ed.). *Laser Systems for Applications*, pp. 119–136, InTech, London (2011).

45. Y. Oyama, T. Yamagata, H. Kariya, T. Tanabe and K. Saito. Non-destructive inspection of copper corrosion via coherent Terahertz light source. *ECS Trans.* **50**, 89–98 (2013).

46. K. Kuroo, R. Hasegawa, T. Tanabe and Y. Oyama. Terahertz Application for non-destructive inspection of coated Al electrical conductive wires. *J. Imaging* **3**, 27 (2017).

47. H. Kariya, A. Sato, T. Tanabe, K. Saito, K. Nishihara, A. Taniyama and Y. Oyama. Non-destructive evaluation of a corroded metal surface using Terahertz wave. *ECS Trans.* **50**, 81–88 (2013).

48. T. Kimura, Y. Nakasato, K. Maeda and Y. Oyama. Nondestructive and remote inspection applications by terahertz spectrum imaging. *IEICE Technical Report* **115**, 387 (2015).

Section III

Systems Design and Applications

11

Low Power Biosensor Design Techniques Based on Information Theoretic Principles

Nicole McFarlane

CONTENTS

11.1 Introduction

Mixed signal complementary metal oxide semiconductor (CMOS)technology has become a popular research area for integrated biosensing applications. However, while modern CMOS processes, through the fulfillment of Moore's Law, realize decreasing minimum sizing, this is accompanied by a lessening power supply. Further, the inherent physical noise still remains the same. This trend leads to poor signal to noise ratios, and dynamic range performance being a significant challenge to sensitive and accurate low power biosensing. The application of information theory to circuits has been introduced to model various topologies. These topologies include chopper stabilized amplifiers, active pixel sensors, and single photon avalanche diodes [1–6]. By using circuit design methodologies based on information theory, it is possible to create mixed signal systems that can operate at lower power, while efficiently transmitting information in the presence of high intrinsic physical and environmental noise. The methodology and its implications in this chapter were previously presented as part of [7].

11.2 Noise and Information Rates in Amplifiers

Noise, from a variety of sources, distorts signals of interest, setting a lower bound on the minimum detectable signal (Figure 11.1). Additionally, biosignals tend to be extremely small and hover around the levels of intrinsic system noise. The intrinsic transistor noise sources include thermal noise, a white band noise source with a constant value spectral

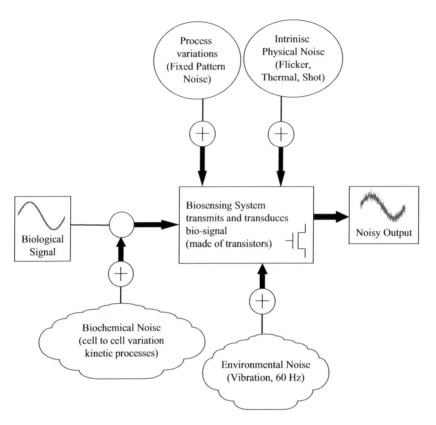

FIGURE 11.1
Overview of possible noise sources in biosensing microsystems.

density, and flicker noise, which varies inversely with the frequency, gate current noise, and shot noise. The flicker noise and thermal noise sources are typically dominant, particularly at the lower frequencies where biosignals are found. The output noise spectral density for the thermal and flicker noise current sources is,

$$S_{I_d} = \gamma 4KTg_m + \frac{K_f I_d^{A_f}}{f^{E_f} C_{ox} L_{eff}^2} \tag{11.1}$$

where K_f, A_f, and E_f parameters are dependent on the fabrication process [8–12]. γ depends on the region of operation. In strong inversion it has a value of $2/3$, while in weak inversion it has a value of $1/2$. The other intrinsic noise sources, gate leakage current noise, and shot noise, are typically non-dominant sources of noise for MOSFETs. p-MOSFETs typically have lower noise than n-MOSFETs, in a given process. The output current noise is tranformed to voltage noise through the equivalent output impedance [13–15],

$$S_{out_v} = S_{I_d} Z_{out}^2. \tag{11.2}$$

For any mixed signal, multi-stage system, including amplifiers, the voltage noise at the input is more important to consider than the output noise. The noise at the input node of any stage is determined from the gain of the output signal with respect to the input signal.

For multiple stages, the noise is divided by the gain for each stage until the input node is reached. Process parasitics, including the gate-to-source and gate-to-drain capacitances (C_{gd} and C_{gs}), may need to be considered to accurately account for frequency effects. Since each device is assumed to have an independent noise generator, the total noise density is simply the sum of all the noise sources. Thus, the total input noise at the input of the amplifier is determined by developing the effect of each noise source on the output divided by the frequency dependent gain of the amplifier.

A method for extracting and measuring noise characteristics of transistors is shown in Figure 11.2 [16,17]. Amplifier noise is measured at the output directly using the spectrum analyzer and a voltage buffering circuit to prevent the analyzer from loading the amplifier. The voltage buffer and amplifier should be supplied using a noiseless power source, such as a battery with voltage regulation. The measured experimental noise for an n-channel transistor in a standard CMOS process is shown in Figure 11.3 [7].

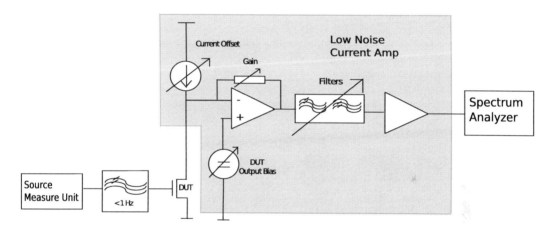

FIGURE 11.2
Setup for noise parameters, K_f and A_f, extraction. DUT is the Device Under Test.

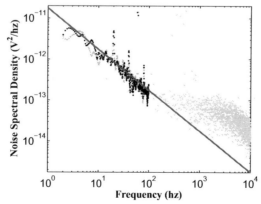

FIGURE 11.3
Flicker (1/f) noise profile of an NMOS transistor with $K_f = 10^{-26}$. (From McFarlane, N. "Information power efficiency tradeoffs in mixed signal CMOS circuits". 2010, PhD dissertation, University of Maryland, College Park, MD.)

11.3 A First Order Model for the Information Rate of an Amplifier

For differential amplifiers, the noise referred to the input terminals are dominated by the initial stage and specifically by the input differential pair. The voltage noise source at the input of the amplifier can be represented as,

$$S_{Vg} = \frac{4KT\gamma}{g_m} + \frac{K_f}{WLC_{ox}fg_m^2} \tag{11.3}$$

More generically, following [18], this has the form,

$$S_{Vg} = S_0\left(1 + \frac{f_k}{f}\right) \tag{11.4}$$

where S_0 is the amount of frequency independent white noise from thermal sources, and f_k is the corner frequency defined as the point where the frequency at which the thermal noise is equal to the flicker noise. The output noise is then shaped by the, typically, low pass transfer function of the amplifier. This leads to an input referred noise for a generic low pass system with single pole characteristics, [19],

$$S_{n_{in}} = \frac{S_0}{A_0^2}\left(1 + \frac{f_k}{f}\right)\left(1 + \left(\frac{f}{f_c}\right)^2\right) \tag{11.5}$$

f_k is the corner frequency as previously defined, S_0 is the noise from thermal sources, A_0 is the midband gain of the system, and f_c is the bandwidth of the system. These parameters are functions of the physical makeup, including the fabrication process, of the transistors. This includes the noise parameters for flicker noise, K_f, A_f, and E_f, the transconductance, g_m, the input differential pair transistors aspect ratio, W/L, the output resistance, r_o, and the bias current from the tail current source, I.

In biological applications, for example in electrophysiological experiments, the input biological signal presented to the amplifier is extremely weak. The assumption is made that the noise sources are Gaussian in nature. Following classic information theory, the amplifier can be appropriately modeled as a Gaussian communication channel. The solution to maximizing the information efficiency of this channel is well known as the water-filling technique, where the input signal is spectrally diffused over frequency locations or bins where the noise is at a minimum [20,21]. This may be mathematically expressed as,

$$C = \int_{f_1}^{f_2} \log_2\left(\frac{v}{S_{n_{in}}(f)}\right)df \tag{11.6}$$

v is the power spectral density of the input signal power plus the spectral density of the flicker and thermal noise signals for the system's bandwidth. $\Delta f = f_2 - f_1$ is a fixed number, and the input signal power is defined as,

$$P = \int_{f_1}^{f_2}\left(v - S_{n_{in}}(f)\right)df$$

$$v = S_{n_{in}}(f_1) = S_{n_{in}}(f_2) \tag{11.7}$$

Assuming an ideal, single pole, low pass transfer function for the filter the information rate is,

$$v = S_{n_{in}}(f_1) \text{ and } f_2 = f_c = f_{3dB}$$

$$I = \frac{1}{\ln 2}\left[2(f_c - f_1) + f_c \ln \frac{\left(1 + \frac{f_k}{f_1}\right)\left(1 + \frac{f_1^2}{f_c^2}\right)}{\left(1 + \frac{f_k}{f_c}\right)2} \right.$$

$$\left. + f_k \ln \frac{f_1 + f_k}{f_c + f_k} - 2f_c\left(\frac{\pi}{4} - \tan^{-1}\frac{f_1^2}{f_c^2}\right) \right] \tag{11.8}$$

The bandwidth and the signal power are given by,

$$P_{sig} = \frac{S_0}{A_0^2}\left[\frac{2}{3}\frac{f_2^3 - f_1^3}{f_c^2} + \frac{f_k}{2f_c^2}\left(f_2^2 - f_1^2\right) - f_k \ln \frac{f_1}{f_2} \right] \tag{11.9}$$

$$0 = \left(1 + \frac{f_k}{f_2}\right)\left(1 + \frac{f_2^2}{f_c^2}\right) - \left(1 + \frac{f_k}{f_1}\right)\left(1 + \frac{f_1^2}{f_c^2}\right) \tag{11.10}$$

As defined previously, in waterfilling, the power of the input signal is first allocated to frequency bins where the thermal + flicker noise power is at a minimum, before being allocated to frequencies where these noise sources are higher. In this manner, as the signal power increases, so does the information rate. Assuming the flicker-thermal corner frequency lies within the bandwidth, the first two quantities rely only on the frequency where the noise corner occurs and system bandwidth. Increasing the bandwidth or the frequency of the noise corner results in an information rate increase. Typically, $f_1 \ll f_c$ and $f_2 \approx f_c$, therefore as the bandwidth, f_c, increases $arctan(f_2 / f_c)$ remains almost at a fixed value, while $arctan(f_1 / f_c)$ linearly increases. For a fixed noise corner frequency, f_k, larger bandwidth systems will have greater information rates. As the bandwidth increases, $f_c \gg f_{1,2}$, the last two terms cancel. With lowered corner frequencies, $f_c < f_{1,2}$, the sum approaches $2(f_2 - f_1)$.

Most systems will have multiple poles and zeroes; for example, most two stage amplifiers will have at least two dominant poles and a zero. Other poles and zeroes are generally high enough to be neglected for the amplifier performance. The information rate can be generalized for a system with poles and zeroes as,

$$C = \frac{1}{\ln 2}\left[2(f_2 - f_1) + f_k \ln \frac{f_1 + f_k}{f_2 + f_k} \right.$$

$$- \sum_{i=1}^{n} 2f_{pi}\left(\tan^{-1}\frac{f_2}{f_{pi}} - \tan^{-1}\frac{f_1}{f_{pi}} \right)$$

$$\left. + \sum_{j=1}^{m} 2f_{zj}\left(\tan^{-1}\frac{f_2}{f_{zj}} - \tan^{-1}\frac{f_1}{f_{zj}} \right) \right] \tag{11.11}$$

where $f_{1,2}$ are determined by the shape of the spectrum for the power density of the noise sources and the quantity of available signal power. Clearly, higher signal powers will require higher bandwidth (subject to the same of the spectral density function). Each additional pole decreases the information rate, while zeroes increase the information rate. A decrease in information capacity will be dominated by the dominant pole, while any increase will be dominated by the first zero. From this result, signal power should be placed at frequencies above the dominant pole where any amplification is typically minimal. Thus, this requires careful design strategies to optimize signal power, noise power, and power supply using information rate and typical application design constraints.

The input referred noise is lowered with decreasing bias currents due to reduced bandwidth and increased midband gain. The dominant pole location is a function of the bias current; thus, lowered bias current also influences the frequency location of the noise minimum. Thus, the waterfilling frequency allocation of signal power may vary significantly with the bias current. If the bandwidth is about the same or smaller than the frequency of the thermal-flicker noise intersection, the frequency location of the minimum value of noise may increase.

11.4 Metrics for Tradeoffs in Power and Noise

There are many other metrics for characterization of the interplay noise of a system and the available power specifications. Specifically, the noise efficiency factor (NEF) is a popular metric for characterizing noise and power of a system. It compares the noise of the given system to an ideal bipolar transistor. The ideal bipolar is assumed to have no base resistance and only have thermal noise sources [22]. Given some frequency bandwidth, Δf, bipolar collector current, I_c, and total current draw from the power supplies of the amplifier, I_{tot}, the NEF is given by,

$$\text{NEF} = \sqrt{\Delta f \frac{\pi}{2} \frac{4kTV_T}{I_c}} \sqrt{\frac{2I_{tot}}{\pi 4kTV_T \Delta f}} \tag{11.12}$$

where V_T is the thermal voltage, 0.025 V. A less noise and power efficient amplifier will have a greater NEF value. Improved system noise performance is implied by having a lowered noise efficiency factor.

Another metric known as the bit energy (BE) measures the energy cost of using the system with optimized amplifier performance [23]. This is measured by the power consumed by the amplifier, P_{amp}, and maximum information rate or information capacity, I, such that

$$BE = \frac{P_{amp}}{I} \tag{11.13}$$

Lower bit energy implies a more efficient amplifier in terms of cost to utilize the system and performance. That is the cost of maximizing amplifier performance will be minimized.

The total power in an amplifier is a function of the power supplies, V_{dd} and V_{ss}, as well as the total current for the amplifier and any biasing an start-up circuitry,

$$P_{amp} = I_{tot}\left(V_{dd} - V_{ss}\right) \tag{11.14}$$

The power increases as the current increases. However, the dominant pole also increases with the current. This leads to less efficient operation, as the maximum information transfer rate decreases and the cost of using the system increases. The noise efficiency factor is related to the bit energy as they both include the total current and noise. However, the bit energy takes the available power resources explicitly into account in its formulation, making it a more accurate measure.

11.5 Tradeoffs in a Simple Amplifier Design Example

A simple operational transconductance amplifier (OTA) is shown in Figure 11.4. The major design parameters and constraints for an amplifier using the EKV model are the bandwidth, midband gain, and intersection of the flicker and thermal spectral density plots. In the EKV model, these are all formulated in terms of the inversion coefficient, IC, is [24]

$$A_o = \frac{\kappa}{U_T} \frac{1 - e^{-\sqrt{IC}}}{\sqrt{IC}} V_{A2,4}$$

$$S_o = \frac{4kT\gamma}{\kappa I_D \frac{1 - e^{-\sqrt{IC}}}{U_T \sqrt{IC}}}$$

$$f_k = \frac{K_f}{WLC_{ox} 4kT \frac{\kappa}{U_T} \frac{1 - e^{-\sqrt{IC}}}{\sqrt{IC}}}$$

$$f_{3dB} = \frac{IC I_o W / L}{2\pi V_{A2,4} C_{out}}$$

$$V_{A2,4} = \frac{V_{A2} V_{A4}}{V_{A4} + V_{A2}}$$

where the inversion coefficient is defined as the ratio of the drain current to the reverse saturation current, I_d / I_s. W / L is the transistor aspect ratio, and V_A is the early voltage (reciprocal of λ the transistor channel length modulation parameter). $U_T = V_T$ is the thermal voltage, 0.025 V, C_{ox} is the capacitance of the gate oxide, I_o is the characteristic current, and κ is the slope in subthreshold. The simplified model of γ uses a constant of $1/2$ for transistors operating above threshold, and a value of $2/3$ for transistors operating in strong inversion. Over weak, moderate, and strong inversion regions, a more complete model is [24],

$$\gamma = \frac{1}{1 + IC} \left(\frac{1}{2} + \frac{2}{3} IC \right). \tag{11.15}$$

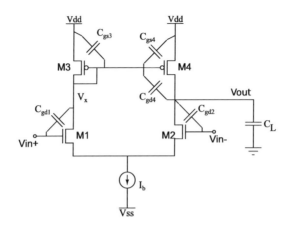

FIGURE 11.4
Simple differential operational transconductance amplifier explicitly showing the parasitic capacitances between the terminals of the transistor. (From McFarlane, N. "Information power efficiency tradeoffs in mixed signal CMOS circuits". 2010, PhD dissertation, University of Maryland, College Park, MD.)

The design parameters to be determined are the bias tail current, typically implemented as a current mirror source with M_5 and M_6, and transistor aspect ratios. The midband gain at lowered frequencies is traditionally written as,

$$A_0 = g_{m1}\left(r_{02} \parallel r_{04}\right) = \sqrt{\frac{\mu C_{ox} W / L}{I_{bias}}} \frac{1}{\lambda_2 + \lambda_4} \tag{11.16}$$

in saturation. The MOSFET square law equations are assumed, and λ is the inverse of the early voltage. The dividing line that separates the weak, moderate, and strong inversion lie at inversion coefficient values of 0.1, 1, and 10.

The minimum inputs occur due to the tail current transistor entering into triode (assuming strong inversion)

$$V_{I_{min}} \geq \sqrt{\frac{I_B}{\beta_1}} + V_{th1} + \sqrt{\frac{2I_B}{\beta_5}} + V_{ss} \tag{11.17}$$

The maximum input signal occurs when M_4 enters triode and eventually turns off, so that,

$$V_{G1} = V_{DD} - \sqrt{\frac{I_b}{\mu C_{ox} W / L}} - V_{th3} + V_{th1} \tag{11.18}$$

The common mode gain is a function of the tail current output resistance,

$$A_c = \frac{1}{2g_{m4}r_{o5}} = \frac{1}{\dfrac{4\kappa 1 - e^{-\sqrt{IC}}}{U_T\sqrt{IC}}} \tag{11.19}$$

And the common mode rejection ratio (CMRR)is given by $|A_v/A_c|$. The slew rate increases with increasing bias current and lowered load capacitances. The load capacitance consists of parasitic capacitances and any explicit capacitances at the output node. Increased current increases both the thermal and flicker noise levels. These noise levels decrease with transistor area. Thus, increasing inversion coefficient increases the noise as well as the thermal and flicker noise corner frequency, system bandwidth, and cost of using the system (bit energy). Gain decreases with decreasing inversion coefficient. However, the system information rate shows a complex response to design parameters [1].

For the amplifier, ignoring process variations means $g_{m1} = g_{m2}$ and $g_{m3} = g_{m4}$. The input referred noise, assuming no frequency effect is [25],

$$v_{eq}^2 = \frac{i_{n1}^2}{g_{m1}^2} + \frac{i_{n2}^2}{g_{m1}^2} + \left(\frac{g_{m3}}{g_{m1}}\right)^2 \left(\frac{i_{n3}^2}{g_{m3}^2} + \frac{i_{n4}^2}{g_{m3}^2}\right) \tag{11.20}$$

Figure 11.5 shows the parasitic capacitances, which affect the transfer function. The capacitances of the NMOS and PMOS pair, from the gate to source and from the gate to the drain, are the assumed to be equal, the sources of differential pair, M1 and M2, are considered to be at virtual ground. Since the bias current adds equally to both sides, it is not considered to add to the system noise. The system transfer function from input to output nodes and from the gate of the current mirror active load to the input is,

$$H(f) = \frac{-1/2\left(sC_{gd1} - g_{m1}\right)\left(sC_{gd1} + sC_1 + 1/R_1 + g_{m4}\right)}{\left(sC_{out} + 1/R_{out}\right)\left(sC_{gd1} + 1/R_1 + sC_1 + sC_{gd4}\right) - sC_{gd4}\left(sC_{gd4} - g_{m4}\right)}$$

$$\frac{V_x}{V_{in}} = \frac{1/2\left(sC_{gd1} - g_{m1}\right)\left(sC_{gd4} + sC_out + 1/R_{out}\right)}{sC_{gd4}\left(sC_{gd4} - g_{m4}\right) - \left(sC_{out} + 1/R_{out}\right)\left(sC_{gd1} + 1/R_1 + sC_1 + sC_{gd4}\right)}$$

the poles and zeroes of the amplifier, assuming C_{gd4} is small, are,

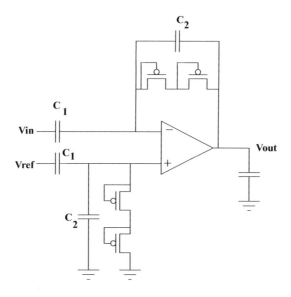

FIGURE 11.5
Frequently used capacitive feedback amplifier topology used for biosignal acquisition. (From Harrison, R.R. and Charles, C., *IEEE J. Solid State Circ.*, 38, 958–965, 2003.)

$$z_1 = \frac{g_{m1}}{C_{gd2}}$$

$$z_2 = \frac{2g_{m4}}{C_1 + C_{gd1}}$$

$$p_1 = \frac{1}{R_o\left(C_L + C_{gd2}\right)}$$

$$p_2 = \frac{g_{m4}}{C_1 + C_{gd1}} \tag{11.21}$$

where

$$C_1 = C_{gs3} + C_{gs4}$$

$$C_{out} = C_L + C_{gd2} + C_{gd4}$$

$$R_1 = r_{o1} \parallel r_{o3} \parallel 1/g_{m3}$$

$$R_{out} = r_{o2} \parallel r_{o4} \tag{11.22}$$

Given that $i = g_m v_{gs}$, the noise voltage is reflected back to the gate of the transistor as i_n^2 / g_m^2 at low frequencies. At mid frequencies, the gate-to-drain capacitance may not necessarily be ignored, and the noise voltage at the gate is $i_n^2 / (g_m + sC_{gd})^2$.

The width of a transistor may be taken out of the design space by considering instead that the bias current, inversion level, and transistor length are known [26,27]. Bias current and length of the transistor both change linearly with the inversion level. Both aspect ratio, W / L and area $W \times L$ impact the system and should be explored first, after which the effects of length and bias current are explored [26].

Increased inversion coefficient, that is moving from weak to moderate to strong inversion, affects the information rate and energy cost of using the system (bit energy). The stronger the inversion level, the better the information rate, implying that above threshold operation will give a cleaner transfer of the input signal to the measured output signal. The cost of using the amplifier is lowered at the lower inversion levels. Since the input differential pair should contribute the most noise, a generalization can be made of the design specifications as functions of the transistor length and region of operation (Table 11.1).

TABLE 11.1
Variation of Design Specification with Inversion Coefficient (IC) and Length (L)

Specification	Name	IC ↑	L ↑
A_o	Low frequency gain	↓	↑
f_k	Noise corner frequency	↑	↓
S_o	White noise value	↑	↑
A_c	Common mode gain	↑	↓
BE	Bit energy	↑	↓
I	Information rate	↑	↓

11.6 Information Rate of Amplifier

Capacitive feedback with an OTA is a popular architecture for weak biosignal acquisition [3,28,29]. The design has been widely implemented in various CMOS processes. A wide range OTA is used as the OTA for the system shown in Figure 11.5. For a typical design in a 500 nm process, the input differential pair transistors have $W = 35$ m, $L = 2.1$ m, $C_2 = 200$ fF,

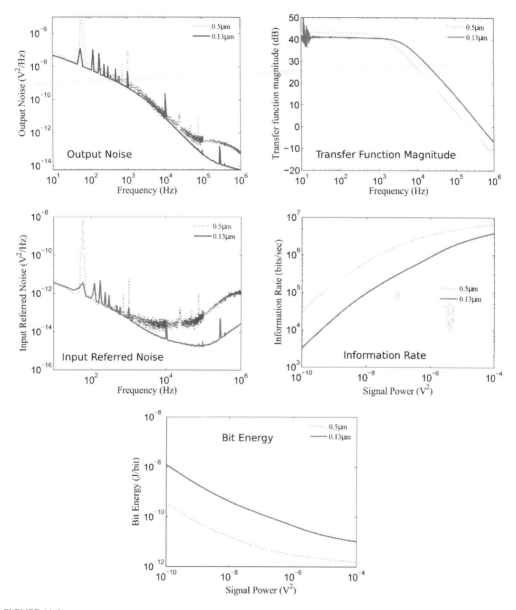

FIGURE 11.6

Comparison of larger and submicron process output voltage noise, input referred noise, transfer function, information rates, and bit energy. (From McFarlane, N. "Information power efficiency tradeoffs in mixed signal CMOS circuits". 2010, PhD dissertation, University of Maryland, College Park, MD.)

and $C_1 = 20$ pF. For a 130-nm process, the input differential pair have $W = 24\,\mu m$, $L = 1.2\,\mu m$, $C_2 = 98.3$ fF, and $C_1 = 10$ pF. The power supply goes from 5 V to 1.8 V between the two processes. The psuedo resistors have a large resistance value. The noise spectrum referred to the input terminals of the OTA is a function of the input capacitance (C_{in}) and the output noise (S_{OTA}), as is [28],

$$S_{amp} = \left(\frac{C_1 + C_2 + C_{in}}{C_1}\right)^2 S_{OTA} \qquad (11.23)$$

Figure 11.6 shows the measured noise at the output, frequency dependent magnitude of the transfer function, spectral density of the noise referred to the input terminals, information rate, and energy cost of using the amplifier [7]. Due to a wider minimal noise spectral density bandwidth, the larger process shows a higher information rate than the shorter channel process. There is a slight decrease in voltage between the processes; however, it is unable to compensate for the noise driven information rate measure. With the fundamental noise sources actually increasing with decreasing process lengths, this result shows that the expected improvement in the energy costs (due to lowered voltage supplies) of sub-micron and deep sub-micron processes does not materialize.

11.7 Conclusion

The input referred thermal and flicker noise spectral density for a amplifier used for weak biological signals along with the information rate and energy cost of using the amplifier was theoretically calculated. The information rate and bit energy derives from treating the circuits as communication channels according to information theory principles. This has been incorporated into an inversion coefficient based design methodology. The use of information rate to optimize amplifiers suggests that optimum operation occurs at frequencies where gain is not maximum. The information derived from this methodology has the potential to be incorporated into the mixed signal design flow and can be particularly important biological applications, where weak signals need to be detected in the presence of fixed system noise.

References

1. N. McFarlane and P. Abshire, "Comparative analysis of information rates of simple amplifier topologies," *IEEE International Symposium of Circuits and Systems*, pp. 785–788, 2011.
2. M. Loganathan, S. Malhotra, and P. Abshire, "Information capacity and power efficiency in operational transconductance amplifier," *IEEE International Symposium on Circuits and Systems*, vol. 1, pp. 193–196, 2004.
3. N. M. Nelson and P. A. Abshire, "An information theoretic approach to optimal amplifier operation," *IEEE Midwest Symposium on Circuits and Systems*, vol. 1, pp. 13–16, 2005.
4. N. Nelson and P. Abshire, "Chopper modulation improves OTA information transmission," *IEEE International Symposium on Circuits and Systems*, pp. 2275–2278, 2007.

5. D. Sander, N. Nelson, and P. Abshire, "Integration time optimization for integrating photosensors," *IEEE International Symposium on Circuits and Systems*, pp. 2354–2357, 2008.

6. J. Gu, M. H. U. Habib, and N. McFarlane, "Perimeter-gated single-photon avalanche diodes: An information theoretic assessment," *IEEE Photonics Technology Letters*, vol. 28, no. 6, pp. 701–704, 2016.

7. N. McFarlane. "Information power efficiency tradeoffs in mixed signal CMOS circuits," PhD dissertation, University of Maryland, College Park, MD, 2010.

8. M. J. Kirton and M. J. Uren, "Noise in solid-state microstructures: A new perspective on individual defects, interface states and low-frequency (1/f) noise," *Advances in Physics*, vol. 38, no. 4, pp. 367–468, 1989.

9. I. Bloom and Y. Nemirovsky, "1/f noise reduction of metal-oxide-semiconductor transistors by cycling from inversion to accumulation," *Applied Physics Letters*, vol. 58, pp. 1664–1666, 1991.

10. E. A. M. Klumperink, S. L. J. Gierink, A. P. Van der Wel, and B. Nauta, "Reduction MOSFET 1/f noise and power consumption by switched biasing," *IEEE Journal of Solid State Circuits*, vol. 35, no. 7, pp. 994–1001, 2000.

11. Y. Isobe, K. Hara, D. Navarro, Y. Takeda, T. Ezaki, and M. Miura-Mattausch, "Shot noise modelling in metal oxide semiconductor field effect transistors under sub threshold condition," *IEICE Transactions on Electronics*, vol. E90-C, no. 4, pp. 885–894, 2007.

12. L. Callegaro, "Unified derivation of Johnson and shot noise expressions," *American Journal of Physics*, vol. 74, pp. 438–440, 2006.

13. S.-C. Liu, J. Kramer, G. Indiveri, T. Delbruck, and R. Douglas, *Analog VLSI: Circuits and Principles*. Cambridge, MA: MIT Press, 2002.

14. C. Jakobson, I. Bloom, and Y. Nemirovsky, "1/f noise in CMOS transistors for analog applications from subthreshold to saturation," *Solid State Electronics*, vol. 42, pp. 1807–1817, 1988.

15. Y. Nemirovsky, I. Brouk, and C. G. Jakobson, "1/f noise in CMOS transistors for analog applications," *IEEE transactions on Electron Devices*, vol. 48, no. 5, pp. 212–218, 2001.

16. R. Tinti, F. Sischka, and C. Morton, "Proposed system solution for 1/f noise parameter extraction," http://literature.cdn.keysight.com/litweb/pdf/5989-9087EN.pdf.

17. A. Blaum, O. Pilloud, G. Scalea, J. Victory, and F. Sischka, "A new robust on-wafer 1/f noise measurement and characterization system," *International Conference on Microelectronic Test Structures*, pp. 125–130, 2001.

18. C. C. Enz and G. C. Temes, "Circuit techniques for reducing the effects of op-amp imperfections: Autozeroing, correlated double sampling, and chopper stabilization," *Proceedings of the IEEE*, vol. 84, no. 11, pp. 1584–1614, 1996.

19. C. C. Enz, E. A. Vittoz, and F. Krummenacher, "A CMOS chopper amplifier," *IEEE Journal of Solid-State Circuits*, vol. 22, no. 3, pp. 335–342, 1987.

20. T. M. Cover and J. A. Thomas, *Elements of Information Theory*. New York: John Wiley & Sons, 1991.

21. P. A. Abshire, "Implicit energy cost of feedback in noisy channels," *IEEE Conference on Decision and Control*, vol. 3, pp. 3217–3222, 2002.

22. M. S. J. Steyaert, W. M. C. Sansen, and C. Zhongyuan, "A micropower low-noise monolithic instrumentation amplifier for medical purposes," *IEEE Journal Solid State Circuits*, vol. 22, no. 6, pp. 1163–1168, 1987.

23. A. G. Andreou, *Low-Voltage/Low Power Integrated Circuits and Systems*. IEEE Press, 1999, ch. An Information Theoretic Framework for Comparing the Bit Energy of Signal Representations at the Circuit Level, pp. 519–540.

24. C. C. Enz, F. Krummenacher, and E. A. Vittoz, "An analytical MOS transistor model valid in all regions of operation and dedicated to low-voltage and low-current applications," *Analog Integrated Circuits and Signal Processing*, vol. 8, no. 1, pp. 83–114, 1995.

25. P. R. Gray, P. J. Hurst, S. H. Lewis, and R. G. Meyer, *Analysis and Design of Analog Integrated Circuits*. New York: John Wiley & Sons, 2001.

26. D. Binkley, B. Blalock, and J. Rochelle, "Optimizing drain current, inversion level, and channel length in analog CMOS design," *Analog Integrated Circuits and Signal Processing*, vol. 47, pp. 137–163, 2006.

27. D. Binkley, C. Hopper, S. Tucker, B. Moss, J. Rochelle, and D. Foty, "A CAD methodology for optimizing transistor current and sizing in analog CMOS design," *IEEE Transaction on Computer-Aided Design*, vol. 22, pp. 225–237, 2003.

28. R. R. Harrison and C. Charles, "A low-power low-noise CMOS amplifier for neural recording applications," *IEEE Journal of Solid State Circuits*, vol. 38, no. 6, pp. 958–965, 2003.

29. S. B. Prakash, N. M. Nelson, A. M. Haas, V. Jeng, P. Abshire, M. Urdaneta, and E. Smela, "Biolabs-on-a-chip: Monitoring cells using CMOS biosensors," *IEEE/NLM Life Science Systems and Applications Workshop*, pp. 1–2, 2006.

12

Low-Power Processor Design Methodology: High-Level Estimation and Optimization via Processor Description Language

Zheng Wang and Anupam Chattopadhyay

CONTENTS

12.1 Introduction

Computing is an integral part of daily life. We encounter two types of computing devices everyday: desktop-based computing devices and embedded computer systems. Desktop-based computing systems encompass traditional "computers," including personal computers, notebook computers, workstations, and servers. Embedded computing systems are ubiquitous—they run the devices hidden inside a vast array of everyday products and appliances, such as smartphones, toys, intelligent sensors as part of Internet-of-Things (IoT), surveillance cameras, and autonomous vehicles. Both types of computing devices use programmable components, such as processors, co-processors, and memories, to execute the application programs. These programmable components are also referred as *programmable accelerators*. Figure 12.1 shows an example of an embedded system with

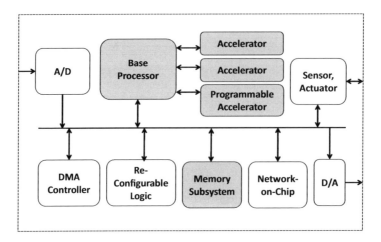

FIGURE 12.1
An exemplary embedded system.

programmable accelerators. Depending on the application domain, the embedded system can have application-specific accelerators, interfaces, controllers, and peripherals. The complexity of the programmable accelerators is increasing at an exponential rate due to technological advances as well as demand for the realization of ever more complex applications in communication, multimedia, networking, and entertainment. Shrinking time-to-market coupled with short product lifetimes create a critical need for design automation of increasingly sophisticated and complex programmable accelerators.

Modeling plays a central role in the design automation of processors. It is necessary to develop a specification language that can model complex processors at a higher level of abstraction and enable automatic analysis and generation of efficient prototypes. The language should be powerful enough to capture a high-level description of the programmable architectures. On the other hand, the language should be simple enough to allow correlation of the information between the specification and the architecture manual.

Specifications widely in use today are still written informally in natural languages, like English. Since natural language specifications are not amenable to automated analysis, there are possibilities of ambiguity, incompleteness, and contradiction: all problems that can lead to different interpretations of the specification. Clearly, formal specification languages are suitable for analysis and verification. Some have become popular because they are input languages for powerful verification tools such as a model checker. Such specifications are popular among verification engineers with expertise in formal languages. However, these specifications are not acceptable by designers and other tool developers. Therefore, the ideal specification language should have formal (unambiguous) semantics as well as easy correlation with the architecture manual.

Architecture Description Languages (ADLs) have been successfully used as a specification language for processor development. Development of a processor is associated with multiple steps as embodied in the Mescal design methodology [1]. These steps— *benchmarking, architectural space identification, design evaluation, design exploration,* and *deployment* are correlated with corresponding tools and implementations. Mescal methodology, in practice, is represented by the ADL-based processor design flow. The ADL specification is used to perform design representation, design evaluation, design validation, and synthesis to more detailed abstraction, such as Register Transfer Level (RTL). This is shown in the Figure 12.2.

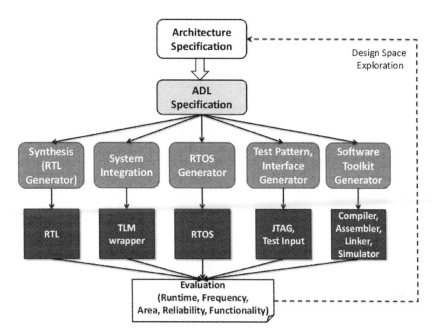

FIGURE 12.2
ADL-driven exploration, synthesis, and validation of programmable architectures.

More specifically, the ADL specification is used to derive a processor toolsuite, such as instruction-set simulator, compiler, debugger, profiler, assembler, and linker. The specification is used to generate detailed and an optimized hardware description in a RTL [2,3]. The ADL specification is used to validate the design by formal, semi-formal, and simulation-based validation flows [4], as well as for the generation of test interfaces [5]. The specification can also be used to generate device drivers for real-time operating systems [6].

The rich modeling capability of ADL is used to design various kinds of processor architectures ranging from programmable coarse-grained reconfigurable architectures to superscalar processors [2,7,8]. The design exploration capability of an ADL-driven toolsuite is extended to cover high-level power and reliability estimations [9]. The prominence of ADL is also palpable in its widespread market acceptance [10,11]. We attempt to cover the entire spectrum of processor design tools, from the perspective of ADL-based design methodology, in this chapter.

Furthermore, ADL-based processor design methodology enables high-level modeling of physical parameters, such as power consumption and reliability, which are traditionally analyzed during the phases of physical design and chip testing. By automating the modeling process, the users can explore the physical parameters in the very early phases of processor design, and therefore customize the architecture according to various domains of applications.

The rest of the chapter is organized as follows. Section 12.2 describes processor modeling using ADLs. Section 12.3 presents ADL-driven methodologies for software toolkit generation, hardware synthesis, exploration, and validation of programmable architectures. Section 12.4 illustrates the ADL-based physical modeling framework, which estimates power consumption, thermal footprint, and logic timing variation. Finally, Section 12.5 concludes the chapter.

12.2 Processor Modeling Using ADLs

The phrase "Architecture Description Language" (ADL) has been used in the context of designing both software and hardware architectures. Software ADLs are used for representing and analyzing software architectures [12]. They capture the behavioral specifications of the components and their interactions that comprises the software architecture. However, hardware ADLs (also known as Processor Description Languages) capture the structure; that is, hardware components and their connectivity, and the behavior (instruction-set) of processor architectures. The concept of using high-level languages for specification of architectures has been around for a long time. Early ADLs such as Instruction Set Processor Specifications (ISPS) [13] were used for simulation, evaluation, and synthesis of computers and other digital systems. This section gives a short overview of prominent ADLs and tries to define a taxonomy of ADLs.

12.2.1 ADLs and Other Languages

How do ADLs differ from programming languages, hardware description languages, modeling languages, and the like? This section attempts to answer this question. However, it is not always possible to answer the following question: Given a language for describing an architecture, what are the criteria for deciding whether it is an ADL or not?

In principle, ADLs differ from programming languages because the latter bind all architectural abstractions to specific point solutions whereas ADLs intentionally suppress or vary such binding. In practice, architecture is embodied and recoverable from code by reverse engineering methods. For example, it might be possible to analyze a piece of code written in C and figure out whether it corresponds to a *Fetch* unit or not. Many languages provide architecture level views of the system. For example, C++ offers the ability to describe the structure of a processor by instantiating objects for the components of the architecture. However, C++ offers little or no architecture-level analytical capabilities. Therefore, it is difficult to describe architecture at a level of abstraction suitable for early analysis and exploration. More importantly, traditional programming languages are not a natural choice for describing architectures due to their unsuitability for capturing hardware features such as parallelism and synchronization.

ADLs differ from modeling languages (such as UML) because the latter are more concerned with the behaviors of the whole rather than the parts, whereas ADLs concentrate on representation of components. In practice, many modeling languages allow the representation of cooperating components and can represent architectures reasonably well. However, the lack of an abstraction would make it harder to describe the instruction-set of the architecture. Traditional Hardware Description Languages (HDL), such as VHDL and Verilog, do not have sufficient abstraction to describe architectures and explore them at the system level. It is possible to perform reverse-engineering to extract the structure of the architecture from the HDL description. However, it is hard to extract the instruction-set behavior of the architecture.

12.2.2 Prominent ADLs

This section briefly surveys some of the prominent ADLs in the context of processor modeling and design automation. There are many comprehensive ADL surveys available in the literature including ADLs for retargetable compilation [14], programmable embedded systems [15], and System-on-Chip (SoC) design [16]. A definitive compilation of the ADLs can be found in [17].

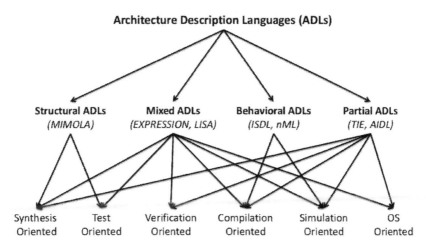

FIGURE 12.3
Taxonomy of ADLs.

Figure 12.3 shows the classification of ADLs based on two aspects: *content* and *objective*. The content-oriented classification is based on the nature of the information an ADL can capture, whereas the objective-oriented classification is based on the purpose of an ADL. ADLs can be classified into six categories based on the objective: simulation-oriented, synthesis-oriented, test-oriented, compilation-oriented, validation-oriented, and operating system (OS) oriented.

ADLs can be classified into four categories based on the nature of the information: structural, behavioral, mixed, and partial. This classification can also be linked with the objective and origin of the different ADLs. The structural ADLs capture the structure in terms of architectural components and their connectivity. The behavioral ADLs, such as nML [19] and Instruction Set Description Language (ISDL) [20], capture the instruction-set behavior of the processor architecture. The mixed ADLs capture both structure and behavior of the architecture. These ADLs capture complete description of the structure or behavior or both. However, the partial ADLs capture specific information about the architecture for the intended task. For example, an ADL intended for interface-synthesis does not require internal structure or behavior of the processor. In the following, few prominent ADLs are discussed.

12.2.2.1 nML

nML was designed at Technical University of Berlin, Germany. nML has been used by code generators CBC [21] and CHESS [22] and instruction set simulators Sigh/Sim [23] and CHECKERS. CHESS/CHECKERS environment is used for automatic and efficient software compilation and instruction-set simulation [23].

nML developers recognized the fact that several instructions share common properties. The final nML description would be compact and simple if the common properties are exploited. Consequently, nML designers used a hierarchical scheme to describe instruction sets. The instructions are the topmost elements in the hierarchy. The intermediate elements of the hierarchy are partial instructions (PI). The relationship between elements can be established using two composition rules: AND-rule and OR-rule. The AND-rule groups several PIs into a larger PI, and the OR-rule enumerates a set of alternatives for one PI. Therefore, instruction definitions in nML can be in the form of an and–or tree. Each possible derivation of the tree corresponds to an actual instruction.

To achieve the goal of sharing instruction descriptions, the instruction set is enumerated by an attributed grammar. Each element in the hierarchy has a few attributes. A non-leaf element's attribute values can be computed based on its children's attribute values. Attribute grammar is also adopted by other ADLs such as ISDL [20]. The following nML description shows an example of instruction specification [19]:

```
op numeric_instruction(a:num_action, src:SRC, dst:DST)
action {
    temp_src = src; temp_dst = dst;
    a.action;
    dst = temp_dst;
}
op num_action = add | sub op add()
action = {
    temp_dst = temp_dst + temp_src
}
```

The definition of *numeric instruction* combines three partial instructions (PI) with the AND rule: *num action*, SRC, and DST. The first PI, *num action*, uses the OR-rule to describe the valid options for actions: *add* or *sub*. The number of all possible derivations of *numeric instruction* is the product of the size of *num action*, *SRC*, and *DST*. The common behavior of all these options is defined in the *action* attribute of *numeric instruction*. Each option for *num action* should have its own action attribute defined as its specific behavior, which is referred by the *a.action* line. For example, the above code segment has an action description for *add* operation. Object code image and assembly syntax can also be specified in the same hierarchical manner. nML also captures the structural information used by instruction-set architecture (ISA). For example, storage units should be declared since they are visible to the instruction-set. nML supports three types of storages: RAM, register, and transitory storage. Transitory storage refers to machine states that are retained only for a limited number of cycles, for example, values on buses and latches. Computations have no delay in nML timing model—only storage units have delay. Instruction delay slots are modeled by introducing storage units as pipeline registers. The result of the computation is propagated through the registers in the behavior specification.

nML models constraints between operations by enumerating all valid combinations. The enumeration of valid cases can make nML descriptions lengthy. More complicated constraints, which often appear in DSPs with irregular instruction level parallelism (ILP) constraints, or VLIW processors with multiple issue slots are hard to model with nML. For example, nML cannot model the constraint that operation *I1* cannot directly follow operation *I0*. nML explicitly supports several addressing modes. However, it implicitly assumes an architecture model that restricts its generality. As a result, it is hard to model multi-cycle or pipelined units and multi-word instructions explicitly. A good critique of nML is given in [24].

Several ADLs endeavored to capture both structural and behavioral details of the processor architecture. We briefly describe two such *mixed* ADLs: EXPRESSION [25] and LISA [26].

12.2.2.2 Expression

The above mixed ADLs require explicit description of Reservation Tables (RT). Processors that contain complex pipelines, large amounts of parallelism, and complex storage subsystems, typically contain a large number of operations and resources (and hence RTs).

Thus, manual specification of RTs on a per-operation basis becomes cumbersome and error-prone. The manual specification of RTs (for each configuration) becomes impractical during rapid architectural exploration. The EXPRESSION ADL [25] describes a processor as a netlist of units and storages to automatically generate RTs based on the netlist [27]. Unlike MIMOLA, the netlist representation of EXPRESSION is of a coarse granularity. It uses a higher level of abstraction similar to a block-diagram level description in an architecture manual.

EXPRESSION ADL was developed at University of California, Irvine. The ADL has been used by the retargetable compiler (EXPRESS [28]) and simulator (SIMPRESS [29]) generation framework. The framework also supports a graphical user interface (GUI) and can be used for design space exploration of programmable architectures consisting of processor cores, coprocessors, and memories [30]. An EXPRESSION description is composed of two main sections: behavior (instruction-set) and structure. The behavior section has three subsections: operations, instruction, and operation mappings. Similarly, the structure section consists of three subsections: components, pipeline/data–transfer paths, and memory subsystem.

The operation subsection describes the instruction-set of the processor. Each operation of the processor is described in terms of its opcode and operands. The types and possible destinations of each operand are also specified. A useful feature of EXPRESSION is operation group that groups similar operations together for the ease of later reference. For example, the following code segment shows an operation group (*alu ops*) containing two ALU operations: *add* and *sub*.

```
(OP_GROUP alu_ops
    (OPCODE add
      (OPERANDS (SRC1 reg) (SRC2 reg/imm) (DEST reg))
      (BEHAVIOR DEST = SRC1 + SRC2)
    ...)
    (OPCODE sub
      (OPERANDS (SRC1 reg) (SRC2 reg/imm) (DEST reg))
      (BEHAVIOR DEST = SRC1 - SRC2)
    ...)
)
```

The instruction subsection captures the parallelism available in the architecture. Each instruction contains a list of slots (to be filled with operations), with each slot corresponding to a functional unit. The operation mapping subsection is used to specify the information needed by instruction selection and architecture-specific optimizations of the compiler. For example, it contains mapping between generic and target instructions.

The component subsection describes each RT-level component in the architecture. The components can be pipeline units, functional units, storage elements, ports, and connections. For multi-cycle or pipelined units, the timing behavior is also specified.

The pipeline/data-transfer path subsection describes the netlist of the processor. The *pipeline path description* provides a mechanism to specify the units, which comprise the pipeline stages, while the *data-transfer path description* provides a mechanism for specifying the valid data-transfers. This information is used to both retarget the simulator and to generate reservation tables needed by the scheduler [27]. An example path declaration for the DLX architecture [31] (Figure 12.4) is shown as follows.

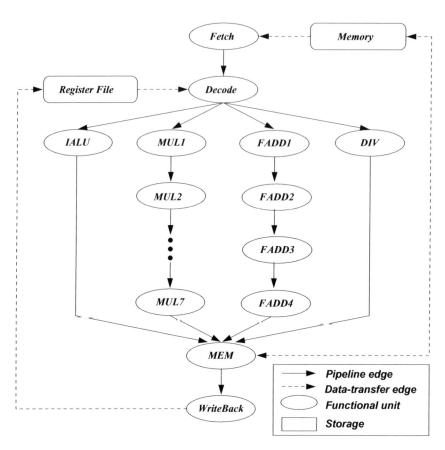

FIGURE 12.4
The DLX architecture.

It describes the processor as having five pipeline stages. It also describes the *Execute* stage a having four parallel paths. Finally, it describes each path; for example, it describes that the *FADD* path has four pipeline stages).

```
(PIPELINE Fetch Decode Execute MEM WriteBack)
(Execute (ALTERNATE IALU MULT FADD DIV))
(MULT (PIPELINE MUL1 MUL2 ... MUL7))
(FADD (PIPELINE FADD1 FADD2 FADD3 FADD4))
```

The memory subsection describes the types and attributes of various storage components (such as register files, SRAMs, DRAMs, and caches). The memory netlist information can be used to generate memory aware compilers and simulators [32]. Memory-aware compilers can exploit the detailed information to hide the latency of the lengthy memory operations. EXPRESSION captures the data path information in the processor. The control path is not explicitly modeled. The instruction model requires extension to capture inter-operation constraints such as sharing of common fields. Such constraints can be modeled by ISDL through cross-field encoding assignment.

12.2.2.3 *Language for Instruction Set Architecture (LISA)*

Language for Instruction Set Architecture (LISA) [26] was developed at RWTH Aachen University, Germany with the original goal of developing fast simulators. The language has been used to produce production-quality simulators [33]. Depending on speed and accuracy constraints, different modes of instruction-set simulator, such as compiled, interpretive, and just-in-time cache-compiled (JIT-CC), can be generated. An important aspect of LISA is its ability to stepwise increase the abstraction details. A designer may start from an instruction-accurate LISA description, perform early design exploration, then move towards a detailed, cycle-accurate model. In this stepwise improvement of architectural details, application profiler, automatic instruction-set encoding [34], as well as custom instruction identification [35] plays an important role. From a cycle-accurate LISA description, optimized, low-power RTL [36,37] generation is permitted. LISA also provides a methodology for automated test pattern and assertion generation [38]. LISA has been used to generate retargetable C compilers [39,40]. LISA descriptions are composed of two types of declarations: Resource and Operation. The resource declarations cover hardware resources, such as registers, pipelines, and memory hierarchy. An example pipeline description for the architecture shown in Figure 12.4 is as follows:

```
PIPELINE int = {Fetch; Decode; IALU; MEM; WriteBack}
PIPELINE flt = {Fetch; Decode; FADD1; FADD2;
               FADD3; FADD4; MEM; WriteBack}
PIPELINE mul = {Fetch; Decode; MUL1; MUL2; MUL3; MUL4;
               MUL5; MUL6; MUL7; MEM; WriteBack}
PIPELINE div = {Fetch; Decode; DIV; MEM; WriteBack}
```

Operations are the basic objects in LISA. They represent the designer's view of the behavior, the structure, and the instruction set of the programmable architecture. Operation definitions capture the description of different properties of the system, such as operation behavior, instruction set information, and timing. These operation attributes are defined in several sections. *Coding* and *Syntax* sections cover the instruction encoding and semantics. The *Behavior* section contains the datapath of the instruction, and the *Activation* section dictates the timing behavior of the instruction across the pipeline stages. LISA exploits the commonality of similar operations by grouping them into one. The following code segment describes the decoding behavior of two immediate-type (i type) operations (ADDI and SUBI) in the DLX *Decode* stage. The complete behavior of an instruction can be obtained by combining the behavior definitions along the operation-activation chain. In this regard, an entire LISA model can be conceptualized as a Directed Acyclic Graph (DAG), where the nodes are the LISA operations and the edges are formed by the LISA Activation section.

```
OPERATION i_type IN pipe_int.Decode {
    DECLARE {
        GROUP opcode={ADDI || SUBI}
        GROUP rs1, rd = {register};
    }
    CODING {opcode rs1 rd immediate}
    SYNTAX {opcode rd ``,'' rs1 ``,'' immediate}
    BEHAVIOR { rd = rs1; imm = immediate; cond = 0;}
    ACTIVATION {opcode, writeback}
}
```

Recently, the language has been extended to cover a wide range of processor architectures, such as weakly programmable ASICs, Coarse-Grained Reconfigurable Architectures (CGRAs) [41], and partially reconfigurable ASIPs (rASIPs) [42]. One particular example of the language extension is to efficiently represent VLIW architectures. This is done by template operation declaration, as shown in the following code. Using this description style, multiple datapaths can be described in a very compact manner. The actual instantiation of the datapath with different values of *tpl* for, say different VLIW slots, take place during simulator/RTL generation.

```
OPERATION alu_op<tpl> IN pipe_int.EX {
    DECLARE {
        GROUP opcode={ADDI<tpl> || SUBI<tpl>}
        GROUP rs1, rd = {register};
    }
    CODING {opcode<tpl> rs1 rd immediate}
    SYNTAX {opcode rd '',`' rs1 '',`' immediate}
    BEHAVIOR { rd = alu_op<tpl>(rs1, immediate); }
    ACTIVATION {writeback}
}
```

Within the class of partial ADLs, an important entry is Tensilica Instruction Extension (TIE) ADL. TIE [43] captures the details of a processor to the extent it is required for the customization of a base processor.

12.2.2.4 Tensilica Instruction Extension (TIE)

To manage the complexity of processor design, customizable processor cores are provided by Tensilica [11] and ARC [44]. For Tensilica's XTensa customizable cores, the design space is restricted by discrete choices such as number of pipeline stages, width, and size of the basic address register file. On the other hand, users can model, for example, arbitrary custom instruction set extensions, register files, VLIW formats, vectorization rules by an ADL, known as Tensilica Instruction Extension (TIE). Exemplary TIE description for a vector add instruction is shown in the following code. Following the automatic parsing of TIE by XTensa C compiler generation, the instruction *add v* can be used in the application description as a C intrinsic call. Detailed description of TIE is available at [43] and [45] (Chapter 6).

```
// large register file
Regfile v 128 16

// custom vector instruction
operation add_v {out v v_out, in v v_a, in v v_b} {}
{
assign v_out = {v_a[127:96] + v_b[127:96],
                v_a[95:64] + v_b[95:64],
                v_a[64:32] + v_b[64:32],
                v_a[31:0] + v_b[31:0];}
}
```

12.3 ADL-Driven Methodologies

This section describes the ADL-driven methodologies used for processor development. It presents the following three methodologies that are used in academic research as well as industry:

- Software toolsuite generation
- Optimized generation of hardware implementation
- Top-down validation

12.3.1 Software Toolsuite Generation

Embedded systems present a tremendous opportunity to customize designs by exploiting the application behavior. Rapid exploration and evaluation of candidate architectures are necessary due to time-to-market pressure and short product lifetimes.

ADLs are used to specify processor and memory architectures and generate software toolkit including compiler, simulator, assembler, profiler, and debugger. Figure 12.5 shows an ADL-based design space exploration flow. The application programs are compiled to machine instructions and simulated, and the feedback is used to modify the ADL specification with the goal of finding the best possible architecture for the given set of application programs under various design constraints, such as area, power, performance, and reliability.

An extensive body of recent work addresses ADL-driven software toolkit generation and design space exploration of processor-based embedded systems, in both academia: ISDL [20], Valen-C [46], MIMOLA [18], LISA [26], nML [19], Sim- nML [47], EXPRESSION [25], and industry: ARC [44], RADL [48], Target [23], Processor Designer [10], Tensilica [11], and MDES [49].

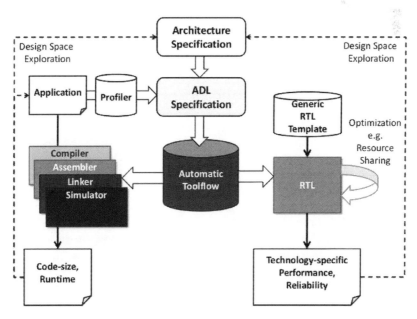

FIGURE 12.5
ADL driven design space exploration.

One of the main purposes of an ADL is to support automatic generation of a high-quality software toolkit including a cycle-accurate simulator. For supporting fast design space exploration, the simulator needs to balance speed and accuracy. In the same manner, the C/C++ compiler to be generated from the ADL specification need to support advanced features, such as specification of ILP. This section describes some of the challenges in the automatic generation of software tools (focusing on compilers and simulators) and surveys some of the approaches adopted by current tools.

1. *Compilers*: Traditionally, software for embedded systems was hand-tuned in assembly. With increasing complexity of embedded systems, it is no longer practical to develop software in assembly language or to optimize it manually, except for critical sections of the code. Compilers that produce optimized machine specific code from a program specified in a high-level language (HLL), such as C/C++ and Java, are necessary in order to produce efficient software within the time budget. Compilers for embedded systems have been the focus of several research efforts recently [50,51].

   The compilation process can be broadly broken into two steps: analysis and synthesis. During analysis, the program (in HLL) is converted into an intermediate representation (IR) that contains all the desired information, such as control and data dependences. During synthesis, the IR is transformed and optimized in order to generate efficient target specific code. The synthesis step is more complex and typically includes the following phases: instruction selection, scheduling, resource allocation, code optimizations/transformations, and code generation. The effectiveness of each phase depends on the algorithms chosen and the target architecture. A further problem during the synthesis step is that the optimal ordering between these phases is highly dependent on the target architecture and the application program. As a result, traditionally, compilers have been painstakingly hand-tuned to a particular architecture (or architecture class) and application domain(s). However, stringent time-to-market constraints for SoC designs no longer make it feasible to manually generate compilers tuned to particular architectures. Automatic generation of an efficient compiler from an abstract description of the processor model becomes essential.

   A promising approach to automatic compiler generation is the "retargetable compiler" approach. A compiler is classified as retargetable if it can be adapted to generate code for different target processors with significant reuse of the compiler source code. Retargetability is typically achieved by providing target machine information (in an ADL) as input to the compiler along with the program corresponding to the application. The complexity in retargeting the compiler depends on the range of target processors it supports and on its optimizing capability. Due to the growing amount of ILP features in modern processor architectures, the difference in quality of code generated by a naive code conversion process and an optimizing ILP compiler can be enormous. Recent approaches on retargetable compilation have focused on developing optimizations/transformations that are "retargetable" and capturing the machine specific information needed by such optimizations in the ADL. The retargetable compilers can be classified into three broad categories, based on the type of the machine model accepted as input.

   Architecture template-based: Such compilers assume a limited architecture template that is parameterizable for customization. The most common parameters include operation latencies, number of functional units, number of registers, etc. Architecture template-based compilers have the advantage that both optimizations and the phase ordering between them can be manually tuned to produce highly efficient code for the limited architecture space. Examples of such compilers include the Valen-C compiler [46] and the GNU-based C/C++ compiler from Tensilica Inc. [11]. The Tensilica GNU-based C/C++ compiler is geared towards the Xtensa parameterizable processor architecture. One important feature of this system is the ability to add new instructions (described through an Instruction Extension Language) and automatically generate software tools tuned to the new instruction-set.

Explicit behavioral information-based: Most compilers require a specification of the behavior in order to retarget their transformations (e.g., instruction selection requires a description of the semantics of each operation). Explicit behavioral information based retargetable compilers require full information about the instruction-set as well as explicit resource conflict information. Examples include the AVIV [52] compiler using ISDL, CHESS [22] using nML, and Elcor [49] using MDES. The AVIV retargetable code generator produces machine code, optimized for minimal size, for target processors with different instruction-set. It solves the phase ordering problem by performing a heuristic branch-and-bound step that performs resource allocation/assignment, operation grouping, and scheduling concurrently. CHESS is a retargetable code generation environment for fixed-point DSPs. CHESS performs instruction selection, register allocation, and scheduling as separate phases (in that order). Elcor is a retargetable compilation environment for VLIW architectures with speculative execution. It implements a software pipelining algorithm (modulo scheduling) and register allocation for static and rotating register files.

Behavioral information generation-based: Recognizing that the architecture information needed by the compiler is not always in a form that may be well suited for other tools (such as synthesis) or does not permit concise specification, some research has focused on extraction of such information from a more amenable specification. Examples include the MSSQ and RECORD compiler using MIMOLA [18], retargetable C compiler based on LISA [39], and the EXPRESS compiler using EXPRESSION [25]. MSSQ translates Pascal-like HLL into microcode for micro-programmable controllers, while RECORD translates code written in a DSP-specific programming language, called data flow language (DFL), into machine code for the target DSP. The EXPRESS compiler tries to bridge the gap between explicit specification of all information (e.g., AVIV) and implicit specification requiring extraction of instruction-set (e.g., RECORD), by having a mixed behavioral/structural view of the processor. The retargetable C compiler generation using LISA is based on reuse of a powerful C compiler platform with many built-in code optimizations and generation of mapping rules for code selection using the instruction semantics information [39]. The commercial Processor Designer environment based on LISA [10] is extended to support a retargetable compiler generation backend based on LLVM [53] or CoSy [54] (see Figure 12.6).

Custom Instruction Synthesis: Although there are embedded processors being designed completely from scratch to meet stringent performance constraints, there is also a trend toward partially pre-defined, configurable embedded processors [11], which can be quickly tuned to given applications by means of *custom instruction* and/or custom feature synthesis. The custom instruction synthesis tool needs to have a frontend, which can identify the custom instructions from an input application under various architectural constraints and

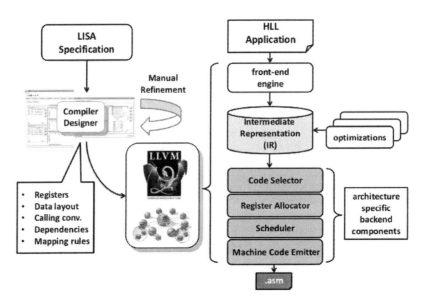

FIGURE 12.6
Retargetable compiler generation from LISA.

a flexible back-end, which can re-target the processor tools and generate the hardware implementation of the custom instruction quickly. Custom instruction, as a standalone problem, has been studied in depth [55–57]. In the context of processor, custom instruction synthesis with architectural backend is provided in [11,35]. Custom instruction synthesis with hardware-oriented optimizations are proposed in [58], while custom instructions are mapped onto a reconfigurable fabric in [59].

2. *Simulators*: Simulators are critical components of the exploration and software design toolkit for the system designer. They can be used to perform diverse tasks, such as verifying the functionality and/or timing behavior of the system (including hardware and software), and generating quantitative measurements (e.g., power consumption [60]) that can aid the design process.

Simulation of the processor system can be performed at various abstraction levels. At the highest level of abstraction, a functional simulation of the processor can be performed by modeling only the instruction-set (IS). Such simulators are termed instruction-accurate (IA) simulators. At lower-levels of abstraction are the cycle-accurate and phase-accurate simulation models that yield more detailed timing information. Simulators can be further classified based on whether they provide bit-accurate models, pin-accurate models, exact pipeline models, and structural models of the processor.

Typically, simulators at higher levels of abstraction are faster but gather less information as compared to those at lower levels of abstraction (e.g., cycle-accurate, phase-accurate). Retargetability (i.e., ability to simulate a wide variety of target processors) is especially important in the arena of embedded SoC design with emphasis on the design space exploration and co-development of hardware and software. Simulators with limited retargetability are very fast but may not be useful in all aspects of the design process. Such simulators typically incorporate a fixed architecture template and allow only limited retargetability in the form of parameters, such as number of registers and ALUs. Examples of such simulators are numerous in the industry and include the HPL-PD [49] simulator using the MDES ADL. The model of simulation adopted has significant impact on the simulation speed and flexibility of the simulator. Based on the simulation model, simulators can be classified into three types: interpretive, compiled, and mixed.

Interpretation based: Such simulators are based on an interpretive model of the instruction-set of the processors. Interpretive simulators store the state of the target processor in host memory. It then follows a fetch, decode, and execute model: instructions are fetched from memory, decoded, then executed in serial order. Advantages of this model include ease of implementation, flexibility, and the ability to collect varied processor state information. However, it suffers from significant performance degradation as compared to the other approaches primarily due to the tremendous overhead in fetching, decoding, and dispatching instructions. Almost all commercially available simulators are interpretive. Examples of interpretive retargetable simulators include SIMPRESS [29] using EXPRESSION, and GENSIM/XSIM [61] using ISDL.

Compilation based: Compilation based approaches reduce the runtime overhead by translating each target instruction into a series of host machine instructions that manipulate the simulated machine state. Such translation can be done either at compile time (static compiled simulation), where the fetch-decode-dispatch overhead is completely eliminated, or at load time (dynamic compiled simulation), which amortizes the overhead over repeated execution of code. Simulators based on the static compilation model are presented by Zhu et al. [62] and Pees et al. [63]. Examples of dynamic compiled code simulators include the Shade simulator [64] and the Embra simulator [65].

Interpretive+Compiled: Traditional interpretive simulation is flexible but slow. Instruction decoding is a time-consuming process in a software simulation. Compiled simulation performs compile time decoding of application programs to improve the simulation performance. However, all compiled simulators rely on the assumption that the complete program code is known before the simulation starts and is furthermore runtime static. Due to the restrictiveness of the compiled technique, interpretive simulators are typically used in embedded systems design flow. Several simulation techniques (JIT-CCS [33] and IS-CS [66]) combine the flexibility of interpretive simulation with the speed of the compiled simulation.

The *just-in-time cache compiled simulation* (JIT-CCS) technique compiles an instruction during runtime, *just-in-time* before the instruction is going to be executed. Subsequently, the extracted information is stored in a simulation cache for direct reuse in a repeated execution of the program address. The simulator recognizes if the program code of a previously executed address has changed and initiates a re-compilation. The *instruction set compiled simulation* (IS-CS) technique performs time-consuming instruction decoding during compile time. If an instruction is modified at runtime, the instruction is re-decoded prior to execution. It also uses an *instruction abstraction* technique to generate aggressively optimized decoded instructions that further improves simulation performance [66].

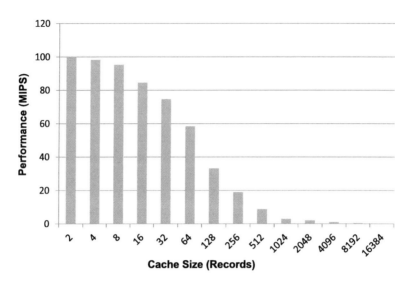

FIGURE 12.7
Performance of the just-in-time cache compiled simulation. (From Nohl, A. et al., A universal technique for fast and flexible instruction-set architecture simulation, In *Proceedings of the 39th Annual Design Automation Conference*, pages 22–27, 2002.)

Hybrid Simulation: Normally, a software developer does not need high simulation accuracy all the time. Some parts of an application just need to be functionally executed to reach a zone of interest. For those parts, a fast but inaccurate simulation is completely sufficient. For the region on interest, however, a detailed and accurate simulation might be required to understand the behavior of the implementation of a particular function. The *Hybrid Simulation* (HySim) [67] concept—a simulation technique—addresses this problem. It allows the user to switch between a detailed simulation, using previously referred ISS techniques, and direct execution of the application code on the host processor. This gives the designer the possibility to trade simulation speed against accuracy. The tricky part in hybrid simulation is to keep the application memory synchronous between both simulators. By limiting the switching to function borders, this problem becomes easier to handle (Figure 12.7).

12.3.2 Generation of Hardware Implementation

For a detailed performance evaluation of the processor as well as the final deployment on a working silicon, a synthesizable RTL description is required. There are two major approaches in the literature for synthesizable HDL generation. The first one is based on a parameterized processor core. These cores are bound to a single processor template whose architecture and tools can be modified to a certain degree. The second approach is based on high-level processor specification; that is, ADLs.

Template-based RTL generation: Examples of template-based RTL generation approaches are Xtensa [11], Jazz [68], and PEAS [69]. Xtensa [11] is a scalable RISC processor core. Configuration options include the width of the register set, caches, and memories. New functional units and instructions can be added using the Tensilica Instruction Language (TIE) [11]. A synthesizable hardware model along with software toolkit can be generated for this class of architectures. Improv's Jazz [68] processor was supported by a flexible design methodology to customize the computational resources and instruction set of the processor. It allows modifications of data width, number of registers, depth of hardware task queue, and addition of custom functionality in Verilog. PEAS [69,70] is a GUI based hardware/software codesign framework. It generates HDL code along with a software toolkit. It has support for several architecture types and a library of configurable resources.

ADL-based RTL generation: Figure 12.5 includes the flow for HDL generation from processor description languages. Structure-centric ADLs, such as MIMOLA, are suitable for hardware generation. Some of the behavioral languages (such as ISDL and nML) are also used for hardware generation. For example, the HDL generator HGEN [61] uses ISDL description, and the synthesis tool GO [23] is based on nML. Itoh et al. [71] have proposed a micro-operation description-based synthesizable HDL generation. It can handle simple processor models with no hardware interlock mechanism or multi-cycle operations.

The synthesizable HDL generation approach based on LISA [3] produces an HDL model of the architecture. The designer has the choice to generate a VHDL, Verilog, or SystemC representation of the target architecture [3]. The commercial offering [10], based on LISA [26], allows the designer to select between a highly optimized code with poor readability or an un-optimized code. Different design options like resource sharing [72], localization of storage, and decision minimization [73] can be enabled. The HDL generation methodology, based on EXPRESSION ADL, is demonstrated to have excellent performance [2].

12.3.3 Top-Down Validation

Validation of microprocessors is one of the most complex and important tasks in the current SoC design methodology. Traditional top-down validation methodology for processor architectures would start from an architectural specification and ensure that the actual implementation is in sync with the specification. The advent of ADL provided an option to have an *executable specification*, thereby allowing top-down validation flow [38,74]. This is shown graphically in Figure 12.8.

This methodology is enabled in two phases. First, the ADL specification is validated for completeness. Second, the ADL specification is used for driving simulation-based verification with increasingly detailed abstraction level.

Validation of ADL Specification: It is important to verify the ADL specification to ensure the correctness of the architecture specified and the generated software toolkit. Both static and dynamic behavior need to be verified to ensure that the specified architecture is well-formed. The static behavior can be validated by analyzing several static properties such as, connectedness, false pipeline, and data-transfer paths, and completeness using a graph-based model of the pipelined architecture [75]. The dynamic behavior can be validated by analyzing the instruction flow in the pipeline using a Finite State Machine (FSM)-based model to verify several important architectural properties such as determinism and in-order execution in the presence of hazards and multiple exceptions [76]. In [38], assertions are generated from a LISA description for detecting incorrect dynamic behavior, for example, multiple write access to the same storage element.

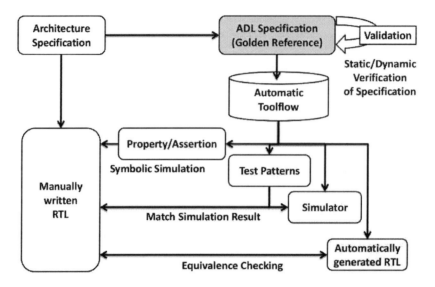

FIGURE 12.8
Top-Down validation flow.

> *Specification-driven Validation*: The validated ADL specification can be used as a golden reference model for top-down validation of programmable architectures. The top-down validation approach has been demonstrated in two directions: functional test program generation and design validation using a combination of equivalence checking and symbolic simulation.

Test generation for functional validation of processors has been demonstrated using MIMOLA [18], EXPRESSION [77], LISA [38], and nML [23]. A model checking based approach is used to automatically generate functional test programs from the processor specification using EXPRESSION [77]. It generates a graph model of the pipelined processor from the ADL specification. The functional test programs are generated based on the coverage of the pipeline behavior. Further test generation covering pipeline interactions and full-coverage test-pattern generations have been demonstrated for EXPRESSION [78] and LISA [38], respectively. ADL-driven design validation using equivalence checking is proposed in [74]. This approach combines ADL-driven hardware generation and validation. The generated hardware model (RTL) is used as a reference model to verify the hand-written implementation (*RTL design*) of the processor. To verify that the implementation satisfies certain properties, the framework generates the intended properties. These properties are applied using symbolic simulation [74].

Note that the functional validation problem is relatively simpler for configurable processor cores. Therefore, in each possible configuration, which is a finite set, the processor specification needs to be validated. This is presented for Tensilica configurable cores [79].

12.4 ADL-Driven Physical Modeling Framework

Processor power estimation techniques have been continuously a hot topic in both research and industry. An instruction-level power model is proposed by Tiwari et al. [80,81], where each instruction is provided with an individual power model. The runtime power can be determined through the profiling of executed instructions. Wattch [82] introduces architecture-level power model that decompose main processor units into categories based on their structures and separates each of the units into stages and forms resistor-capacitor (RC) circuits for each stage. McPAT [83] models all dynamic, static, and short-circuit power while providing a joint modeling capability of area and timing. To increase modeling accuracy, a hybrid FLPA (functional level power analysis) and ILPA (instruction level power analysis) model [84] is elaborated that advantageously combines the lower modeling and computational efforts of an FLPA model and the higher accuracy of an ILPA model. The trade-off is further explained in [85] with a 3-dimensional Look-up Table (LUT) and a tripartite hyper-graph.

The heat dissipation from power consumption leads to increased and unevenly distributed temperature that causes potential reliability problems [86,87], where there is high demand for research into architecture-level thermal management techniques. Consequently, accurate architecture level thermal modeling has received huge interests. In this domain, HotSpot [88] is the de facto standard, where the thermal effects for individual architecture blocks can be fast estimated. HotSpot is easy to integrate with any source level power simulator, which spreads its appliance into huge research bodies [89,90].

Recently, there is an emerging research trend for multi-domain simulation, where physical factors in more than one system, such as electrical, chemical and mechanical, are jointly simulated [91]. In the domain of digital processor design, Cacti [92] estimates power, area, and timing specifically for the memory system. McPAT [83] jointly models power, area, and timing for individual system-level blocks, including cores and memories [93], applies a joint performance, power, and thermal simulation framework for the design of network-on-chip [94], and extends the work with the ability to simulate optimization techniques, such as Dynamic Voltage Frequency Scaling (DVFS) and Power Gating.

However, the previous work simulates the physical behaviors using off the shelf libraries on a higher abstraction level for individual blocks, which did not deal with the complexity of processor architecture itself. An Application-specific Integrated Processor (ASIP) can have arbitrary logic blocks, which need detailed block level modeling of physical parameters. Previous work also lacks the ability to accurately estimate power/temperature with application-specific switching activities. The reason is that modeling and simulation are treated as separate issues, where the modeling part is more tent to be provided from IP vendors as technology dependent databases.

In this section, a joint physical modeling framework is demonstrated by integrating power, thermal, and logic timing in an ADL LISA-based processor design environment, where both accurately model through low-level characterization and cross-domain simulation using instruction-set simulator are fast realized. The reliability simulation is achieved as an extension to the high-level fault injection technique [95], where faults are modeled as delay variation on logic paths resulted from instantaneous power and thermal footprints. By automating the complete modeling and simulation flow, the processor designer can easily perform architectural and application-level design space exploration with power, temperature, and reliability issues.

The work is organized in the following manner. Section 12.4.1 discusses the approach of high-level power modeling and estimation for a LISA-based processor design framework. Section 12.4.2 illustrates the thermal modeling and integration using HotSpot package. Section 12.4.3 introduces the approach of high-level logic timing simulation. Section 12.4.4 focuses on the automation flow and analyses its runtime overhead.

12.4.1 ADL-Driven Power Modeling and Estimation

The runtime power consumption for LISA units (operations and resources) can be characterized from power simulation of low levels, such as layout or gate-level. The power models are integrated into the instruction-set simulator to estimate power with significant speed-up. The accuracy of the simulation is related with the level of details during characterization. This section explains the proposed modeling flow and features. First, an overview on the power modeling flow is presented. Second, construction of architectural power models is illustrated. Afterward, approaches to handle power related factors are introduced.

1. *Flow Overview*: Figure 12.9 illustrates the power estimation flow consisting of the simulation, characterization, estimation, and exploration phases.

   a. *Simulation*: This phase collects cycle-accurate power data from low-level power simulation. The testbenches for each simulation is constructed with single instruction type with random operands, which is used to create power models for individual instruction. To improve modeling accuracy, one group of instructions (e.g., ALU instruction) can be further differentiated as addition, subtraction, and multiplication to collect corresponding power traces due to their obvious difference in power. Besides power data, the trace of switching activity on interface signals of individual architecture unit is also gathered.

   b. *Characterization*: Both traces of power and switching activity are required to characterize the power coefficients for interface signals of architecture units. Multivariate regression analysis is utilized for coefficient extraction. Enhanced

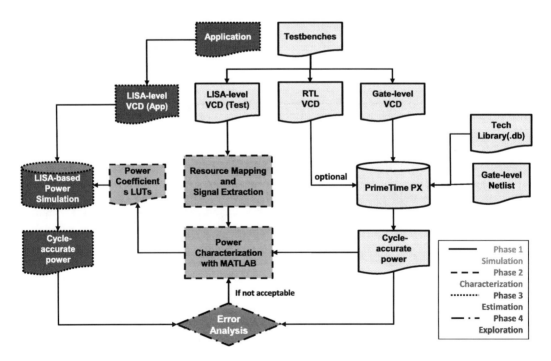

FIGURE 12.9

LISA-based power modeling and simulation flow. (From Wang, Z. et al., Power modeling and estimation during ADL-driven embedded processor design, In *2013 4th Annual International Conference on Energy Aware Computing Systems and Applications* (*ICEAC*), pages 97–102, 2013. © 2013 IEEE. With permission.)

regression orders, such as polynomial order instead of linear one, can be used to improve regression accuracy while increasing extraction efforts.

c. *Estimation*: Applications running in instruction set simulator (ISS) can calculate runtime power on each unit with extracted power coefficient. The estimated power is recorded in PrimeTime power format for benchmarking purpose.

d. *Exploration*: The estimated runtime power from ISS is benchmarked with low-level power simulation to check for modeling accuracy. The estimation accuracy can be further improved by the techniques in the simulation and characterization phases.

2. *Power Modeling*: Finer power modeling from granularity of instruction to architecture units improves the accuracy of estimation. ADL-based architecture design environment facilitates the construction of power models on different granularities. Figure 12.10 lists the hierarchical representation of an RISC processor with 5 pipeline stages. The instruction level power model can be constructed by power for various

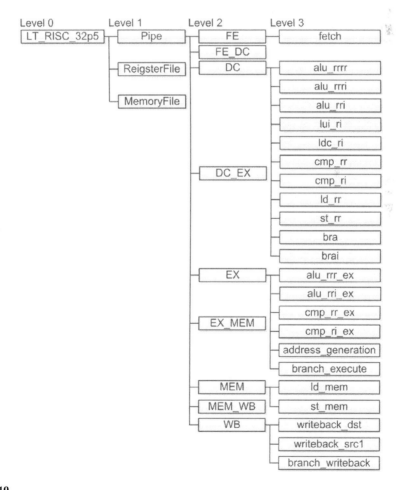

FIGURE 12.10
Hierarchical representation of RISC processor architecture. (From Wang, Z. et al., Power modeling and estimation during ADL-driven embedded processor design, In *2013 4th Annual International Conference on Energy Aware Computing Systems and Applications* (*ICEAC*), pages 97–102, 2013. © 2013 IEEE. With permission.)

instructions running on the core at level 0. Furthermore, architectural power models are created for the hardware modules at level 1 and 2 for each instruction. The proposed unit-level modeling further extends the pipeline module into operation units in level 3, which shows significant power difference among operation units. Such an approach addresses the modeling accuracy through model decomposition, while also introducing a systematic modeling approach.

Figure 12.11 illustrates the concept of unit-based power model with a block representing either a logic operation or a storage unit. The power model for each unit is constructed using the cycle-wise toggling information, or *Hamming Distances*, of interfacing signals. The runtime power of the individual unit is estimated according to the weighted summation of the power of all interfacing signals, where the weights are the power coefficient extracted from characterization phase.

Figure 12.11 gives an example of coefficient extraction for single variable using interpolation technique [97]. Both linear and polynomial curves are shown with formulas. As explained previously, the order of the formula is in proportion to the estimation accuracy. This work demonstrates power models under linear curve fitting for all units.

The power coefficient-based method gives the desired accuracy only for circuits under the same activation modes. However, an inactivated logic operation consumes one order less power than the activated one provided same hamming distance.

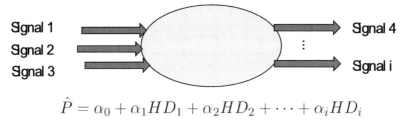

$$\hat{P} = \alpha_0 + \alpha_1 H D_1 + \alpha_2 H D_2 + \cdots + \alpha_i H D_i$$

$\hat{P}$: estimated power, HD_i: Hamming Distance for signal i,
α_i: Power coefficient for signal i, α_0: Base power coefficient

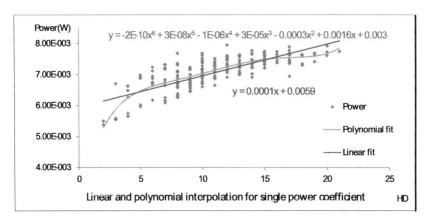

FIGURE 12.11
Unit-level power model. (From Wang, Z. et al., Power modeling and estimation during ADL-driven embedded processor design, In *2013 4th Annual International Conference on Energy Aware Computing Systems and Applications (ICEAC)*, pages 97–102, 2013. © 2013 IEEE. With permission.)

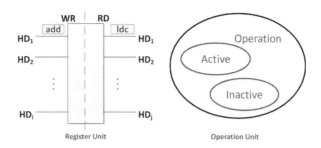

FIGURE 12.12
Separate modes for power models. (From Wang, Z. et al., Power modeling and estimation during ADL-driven embedded processor design, In *2013 4th Annual International Conference on Energy Aware Computing Systems and Applications (ICEAC)*, pages 97–102, 2013. © 2013 IEEE. With permission.)

This happens because logic switching does not propagate inside the circuit when it is inactive. With regard to storages, two modes, which are written and read, need to be separately considered, where register write can cause one order more power than register read. Such effects are included in the proposed modeling approach, as shown in Figure 12.12. According to the cycle-wise activation condition of operation units, the simulator applies either an activate or inactivate power model for calculation. Register power equals the summation of its read power and writes power.

3. *Power Related Factors*:

   a. *Inter-instruction Effect*: One key challenge for instruction level power models is the handling of inter-instruction effect (IIE), which is to quantify power caused by adjacent instruction pairs. Previous work characterizes power consumption for all instruction pairs, which demands significant efforts.

      The unit level power model handles the IIE through architecture decomposition. Adjacent instructions tend to activate different architecture units in each pipeline stage, where each one has individual power models. IIE power can be considered as the deactivating power for previously active operations, which is difficult to be included for any instruction-level power models.

   b. *Custom Instructions*: For a processor supporting N instructions in ISA, another custom instruction increases the number of instruction pairs by 2N, which is a number of pairs to be characterized under instruction-level power models. However, only a few additional units (which should be usually less than the number of pipeline stages) are added into processor pipeline to support the new instruction. This greatly reduces the effort of characterization, which enhances the flexibility of proposed flow.

   c. *Static Power*: The proposed modeling approach realizes the contribution of static power as the constant value of base power coefficient in Figure 12.11, which does not scale with any hamming distances.

4. *Experimental Results in Power Modeling*:

   a. *Instruction-level Power*: The accuracy of instruction-level power is demonstrated for a synthesized RISC processor under 90 nm technology node at 500 MHz. The average error of power by the proposed method and that with gate-level power simulation by PrimeTime PX are present in Table 12.1. Except load instruction *ld rr*, inaccuracy for all instruction groups are below 5%.

TABLE 12.1

Power Estimation Accuracy for Each Instruction Group

Instruction	Difference	Instruction	Difference
alu rrri	1.62%	st rr	2.86%
alu rrrr	0.36%	cmp rr	3.63%
alu rri	2.55%	cmp ri	3.03%
alu rrr	1.99%	ld rr	7.89%
ldc ri	3.48%	bra	2.28%
lui ri	3.64%	brau	0.53%

Source: Wang, Z. et al., Power modeling and estimation during ADL-driven embedded processor design, In *2013 4th Annual International Conference on Energy Aware Computing Systems and Applications (ICEAC)*, pages 97–102, 2013. © 2013 IEEE. With permission.

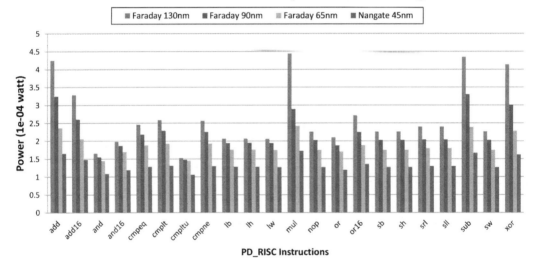

FIGURE 12.13
Instruction-level power for RISC processor.

The load instruction is relatively inaccurate (7.89%) due to the memory interface module, which is not included in the core architecture.

Instruction-level power models are straightforward to extend to other processor models. Power models for another RISC processor with mixed 16/32 bits ISA of 6 pipeline stages, synthesized under four different technologies, are further characterized. The processor is synthesized at 25 MHz. Figure 12.13 presents the instruction level power consumption.

b. *Application-level Power*: Six embedded applications are applied to demonstrate the speed and accuracy of proposed power modeling flow. Figure 12.14 shows the instruction profiling for applications in top half and accuracy/speed comparison in the bottom half. LISA-based power estimation shows a close match with the gate-level simulation for all applications, out of which Sieve has relatively larger mismatch due to its larger percentage of memory load

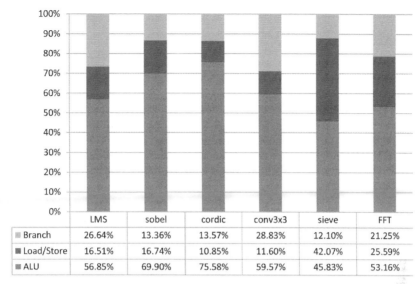

App	Average Power(mW)		Error	Speed(s)		Faster
	LISA-based Simulation	Gate level Simulation		LISA-based Simulation	Gate level Simulation	
LMS	8.70	8.80	1.40%	1.23	40.57	X 32.97
sobel	9.20	9.70	5.04%	2.56	66.35	X 25.87
cordic	8.20	8.40	2.75%	0.79	32.59	X 41.02
conv3x3	8.70	8.84	1.58%	13.60	219.29	X 16.13
sieve	8.30	9.61	13.99%	11.40	215.81	X 18.92
FFT	8.80	9.30	5.38%	22.00	726.27	X 33.01
Average						X 27.99

FIGURE 12.14
Application profiling and average power. (From Wang, Z. et al., Power modeling and estimation during ADL-driven embedded processor design, In *2013 4th Annual International Conference on Energy Aware Computing Systems and Applications (ICEAC)*, pages 97–102, 2013. © 2013 IEEE. With permission.)

instructions, which is shown as less accurate in Table 12.1. It is worth noticing that high-level power estimation in ISS has its limitation to further improve accuracy since a lot of signals existing in RTL/gate-level are trimmed out in high-level representation. Such signals, which are neglected in high-level models, still contribute to overall power consumption. With regard to estimation speed, 28x speed-up is achieved in average for all applications, which is mainly caused by the intrinsically fast speed in ISS simulation compared with low-level ones [98]. Compared with pure behavioral simulation in ISS, the simulation overhead is caused by the cycle-based activation analysis of operation units, the calculation of hamming distance, and unit-level power consumption.

Figure 12.15 visualizes the instantaneous power consumption for both levels of four applications, which shows a dynamically close match. The resolution of proposed accurate power modeling is in one clock cycle, which is quite advanced among state-of-the-art technologies.

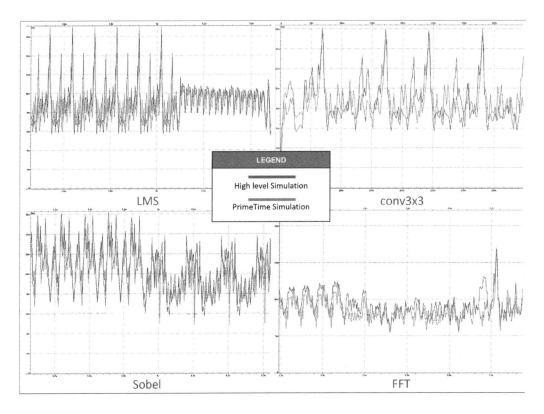

FIGURE 12.15
Instantaneous power for selected applications. (From Wang, Z. et al., Power modeling and estimation during ADL-driven embedded processor design, In *2013 4th Annual International Conference on Energy Aware Computing Systems and Applications (ICEAC)*, pages 97–102, 2013. © 2013 IEEE. With permission.)

TABLE 12.2

Power Estimation for Custom Instruction

Instruction	LISA-based Power	Gate-level Power	Difference
ZOL	5.1mw	5.3mw	3.63%

Source: Wang, Z. et al., Power modeling and estimation during ADL-driven embedded processor design, In *2013 4th Annual International Conference on Energy Aware Computing Systems and Applications (ICEAC)*, pages 97–102, 2013. © 2013 IEEE. With permission.

c. *Power for Custom Instruction*: The Zero Overhead Loop (ZOL) instruction is implemented on the RISC processor to demonstrate the flexibility of proposed modeling approach. ZOL is used to replace explicitly specified jump instructions to create a loop of subsequent instructions with a fixed amount of iterations. In the processor model, one additional *ZOL control* operation is created in *F ET CH* pipeline stage for setting program counter and detect end of iteration. Another modification is in the *DECODE* stage where ZOL instruction needs to be included in the ISA. Power modeling for updated architecture characterizes *ZOL control* and the updated *DECODE* operations. Table 12.2 presents a modeling error of 3.63%.

12.4.2 ADL-Driven Thermal Modeling

1. *Thermal Modeling using HotSpot*: HotSpot is an open source package for temperature estimation of architecture-level units. It has been applied in both academia and industry for architecture-level thermal modeling and management. HotSpot is easily integrated into any performance/power simulator by providing the floorplan and instantaneous power information. By transforming the floorplan into an equivalent thermal RC circuit, which is called *compact models*, HotSpot calculates instantaneous temperature by solving the thermal differential equation using a fourth-order Runge-Kutta method. The temperature for each block is updated by each call to the RC solver. For details of applying HotSpot for thermal modeling, please refer to [89].

2. *Integration of Power Simulator with HotSpot*: The integration of LISA power simulator with HotSpot generally follows the guideline in [99]. Two phases are required, the initialization and runtime phases, which are briefly explained in the following.

   - The initialization phase, where the RC equivalent circuits are first initialized based on user provided floorplan and thermal configurations, such as parameters for heat sink and heat spread. Afterward, the initial temperature is set by the user. For instance, 60°C is initialized for the starting temperature while 45°C is set as ambient temperature. The floorplan information could be obtained from commercial physical synthesis tool, such as Cadence SoC Encounter, or derived according to the area report from the logic synthesizer.

   - The runtime phase, where the simulator iteratively calls the temperature computing routine to update the block temperatures. Such routine does not need to be called during each clock cycle due to the nature of slow changing temperature. In practice, a sampling interval of 3.33 microseconds is adopted. The power values provided to HotSpot are the average values of the previous sampling interval.

3. *Temperature Simulation and Analysis*: Figure 12.16 shows an example of the runtime temperature simulation for Bose–Chaudhuri–Hocquenghem (BCH) application under a synthesis frequency of 500 MHz. The unit of time is in the nanosecond while the temperature is in degree Celsius.

Table 12.3 shows the temperature and power consumption for architectural units with different design frequencies, where BCH application runs on the processor. As the power increases dramatically with frequency, the temperature shows slightly increment for most of the units, such as DC, MEM, and WB.

Rapid increment lies between 100 MHz and 500 MHz for units FE and FE DC, even though their power consumptions are relatively small compared with other units. On the contrary, units such as RegisterFile, which incur higher power consumption, shows only a slight increment in temperature. The reason behind this is the high-power density of such units due to their small area provided by the floorplan.

Table 12.3 shows the temperature for BCH application using different floorplans. The first floorplan adopts the ratio of unit size from logic synthesis tools. However, the runtime temperature shows strong differences among different architectural units, which has the potential to incur temperature related reliability issues. Floorplan 2 tries to increase the sizes of units with high power density so that the power density will be significantly reduced. As seen from the thermal simulation, the temperature of hot units reduces dramatically so that the thermal footprints of pipeline registers and RegisterFile are finalizing at similar values. To prevent large area overhead, a slight increment to the area of registers

```
Time                              Temperature (Degree Celsius)
(nanosec)  FE      DC      EX      MEM     WB      FE_DC   DC_EX   EX_MEM  MEM_WB  RegisterFile
3320       60.01   60.00   60.00   60.00   60.00   60.01   60.00   60.00   60.00   60.01
6680       60.03   60.00   60.00   60.00   60.00   60.03   60.01   60.01   60.01   60.01
10040      60.04   60.00   60.00   60.00   60.00   60.04   60.01   60.01   60.01   60.02
13400      60.06   60.00   60.01   60.00   60.00   60.05   60.01   60.01   60.01   60.02
16760      60.07   60.00   60.01   60.00   60.00   60.06   60.02   60.02   60.02   60.03
20120      60.09   60.00   60.01   60.00   60.00   60.07   60.02   60.02   60.02   60.03
23480      60.10   60.01   60.01   60.01   60.00   60.08   60.02   60.02   60.02   60.03
26840      60.12   60.01   60.01   60.01   60.00   60.09   60.02   60.03   60.03   60.04
30200      60.13   60.01   60.01   60.01   60.00   60.10   60.03   60.03   60.03   60.04
33560      60.14   60.01   60.01   60.01   60.00   60.11   60.03   60.03   60.03   60.05
36920      60.16   60.01   60.01   60.01   60.00   60.12   60.03   60.03   60.04   60.05
40280      60.17   60.01   60.02   60.01   60.01   60.13   60.04   60.04   60.04   60.06
43640      60.18   60.01   60.02   60.01   60.01   60.14   60.04   60.04   60.04   60.06
47000      60.19   60.01   60.02   60.01   60.01   60.15   60.04   60.04   60.04   60.07
50360      60.21   60.01   60.02   60.01   60.01   60.16   60.04   60.05   60.05   60.07
53720      60.22   60.01   60.02   60.01   60.01   60.17   60.05   60.05   60.05   60.07
57080      60.23   60.01   60.02   60.01   60.01   60.18   60.05   60.05   60.05   60.08
60440      60.24   60.01   60.02   60.02   60.01   60.19   60.05   60.05   60.06   60.08
63800      60.26   60.01   60.02   60.02   60.01   60.20   60.05   60.06   60.06   60.09
67160      60.27   60.01   60.03   60.02   60.01   60.21   60.06   60.06   60.06   60.09
70520      60.28   60.01   60.03   60.02   60.01   60.22   60.06   60.06   60.07   60.09
73880      60.29   60.02   60.03   60.02   60.01   60.23   60.06   60.06   60.07   60.10
77240      60.31   60.02   60.03   60.02   60.01   60.24   60.06   60.07   60.07   60.10
80600      60.32   60.02   60.03   60.02   60.01   60.25   60.07   60.07   60.07   60.11
83960      60.33   60.02   60.03   60.02   60.01   60.26   60.07   60.07   60.08   60.11
87320      60.34   60.02   60.03   60.02   60.01   60.27   60.07   60.07   60.08   60.11
90680      60.35   60.02   60.03   60.02   60.01   60.28   60.07   60.08   60.08   60.12
94040      60.36   60.02   60.03   60.02   60.01   60.29   60.08   60.08   60.08   60.12
97400      60.38   60.02   60.04   60.02   60.01   60.30   60.08   60.08   60.09   60.13
100760     60.39   60.02   60.04   60.03   60.01   60.30   60.08   60.08   60.09   60.13
```

FIGURE 12.16
Instantaneous temperature generated by HotSpot.

TABLE 12.3

Temperature and Power of LT_RISC at Different Frequencies Running BCH Application

Frequencies	25 MHz		100 MHz		500 MHz	
Units	Temp(°C)	Power(mW)	Temp(°C)	Power(mW)	Temp(°C)	Power(mW)
FE	63.90	3.19e-3	69.39	9.08e-3	84.25	3.88e-2
DC	60.16	2.56e-2	60.43	7.18e-2	61.15	2.99e-1
EX	60.23	4.91e-2	60.60	1.40e-1	62.11	7.06e-1
MEM	60.17	3.88e-3	60.63	9.53e-3	61.48	3.72e-2
WB	60.05	2.69e-3	60.20	8.47e-3	60.66	3.86e-2
FE_DC	62.16	2.71e-2	68.44	1.05e-1	89.04	4.93e-1
DC_EX	60.74	5.72e-2	62.66	2.08e-1	67.68	1.08
EX_MEM	60.74	4.09e-2	62.40	1.50e-1	67.66	7.61e-1
MEM_WB	60.45	2.32e-2	61.74	8.73e-2	66.89	4.36e-1
RegisterFile	61.09	2.39e-1	63.25	7.54e-1	69.79	3.52

is introduced due to their initially large size. The area of FE is also increased to achieve uniform temperature for all logic between pipeline stages. Overall, a 38.6% area overhead is incurred to achieve the thermal footprints where all units show temperature under 68°C. In other words, it reflects a maximal power density around 2.00 W/m^2 for arbitrary logic units. According to the strong relationship of temperature with power density, further thermal optimization techniques could be purposed (Table 12.4).

TABLE 12.4

Temperature of LT_RISC Running BCH Application Using Different Floorplans

Units	Power @500 MHz (mW)	Floorplan 1			Floorplan 2		
		Size (mm²)	Power Density (W/m²)	Temp (°C)	Size (mm²)	Power Density (W/m²)	Temp (°C)
FE	3.88e-2	0.01	4.95	84.25	1.00	0.04	60.19
DC	2.99e-1	1.24	0.24	61.15	1.24	0.24	61.15
EX	7.06e-1	1.50	0.47	62.11	1.50	0.47	62.11
MEM	3.72e-2	0.28	0.13	61.48	0.28	0.13	61.37
WB	3.86e-2	0.27	0.14	60.66	0.27	0.14	60.66
FE_DC	4.93e-1	0.08	6.03	89.04	1.00	0.49	62.40
DC_EX	1.08	0.69	1.57	67.68	0.76	1.43	67.02
EX_MEM	7.61e-1	0.48	1.59	67.66	0.49	1.55	67.42
MEM_WB	4.36e-1	0.26	1.65	66.89	0.30	1.45	66.17
RegisterFile	3.52	1.74	3.52	69.79	2.25	2.02	67.57
Total	—	6.55	—	—	9.08	—	—

TABLE 12.5

Temperature of LT_RISC at 500 MHz for Different Applications Under Floorplan 1

Units	Temperature (°C) for Different Applications									
	bch	cordic	crc32	fft	idct	median	qsort	sieve	sobel	viterbi
FE	84.25	60.38	61.10	60.57	61.70	73.67	72.62	61.27	60.73	72.65
DC	61.15	60.02	60.06	60.03	60.09	60.70	60.65	60.06	60.04	60.69
EX	62.11	60.05	60.15	60.06	60.20	61.50	61.51	60.07	60.10	61.69
MEM	61.48	60.03	60.07	60.03	60.09	60.94	60.95	60.02	60.04	60.80
WB	60.66	60.02	60.05	60.02	60.06	60.50	60.52	60.03	60.03	60.48
FE_DC	89.04	60.44	61.29	60.66	62.04	76.01	75.13	61.51	60.86	75.50
DC_EX	67.68	60.12	60.34	60.17	60.55	64.25	64.02	60.37	60.23	64.06
EX_MEM	67.66	60.12	60.36	60.18	60.54	64.32	64.14	60.38	60.25	64.22
MEM_WB	66.89	60.14	60.39	60.19	60.57	64.42	64.31	60.39	60.27	64.22
RegisterFile	69.79	60.15	60.44	60.22	60.70	65.33	65.17	60.48	60.30	65.10
Finish time (μs)	900.2	6.7	20.0	10.0	33.3	350.1	333.4	23.3	13.3	320.1

Table 12.5 shows the temperature of processor units by end of the simulation time for 10 embedded benchmarks using the initial floorplan 1. The temperature differs among applications mainly due to the difference in execution time of the applications. For instance, the BCH application that runs for 900 μs is significantly hotter on most of the units than other short applications. For applications with similar execution time, such as CRC32 and Sieve, no huge differences in temperature among all units is detected. Note that change in temperature is a slow process compared with power consumption, where application dependent thermal effects will exhibit for long execution time. For instance, with 91.4% execution time of Median application, Viterbi achieves a slightly higher temperature in EX units, which is due to the nature of more ALU instructions. Assembly level profiling shows that Viterbi incurs 59,739 ALU instructions (37.12% of all instructions) while median has the amount of 46,301 (26.39% of all instructions), which verifies Viterbi's hotter temperature in EX pipeline unit than that for Median.

12.4.3 Thermal-Aware Logic Delay Simulation

The effects of temperature on the logic delay of nanoscale CMOS technology have been heavily investigated such as Negative-bias Temperature Instability (NBTI) [86] and Inverted Temperature Dependence (ITD) [87]. Most of the previous work has focused on device and gate-level. Such effects can be modeled using the architectural level thermal simulation framework proposed in this chapter so that a thermal delay simulator for generic processor architecture could be easily generated and explored.

Figure 12.17 shows the integration framework with power and thermal simulator to model the delay fault. As discussed in Section 12.4.2, the LISA-level temperature simulator is generated using power simulator, HotSpot package, and architectural floorplan. Thermal directed delay fault is modeled by combining the thermal simulator and the high-level timing fault injection, where the runtime delay of individual logic paths is updated using temperature and a user provided delay variation model. In this section, the effects of delay change with temperature are modeled according to a second order polynomial model for 65 nm technology. The effect of ITD for different applications running on an RISC processor is also presented.

1. *Inverted Temperature Dependence*: Propagation delay of CMOS transistor is widely modelled using the *Alpha-power*

   law [100] as:

$$Delay \propto \frac{C_{out}V_{dd}}{I_d} = \frac{C_{out}V_{dd}}{\mu(T)(V_{dd} - V_{th}(T))^{\alpha}} \tag{12.1}$$

   where C_{out} is the load capacitance, α is a constant, $\mu(T)$ is the temperature-dependent carrier mobility, $V_{th}(T)$ is the temperature dependent threshold voltage. The temperature affects the delay in two ways: at high voltage V_{dd}, the delay is less sensitive to the term $V_{th}(T)$ but to the mobility, while at low temperature the thermal effects on threshold voltage dominate the delay change. As a consequence of advanced technology with small driving voltage, the increment in temperature could reduce the propagation delay rather than increase it for technologies with higher voltage. Such effect is named as *Inverted Temperature Dependence* (ITD), and the voltage that inverts the trend of thermal dependent is the *Zero-temperature coefficient* (ZTC) voltage.

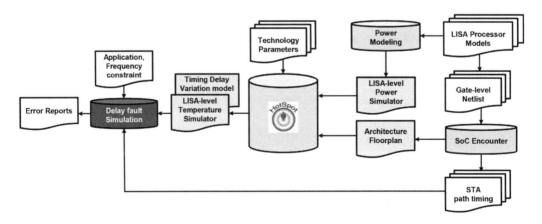

FIGURE 12.17
Thermal-aware fault injection.

2. *Timing Variation Function for Inverted Temperature Dependence*: The effects of ITD for 65 nm technology are modeled using the trend of delay change for clock tree network in [101]. Two assumptions are made to simplify the high-level modeling:

   a. The delay of logic path follows the same ratio of temperature/voltage dependency of the individual logic buffer.

   b. The temperature within one architecture block is uniform.

   c. Other thermal effects on the change of threshold voltage such as NBTI is not modeled currently.

Figure 12.18 shows two critical paths for the RISC processor and their transverse architectural blocks, which are generated by the STA tools. The delay of the complete logic path equals to the sum of path delay of individual logic blocks on the path. For instance, the critical path 1 gets two operands from pipeline register and RegisterFile transverse in order the following block: *MEM WB, DC, BYPASS DC, DC, RegisterFile, DC, ALU DC, DC, DC EX*. The critical path 2 transverses *EX MEM, EX, BYPASS EX, EX, ALU EX, EX, EX MEM*. The delay within individual architectural units are updated using its own running temperature, which is generated from the thermal simulation. In extreme cases, each cell uses its own running thermal footprints to update its delay, which can only be simulated using gate-level thermal analysis.

With the above assumption and the referred data for 65 nm technology in [101], the second order polynomials shown in Figure 12.19 are interpolated to represent the relationship between the supply voltage, instantaneous temperature, and propagation delay.

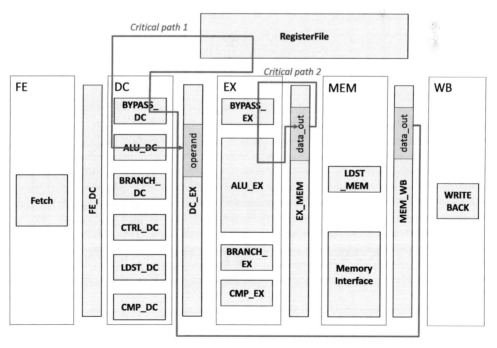

FIGURE 12.18
Critical paths and transverse blocks.

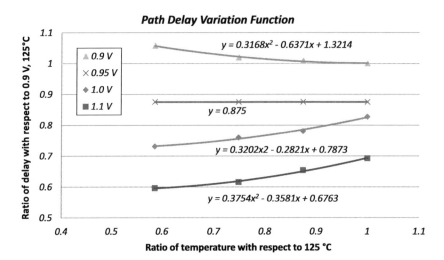

FIGURE 12.19
Delay variation function under several conditions.

It is observed that the trend of propagation delay with temperature differs with supply voltage. For 1.0V and 1.1V the delay increases with temperature while decreases at 0.9V. In [101] the ZTC voltage is known to be 0.95 V for 65 nm technology from STMicroelectronics, which proves the effect of ITD for advanced technology.

3. *ITD Logic Delay Estimation*: The polynomials are used as the path timing variation models for the RISC processor and to test the change of critical path running embedded applications. Figure 12.20 shows the runtime delay of the critical path

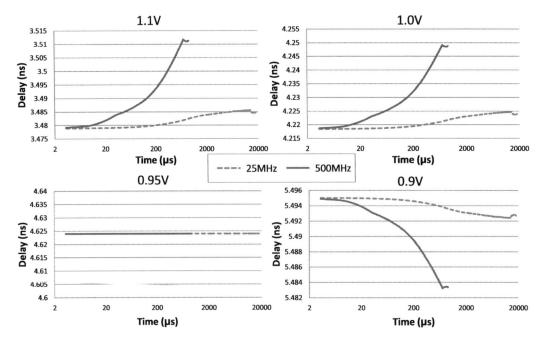

FIGURE 12.20
Runtime delay of critical path for BCH application.

for the RISC processor running BCH application. Curves are plotted for both frequencies of 25 and 500 MHz. The supply voltage is simulated using 0.9 V, 0.95 V, 1.0 V and 1.1 V. The initial delay of critical path extracted out of the timing analysis tool is for the worst case condition under 125°C, 0.9 V. It is observed that for high supply voltage such as 1.1 V and 1.0 V, the delay increases with temperature till a saturation point then slightly decreases according to the characteristics of the application. For a low voltage of 0.9 V, the inverse trend is shown where the delay decreases with temperature till the saturation point and then slightly increases. Under the ZTC voltage, which is 0.95 V, the delay is not affected by the temperature as expected. The effect of ITD shows the potential of frequency over-scaling under lower voltage, which is predicted for 65 nm and further technologies in [102]. With regard to different running frequencies, the processor running at 500 MHz consumes higher power, which leads to higher temperature compared to the data at 25 MHz. Consequently, the speed of delay change shows more significant dependence on temperature for higher frequencies.

12.4.4 Automation Flow and Overhead Analysis

In this section, the purposed automated estimation flow for Power/Thermal/Delay is briefly documented, which functions as a simulator wrapper to the Synopsys Processor Designer [103]. Furthermore, the overheads for both characterization and simulation are discussed.

1. *Flow Summary*: Figure 12.21 illustrates the complete analysis framework where the architecture description and application of interests are provided as inputs. The framework consists of characterization and simulation phase. The power characterization phase consists of four modules, which are briefly explained:

   a. *Testbench generation*: is used to generate processor-specific testbenches for power characterization. This module parses the syntax section of processor description to produce instructions with random operands. One testbench is generated for each type of instruction, which runs for a predefined simulation clock cycles.

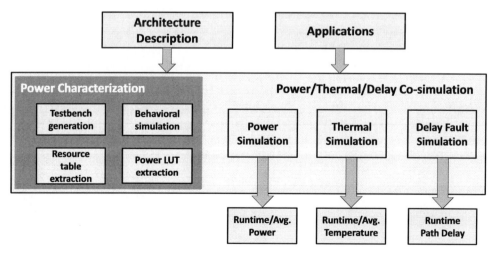

FIGURE 12.21
Automation flow of power/thermal/logic delay co-simulation.

b. *Resource table extraction*: gets the hierarchical information of the architecture and extracts input and output signals for each architecture unit. Read and write power models in the form of interpolated polynomial will be generated to each unit.

c. *Behavioral simulation*: dumps the runtime hamming distance of input/output signals per architecture unit, which is used for power coefficient extraction.

d. *Power LUT extraction*: interpolates power coefficients in the form of LUT using hamming distance and data from low-level power simulation. The interpolation itself is carried out using MATLAB tool.

e. *Power simulation*: takes loops to simulate processor behavior and power consumption until the end of the simulation cycles. In each control step, the simulator calculates power consumption based on the architecture unit specific instruction type, runtime hamming distances of the pins, and power coefficient of the architecture units. Instead of list-based implementation of power LUT, the hash container is applied to increase the speed of instruction-architecture specific LUT addressing. The hierarchical power data according to Figure 12.10 is dumped during simulation. More modeling architecture units lead to a higher overhead of power estimation.

f. *Thermal and delay simulations*: are automatically generated once the power simulator is ready since no further characterization steps are required for thermal and delay simulation.

The proposed flow is demonstrated by using Synopsys Processor Designer and is portable to any high-level architecture simulation environment and architectures. Further work includes the porting of the framework into other ADL such as SystemC.

2. *Overhead Analysis*: Table 12.6 shows the timing and accuracy for power characterization phase under two groups of testbenches where 10 architecture units are modeled. The first group consists of 14 types of instructions to cover the most generalized processor instructions. For instance, ALU instructions such as add and sub, which operate on 2 register operands and 1 immediate are grouped together in one instruction type. The second group consists of 33 types of instructions where each instruction type consists of exact one operational mode. The characterization is performed on the machine with Intel Core i7 CPU at 2.8 GHz. Each instruction file is running for 2,000 clock cycle.

As shown in Table 12.6, group one achieves faster characterization time than group two. However, group two achieves higher estimation accuracy when benchmarked with gate-level power estimation. Generally, the power characterization time in the range of several minutes is acceptable for power modeling of embedded processors.

TABLE 12.6

Time and Accuracy of Power Characterization for Testbench Groups

Number of Testbenches	14 instructions	33 instructions
Time (minutes)	3	8
Average error (%)	21.3	8.6

TABLE 12.7

Runtime Overhead for Different Simulation Modes

Simulator Applications	Behavior (sec)	Power (sec)	Times	Thermal (sec)	+%	Delay (sec)	+%
BCH	2.04	124.94	61x	125.47	0.4	129.72	3.4
Viterbi	0.82	43.49	53x	44.37	2.0	47.86	7.9
Median	0.87	49.40	57x	49.45	0.1	53.00	7.2
Qsort	0.81	45.45	56x	46.65	2.6	48.53	4.0
IDCT	0.19	5.17	27x	5.22	1.0	5.69	9.0
Average	—	—	51x	—	1.2	—	6.3

Table 12.7 represents the runtime overhead of different simulation mode including pure behavioral simulation, power estimation, thermal estimation and delay simulation, where 10 architecture units are modeled. It is observed that the runtime overhead significantly lies in the power estimation compared with behavioral simulation, on which details have been discussed in Section 12.4.4.4b. The thermal simulator achieves only 1.2% of overhead compared with power simulator, which is due to the light weight implementation of HotSpot package and smooth integration with power simulator. The delay simulation achieves in average 6.3% of overhead compared with the thermal simulator, which is mainly due to the parsing of delay information from timing analysis file that contains delay of the longest 1,000 paths.

12.5 Conclusions

The emergence of heterogeneous Multiprocessor SoCs has increased the importance of application-specific processors/accelerators. The complexity and tight time-to-market constraints of such accelerators require the use of automated tools and techniques.

Consequently, over the last decade, ADLs have made a successful transition from pure academic research to widespread acceptance in industry [11,10,23,44,104].

Indeed, the academic evolution and stepwise transition of ADLs to industrial usage makes an interesting study in the history of technology. In the academic research, starting from the early effect of nML [19], Target Compiler Technologies was founded, which was eventually acquired by Synopsys [23]. LISA, another prominent ADL, commercially ventured out as LISATek, before being acquired by CoWare Inc. and finally by Synopsys [10]. The ASIP design environment based on PEAS [69] is commercialized [70], as well. Many other notable ADLs such as MIMOLA [18] and ISDL [20] are not pursued for research or commercial usage anymore. EXPRESSION [30], ArchC [104], and MDES [49] are freely available. Among configurable cores, Tensilica was acquired by Cadence [11], and ARC configurable cores are now available via Synopsys [44]. While the basic design methodology with ADL has become commercially mature and, therefore, has been adopted heavily, the research on accurate high-level estimation of physical parameters continues to be a challenging task. This is particularly due to the reliance of physical parameters on the technology platforms, which depends on rigorous benchmarking. On the other hand, growing issues like energy-efficiency and reliability are unanimously accepted as a system-level design goal, thus must be handled at an early design phase. To that effect, ADL-based power and thermal modeling strategies evolved in the last few years, which is briefly discussed in this chapter.

This chapter provided a short overview of the high-level processor architecture design methodologies. Detailed treatment of ADLs can be found in [17]. Custom processor architectures and language-based partially reconfigurable processor architectures are discussed in detail in [105] and [42], respectively. Technically inquisitive readers can fiddle with the open source and academic ADLs [30,104] or commercial ADLs [10,23], as well as commercial template-based processor design flows [11].

References

1. M. Gries and K. Keutzer. *Building ASIPs: The Mescal Methodology*. Boston, MA: Springer, 2005.
2. P. Mishra, A. Kejariwal, and N. Dutt. Synthesis-driven exploration of pipelined embedded processors. In *VLSI Design, 2004. Proceedings. 17th International Conference on*, pages 921–926, 2004.
3. O. Schliebusch, A. Chattopadhyay, R. Leupers, G. Ascheid, H. Meyr, M. Steinert, G. Braun, and A. Nohl. RTL processor synthesis for architecture exploration and implementation. In *Design, Automation and Test in Europe Conference and Exhibition, 2004. Proceedings*, volume 3, pages 156–160, 2004.
4. P. Mishra and N. Dutt. Functional coverage driven test generation for validation of pipelined processors. In *Proceedings of the Conference on Design, Automation and Test in Europe - Volume 2*, DATE'05, pages 678–683, 2005.
5. O. Schliebusch, D. Kammler, A. Chattopadhyay, R. Leupers, G. Ascheid, H. Meyr. Automatic generation of JTAG interface and debug mechanism for ASIPs. In *GSPx, 2004. Proceedings*, 2004.
6. S. Wang and S. Malik. Synthesizing operating system based device drivers in embedded systems. In *Proceedings of the 1st IEEE/ACM/IFIP International Conference on Hardware/Software Codesign and System Synthesis*, pages 37–44, 2003.
7. Z. E. Rakossy, T. Naphade, and A. Chattopadhyay. Design and analysis of layered coarse-grained reconfigurable architecture. In *Reconfigurable Computing and FPGAs (ReConFig), 2012 International Conference on*, pages 1–6, 2012.
8. Z. E. Rakossy, A. A. Aponte, and A. Chattopadhyay. Exploiting architecture description language for diverse ip synthesis in heterogeneous MPSoC. In *Reconfigurable Computing and FPGAs (ReConFig), 2013 International Conference on*, pages 1–6, 2013.
9. C. Chen Z. Wang, and A. Chattopadhyay. Fast reliability exploration for embedded processors via high-level fault injection. In *Quality Electronic Design (ISQED), 2013 14th International Symposium on*, pages 265–272, 2013.
10. http://www.synopsys.com/systems/blockdesign/processordev/pages/default.aspx (formerly CoWare Processor Designer). Synopsys processor designer.
11. http://ip.cadence.com/ipportfolio/tensilica-ip. Cadence Tensilica Customizable Processor IP.
12. P. C. Clements. A survey of architecture description languages. In *Proceedings of the 8th International Workshop on Software Specification and Design*, pages 16, 1996.
13. M. R. Barbacci. Instruction Set Processor Specifications (ISPS): The notation and its applications. *IEEE Transactions on Computers*, 30(1):24–40, 1981.
14. W. Qin and S. Malik. Architecture description languages for retargetable compilation. In *Compiler Design Handbook: Optimizations & Machine Code Generation*, pages 535–564. Boca Raton, FL: CRC Press, 2002.
15. P. Mishra and N. Dutt. Architecture description languages for programmable embedded systems. In *IEE Proceedings on Computers and Digital Techniques*, 2005.
16. H. Tomiyama, A. Halambi, P. Grun, N. Dutt, and A. Nicolau. Architecture description languages for systems-onchip design. In *The Sixth Asia Pacific Conference on Chip Design Language*, pages 109–116, 1999.

17. P. Mishra and N. Dutt (Eds.). *Processor Description Languages*. Burlington, MA: Morgan Kaufmann Publishers, 2008.
18. R. Leupers and P. Marwedel. Retargetable code generation based on structural processor description. *Design Automation for Embedded Systems*, 3(1):75–108, 1998.
19. M. Freericks. The nML machine description formalism. TU Berlin CS Dept. Technical Report TR SM-IMP/DIST/08, 1993.
20. G. Hadjiyiannis, S. Hanono, and S. Devadas. ISDL: An instruction set description language for retargetability. In *Proceedings of the 34th Annual Design Automation Conference*, pages 299–302, 1997.
21. A. Fauth and A. Knoll. Automated generation of DSP program development tools using a machine description formalism. In *Acoustics, Speech, and Signal Processing, 1993 IEEE International Conference on*, volume 1, pages 457–460, 1993.
22. D. Lanneer, J. Praet, A. Kifli, K. Schoofs, W. Geurts, F. Thoen, and G. Goossens. CHESS: Retargetable code generation for embedded DSP processors. *Code Generation for Embedded Processors*, pages 85–102, 1995.
23. Synopsys IP Designer, IP Programmer and MP Designer (formerly Target Compiler Technologies). http://www.synopsys.com/IP/ProcessorIP/asip/ip- mp-designer/Pages/default.aspx.
24. M. R. Hartoog, J. A. Rowson, P. D. Reddy, S. Desai, D. D. Dunlop, E. A. Harcourt, and N. Khullar. Generation of software tools from processor descriptions for hardware/software codesign. In *Proceedings of the 34th Annual Design Automation Conference*, pages 303–306, 1997.
25. A. Halambi, P. Grun, V. Ganesh, A. Khare, N. Dutt, and A. Nicolau. EXPRESSION: A language for architecture exploration through compiler/simulator retargetability. In *Proceedings of the Design, Automation and Test in Europe Conference and Exhibition 1999*, pages 485–490, 1999.
26. H. Meyr A. Chattopadhyay, and R. Leupers. LISA: A uniform ADL for embedded processor modelling, implementation and software toolsuite generation. *Processor Description Languages*, P. Mishra and N. Dutt (Eds.), pages 95–130, 2008.
27. P. Grun, A. Halambi, N. Dutt, and A. Nicolau. RTGEN: An algorithm for automatic generation of reservation tables from architectural descriptions. In *System Synthesis, 1999. Proceedings. 12th International Symposium on*, pages 44–50, 1999.
28. A. Halambi, A. Shrivastava, N. Dutt, and A. Nicolau. A customizable compiler framework for embedded systems. In *Proceedings of Software and Compilers for Embedded Systems (SCOPES)*, 2001.
29. A. Khare, N. Savoiu, A. Halambi, P. Grun, N. Dutt, and A. Nicolau. V-SAT: A visual specification and analysis tool for system-on-chip exploration. In *Proceedings of the 25th EUROMICRO Conference*, volume 1, pages 196–203, 1999.
30. Exploration framework using EXPRESSION. http://www.ics.uci.edu/~express.
31. J. Hennessy and D. Patterson. *Computer Architecture: A Quantitative Approach*. Burlington, MA: Morgan Kaufmann Publishers, 1990.
32. P. Mishra, M. Mamidipaka, and N. Dutt. Processor-memory coexploration using an architecture description language. *ACM Transactions on Embedded Computing Systems*, 3(1):140–162, 2004.
33. A. Nohl, G. Braun, O. Schliebusch, R. Leupers, H. Meyr, and A. Hoffmann. A universal technique for fast and flexible instruction-set architecture simulation. In *Proceedings of the 39th Annual Design Automation Conference*, pages 22–27, 2002.
34. A. Nohl, V. Greive, G. Braun, A. Hoffman, R. Leupers, O. Schliebusch, and H. Meyr. Instruction encoding synthesis for architecture exploration using hierarchical processor models. In *Design Automation Conference, 2003. Proceedings*, pages 262–267, 2003.
35. R. Leupers, K. Karuri, S. Kraemer and M. Pandey. A design flow for configurable embedded processors based on optimized instruction set extension synthesis. In *DATE'06: Proceedings of the Conference on Design, Automation and Test in Europe*, pages 581–586, Belgium, 2006. European Design and Automation Association.

36. A. Chattopadhyay, D. Kammler, E. M. Witte, O. Schliebusch, H. Ishebabi, B. Geukes, R. Leupers, G. Ascheid, and H. Meyr. Automatic low power optimizations during ADL-driven ASIP design. In *VLSI Design, Automation and Test, 2006 International Symposium on*, pages 1–4, 2006.

37. A. Chattopadhyay, B. Geukes, D. Kammler, E. M. Witte, O. Schliebusch, H. Ishebabi, R. Leupers, G. Ascheid, and H. Meyr. Automatic ADL-based operand isolation for embedded processors. In *Proceedings of the Conference on Design, Automation and Test in Europe: Proceedings*, pages 600–605, 2006.

38. A. Chattopadhyay, A. Sinha, Diandian Zhang, R. Leupers, G. Ascheid, and H. Meyr. Integrated verification approach during ADL-driven processor design. In *Rapid System Prototyping, 2006. Seventeenth IEEE International Workshop on*, pages 110–118, 2006.

39. M. Hohenauer, H. Scharwaechter, K. Karuri, O. Wahlen, T. Kogel, R. Leupers, G. Ascheid, H. Meyr, G. Braun, and H. van Someren. A methodology and tool suite for C compiler generation from ADL processor models. In *Proceedings of the Conference on Design, Automation and Test in Europe - Volume 2*, 2004.

40. O. Wahlen, M. Hohenauer, R. Leupers, and H. Meyr. Instruction scheduler generation for retargetable compilation. *Design Test of Computers*, 20(1):34–41, 2003.

41. A. Chattopadhyay, X. Chen, H. Ishebabi, R. Leupers, G. Ascheid, and H. Meyr. High-level modelling and exploration of coarse-grained re-configurable architectures. In *Proceedings of the Conference on Design, Automation and Test in Europe*, pages 1334–1339, 2008.

42. A. Chattopadhyay, R. Leupers, H. Meyr, and G. Ascheid. *Language-driven Exploration and Implementation of Partially Re-configurable ASIPs*. Dordrecht, the Netherlands: Springer, 2009.

43. H. Sanghavi and N. Andrews. TIE: An ADL for designing application-specific instruction-set extensions. *Processor Description Languages*, P. Mishra and N. Dutt (Eds.), pages 183–216, 2008.

44. Synopsys DesignWare ARC Processor Cores. http://www.synopsys.com/IP/ProcessorIP/ARCProcessors/Pages/default.aspx.

45. B. Bailey and G. Martin. *ESL Models and Their Application*. Boston, MA: Springer, 2010.

46. A. Inoue, H. Tomiyama, E. Fajar, N. H. Yasuura, and H. Kanbara. A programming language for processor based embedded systems. In *Proceedings of APCHDL*, pages 89–94, 1998.

47. V. Rajesh and R. Moona. Processor modeling for hardware software codesign. In *VLSI Design, 1999. Proceedings. Twelfth International Conference On*, pages 132–137, 1999.

48. C. Siska. A processor desription language supporting retargetable multi-pipeline DSP program development tools. In *Proceedings of the 11th International Symposium on System Synthesis*, pages 31–36, 1998.

49. The MDES User Manual. http://www.trimaran.org.

50. M. Hohenauer, F. Engel, R. Leupers, G. Ascheid, H. Meyr, G. Bette, and B. Singh. Retargetable code optimization for predicated execution. In *Proceedings of the Conference on Design, Automation and Test in Europe*, pages 1492–1497, 2008.

51. M. Hohenauer, C. Schumacher, R. Leupers, G. Ascheid, H. Meyr and H. v. Someren. Retargetable code optimization with SIMD instructions. In *Proceedings of the 4th International Conference on Hardware/Software Codesign and System Synthesis*, pages 148–153, New York, 2006. ACM.

52. S. Hanono and S. Devadas. Instruction selection, resource allocation, and scheduling in the aviv retargetable code generator. In *Proceedings of the 35th Annual Design Automation Conference*, pages 510–515, 1998.

53. The LLVM Compiler Infrastructure. llvm.org.

54. Associated Compiler Experts. http://www.ace.nl.

55. K. Atasu, W. Luk, O. Mencer, C. Ozturan, and G. Dundar. FISH: Fast instruction synthesis for custom processors. *IEEE Transactions on Very Large Scale Integration Systems*, 20(1):52–65, 2012.

56. N. Pothineni, A. Kumar, and K. Paul. Exhaustive enumeration of legal custom instructions for extensible processors. In *VLSI Design, 2008. VLSID 2008. 21st International Conference on*, pages 261–266, 2008.

57. P. Biswas, N. D. Dutt, L. Pozzi, and P. Ienne. Introduction of architecturally visible storage in instruction set extensions. *Computer-Aided Design of Integrated Circuits and Systems, IEEE Transactions on*, 26(3):435–446, 2007.

58. K. Karuri, A. Chattopadhyay, M. Hohenauer, R. Leupers, G. Ascheid and H. Meyr. Increasing data-bandwidth to instruction-set extensions through register clustering. In *IEEE/ACM International Conference on Computer-Aided Design (ICCAD)*, 2007.

59. K. Karuri, A. Chattopadhyay, X. Chen, D. Kammler, L. Hao, R. Leupers, H. Meyr, and G. Ascheid. A design flow for architecture exploration and implementation of partially reconfigurable processors. *IEEE Transactions on Very Large Scale Integration Systems*, 16(10):1281–1294, 2008.

60. H. Xie Z. Wang, L. Wang and A. Chattopadhyay. Power modeling and estimation during ADL-driven embedded processor design. In *Energy Aware Computing Systems and Applications (ICEAC), 2013 4th Annual International Conference on*, pages 97–102, 2013.

61. G. Hadjiyiannis, P. Russo, and S. Devadas. A methodology for accurate performance evaluation in architecture exploration. In *Proceedings of the 36th Annual ACM/IEEE Design Automation Conference*, pages 927–932, 1999.

62. J. Zhu and D. D. Gajski. A retargetable, ultra-fast instruction set simulator. In *Proceedings of the Conference on Design, Automation and Test in Europe*, 1999.

63. S. Pees, A. Hoffmann, and H. Meyr. Retargetable compiled simulation of embedded processors using a machine description language. *ACM Transactions on Design Automation of Electronic Systems*, 5(4):815–834, 2000.

64. B. Cmelik and D. Keppel. Shade: A fast instruction-set simulator for execution profiling. In *Proceedings of the 1994 ACM SIGMETRICS Conference on Measurement and Modeling of Computer Systems*, pages 128–137, 1994.

65. E. Witchel and M. Rosenblum. Embra: Fast and flexible machine simulation. In *Proceedings of the 1996 ACM SIGMETRICS International Conference on Measurement and Modeling of Computer Systems*, pages 68–79, 1996.

66. M. Reshadi, P. Mishra, and N. Dutt. Instruction set compiled simulation: A technique for fast and flexible instruction set simulation. In *Proceedings of the 40th Annual Design Automation Conference*, pages 758–763, 2003.

67. S. Kraemer, L. Gao, J. Weinstock, R. Leupers, G. Ascheid and H. Meyr. HySim: A fast simulation framework for embedded software development. In *CODES+ISSS'07: Proceedings of the 5th IEEE/ACM International Conference on Hardware/Software Codesign and System Synthesis*, pages 75–80, 2007.

68. Improv Inc. http://www.improvsys.com (now defunct).

69. M. Itoh, S. Higaki, J. Sato, A. Shiomi, Y. Takeuchi, A. Kitajima, and M. Imai. PEAS-III: an ASIP design environment. In *Computer Design, 2000. Proceedings. 2000 International Conference on*, pages 430–436, 2000.

70. ASIP Solutions. http://www.asip-solutions.com/.

71. M. Itoh and Y. Takeuchi and M. Imai and A. Shiomi. Synthesizable HDL generation for pipelined processors from a micro-operation description. *IEICE Transactions on Fundamentals of Eletronics, Communications and Computer Sciences*, 2000.

72. E. M. Witte, A. Chattopadhyay, O. Schliebusch, D. Kammler, R. Leupers, G. Ascheid, and H. Meyr. Applying resource sharing algorithms to ADL-driven automatic ASIP implementation. In *Computer Design: VLSI in Computers and Processors, 2005. ICCD 2005. Proceedings. 2005 IEEE International Conference on*, pages 193–199, 2005.

73. O. Schliebusch, A. Chattopadhyay, E. M. Witte, D. Kammler, G. Ascheid, R. Leupers, and H. Meyr. Optimization techniques for ADL-driven RTL processor synthesis. In *Proceedings of the 16th IEEE International Workshop on Rapid System Prototyping*, pages 165–171, 2005.

74. P. Mishra. Processor validation: A top-down approach. *IEEE Potentials*, 24(1):29–33, 2005.

75. P. Mishra and N. Dutt. Modeling and validation of pipeline specifications. *ACM Transactions on Embedded Computing Systems*, 3(1):114–139, 2004.

76. P. Mishra, N. Dutt, and H. Tomiyama. Towards automatic validation of dynamic behavior in pipelined processor specifications. *Design Automation for Embedded Systems*, 8(2–3):249–265, 2003.

77. P. Mishra and N. Dutt. Graph-based functional test program generation for pipelined processors. In *Design, Automation and Test in Europe Conference and Exhibition, 2004. Proceedings*, volume 1, pages 182–187, 2004.

78. T. N. Dang, A. Roychoudhury, T. Mitra, and P. Mishra. Generating test programs to cover pipe-line interactions. In *Proceedings of the 46th Annual Design Automation Conference*, pages 142–147, 2009.

79. M. Puig-Medina, G. Ezer, and P. Konas. Verification of configurable processor cores. In *Design Automation Conference, 2000. Proceedings 2000*, pages 426–431, 2000.

80. V. Tiwari, S. Malik, and A. Wolfe. Power analysis of embedded software: A first step towards software power minimization. *Very Large Scale Integration (VLSI) Systems, IEEE Transactions on*, 2(4):437–445, 1994.

81. V. Tiwari, S. Malik, A. Wolfe, and M. Tien-Chien Lee. Instruction level power analysis and optimization of software. *The Journal of VLSI Signal Processing*, 13(2):223–238, 1996.

82. D. Brooks, V. Tiwari, M. Martonosi. Wattch: A framework for architectural-level power analysis and optimizations. In *ISCA'00: Proceedings of the 27th annual international symposium on Computer architecture*, pages 83–94, New York, 2000. ACM.

83. S. Li, J. H. Ahn, R. D. Strong, J. B. Brockman, D. M. Tullsen, and N. P. Jouppi. McPAT: An integrated power, area, and timing modeling framework for multicore and manycore architectures. In *Microarchitecture, 2009. MICRO-42. 42nd Annual IEEE/ACM International Symposium on*, pages 469–480. IEEE, 2009.

84. H. Blume, D. Becker, M. Botteck, J. Brakensiek, and T. Noll. Hybrid functional and instruction level power modeling for embedded processors. *Embedded Computer Systems: Architectures, Modeling, and Simulation*, pages 216–226, 2006.

85. Y. H. Park, S. Pasricha, F. J. Kurdahi, and N. Dutt. A multi-granularity power modeling methodology for embedded processors. *Very Large Scale Integration (VLSI) Systems, IEEE Transactions on*, 19(4):668–681, 2011.

86. M. A. Alam and S. Mahapatra. A comprehensive model of PMOS NBTI degradation. *Microelectronics Reliability*, 45(1):71–81, 2005.

87. K. Kanda, K. Nose, H. Kawaguchi, and T. Sakurai. Design impact of positive temperature dependence on drain current in sub-1-v CMOS VLSIs. *Solid-State Circuits, IEEE Journal of*, 36(10):1559–1564, 2001.

88. K. Skadron, M. R. Stan, W. Huang, S. Velusamy, K. Sankaranarayanan, and D. Tarjan. Temperature-aware microarchitecture. *ACM SIGARCH Computer Architecture News*, 31(2):2–13, 2003.

89. W. Huang, S. Ghosh, K. Sankaranarayanan, K. Skadron, and M. R. Stan. Hotspot: Thermal modeling for CMOS VLSI systems. *IEEE Transactions on Component Packaging and Manufacturing Technology*, pages 200–205, 2005.

90. J. Donald and M. Martonosi. Techniques for multicore thermal management: Classification and new exploration. In *ACM SIGARCH Computer Architecture News*, volume 34, pages 78–88, 2006. IEEE Computer Society.

91. M. O. Faruque, V. Dinavahi, M. Steurer, A. Monti, K. Strunz, J. A. Martinez, G. W. Chang, J. Jatskevich, R. Iravani, and A. Davoudi. Interfacing issues in multi-domain simulation tools. *Power Delivery, IEEE Transactions on*, 27(1):439–448, 2012.

92. N. Muralimanohar, R. Balasubramonian, and N. P. Jouppi. CACTI 6.0: A tool to model large caches. *HP Laboratories*, pages 22–31, 2009.

93. M. Hsieh, A. Rodrigues, R. Riesen, K. Thompson, and W. Song. A framework for architecture-level power, area, and thermal simulation and its application to network-on-chip design exploration. *ACM SIGMETRICS Performance Evaluation Review*, 38(4):63–68, 2011.

94. F. Terraneo, D. Zoni, and W. Fornaciari. An accurate simulation framework for thermal explorations and optimizations. In *Proceedings of the 2015 Workshop on Rapid Simulation and Performance Evaluation: Methods and Tools*, page 5, 2015. ACM.

95. Z. Wang, C. Chen, and A. Chattopadhyay. Fast reliability exploration for embedded processors via high-level fault injection. In *ISQED*, pages 265–272, 2013.

96. Z. Wang, L. Wang, H. Xie, and A. Chattopadhyay. Power modeling and estimation during ADL-driven embedded processor design. In *2013 4th Annual International Conference on Energy Aware Computing Systems and Applications (ICEAC)*, pages 97–102, 2013. IEEE.

97. J. Kiusalaas. *Numerical Methods in Engineering with MATLAB®*. Cambridge: Cambridge University Press, 2010.

98. R. Leupers and O. Temam. *Processor and System-on-Chip Simulation*. Heidelberg, Germany: Springer, 2010.

99. HotSpot 6.0. Documentation http://lava.cs.virginia.edu/HotSpot/documentation.htm.

100. T. Sakurai et al. Alpha-power law mosfet model and its applications to cmos inverter delay and other formulas. *Solid-State Circuits, IEEE Journal of,* 25(2):584–594, 1990.

101. A. Sassone, A. Calimera, A. Macii, E. Macii, M. Poncino, R. Goldman, V. Melikyan, E. Babayan, and S. Rinaudo. Investigating the effects of inverted temperature dependence (ITD) on clock distribution networks. In *2012 Design, Automation & Test in Europe Conference & Exhibition,* pages 165–166, 2012.

102. W. Zhao and Y. Cao. New generation of predictive technology model for sub-45 nm early design exploration. *Electron Devices, IEEE Transactions on,* 53(11):2816–2823, 2006.

103. Synopsys. Synopsys Processor Designer http://www.synopsys.com/Systems/BlockDesign/processorDev.

104. ArchC Architecture Description Language. http://archc.sourceforge.net/.

105. P. Ienne and R. Leupers. *Customizable Embedded Processors*. Burlington, MA: Morgan Kaufmann Publishers, 2006.

Spatio-Temporal Multi-Application Request Scheduling in Energy-Efficient Data Centers

Haitao Yuan, Jing Bi, and MengChu Zhou

CONTENTS

13.1 Introduction

Infrastructure resources in cloud data centers (CDCs) are shared to concurrently operate multiple applications that provide services to global users [1,2]. Each CDC typically consumes tens of megawatts of power for running and cooling tens of thousands of servers [3]. To achieve low latency and high availability, applications are replicated and deployed in multiple CDCs distributed in different locations [4]. To achieve both desired cost and performance objectives, each CDC connects to multiple Internet service providers (ISPs) that carry gigantic traffic between millions of users and distributed CDCs.

It has been shown that the energy and bandwidth costs account for a majority of the operational expenses (OPEX) of CDC providers [3]. As requests of applications in distributed CDCs soar, the energy cost of CDC providers is skyrocketing. Recently, there have been studies from both academia and industry focusing on the energy minimization problem [5,6]. However, user requests must first go through a wide-area network (WAN) consisting of multiple available ISPs and then arrive at distributed CDCs. For example, Google's WAN provides a great number of applications including mail, search, and video to global users [4]. Existing cloud providers deliver at least a petabyte of traffic per day [7]. Thus, a CDC provider suffers huge ISP bandwidth cost to deliver traffic. Besides, the bandwidth cost of each ISP is specified based on the service-level agreement (SLA) signed with a CDC provider, with some ISPs being much cheaper than others. However, recent investigations ignore the diversity in the bandwidth cost and capacities of ISPs and cause high cost and request loss. In addition, CDCs are located in different areas where the energy cost is only regionally valid; that is, the energy cost in distributed CDCs also exhibits geographical diversity. Therefore, it is challenging to minimize the total cost of CDC providers in a market where the bandwidth and energy cost show significant geographical diversity. Therefore, to solve the problem, the **first objective** of this chapter aims to present a cost-aware workload scheduling method that jointly optimizes the number of active servers in each CDC and the selection of Internet service providers for a CDC provider.

In addition, a growing number of Internet services, e.g., social networking and e-commerce, are running in large-scale cloud infrastructure in a cost-effective way [8]. However, the usage of electric power in CDCs (e.g., Amazon and Google) becomes an increasingly important concern to cloud providers in recent years. In 2011, the electric power consumed by these centers comprised roughly 3% of the total power consumption in the U.S., and the percentage might reach 15% in the future [9]. Therefore, such power consumption significantly increases the cost of CDCs. Besides, the usage of brown energy leads to severe damages to the environment. Over 57% of the electricity power in the U.S. are produced with coal in 2009 [10]. Therefore, an increasing number of large green cloud data centers (GCDCs) (e.g., Google and Microsoft) intend to reduce the usage of brown energy by adopting renewable energy sources, such as wind and solar energy. In this way, their carbon footprint can be reduced. For example, to produce solar energy, Apple has recently built a solar farm close to its existing iCloud data center.

Typically, a GCDC aims to serve all user requests in a cost-effective way while guaranteeing a user-specified delay bound. Similar to [11,12], this chapter focuses on delay-bounded requests that have a relaxed and relatively long delay bound. Typical requests in a cloud include massive-scale data analysis [13], scientific computing [14], and so on. During the delay bound of user requests, multiple factors show the temporal diversity. In the real-life market, the price of grid energy often varies during the delay bound of requests. In addition, wind speed and solar irradiance change with time [14]. Therefore, temporal diversity in these factors brings a big challenge of how to minimize the cost of grid energy consumed by a GCDC while guaranteeing a given delay bound for user requests. Therefore, to solve the problem, the **second objective** of this chapter aims to achieve the minimization of the grid energy cost by jointly considering the temporal variation of grid price, wind speed, and solar irradiance during the delay bound of requests.

The remainder of this chapter is organized as follows. Section 13.2 presents the cost-aware workload scheduling in distributed CDCs. Then, the cost-aware workload scheduling problem is formulated and the solution method is then proposed. Section 13.3 presents a temporal request scheduling framework in a GCDC. Based on it, a temporal task scheduling problem in a GCDC is further formulated and solved by the designed temporal request scheduling (TRS) algorithm. Section 4 evaluates the methods proposed in Sections 13.2 and 13.3 with the real-life workload in a Google production cluster and the realistic trace from the 1998 World Cup website, respectively. Finally, Section 13.5 concludes this chapter and discusses future directions.

13.2 Cost-Aware Workload Scheduling in Distributed CDCs

The arrival of requests in distributed CDCs is dynamic and hard to accurately predict [14]. To solve the problem, this section first proposes a revenue-based workload admission control method that considers priority, revenue, and the expected response time of every request flow. This method inclines to accept higher priority requests while meeting the bandwidth constraint of ISPs. Furthermore, this section proposes cost-aware workload scheduling to minimize the total cost of the CDCs provider by exploiting the geographical diversity of the bandwidth and energy cost. The proposed scheduling method can provide joint optimization of the ISPs selection and the number of active servers in distributed CDCs for multiple competing applications.

This section comprises two major stages shown in Figure 13.1. The first stage executes the revenue-based workload admission control method according to outside arrival requests and provides the input for the second stage. The second stage runs the cost-aware workload scheduling to specify the optimal workload assignment that can minimize the total cost of the CDCs provider.

The remainder of this section is organized as follows. Section 13.2.1 reviews the related work in the literature. Section 13.2.2 proposes the architecture of cost-aware workload scheduling in distributed CDCs. Section 13.2.3 proposes the revenue-based workload admission control method. Section 13.2.4 formulates the cost-aware workload scheduling problem and then proposes the solution method.

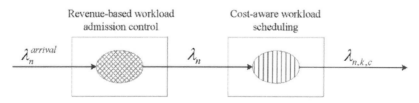

FIGURE 13.1
Two-stage design.

13.2.1 Related Work

Here, this section discusses the related work and presents the contribution of the workload admission control and cost-aware workload scheduling in comparison to existing works.

13.2.1.1 Performance Modeling

There are several works focusing on the performance modeling and analysis of cloud infrastructure by considering virtual machines (VMs) [1,14–16]. The authors in [1] present an analytical approach to evaluate the performance of cloud infrastructure by considering several metrics including system overhead rate, the rejection probability and the expected completion time. Authors in [14] propose a stochastic analytic model to quantify the performance of cloud infrastructure. Compared with traditional analytic models, their interacting iterations of multilevel submodels can obtain the solution of the overall model. The authors in [15] formulate the resource allocation problem as an integer programming and propose a heuristic algorithm to maximize the profit of cloud infrastructure. The authors in [16] propose a stochastic reward nets-based analytical model to evaluate the performance of cloud infrastructure. The behavior of a cloud system is quantified and evaluated using the defined performance metrics. However, these works cannot accurately model the energy cost of distributed CDCs. This chapter models the energy cost based on servers and consider its geographical diversity for minimizing the total cost of the CDCs provider.

13.2.1.2 Admission Control

The objective of admission control is to protect servers from overload and to guarantee the performance of applications. In [17], the authors provide joint response routing and request mapping in geographically distributed CDCs. The authors in [18] propose a coordinated method to provide admission control and to provision resources for multi-tier services in a shared platform. The reinforcement learning method and cascade neural networks are integrated to improve the scalability and agility of the system. In [11], the authors present a simple algorithm that drops excessive workload provided that the performance is met. However, these works only consider a single application and ignore the resource conflict among multiple applications. The revenue-based workload admission control method can judiciously admit requests by considering priority, revenue and the expected response time of different request flows.

13.2.1.3 Traffic Engineering

Recently, there have been several works on traffic engineering algorithms [19–21]. In [19], the authors adopt an approximation algorithm to solve the virtual local area network (VLAN) assignment problem in different network topologies. Their result shows that the approximation algorithm can provide close-to-optimal traffic engineering and performance guarantee. In [20], the authors present a joint optimization of the workload scheduling and the virtual machine placement to improve traffic engineering in CDCs. However, these works do not adopt a centralized controller to provide traffic engineering. Authors in [21] apply a centralized controller to realize traffic engineering in a network where software-defined networking (SDN) is incrementally deployed. However, their work only applies to a network where only a few switches can be controlled by a centralized SDN controller. This chapter considers a network where all switches can be remotely controlled. In addition, the proposed cost-aware workload scheduling can provide the joint optimization of the ISPs selection and the number of servers in each CDC.

13.2.2 Architecture of Distributed CDCs

This section presents the architecture of distributed CDCs in Figure 13.2. Every CDC hosts a great number of servers ranging from several hundred to several thousands. Besides, for robustness and performance, multiple ISPs that deliver traffic between distributed CDCs and users connect to each CDC.

In Figure 13.2, it is assumed that users around the world send hybrid requests to distributed CDCs where applications run. Then, users' requests are processed by three modules that are **Request Classifier, Admission Control**, and **Cost-Aware Scheduling**, respectively. The **Request Classifier** module classifies users' hybrid requests into request flows and determines the request arrival rate for each application. The **Admission Control** module executes the revenue-based workload admission control to judiciously admit requests. Then, based on the admitted requests, the *Cost-Aware Scheduling module* can minimize the total cost of the CDCs provider by specifying the workload assignment between ISPs and the number of active servers in every CDC. In addition, it is assumed that there are available ISPs delivering traffic between users and distributed CDCs. What's more, similar to the work in [22], it is assumed that replicas including programs and data that are indispensable for each application have been distributed across all CDCs. Therefore, applications and their corresponding essential data are strictly consistent with each other.

It has been shown that the centralized control of workload in distributed CDCs is feasible [23]. Therefore, similar to the work in [23], as shown in Figure 13.2, it is assumed that

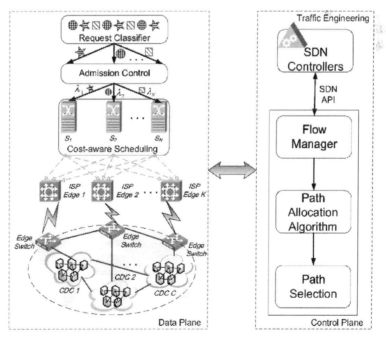

FIGURE 13.2
Architecture of distributed CDCs.

SDN controllers in the control plane can realize traffic engineering and specify routing paths for each request flow. In addition, it is assumed that SDN controllers can allocate the bandwidth resource of ISPs in the data plane among multiple competing applications. Figure 13.2 illustrates that there are multiple available paths to each CDC by specifying routing paths and that there are multiple available paths to each CDC for every request flow corresponding to each application. Therefore, SDN controllers can realize the cost-aware workload scheduling by sending OpenFlow messages [24] and installing flow entries in OpenFlow-enabled switches. Based on the architecture, Section 13.2.3 proposes the revenue-based workload admission control method. Section 13.2.4 presents the cost-aware workload scheduling that aims to minimize the total cost of the CDCs provider.

13.2.3 Workload Admission Control Problem

This section formulates the workload admission control problem and further presents the solution. The performance of the workload admission control plays an important role in the cost-aware workload scheduling in distributed CDCs. For clarity, main notations are summarized in Table 13.1.

Higher priority requests can bring more revenue to the CDCs provider than lower priority ones. Therefore, the revenue-based workload admission control method inclines to admit higher priority requests. However, this does not mean that lower priority requests cannot be admitted until higher priority ones have been completely admitted. For example, if there are not enough servers to execute higher priority requests, these requests may experience extremely long response time and bring less or no revenue to the CDCs provider. In this case, to maximize the total revenue, the revenue-based workload admission control method can refuse some of higher priority requests and intelligently admit lower priority ones that can bring more revenue.

Then, the revenue-based workload admission control problem is formulated as follows. It is assumed that requests of application n arrive in a Poisson process with rate of λ_n. The distributed CDCs are modeled as an $M/M/m$ queueing system [25]. For application n, the average serving rate of each server is defined as the average serving rate of all servers in distributed CDCs, i.e., $\sum_{c=1}^{C} M_{c,n}\mu_{c,n}/\sum_{c=1}^{C} M_{c,n}$. It is assumed that active servers are busy all the time; that is, there are always requests that wait in the queue. Therefore, the expected average response time for the arrival requests of application n, ERT_n, is calculated as follows:

$$ERT_n = \frac{1}{\left(\sum_{c=1}^{C} M_{c,n}\mu_{c,n}\right) - \lambda_n} + \frac{1}{\dfrac{\sum_{c=1}^{C} M_{c,n}\mu_{c,n}}{\sum_{c=1}^{C} M_{c,n}}} \tag{13.1}$$

An $M/M/m$ queueing system can keep stable if the traffic intensity of application n, ρ_n, must be less than 1, i.e., $\rho_n = (\lambda_n / \sum_{c=1}^{C} M_{c,n}\mu_{c,n}) < 1$. The utility (revenue) of executing a request from application n in time t_n is denoted by $u_n(t_n)$. Then, we define the utility function $u_n(t_n)$ as

$$u_n(t_n) = \begin{cases} R_n, & t_n \leq T_n^{\min} \\ R_n - \alpha_n\left(t_n - T_n^{\min}\right), & T_n^{\min} < t_n \leq T_n^{\max} \\ 0, & t_n > T_n^{\max}. \end{cases} \tag{13.2}$$

TABLE 13.1

Notations

Notation	Definition
N	Number of applications
K	Number of ISPs
C	Number of CDCs
ISP_k^{BWCap}	Bandwidth capacity of ISF k
$price_k$	Price of unit bandwidth of ISP k
$\lambda_n^{arrival}$	Request arrival rate of application n
$\lambda_{c,n}$	Request rate of application n allocated to CDC c
$\mu_{c,n}$	Serving rate of a server for application n in CDC c
λ_n	Admitted request rate of application n
$\lambda_{n,k}$	Part of λ_n allocated to ISP k
$\lambda_{n,k,c}$	Part of $\lambda_{n,k}$ allocated to CDC c
RT_n^{user}	User-defined response time constraint for application n
$ART_{c,n}$	Average response time of application n in CDC c
$m_{c,n}$	Average number of active servers for application n in CDC c
$M_{c,n}$	Total number of servers for application n in CDC c
M_c	Total number of servers in CDC c
$b_{c,n}^a$	Energy cost of an active server for application n in CDC c
$b_{c,n}^s$	Energy cost of a spare server for application n in CDC c
s_n	Size of every request of application n
Λ_c	Request rate admitted in CDC c
$penalty_n$	Penalty of unit bandwidth corresponding to refused requests of type n
ERT_n	Expected average response time for the arrival requests of application n
T_n^{max}	Maximum acceptable average service response time for request flow of application n
T_n^{min}	Minimum acceptable average service response time for request flow of application n
$u_n(t_n)$	Utility (revenue) of executing a request from application n in time t_n
α_n	Revenue decay rate for the average response time of application n
r_n	Average revenue of a request of application n

The time-varying utility function is $u_n(t_n)$ negatively correlated with t_n. R_n denotes the maximum revenue brought by executing a request of application. Note that higher priority requests bring more revenue to the CDCs provider. The request flow of application, n, has a maximum acceptable average service response time, T_n^{max}; that is, requests expect to be executed within the time limit T_n^{max}. T_n^{min} denotes the minimum response time required to execute a request of application n. Here, the value of T_n^{min} is equal to that of RT_n^{user}, which denotes the user-defined response time constraint for application$n$. α_n denotes the revenue decay rate of application n. In this way, the revenue brought by a request is proportional to its average response time. Equation (13.2) shows that, if the actual average response time of application n, t_n, is less than T_n^{min}, the request flow will bring maximum revenue, R_n, to the CDCs provider. However, if t_n is greater than T_n^{min}, the revenue brought by the request flow decreases. Furthermore, if t_n is greater than $T_n^{\max}$, no revenue is brought to the CDCs provider. This means that $u_n(t_n)$ is equal to 0 if t_n is equal to $T_n^{\max}$, i.e., $u_n(T_n^{\max}) = 0$. Therefore, $R_n - \alpha_n(T_n^{\max} - T_n^{\min}) = 0$. Then, α_n can be calculated using

$$\alpha_n = \frac{R_n}{\left(T_n^{\max} - T_n^{\min}\right)} \tag{13.3}$$

Therefore, (13.2) can be rewritten as

$$u_n(t_n) = \begin{cases} R_n, & t_n \leq T_n^{\min} \\ R_n - \dfrac{R_n}{\left(T_n^{\max} - T_n^{\min}\right)}\left(t_n - T_n^{\min}\right), & T_n^{\min} < t_n \leq T_n^{\max} \\ 0, & t_n > T_n^{\max}. \end{cases} \tag{13.4}$$

Let ISR_k^{BWCap} and $\lambda_n^{arrival}$ denote bandwidth capacity of ISP$k$and request arrival rate of application n, respectively. Let $\lambda_{n,k}$ denote part of λ_n allocated to ISP k. Let $\lambda_{n,k,c}$ denote part of $\lambda_{n,k}$ allocated to CDC c. Let r_n denote the average revenue of a request of application n.

Our objective is to maximize the total revenue of the CDCs provider. Then, the revenue-based workload admission control problem can be formulated as follows:

$$\max_{\lambda_n} Revenue = \sum_{n=1}^{N} \left(r_n \cdot \lambda_n\right)$$

subject to

$$\lambda_n \leq \lambda_n^{arrival} \tag{13.5}$$

$$\sum_{n=1}^{N} \left(r_n \cdot S_n\right) \leq \sum_{k=1}^{K} ISP_k^{BWCap} \tag{13.6}$$

$$\lambda_n < \sum_{c=1}^{C} M_{c,n}\mu_{c,n} \tag{13.7}$$

$$r_n = u_n\left(ERT_n\right) \tag{13.8}$$

$$ERT_n = \frac{1}{\left(\displaystyle\sum_{c=1}^{C} M_{c,n}\mu_{c,n}\right) - \lambda_n} + \frac{\displaystyle\sum_{c=1}^{C} M_{c,n}}{\displaystyle\sum_{c=1}^{C} M_{c,n}\mu_{c,n}}$$

$$n \in \{1, \cdots, N\} \tag{13.9}$$

Constraint (13.5) ensures the admitted request rate of application must be less than corresponding request arrival rate. Constraint (13.6) means that the total occupied bandwidth of all admitted request flows cannot exceed the total bandwidth of all ISPs. Constraint (13.7) ensures an $M/M/m$ queueing system can keep stable. Constraint (13.8) shows the revenue calculated based on the proposed utility function in (13.4). Constraint (13.9) calculates the expected average response time of application n.

In this problem, the decision variables are $\lambda_n.(n \in \{1, \cdots, N\})$. Note that the objective function is nonlinear with respect to λ_n. In addition, constraints (13.5) through (13.7) are linear constraints with respect to λ_n. Therefore, this problem is a constrained nonlinear programming. There are several traditional deterministic algorithms to solve this problem, including branch and bound [26], dynamic programming [27], and dynamic back-tracking [28]. These algorithms usually rely on the structure of a specific problem and transform it into another one that can be directly solved. However, these algorithms usually obtain optimal solution at the cost of relatively long execution time depending on the complexity of problems.

Recently, stochastic optimization algorithms have been demonstrated to be an efficient tool for tackling constrained nonlinear programming problems. They do not require any knowledge about the mathematical structure of the problems. Besides, the robustness and easy implementation of stochastic algorithms make them widely adopted to solve constrained nonlinear problems. Therefore, to tackle the drawbacks of deterministic algorithms, this section adopts a hybrid heuristic algorithm based on simulated annealing (SA) [29] and particle swarm optimization (PSO) [30] to solve the formulated problem. This section first applies the penalty function method to transform the formulated problem into an unconstrained one. Let *Penalty* denote the value of penalty function defined in

$$Penalty = \sum_{v=1}^{p} \left(\max \left\{ 0, -g_v(x) \right\} \right)^{\gamma} + \sum_{w=1}^{q} \left| h_w(x) \right|^{\delta} \tag{13.10}$$

Each equality or inequality constraint in the formulated problem corresponds to a penalty added to the objective function, *Revenue*.

In (13.10), x denotes a vector of decision variables consisting of λ_n ($n \in \{1, \cdots, N\}$). In addition, γ and δ are two constant parameters. Given p inequality constraints, constraint v can be converted into $g_v(x) \geq 0$, $1 \leq v \leq p$. Similarly, given q equality constraints, constraint w can be converted into $h_w(x) = 0$, $1 \leq w \leq q$. In this way, the formulated problem can be converted into an unconstrained one described as follows:

$$\min_{\lambda_n} aug\,Revenue = \max_{\lambda_n} \left(-Revenue + \sigma \cdot Penalty \right)$$

augRevenue denotes the augmented objective function. In addition, parameter σ is an extremely large positive number, which exaggerates the impact of *Penalty* on *augRevenue*. If a solution is not valid, *Penalty* is greater than 0. Therefore, the minimization of *augRevenue* can not only find a valid solution that does not cause any penalty, but also maximize *Revenue*.

SA provides a chance to escape from local optima by allowing moves that worsen the objection value in the hope of finding global optima. However, the running time of SA is relatively long especially when the searching space is large [29]. PSO converges to its final solution much quicker than SA. Nevertheless, PSO easily traps into local optima in solving constrained nonlinear programming [30]. Therefore, this section applies a hybrid heuristic algorithm, which integrates strengths of SA and PSO. In this hybrid algorithm, old solution of each particle and new one are compared. Better solutions are immediately accepted while inferior solutions are accepted according to the Metropolis criterion that is a typical characteristic of SA. Therefore, the hybrid algorithm can escape from local optima and eventually find global optima in solving the converted unconstrained problem.

13.2.4 Cost-Aware Workload Scheduling

This section describes the cost-aware workload scheduling problem in distributed CDCs. Similar to the work [2], each application in a specific CDC is modeled as an $M/M/m$ queueing system. It is assumed that the service time of every server conforms to exponential distribution. Besides, it is assumed that the arrival process is Poisson.

In addition, the average serving rate and the average request arrival rate of a server for application n in CDC c is denoted by $\mu_{c,n}$ and $\lambda_{c,n}$, respectively. Besides, the average number of active servers for application n in CDC c is denoted by $m_{c,n}$. The workload intensity in a queueing system is denoted by, i.e., $\rho_{c,n} = (\lambda_{c,n}/m_{c,n}\mu_{c,n})$. Based on the queueing theory [25], the condition that an $M/M/m$ queueing system can keep stable is $\rho_{c,n} < 1$. This means that $\lambda_{c,n}$ must be less than $m_{c,n}\mu_{c,n}$. Let $P_{c,n}^Q$ denote the probability of requests waiting in the queue. Let $ART_{c,n}$ denote the average response time in application n of CDC c. Therefore, $ART_{c,n}$ can be calculated according to (13.11).

$$ART_{c,n} = \frac{P_{c,n}^Q}{m_{c,n}\mu_{c,n} - \lambda_{c,n}} + \frac{1}{\mu_{c,n}} \tag{13.11}$$

Without loss of generality, it is assumed that active servers are always busy in each CDC. This means that there are requests waiting in a first-come, first-served (FCFS) queue all the time. Therefore, it is assumed that $P_{c,n}^Q$ in (13.11) equals 1. Then,

$$ART_{c,n} = \frac{1}{m_{c,n}\mu_{c.n} - \lambda_{c,n}} + \frac{1}{\mu_{c.n}} \tag{13.12}$$

Let RT_n^{user} denote the user-defined response time constraint for application n. Therefore, $ART_{c,n}$ must be less than RT_n^{user}. Therefore,

$$\frac{1}{m_{c,n}\mu_{c.n} - \lambda_{c,n}} + \frac{1}{\mu_{c.n}} \leq RT_n^{user} \tag{13.13}$$

The request rate admitted in CDC c is denoted by Λ_c, which can be calculated as follows:

$$\Lambda_c = \sum_{n=1}^{N}\sum_{k=1}^{K}\lambda_{n,k,c}, c = 1,2,\cdots,C. \tag{13.14}$$

In addition, the total number of available servers for application n in CDC c is denoted by $M_{c,n}$. Similarly, the total number of available servers for all applications in CDC c is denoted by M_c. Therefore, the total number of available servers for all applications in CDC c must be equal to M_c. Then,

$$\sum_{n=1}^{N} M_{c,n} = M_c, c = 1,2,\cdots,C. \tag{13.15}$$

Let $price_k$ denote the price of unit bandwidth of ISP k. In addition, the energy cost of an active server for application n in CDC c is denoted by $b_{c,n}^a$. Similarly, the energy cost of a spare server for application n in CDC c is denoted by $b_{c,n}^s$. Therefore, the optimization problem **(Problem One)** can be formulated as follows:

$$\min_{\lambda_{n,k,c},m_{c,n}} TotalCost = \left(\sum_{k=1}^{K} \left(price_k \cdot \left(\sum_{n=1}^{N} \sum_{c=1}^{C} \lambda_{n,k,c} \cdot s_n \right) \right) \right)$$
$$+ \sum_{c=1}^{C} \left(\sum_{n=1}^{N} \left(m_{c,n} \cdot b_{c,n}^a \right) + \sum_{n=1}^{N} \left(\left(M_{c,n} - m_{c,n} \right) \cdot b_{c,n}^s \right) \right)$$

subject to

$$\sum_{n=1}^{N} \sum_{c=1}^{C} \lambda_{n,k,c} \cdot s_n \leq ISP_k^{BWCap} \tag{13.16}$$

$$\frac{1}{m_{c,n}\mu_{c,n} - \lambda_{c,n}} + \frac{1}{\mu_{c,n}} \leq RT_n^{user} \tag{13.17}$$

$$m_{c,n} \leq M_{c,n} \tag{13.18}$$

$$\lambda_{c,n} = \sum_{c=1}^{C} \lambda_{n,k,c} < m_{c,n}\mu_{c,n} \tag{13.19}$$

$$\lambda_n = \sum_{c=1}^{C} \lambda_{c,n} = \sum_{c=1}^{C} \sum_{k=1}^{K} \lambda_{n,k,c} \tag{13.20}$$

$$m_{c,n} \in N^+ \tag{13.21}$$

$$\lambda_{n,k,c} \geq 0. \, n = 1,2,\cdots,N, k = 1,2,\cdots,K, c = 1,2,\cdots,C \tag{13.22}$$

In this problem, the objective function *TotalCost* denotes the total cost of the CDCs provider including the ISP bandwidth cost and the energy cost. Constraint (13.16) shows that the total occupied bandwidth of all admitted requests that traverse ISP k must be less than the bandwidth capacity of the corresponding ISP. Constraint (13.17) guarantees that the average response time of requests corresponding to application n must be less than the response time constraint of that application. Constraint (13.18) shows that the number of active servers for application n in CDC c cannot exceed the corresponding limit, $M_{c,n}$. Constraint (13.19) guarantees that the request arrival rate of application n in CDC c must be less than the total capacity of all corresponding servers. Besides, constraint (13.20) guarantees that all arrival requests of application n have been allocated to execute in distributed CDCs. Constraints (13.21) and (13.22) specify the valid ranges of decision variables including $m_{c,n}$ and $\lambda_{n,k,c}$.

Note that in the problem, the objection function and the corresponding constraints are both linear. Besides, decision variables include continuous variables $\lambda_{n,k,c}$ and discrete variables $m_{c,n}$. Thus, **Problem One** is a typical mixed integer linear programming (MILP) [31]. To solve this problem, this section applies the rounding method [32] to tackle the MILP problem by using a safe bound. According to [32], constraint (17) can be converted to $m_{c,n} \geq (1/\mu_{c,n}RT_n^{user} - 1) + (\lambda_{c,n}/\mu_{c,n})$. Besides, $m_{c,n}$ integer varies, and therefore, $m_{c,n} = [(1/\mu_{c,n}RT_n^{user} - 1) + (\lambda_{c,n}/\mu_{c,n})]$. Then, $m_{c,n}$ can be replaced with $[(1/\mu_{c,n}RT_n^{user} - 1) + (\lambda_{c,n}/\mu_{c,n})]$ in constraint (13.18). In this way, $(1/\mu_{c,n}RT_n^{user} - 1) + (\lambda_{c,n}/\mu_{c,n}) \leq [(1/\mu_{c,n}RT_n^{user} - 1) + (\lambda_{c,n}/\mu_{c,n})] \leq M_{c,n}$. Then, it is easy to obtain $(1/\mu_{c,n}RT_n^{user} - 1) + (\lambda_{c,n}/\mu_{c,n}) \leq M_{c,n}$. Therefore, constraint (13.18) is converted to $\lambda_{c,n} \leq M_{c,n}\mu_{c,n} - (1/\mu_{c,n}RT_n^{user} -)(1/\alpha_{c,n})$.

In addition, $m_{c,n}$ can be replaced with $(1/\mu_{c,n}RT_n^{user} - 1) + (\lambda_{c,n}/\mu_{c,n})$ in constraint (13.19). Then, constraint (13.19) is further convert to $\lambda_{c,n} \leq \mu_{c,n}((1/\mu_{c,n}RT_n^{user} - 1) + (\lambda_{c,n}/\mu_{c,n}))$. Constraint (13.17) shows that RT_n^{user} must be greater than $1/\alpha_{c,n}$. Thus, constraint (13.19) is obviously met and can be directly removed. In this way, **Problem One** can be rewritten as **Problem Two**, which is a linear programming problem described as follows:

$$\min_{\lambda_{n,k,c}} TotalCost = \left(\sum_{k=1}^{K} \left(price_k \cdot \left(\sum_{n=1}^{N} \sum_{c=1}^{C} \lambda_{n,k,c} \cdot s_n \right) \right) \right) + \sum_{c=1}^{C} \left(\sum_{n=1}^{N} \left(\frac{1}{\mu_{c,n}RT_n^{user} - 1} + \frac{\lambda_{c,n}}{\mu_{c,n}} \right) + b_{c,n}^s \right)$$

$$+ \sum_{n=1}^{N} \left(\left(M_{c,n} - \left(\frac{1}{\mu_{c,n}RT_n^{user} - 1} + \frac{\lambda_{c,n}}{\mu_{c,n}} \right) \right) \cdot b_{c,n}^s \right)$$

subject to

$$\sum_{n=1}^{N} \sum_{c=1}^{C} \lambda_{n,k,c} \cdot s_n \leq ISP_k^{BWCap} \tag{13.23}$$

$$\lambda_{c,n} \leq M_{c,n}\mu_{c,n} - \frac{1}{\mu_{c,n}RT_n^{user} - \dfrac{1}{\mu_{c,n}}} \tag{13.24}$$

$$\lambda_n = \sum_{c=1}^{C} \lambda_{c,n} = \sum_{c=1}^{C} \sum_{k=1}^{K} \lambda_{n,k,c} \tag{13.25}$$

$$\lambda_{n,k,c} \geq 0 \tag{13.26}$$

$$m_{c,n} = \left\lceil \frac{1}{\mu_{c,n}RT_n^{user} - 1} + \frac{\lambda_{c,n}}{\mu_{c,n}} \right\rceil$$

$$k = 1, 2, \cdots, K, c = 1, 2, \cdots, C, n = 1, 2, \cdots, N. \tag{13.27}$$

Therefore, $\lambda_{n,k,c}$ can be first obtained by directly solving **Problem Two**. Then, the average number of active servers $m_{c,n}$ for application n in CDC c can be further obtained according to (13.27). In this way, the total cost of the CDCs provider can be minimized by specifying the workload assignment between ISPs and the number of active servers in every CDC.

13.3 Temporal Request Scheduling (TRS) Algorithm in a GCDC

Growing efforts are made to reduce the grid energy cost of a GCDC [11,33–36]. Some papers explore the temporal diversity in electricity price to reduce the energy cost of data centers [11,35]. Other papers adopt the geographical diversity in electricity price of different regions to decrease energy cost of distributed data centers [34,36]. Different from such early studies, this section achieves the minimization of the grid energy cost by jointly considering the temporal variation of grid price, wind speed, and solar irradiance during the delay bound of requests. Besides, most of existing scheduling methods can only meet the average delay bound of all arriving requests [33]. However, the long tail in the delay of real-life requests makes it possible that the delay performance of some requests is not guaranteed [37]. Therefore, different from prior methods, this section seeks to provide delay assurance for all arriving requests by proposing a TRS algorithm.

To avoid overload of data centers, many existing papers selectively admit arriving requests and, therefore, choose to refuse excessive requests, e.g., in [11,38]. However, these papers do not provide an explicit analysis of the relation between the service rate in a data center and the refusal of delay bounded requests. Different from these papers, this section focuses on delay bounded requests and explicitly investigates the mathematical modeling of this relation. Specifically, this section formulates the grid energy cost minimization as a constrained nonlinear optimization problem and solves it with the proposed TRS based on the combination of typical meta-heuristics. It can provide strict delay assurance for each arriving request by smartly scheduling all requests to execute within their corresponding delay bound. The proposed TRS is evaluated by trace-driven simulation based on experimental data including the realistic trace from 1998 World Cup website [39], the grid price and renewable energy resources. Extensive simulation results are presented to assess the effectiveness of TRS and show that it outperforms some existing request scheduling methods in terms of grid energy cost and throughput.

The remainder of the section is organized as follows. Section 13.3.1 gives a brief discussion of related work. Section 13.3.2 presents the TRS framework in a GCDC. Based on the model of GCDC framework, Section 13.3.3 gives the formulation of a problem in a GCDC. Then, Section 13.3.4 presents the design of the TRS.

13.3.1 Related Work

This section gives a summary of some previous studies related to the research issue in this section, and reveals the differences between this section and them.

13.3.1.1 Request Scheduling

Request scheduling in data centers is a challenging topic that was studied in the past [40–44]. In [40], to mitigate interference effects for concurrent data-intensive applications, a framework that adopts control and modeling methods based on statistical machine learning is presented to significantly enhance the system performance. In [41], the problem of task scheduling in the mobile cloud computing is studied. To reduce energy consumption, its proposed algorithm performs task migration and dynamic voltage and frequency scaling after minimal-delay scheduling. In [42], a service-oriented workflow scheduling algorithm that applies a hybrid metric based on recommendation trust and direct trust is proposed to tackle the challenge brought by the unreliability and uncertainty of workflow scheduling in a cloud. In [43], to realize the suboptimal scheduling of multitask jobs over a long period, a method of iterative ordinal optimization is presented and adopted in every iteration. In [44], to lower the execution time of tasks, a fine-grained MapReduce scheduler that divides each task into several phases is introduced. This scheduler considers the high variation in resource requirements of tasks, and schedules tasks in each phase that has a specific profile of resource usage. However, request scheduling methods in these studies do not consider the cost minimization problem of data centers.

13.3.1.2 Green Cloud

Recently, several studies to focus on the application of widely available renewable energy in large-scale cloud [9,10,45–47]. The work [9] proposes a convex optimization-based strategy to maximize the profit of a data center by explicitly considering a service-level agreement. The proposed model integrates multiple factors including the stochastic nature in requests and the availability of renewable energy resources. The work [10] designs a novel online algorithm to jointly perform load scheduling and power management for distributed clouds under highly dynamic user demand. Based on the theory of Lyapunov optimization, the algorithm aims to minimize the eco-aware power cost of cloud providers provided that requests' quality of service (QoS) is ensured. The work [45] investigates the adoption of green energy in a cloud and guarantees that carbon emissions of cloud providers cannot exceed a predefined bound. The proposed framework of resource management aims to reduce operational cost by allocating resources across geo-distributed data centers. In [46], to cope with the intermittency and unpredictability of renewable energy sources, a thermal storage mechanism and load balancing in multiple geographies are exploited to better facilitate the application of green energy in data centers. Then, a stochastic program is formulated and solved by an online algorithm. The work [47] proposes a right-sizing algorithm to achieve power consumption by dynamically turning off servers. Trace-driven experiments show that the proposed algorithm can significantly achieve cost savings. The proposed TRS differs from the early studies in that it aims to minimize the grid energy cost by jointly considering the temporal variation of grid price, wind speed, and solar irradiance during the delay bound of requests.

13.3.1.3 Delay Assurance

Several existing research studies focus on modeling the delay of users' requests and provide delay assurance [48–52]. In [48], a hybrid queueing model is established to calculate the number of resources in each tier of every application in a cloud. Then, the profit maximization problem is formulated as a constrained optimization one and solved using a heuristic algorithm. The work [49] studies the problem of profit maximization by specifying the optimal multiserver setting in a cloud. The multiserver system is modeled as an $M/M/m$ queuing model based on which an optimization problem is formulated and analytically solved. In [50], a stochastic model is constructed for jobs that need to be nonpreemptively scheduled to servers. Then, a load balancing algorithm is proposed to achieve the optimal throughput. The work [51] aims to maximize the long-term profit of a cloud provider by jointly optimizing pricing and scheduling in a wireless cloud system. A dynamic algorithm is developed and applied in an arbitrarily random wireless cloud. In [52], a cost-aware scheduling approach is proposed to jointly optimize the number of servers in geo-distributed clouds and the connection of Internet service providers. However, all these studies can only guarantee the average delay of all requests. It has been demonstrated that there exists a long tail in the delay of requests in a real-life cloud [37]. Therefore, the delay of some requests may be significantly longer than the delay bound. Different from the prior methods, TRS can ensure that each request is handled within a given delay bound.

13.3.2 Motivation

The cloud computing paradigm enables users to flexibly manage their own cloud infrastructure including hundreds of thousands of servers and cooling equipment [45,52]. Many current cloud data centers run based on the electricity power produced by the burn of oil, fuel fossil, and so on. However, an increasing number of companies evolve to GCDCs. In GCDCs, the wind and solar energy and the grid price show the temporal diversity. This section aims to minimize the grid energy cost for a GCDC by jointly exploiting temporal diversity in grid price, wind speed, and solar irradiance. Compared with some existing scheduling methods, the proposed TRS can effectively reduce the grid energy cost of a GCDC and meet the delay requirement of each arriving request.

This section considers a typical realistic framework of a GCDC, as shown in Figure 13.3. A GCDC owned by a cloud provider provides services to global users who send requests from different types of devices including personal computers, laptops, and smartphones. A GCDC may obtain electricity from three types of power supplies including a nonrenewable power grid (G), the wind energy source (w), and the solar energy source (s). Such a GCDC is run as follows. Similar to [33], users' arriving requests are all enqueued into a FCFS queue. The queue information is then periodically reported to the core component, *Requestscheduler*. In addition, *Requestscheduler* periodically collects the information about G, w, and s. This section focuses on the *Requestscheduler* that runs TRS. According to the information of the FCFS queue, it can specify the service rate of requests for each server in a GCDC in each time slot. Then, it sends the setting message of each server to *Resourceallocator* that specifies the service rate of requests for each server in clusters. Similar to [53], it is assumed that servers in a GCDC are homogeneous. It can minimize the grid energy cost of a GCDC provider while guaranteeing each arriving request to be handled in a user-specified delay bound.

13.3.3 Modeling and Formulation

Based on the model in Figure 13.3, this section presents the formulation of the TRS that can minimize the grid energy cost of a GCDC. Similar to [9,11], a GCDC system is modeled as a discrete-time system that evolves in a series of equal-length time slots. In addition,

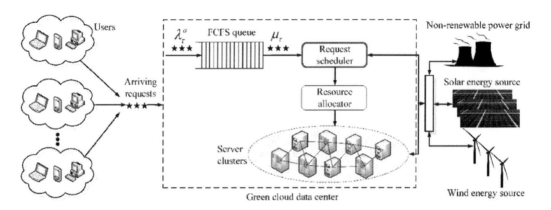

FIGURE 13.3
Illustration of a GCDC framework.

the growing deployment of high-performance clusters in it makes it realistic that an increasing number of arriving requests can be decomposed into many small subrequests executed in parallel. Thus, similar to [11], it is assumed that each arriving request can finish its execution within a single time slot.

The GCDC model shown in Figure 13.3 first enqueues users' requests into an FCFS queue. Let B denote the delay bound of users' arriving requests. λ_τ denotes the arriving rate of requests in time slot τ. μ_τ and $\mu_{\tau+b}$ denote the service rates of requests; that is, the rate at which requests are removed from the FCFS queue, in time slots τ and $\tau + b$ ($b \in \{1,..., B\}$), respectively. Λ_τ denotes the number of accumulated requests that arrive during the period of τ time slots. D_τ denotes the number of accumulated requests scheduled during the period of τ time slots. L denotes the length of each time slot. Then,

$$\Lambda_\tau = \sum_{i=1}^{\tau} \left(\lambda_i L\right) \tag{13.28}$$

$$D_\tau = \sum_{i=1}^{\tau} \left(\lambda_i^a \left(1 - \delta\left(\mu_i\right)\right) L\right) \tag{13.29}$$

In (13.29), λ_τ^a denotes the accumulated arriving rate of requests in time slot τ. Let λ_τ^r denote the remaining arriving rate of requests in time slot τ. Note that TRS can guarantee that by time slot τ, all requests that arrived in time slot τ or earlier have been scheduled in a GCDC, i.e., $\lambda_i^r = 0(i \leq \tau - B - 1)$. Thus, λ_τ^a can be calculated by summing up λ_τ and $\lambda_i^r = 0(\tau - B \leq i \leq \tau - 1)$. Then,

$$\lambda_\tau^a = \lambda_\tau + \sum_{i=\tau-B}^{\tau-1} \lambda_i^r \tag{13.30}$$

In (13.29), $\delta(\mu_i)$ denotes the loss possibility of requests in the GCDC when the service rate of requests is μ_i. Similar to [33,54], it is assumed that requests arrive into the GCDC based on a Poisson process, and the service time in the GCDC is Exponential. In this section, similar to [55,56], the GCDC is modeled as an $M/M/1/N/\infty$ queueing system. Similar to [33,54], the steady-state formulas of a queueing system are used to model a cloud and to provide performance guarantee. The capacity of the GCDC system is only N requests in total. Then,

$$\delta\left(\mu_\tau\right) = \frac{1 - \dfrac{\lambda_\tau^a}{\mu_\tau}}{1 - \left(\dfrac{\lambda_\tau^a}{\mu_\tau}\right)^{N+1}}\left(\frac{\lambda_\tau^a}{\mu_\tau}\right)^N \tag{13.31}$$

Moreover, each request should be scheduled to execute in a GCDC during its delay bound B. Thus, all requests that arrive in time slot τ can only be scheduled to execute in a GCDC from time slot τ to $\tau+B$. All requests that arrived in time slot $\tau-B$ or earlier should have been scheduled to execute in a GCDC by time slot τ. Then,

$$\Lambda_{\tau-B-1} + \lambda_{\tau-B}L \le D_{\tau-1} + \left(\lambda_\tau^a\left(1 - \delta\left(\mu_\tau\right)\right)L\right) \tag{13.32}$$

Besides, all requests that arrived in time slot $\tau-B+b$ or earlier should have been scheduled to execute in a GCDC by time slot $\tau + b$. Then,

$$\Lambda_{\tau-B-1} + \sum_{u=\tau-B}^{\tau-B+b}\left(\lambda_u L\right) \le D_{\tau-1} + \left(\lambda_\tau^a\left(1 - \delta\left(\mu_\tau\right)\right)L\right) + \sum_{u=\tau+1}^{\tau+b}\left(\tilde{\lambda}_u^a\left(1 - \delta\left(\tilde{\mu}_u\right)\right)L\right) \tag{13.33}$$

In addition, the number of requests that have been scheduled to execute in a GCDC by time slot $\tau + B$ should be equal to that of requests that have arrived by time slot τ. Then,

$$\Lambda_{\tau-B-1} + \sum_{u=\tau-B}^{\tau}\left(\lambda_u L\right) = D_{\tau-1} + \left(\lambda_\tau^a\left(1 - \delta\left(\mu_\tau\right)\right)L\right) + \sum_{u=\tau+1}^{\tau+B}\left(\tilde{\lambda}_u^a\left(1 - \delta\left(\tilde{\mu}_u\right)\right)L\right) \tag{13.34}$$

13.3.3.1 Total Energy Model

In this model, similar to [50], it is assumed that all the servers in a GCDC are homogeneous, and therefore the properties of energy consumption are identical. Let m_τ denote the number of active servers in the GCDC system in each time slot τ. In addition, each active server can execute σ requests per minute. Therefore, the service rate of requests in each time slot τ, μ_τ, can be obtained as follows.

$$\mu_\tau = \sigma m_\tau \tag{13.35}$$

The total power consumed in a GCDC includes the power consumed by the server clusters, and the power consumed by the cooling and lighting facilities. Let γ denote the GCDC's power usage effectiveness (PUE), which is the ratio of its total power consumption to the power consumed by its server clusters [57]. γ is usually defined as a metric for the energy efficiency of a GCDC. The smaller γ, the more energy-efficient. Typical γ for most of current enterprise data centers is at least 2.0. However, several studies have shown that the PUE of many existing GCDCs have reached 1.2 [46].

Let $\bar{P}_{idle}$ and $\bar{P}_{peak}$ denote the average power consumption of each server and the average peak power consumption when a server is executing requests, respectively. Let u_τ denote the CPU utilization of each active server of a GCDC in time slot τ. Therefore, the total amount of power consumed by a GCDC in time slot τ, P_τ, can be calculated as follows [58].

$$P_\tau = m_\tau \left[\bar{P}_{idle} + (\gamma - 1)\bar{P}_{peak} + \left(\bar{P}_{peak} - \bar{P}_{idle}\right)u_\tau \right] \tag{13.36}$$

Given loss possibility $\delta(\mu_\tau)$, in each time slot τ, the number of requests that each active server can execute is:

$$\frac{L\left(1 - \delta\left(\mu_\tau\right)\right)\lambda_\tau^a}{m_\tau} \tag{13.37}$$

The busy time of each server is $(L(1 - \delta(\mu_\tau))\lambda_\tau^a)/(\sigma m_\tau)$ minutes. Therefore, u_τ can be obtained by dividing the busy time of each server by L:

$$u_\tau = \frac{\left(1 - \delta\left(\mu_\tau\right)\right)\lambda_\tau^a}{\sigma m_\tau} \tag{13.38}$$

Based on (13.36) through (13.38), the total amount of energy consumed by a GCDC in time slot τ, Γ_τ, can be calculated as follows.

$$\Gamma_\tau = \frac{g\mu_\tau + h\left(\lambda_\tau^a \left(1 - \delta\left(\mu_\tau\right)\right)\right)}{\sigma} \cdot L \tag{13.39}$$

where

$$g \triangleq \bar{P}_{idle} + (\gamma - 1)\bar{P}_{peak}, \quad h \triangleq \bar{P}_{peak} - \bar{P}_{idle}$$

Let Γ_{max} denote the maximum total amount of energy. The total amount of energy consumed by a GCDC in time slots τ and $\tau + b$ should be less than or equal to Γ_{max}, respectively. Then,

$$\frac{g\mu_\tau + h\left(\lambda_\tau^a \left(1 - \delta\left(\mu_\tau\right)\right)\right)}{\sigma} L \leq \Gamma_{max} \tag{13.40}$$

$$\frac{g\tilde{\mu}_{\tau+b} + h\left(\tilde{\lambda}_{\tau+b}^a \left(1 - \delta\left(\tilde{\mu}_{\tau+b}\right)\right)\right)}{\sigma} L \leq \Gamma_{max} \tag{13.41}$$

13.3.3.2 Green Energy Model

Typical GCDCs can support green energy generators, such as wind turbines and solar panels. The adoption of green energy can reduce the amount of grid energy consumed and the carbon emission. This section considers two typical green energy supplies including wind and solar sources. It is assumed that the wind speed and solar irradiance do not vary within each time slot, but may vary from slot to slot. This assumption is reasonable because the interval of each time slot is small enough, e.g., 5 min. Let E_τ^w denote the amount of wind energy produced in time slot τ. Let ϕ denote the conversion efficiency of wind-to-electricity. Let ρ denote the air density. Besides, let v_τ and R denote the wind speed and the rotor area of wind turbines, respectively. Following [59], E_τ^w can be calculated:

$$E_\tau^w = \frac{1}{2} \cdot \phi \cdot R \cdot \rho \cdot \left(v_\tau\right)^3 \cdot L \tag{13.42}$$

Similarly, let E_τ^s denote the amount of solar energy produced in time slot τ. Let ϕ denote the conversion efficiency of solar-to-electricity. Besides, let I_τ and A denote the solar irradiance and the active irradiation area of solar panels, respectively. Following [57], E_τ^w can be calculated:

$$E_\tau^s = \phi \cdot A \cdot I_\tau \cdot L \tag{13.43}$$

Let p_τ and $\tilde{p}_{\tau+b}$ denote the price of grid energy in time slots τ and $\tau + b$, respectively. In addition, it is assumed that the price of grid energy keeps stable within each time slot that is small enough. Besides, the amount of grid energy consumed in time slot τ is $[\tilde{\Gamma}_\tau - E_\tau^s - E_\tau^w, 0]^+$, where $[*,0]^+ = \max(*,0)$. Similarly, the amount of grid energy consumed in time slot $\tau+b$ is $[\tilde{\Gamma}_{\tau+b} - \tilde{E}_{\tau+b}^s - \tilde{E}_{\tau+b}^w, 0]^+$. Let C denote the cost of grid energy consumed by a GCDC from time slots τ to $\tau + B$. Then, C can be obtained as follows.

$$C = p_\tau \left[\Gamma_\tau - E_\tau^s - E_\tau^w, 0\right]^+ + \sum_{b=1}^{B} \left\{ \tilde{p}_{\tau+b} \left[\tilde{\Gamma}_{\tau+b} - \tilde{E}_{\tau+b}^s - \tilde{E}_{\tau+b}^w, 0\right]^+ \right\} \tag{13.44}$$

Besides, there are plenty of existing papers for predicting future information. In addition, prediction models are not the focus of this section and are not considered. Therefore, we simply assume that parameters including p_τ, $\tilde{p}_{\tau+b}$, λ_τ^a, $\tilde{\lambda}_{\tau+b}^a$, v_τ, $\tilde{v}_{\tau+b}$, I_τ, and $\tilde{I}_{\tau+b}$ are all known at the start of time slot τ. Then, based on (13.28) through (13.44), the cost minimization problem of grid energy (**CMPG**) consumed by a GCDC can be formulated as follows.

$$\underset{\mu_\tau, \tilde{\mu}_{\tau+b}}{\text{Minimize}} \quad C = p_\tau \left[\Gamma_\tau - E_\tau^s - E_\tau^w, 0\right]^+ + \sum_{b=1}^{B} \left\{ \tilde{p}_{\tau+b} \left[\tilde{\Gamma}_{\tau+b} - \tilde{E}_{\tau+b}^s - \tilde{E}_{\tau+b}^w, 0\right]^+ \right\}$$

$$\frac{g\mu_\tau + h\left(\lambda_\tau^a \left(1 - \delta\left(\mu_\tau\right)\right)\right)}{\sigma} L \leq \Gamma_{max} \tag{13.45}$$

$$\frac{g\tilde{\mu}_{\tau+b} + h\left(\tilde{\lambda}_{\tau+b}^a \left(1 - \delta\left(\tilde{\mu}_{\tau+b}\right)\right)\right)}{\sigma} L \leq \Gamma_{max} \tag{13.46}$$

$$\Lambda_{\tau-B-1} + \lambda_{\tau-B} L \leq D_{\tau-1} + \left(\lambda_\tau^a \left(1 - \delta\left(\mu_\tau\right)\right)L\right) \tag{13.47}$$

$$\Lambda_{\tau-B-1} + \sum_{u=\tau-B}^{\tau-B+b} \left(\lambda_u L\right) \leq D_{\tau-1} + \left(\lambda_\tau^a \left(1 - \delta\left(\mu_\tau\right)\right)L\right) + \sum_{u=\tau+1}^{\tau+b} \left(\tilde{\lambda}_u^a \left(1 - \delta\left(\tilde{\mu}_u\right)\right)L\right) \tag{13.48}$$

$$\Lambda_{\tau-B-1} + \sum_{u=\tau-B}^{\tau} \left(\lambda_u L\right) \leq D_{\tau-1} + \left(\lambda_\tau^a \left(1 - \delta\left(\mu_\tau\right)\right)L\right) + \sum_{u=\tau+1}^{\tau+B} \left(\tilde{\lambda}_u^a \left(1 - \delta\left(\tilde{\mu}_u\right)\right)L\right) \tag{13.49}$$

$$\mu_\tau \geq 0 \geq \tilde{\mu}_{\tau+b} \geq 0 \tag{13.50}$$

In this problem, (13.50) shows the valid ranges of decision variables μ_τ and $\tilde{\mu}_{\tau+b}$. The solution to the problem can specify the request scheduling strategy and decide the service rate of requests in time slot during the delay bound of all requests. In this way, the grid energy cost of a GCDC can be minimized while the delay bound of requests can be guaranteed.

13.3.4 Temporal Request Scheduling Algorithm

Note that **CMPG** is a constrained optimization problem. To solve it, we first adopt the method of a penalty function to convert it into an unconstrained cost minimization problem of grid energy (**UCMPG**) that can be solved by the meta-heuristic algorithm to be proposed. Let $\vec{X}$ denote a vector of decision variables that include μ_τ and $\tilde{\mu}_{\tau+b}$. **UCMPG** is shown as follows.

$$\underset{\vec{X}}{Min}\left\{\tilde{f} = C + \varepsilon \cdot \varsigma\right\} \tag{13.51}$$

In (13.51), $\tilde{f}$ denotes the new objective function in **UCMPG**. The parameter ϵ is a large enough positive number. The parameter ς denotes the penalty of all constraints, each of which may bring a corresponding penalty to the objective function in **CMPG** if this constraint is not met. Then, ς can be obtained as follows.

$$\varsigma = \sum_{k=1}^{K}\left|g_k\left(\vec{X}\right)\right|^\alpha + \sum_{q=1}^{Q}\left(max\left\{0, -h_q\left(\vec{X}\right)\right\}\right)^\beta \tag{13.52}$$

It is assumed that there are K equality constraints and Q inequality constraints in (13.52). Every equality constraint k in **CMPG** can be transformed into $g_k(\vec{X}) = 0$. Then, its violation causes the penalty of $\left|g_k(\vec{X})\right|^\alpha$ where α is a positive constant. Similarly, every inequality constraint q can be transformed into $h_q(\vec{X}) \geq 0$. Then, its violation causes the penalty of $(max\{0, -h_q(\vec{X})\})^\beta$. For instance, constraint (13.45) is first transformed into $\Gamma_{max} - g\mu_\tau + h((\lambda_\tau^a(1 - \delta(\mu_\tau)))/\sigma)L \geq 0$. Its violation brings the penalty of $(max\{0, -(\Gamma_{max} - (g\mu_\tau + h(\lambda_\tau^a(1 - \delta(\mu_\tau)))/\sigma)L)\})^\beta$. In this way, **CMPG** is transformed into unconstrained problem **UCMPG** that is easier to solve based on the meta-heuristic algorithm proposed next.

Note that in **UCMPG**, μ_τ and $\tilde{\mu}_{\tau+b}$ are continuous variables. The objective function $\tilde{f}$ in **UCMPG** is nonlinear with respective to decision variables. Thus, **UCMPG** is a nonlinear optimization problem. Several present algorithms can be applied to solve this type of problem, these include Sequential Quadratic Programming (SQP) [60], Davidon–Fletcher–Powell (DFP) [61], and Broyden–Fletcher–Goldfarb-Shanno (BFGS) [62]. However, they usually rely on the specific structure of a problem and require the first-order or second-order derivative. In **UCMPG**, the function of max($*$,0) is not differentiable. Thus, **UCMPG** cannot be directly solved by standard methods for nonlinear optimization. Besides, the search for global optima is difficult especially when the space of solution is large.

On the other hand, existing meta-heuristics can avoid drawbacks of the above algorithms and are robust to solve many optimization problems with different types of mathematical structures. In addition, the implementation of meta-heuristics are easy, and therefore they are commonly applied in complicated optimization problems. Nevertheless, existing meta-heuristics all have corresponding pros and cons. For example, simulated annealing (SA) [63] is a typical meta-heuristic that is suitable to solve discrete and continuous problems with different constraints. One main characteristic of SA is that it can escape from local optima by allowing moves worsening the value of the objective function. It has been demonstrated that SA can obtain global optima in theory by meticulously choosing the cooling of temperature. However, the convergence of SA is relatively slow. Particle swarm optimization (PSO) is a population-based meta-heuristic that can quickly converge. However, although the convergence time of PSO is less than that of other meta-heuristics, it is shown that PSO is easy to trap into local optima. Thus, the quality of its solutions to complex problems is sometimes unacceptable [64].

Therefore, we combine the advantages of SA and PSO and apply a hybrid meta-heuristic, Hybrid Simulated-annealing Particle-swarm-optimization (HSP) to tackle **UCMPG**. In HSP, every particle dynamically updates its velocity according to its and other particles' positions in the current swarm. Then, every particle changes its position based on the Metropolis criterion in SA. In each iteration of HSP, the old position of each particle is compared with its new one in terms of the objective function values. Then, for each particle, better position is straightly accepted while worse one is accepted with a certain possibility. In this way, HSP can escape from local optima and likely obtain global optima that can minimize the cost of grid energy consumed by a GCDC.

TRS is shown in Algorithm 1. Line 1 initializes λ_τ ($0 \leq \tau \leq B - 1$) with 0. Line 2 shows that λ_τ^r and λ_τ^a are both initiated to λ_τ. Line 3 initiates $\Lambda - 1$ and D_{B-1} with 0. Let N_S denote the number of time slots. Line 6 updates λ_τ^a based on (30). Line 7 solves the sub-problem **UCMPG** to obtain μ_τ and $\tilde{\mu}_{\tau+b}$ using **HSP**. Line 9 schedules requests in the amount of d_τ and removes them from the head of the FIFO queue. Line 10 updates $\lambda_i^r(\tau - B \leq i \leq \tau)$. Lines 11–12 update D_τ and $\Lambda_{\tau-B}$, respectively. The while loop stops if the number of iterations is larger than N_S.

ALGORITHM 1 TRS.

1: Initialize λ_τ ($0 \leq \tau \leq B - 1$) with 0

2: Initialize λ_τ^r and λ_τ^a ($0 \leq \tau \leq N_S$) with λ_τ

3: Initialize Λ_{-1} and D_{B-1} with 0

4: $\tau \leftarrow B$

5: **while** $\tau \leq N_S$ **do**

6: $\lambda_\tau^a \leftarrow \lambda_\tau + \sum_{i=\tau-B}^{\tau-1} \lambda_i^r$

7: Solve **UCMPG** to obtain μ_τ and $\tilde{\mu}_{\tau+b}$ by using **HSP**

8: Update $\mathfrak{z}(\tau - B \leq i \leq \tau)$

9: $D_\tau \leftarrow D_{\tau-1} + d_\tau$

10: $\Lambda_{\tau-B} \leftarrow \Lambda_{\tau-B-1} + \lambda_{\tau-B}L$

11: $\tau \leftarrow \tau + 1$

12: **end while**

In Line 7 of Algorithm 1, **UCMPG** is solved by HSP whose detail is shown in Algorithm 2. The notations in HSP are first introduced. Let ω denote the inertia weight that is used to avoid the excessive variation of every particle's velocity. w_{min} and w_{max} denote the lower and upper bound of inertia weight, respectively. Let c_1 denote the individual acceleration coefficient that reflects every particle's local search ability. Similarly, let c_2 denote the social acceleration coefficient that reflects the effect of all particles in the current swarm. To avoid unexpected roaming in a solution space, the velocity of every particle can only vary from $-v_{max}$ to v_{max}. Let *pBest* denote the best position of each particle in every swarm. Besides, let *gBest* denote the best position among all particles in every swarm. Let ξ denote the size of each swarm. Let D denote the dimension of each particle in the swarm. Let η denote the total number of iterations in Algorithm 2. In addition, let t^0 and cr denote the initial temperature and the corresponding cooling rate, respectively. Besides, let κ_i denote the percentage of particles with the same fitness value in the swarm corresponding to iteration i.

In Algorithm 2, the first swarm P_0^τ is randomly initialized if HSP is executed for the first time. Otherwise, P_0^τ is initiated with $P_\eta^{\tau-1}$. Line 2 calculates fitness values of all particles in P_0^τ according to (51). Then, *pBest* and *gBest* are updated accordingly. Line 3 initializes w_{min}, w_{max}, c_1, c_2, v_{max}, and cr. Line 4 shows that the inertia weight and temperature are initiated with t^0 and w_{max}, respectively. Lines 5-13 show the **while** loop. The velocities and positions of all particles in the current swarm are updated in Line 6 based on the acceptance criterion of Metropolis. Lines 7-8 calculate the fitness values of all particles in P_i^τ, and update *pBest* and *gBest*. The next swarm in the current time slot, i.e., P_{i+1}^τ, is produced after iteration i in Line 9. The temperature (*tmp*) and inertia weight (ω) are updated in lines 10–11, respectively. The **while** loop terminates if the number of completed iterations is greater than η, or κ_i is more than 95%. Then, *gBest* is chosen as the final solution to **UCMPG** in time slot τ after the **while** loop. Finally, *gBest* can be converted to decision variables μ_τ and $\tilde{\mu}_{\tau+b}$.

ALGORITHM 2 HSP.

1: Randomly initialize velocities and positions of particles in P_0^τ if HSP is executed for the first time: otherwise, $P_0^\tau \leftarrow P_\eta^{\tau-1}$

2: Calculate fitness values of particles in P_0^τ, and update *pBest* and *gBest*

3: Initialize w_{min}, w_{max}, c_1, c_2, v_{max}, and cr

4: $tmp \leftarrow t^0$, $w \leftarrow w_{max}$, $i \leftarrow 0$

5: **while** $i \leq \eta$ and $k_i \leq 95\%$ **do**

6: Update velocities and positions of particles based on Metropolis rule

7: Calculate fitness values of particles in P_i^τ

8: Update *pBest* and *gBest*

9: $P_{i+1}^\tau \leftarrow P_i^\tau$

10: $tmp \leftarrow tmp.cr$

11: $w \leftarrow w_{max} - (w_{max} - w_{min}) \cdot i/\eta$

12: $i \leftarrow i+1$

13: **end while**

14: Output *gBest* as the final solution in time slot τ

We further provide a complexity analysis of Algorithms 1 and 2. In Algorithm 2, most of the execution time is consumed by the **while** loop. In the worst case, the **while** loop terminates after η iterations. Then, Lines 6–9 in Algorithm 2 show that the time complexity of each iteration in its **while** loop is $O(\xi D)$. Therefore, the time complexity of Algorithm 2 is $O(\xi D)$. Similarly, most of the execution time in Algorithm 1 is consumed by its **while** loop in which $N_S - B + 1$ iterations are executed. In the **while** loop of Algorithm 1, the time complexity of each iteration is predominantly determined by HSP in Line 7. In addition, N_S is typically much larger than B. Therefore, the time complexity of Algorithm 1 is $O(N_S \eta \xi D)$.

13.4 Performance Evaluation

This section evaluates the cost-aware workload scheduling in distributed CDCs and the TRS algorithm proposed in Sections 13.2 and 13.3 with the real-life workload in Google production cluster and the realistic trace from 1998 World Cup website, respectively.

13.4.1 Cost-Aware Workload Scheduling in Distributed CDCs

13.4.1.1 Simulation Setting

This section evaluates the effectiveness of the proposed methods by adopting the real-life workload traces in Google production cluster [65] shown in Figure 13.4. The dataset contains arrival requests of four applications in Google production cluster for 370 min in May 2011. In addition, the length of sampling period in Figure 13.4 is set to 5 min, i.e., the arrival rate of each application is calculated every five minutes. For clarity, in the following part of the paper, the sampling period and the time slot can be used interchangeably. There are existing works on workload prediction, therefore, this experiment simply adopts the actual request arrival rates during the 370 minutes. Besides, the priority values of request flows (type 1, 2, 3, and 4) are set to 1, 2, 3, and 4, respectively. A greater priority value means higher priority. A request flow with a greater priority value brings more revenue to the CDCs provider and has the privilege to be admitted to execute.

In this experiment, the arrival requests of four applications are allocated to three distributed CDCs (i.e., $N = 4$, $C = 3$). All the arrival requests can be delivered to distributed CDCs through three available ISPs (i.e., $K = 3$). The size of each request corresponding to application n is denoted by s_n. It is assumed that a request corresponding to application n ($n = 1,2,3,4$) contains 2, 5, 8, and 10 MB of data on average, respectively, i.e., $s_1 = 2$MB, $s_2 = 5$MB, $s_3 = 8$MB and $s_4 = 10$MB. In addition, parameter σ is set to 10^{20}. Parameters γ and δ are both set to 2. T_n^{max} is set to 1.5 times of that of T_n^{min} that is equal to RT_n^{user}. ISP_1^{BWCap}, ISP_2^{BWCap}, and ISP_3^{BWCap} are set to 10^9 Mbps, 0.8×10^9Mbps, and 0.6×10^9Mbps, respectively. $price_1$, $price_2$, and $price_3$ are set to 2×10^{-10}\$/Mbps, 3×10^{-10}\$/Mbps, and 6×10^{-10}\$/Mbps, respectively. In addition, according to the work [3,7], the setting of other parameters are shown in Table 13.2.

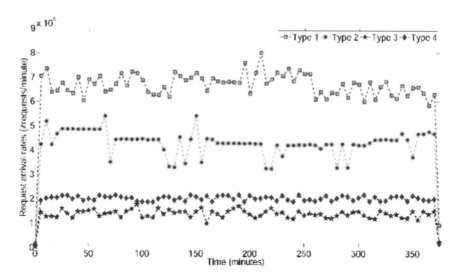

FIGURE 13.4
Arrival requests in Google's workload traces.

TABLE 13.2

Parameters of three CDCs

c	n	$M_{c,n}$	$b_{c,n}^s$ ($/hour)	RT_n^{user} (ms)	$\mu_{c,n}$ (10^3 requests/second)
$c=1$	$n=1$	1000	0.00007	1	9
	$n=2$	800	0.00013	1.25	7.5
	$n=3$	800	0.0002	2	3
	$n=4$	1200	0.00033	2.5	1.5
	$M_1 = 3800$				
$c=2$	$n=1$	1800	0.00001	1	4.5
	$n=2$	1800	0.00002	1.25	3.75
	$n=3$	1200	0.00003	2	1.6
	$n=4$	2500	0.00004	2.5	1
	$M_2 = 7300$				
$c=3$	$n=1$	1000	0.00003	1	10.5
	$n=2$	1000	0.00007	1.25	6
	$n=3$	1200	0.0001	2	3
	$n=4$	2000	0.00017	2.5	1.6
	$M_3 = 5200$				

13.4.1.2 Workload Admission Control

Figure 13.5 illustrates the admitted requests of four applications using the proposed method. It can be clearly observed that requests of application 3 are almost directly admitted into distributed CDCs. However, a large number of requests corresponding to applications 1, 2, and 4 are refused. This result demonstrates the effectiveness of the proposed

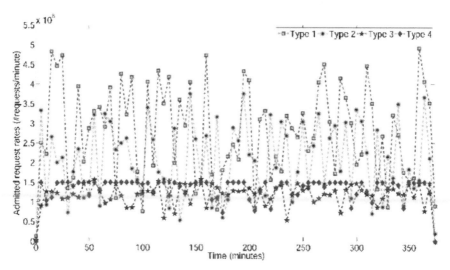

FIGURE 13.5
Admitted request rates of four applications.

method, which first estimates the average expected response time of each request flow and calculates its expected revenue. Higher priority requests, for example, application 3, bring more revenue to the CDCs provider than lower priority ones. Therefore, the proposed method inclines to admit higher priority requests. However, it is worth noting that though the priority of requests corresponding to application 4 is greater than that of requests corresponding to application 3, requests of application 3 are preferred to that of application 4. The reason is that there are not enough servers to execute requests of application 4. In this case, requests of application 4 may experience relatively long response time and will bring less or no revenue to the CDCs provider.

Then, this section evaluates the performance of our revenue-based admission control method by comparing it with the priority-based admission control method [66]. The priority-based admission control method ensures that lower priority requests cannot be admitted until all higher priority ones have been admitted. Figure 13.6 shows the revenue of two methods. It is observed that the revenue of the proposed revenue-based workload control method is obviously greater than that of the priority-based admission control method in each time slot. The reason is that the proposed control method estimates the expected revenue brought by admitted requests and admits requests that can maximize the total revenue of the CDCs provider.

13.4.1.3 Cost-Aware Workload Scheduling

The experiments adopt the same parameter setting and further evaluate the performance of the proposed cost-aware workload scheduling method by comparing it with typical average workload scheduling method [67]. The average workload scheduling method equally allocates requests among multiple ISPs and does not consider the bandwidth capacities of ISPs. If the total occupied bandwidth of requests scheduled to a specific ISP exceeds the bandwidth capacity of the ISP, some requests must be refused and cannot traverse this ISP. Besides, requests admitted by each ISP are also equally allocated to

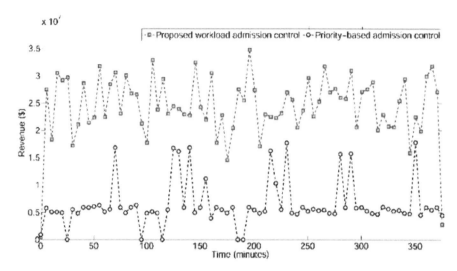

FIGURE 13.6
Revenue comparison.

multiple distributed CDCs. Therefore, requests scheduled to each CDC may exceed the capacity of available servers in that CDC. Figure 13.7 illustrates that the throughput of the cost-aware workload scheduling method is greater than that of the average workload scheduling method by the maximum amount of 10.08%.

Note that the average workload scheduling method may cause that some admitted requests cannot be scheduled to execute in distributed CDCs. Therefore, to fairly compare two workload scheduling methods, this section applies a penalty cost for each request refused to execute in distributed CDCs. Besides, the penalty cost per unit bandwidth corresponding to type n requests is denoted by $penalty_n$. The penalty cost for

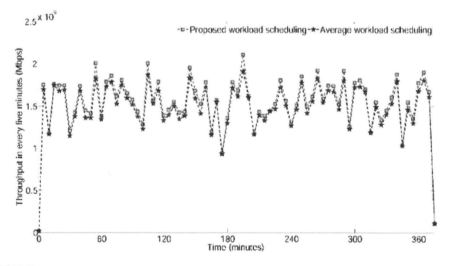

FIGURE 13.7
Throughput comparison between cost-aware and average workload scheduling.

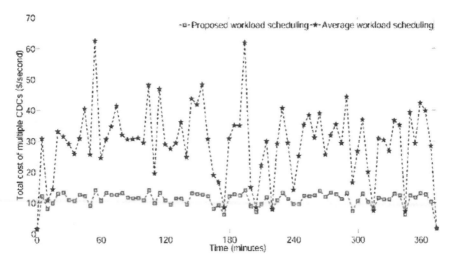

FIGURE 13.8
Total cost comparison between cost-aware and average workload scheduling.

lower priority requests is less than that of higher priority requests. If more admitted requests are allocated to execute in distributed CDCs, these requests will add additional bandwidth and energy cost to the total cost of the CDCs provider. Therefore, the value of $penalty_n$ should be greater than or at least equal to that of $price_k$. In this experiment, $penalty_1$, $penalty_2$, $penalty_3$, and $penalty_4$ are set to 2×10^{-7} \$/Mbps, 2.5×10^{-7} \$/Mbps, 3×10^{-7} \$/Mbps, and 3.5×10^{-7} \$/Mbps, respectively. Figure 13.8 shows the total cost comparison between the proposed cost-aware and the average workload scheduling method. It is observed that compared with the average workload scheduling method, the total cost of the proposed workload scheduling method can be reduced significantly almost in every time slot.

Then, the occupied bandwidth of each ISP connecting to distributed CDCs is further presented. The charging policies of ISPs are heterogeneous, therefore, the unit bandwidth cost of ISPk, $price_k$ is varying. Figure 13.9 shows that the occupied bandwidth of each ISP varies a lot due to the difference in charging policies of ISPs. The reason is that the proposed workload scheduling method can intelligently determine suitable ISPs for admitted requests and minimize the total cost of the CDCs provider. The number of requests that traverse ISP 1 is the largest. Besides, the number of requests that traverse ISP 3 is the smallest among three ISPs. The result in Figure 13.9 is consistent with charging policies of three ISPs, i.e., the unit bandwidth price of ISP 1, $penalty_1$, is the smallest among three ISPs while the unit bandwidth price of ISP 3, $penalty_3$, is the largest among three ISPs.

Figure 13.10 shows the number of requests admitted into each CDC based on the proposed cost-aware workload scheduling method. The number of available servers for each application and the total number of available servers in every CDC are both limited. Thus, the number of requests admitted into each CDC cannot exceed the total capacity of the corresponding servers in that CDC. Besides, the energy cost of active (spare) servers in different CDCs are varying. Therefore, to minimize the total cost of the CDCs provider, the number of requests admitted into each CDC is also varying. It can be clearly observed that the number of requests admitted into CDC 1 is the smallest among three CDCs. The reason is that $b_{1,n}^a (b_{1,n}^s)$ $(n = 1, 2, 3, 4)$ is the largest among three CDCs.

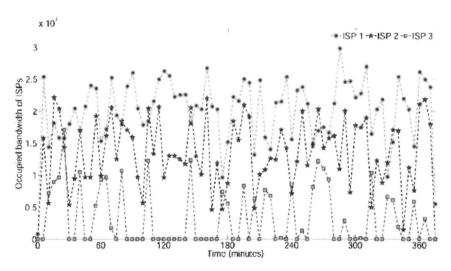

FIGURE 13.9
Occupied bandwidth of ISPs.

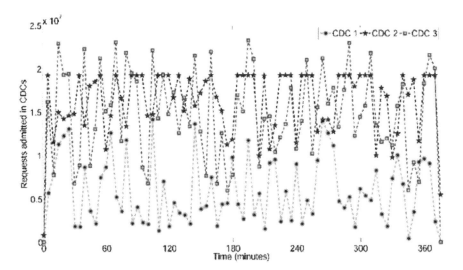

FIGURE 13.10
Requests admitted in distributed CDCs.

13.4.2 Temporal Request Scheduling (TRS) Algorithm in a GCDC

13.4.2.1 Experimental Setting

This section adopts real-life requests and real-life grid price to conduct the experiments. The length of each time slot is 5 minutes. The delay bound is set to 3 time slots, i.e., 15 minutes. To simulate the requests, this section adopts the publicly available data from 1998 World Cup website that spans from the 24 h on June 1, 1998 [39]. Figure 13.11 illustrates the request

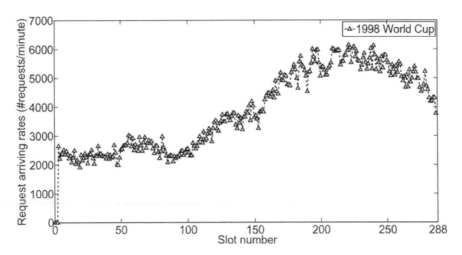

FIGURE 13.11
Requests of the 1998 World Cup website on June 1, 1998. (From Arlitt, M. and Jin, T., *IEEE Netw.*, 14, 30–37, 2000.)

arriving rate. Similar to [8,68], the request arriving rate in this trace is then used as λ_τ in our queueing system. For the grid price, to simulate grid energy cost for a GCDC, this section adopts the historical pricing data that spans from the 24 h on January 1, 2015 in capital region, New York [69]. To guarantee the validity of the real-life grid price, the sampling interval in a GCDC is set to 5 minutes.

Then, the setting of the corresponding parameters is described. For the total energy model, $\Gamma_{max} = 2.5$ (MWH), $\bar{P}_{idle} = 100$(W), $P_{peak} = 200$(W), $\gamma = 1.2$, and $\sigma = 6$ requests/min. Besides, the capacity of a GCDC queueing system is 50, i.e., $N = 50$. For the green energy, this section uses the parameter setting that is similar to that in [10]. The data sets about wind speed [70] and solar irradiance [71] are obtained from the National Renewable Energy Laboratory. Besides, $\varphi = 0.2$, $A = 1.5 * 10^4$(m²), $\phi = 0.3$, $R = 2.5 * 10^4$(m²), and $\rho = 1.225$ (kg/m³). The setting of HSP parameters in Algorithm 2 are: $\xi = 200$, $\eta = 100$, $w_{min} = 0.4$, $w_{max} = 0.95$, $v_{max} = 1$, $c_1 = c_2 = 0.5$, and $k = 0.975$. The initial temperature t^0 in Algorithm 2 is set to 10^{20}, i.e., $t^0 = 10^{20}$.

13.4.2.2 Experimental Results

Figure 13.12 shows the energy of grid, wind, and solar consumed in each time slot, respectively. As shown in Figure 13.12, the maximum total amount of energy is $2.5 * 1^6$(MWH), i.e., $\Gamma_{max} = 2.5 * 10^6$. It can be clearly observed that the total energy consumed in each time slot does not exceed the corresponding capacity Γ_{max}. Figure 13.13 illustrates the accumulative scheduled request, respectively. The delay bound B in this experiment is set to 3 time slots. It can be observed that all arriving requests can be scheduled to execute in a GCDC during its corresponding delay bound. The result demonstrates that TRS can well meet the delay bound requirement of all arriving requests.

13.4.2.3 Comparison Results

TRS is compared with the following three existing algorithms [10,72,15] in terms of throughput and grid energy cost.

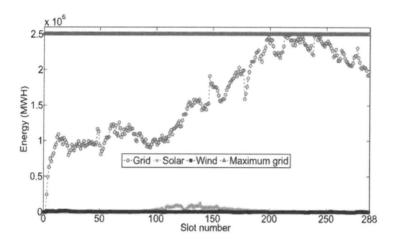

FIGURE 13.12
Energy of grid, solar and wind.

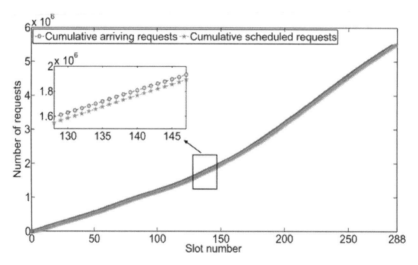

FIGURE 13.13
Accumulative scheduled and arriving requests.

1. A1 presented in [15] denotes the operation of a GCDC that does not consider the temporal variation in grid price and the amount of available green energy from wind and solar sources. Therefore, the arriving requests are not queued and are immediately scheduled to execute in a GCDC in the same time slot when they arrive.

2. A2, which is a variant of the cheap-first scheduling approach presented in [10], denotes the operation of scheduling the arriving requests to execute in the time slot when the grid price is the lowest during the delay bound. However, the total energy in that time slot is limited. Therefore, to meet the delay bound requirement of admitted requests, some arriving requests may be refused.

3. A3, which is a variant of the renewable energy-first scheduling approach pre-sented in [10,72], denotes the operation of scheduling the arriving requests to exe-cute in the time slot when the amount of green energy is the largest during the delay bound. Similar to A2, some arriving requests may be also refused due to the limitation of the total energy.

Figure 13.14 shows the results of four algorithms in terms of the accumulative through-put. Here the accumulative throughput in Figure 13.14 is calculated by summing up the number of scheduled requests in each time slot during the delay bound; that is, the accu-mulative throughput in each time slot τ is the sum of the number of scheduled requests in time slot τ to $\tau+3$. As shown in Figure 13.14, the accumulative throughput of TRS is much larger than that of A1–A3, respectively, in each time slot. The reason is that A1–A3 all have to refuse some of the arriving requests due to the limitation of the total energy at their corresponding time slots. Therefore, the result in Figure 13.14 demonstrates that TRS can significantly increase the accumulative throughput of a GCDC.

However, the refused requests in A1–A3 are not scheduled to execute and, therefore, do not bring the grid energy cost to a GCDC. In real-world cases, to guarantee performance of users' requests, users usually specify an SLA with cloud providers [73]. The SLA specifies many criteria including availability, latency, throughput, security, penalty cost, and so on. Typically, penalty cost in an SLA denotes the price that a GCDC provider has to pay for all requests that violate the SLA. As a result, a GCDC provider may suffer from the penalty cost due to the refusal of some arriving requests. To impartially compare TRS and A1–A3, the penalty cost for each refused request in each time slot τ should be at least more than the maximum grid energy cost per request during the delay bound. Specifically, let pc_τ denote the penalty cost for each refused request in each time slot τ. Let ec_τ denote the grid energy cost per request in each time slot τ. Therefore, $pc_\tau = max_{b \in \{0, 1,..., B\}}(ec_{\tau+b})$. Therefore, Figure 13.15 further illustrates the accumulative total costs of TRS and A1–A3, respec-tively. The accumulative total cost in each time slot is the sum of the accumulative grid

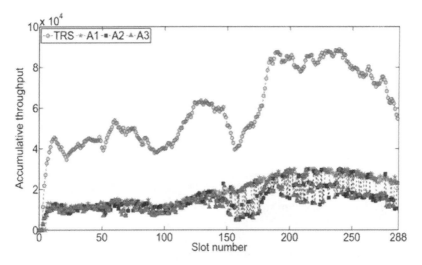

FIGURE 13.14
Accumulative throughput of TRS, A1–A3.

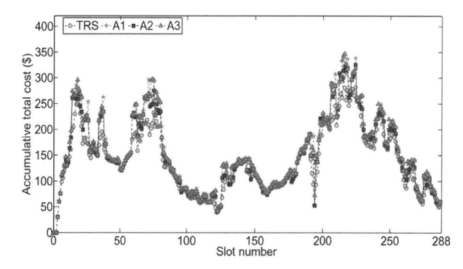

FIGURE 13.15
Accumulative total costs of TRS, A1–A3.

energy cost and the corresponding accumulative penalty cost at that time slot. As shown in Figure 13.15, compared with those of A1–A3, TRS's total cost can be decreased by 9.9%, 7.5%, and 10.2% on average, respectively. Therefore, this comparison result demonstrates that TRS can also decrease the accumulative total costs of a GCDC significantly.

13.5 Conclusion

Distributed CDCs require gigantic bandwidth and energy cost to support multiple applications. Existing works focus on the energy cost minimization problem in CDCs. However, the geographical diversity of bandwidth and energy cost brings a challenge to minimize the total cost of a CDCs provider. Therefore, **firstly**, this chapter studies the total cost minimization problem for the CDCs provider in a market where the bandwidth and energy cost show geographical diversity. To solve this problem, this chapter proposes a revenue-based workload admission control method and a cost-aware workload scheduling method for distributed CDCs.

What's more, the execution of Internet services in current GCDCs requires a huge amount of grid energy cost. The temporal diversity in the grid price, wind speed and solar irradiance makes it challenging to minimize the grid energy cost of a GCDC while meeting the delay bound of all requests. Therefore, **secondly**, this chapter proposes a TRS to efficiently schedule all arriving requests to execute in a GCDC during their delay bound. The long tail in real-life requests' delay is considered, and the delay requirement of each request is strictly met. In each iteration of TRS, a constrained nonlinear optimization problem is formulated and solved by a hybrid meta-heuristic. In this way, a meta-heuristic grid energy cost minimization strategy for a GCDC is developed.

Simulation results with the real-life workload in Google production cluster demonstrate that compared with existing methods, the cost-aware workload scheduling method proposed in Section 13.2 can greatly reduce the total cost and increase the throughput of the

CDCs provider. Besides, the evaluation results based on the realistic trace from 1998 World Cup website demonstrate that TRS proposed in Section 13.3 can achieve higher throughput and lower grid energy cost for a GCDC while delay assurance is strictly provided for all arriving requests.

In the future, the current work would be extended to consider finer-grained metrics (e.g., memory and storage constraints). In addition, the VM scheduling in distributed CDCs where multiple heterogeneous VMs concurrently run in a server would be further considered. Besides, a realistic GCDC would be implemented to evaluate the performance of TRS. Finally, the scheduling of multiple types of requests that need heterogeneous resources (e.g., storage and bandwidth) would be considered and more types of renewable energy sources would be incorporated.

References

1. Y. Xia, M. Zhou, X. Luo, Q. Zhu, J. Li, and Y. Huang, "Stochastic modeling and quality evaluation of Infrastructure-as-a-Service clouds," *IEEE Transactions on Automation Science and Engineering*, 12(1): 1131–1146, 2013.
2. Z. Zhu, J. Bi, H. Yuan, and Y. Chen, "SLA based dynamic virtualized resources provisioning for shared cloud data centers," in *4th IEEE International Conference on Cloud Computing Technology and Science Proceedings*, 2011, pp. 630–637.
3. A. Greenberg, J. Hamilton, D. A. Maltz, and P. Patel, "The cost of a cloud: Research problems in data center networks," *ACM SIGCOMM Computer Communication Review*, 39(1): 68–73, 2009.
4. C. Y. Hong, S. Kandula, R. Mahajan, M. Zhang, V. Gill, M. Nanduri, and R. Wattenhofer, "Achieving high utilization with software-driven WAN," *ACM SIGCOMM Computer Communication Review*, 43(4): 15–26, 2013.
5. Y. Guo and Y. Fang, "Electricity cost saving strategy in data centers by using energy storage," *IEEE Transactions on Parallel and Distributed Systems*, 24(6): 1149–1160, 2013.
6. Z. Abbasi, G. Varsamopoulos, and S. K. S. Gupta, "Thermal aware server provisioning and workload distribution for Internet data centers," in *Proceedings of the 19th ACM International Symposium on High Performance Distributed Computing*, 2010, pp. 130–141.
7. Z. Zhang, M. Zhang, A. Greenberg, Y. C. Hu, R. Mahajan, and B. Christian, "Optimizing cost and performance in online service provider networks," in *7th USENIX Symposium on Networked Systems Design and implementation*, 2010, pp. 33–48.
8. J. Bi, H. Yuan, Y. Fan, W. Tan, and J. Zhang, "Dynamic fine-grained resource provisioning for heterogeneous applications in virtualized cloud data center," in *Proceedings of the 8th IEEE International Conference on Cloud Computing*, 2015, pp. 429–436.
9. M. Ghamkhari and H. Mohsenian-Rad, "Energy and performance management of green data centers: A profit maximization approach," *IEEE Transactions on Smart Grid*, 4(2): 1017–1025, 2013.
10. X. Deng, D. Wu, J. Shen, and J. He, "Eco-aware online power management and load scheduling for green cloud datacenters," *IEEE Systems Journal*, 10(1): 78–87, 2014.
11. J. Luo, L. Rao, and X. Liu, "Temporal load balancing with service delay guarantees for data center energy cost optimization," *IEEE Transactions on Parallel and Distributed Systems*, 25(3): 775–784, 2014.
12. Y. Yao, L. Huang, A. Sharma, L. Golubchik, and M. Neely, "Data centers power reduction: A two time scale approach for delay tolerant workloads," in *Proceedings of the IEEE International Conference on Computer Communications*, 2012, pp. 1431–1439.
13. G. Chen, T. Hu, D. Jiang, P. Lu, K.L. Tan, H.T. Vo, and S. Wu, "Bestpeer++: A peer-to-peer based large-scale data processing platform," *IEEE Transactions on Knowledge and Data Engineering*, 26(6): 1316–1331, 2014.

14. R. Ghosh, F. Longo, V. K. Naik, and K. S. Trivedi, "Modeling and performance analysis of large scale IaaS clouds," vol. *Future Generation Computing Systems*, 29(5): 1216–1234, 2013.

15. X. Zuo, G. Zhang, and W. Tan, "Self-adaptive learning PSO-based deadline constrained task scheduling for hybrid IaaS cloud," *IEEE Transactions on Automation Science and Engineering*, 11(2): 564–573, 2014.

16. D. Bruneo, "A stochastic model to investigate data center performance and QoS in IaaS cloud computing systems," *IEEE Transactions on Parallel and Distributed Systems*, 25(3): 560–569, 2014.

17. H. Xu and B. Li, "Joint request mapping and response routing for geo-distributed cloud services," in Proceedings of the 32nd *IEEE International Conference on Computer Communications*, 2013, pp. 854–862.

18. S. Muppala, G. Chen, and X. Zhou, "Multi-tier service differentiation by coordinated learning-based resource provisioning and admission control," *Journal of Parallel and Distributed Computing*, 74(5): 2351–2364, 2014.

19. Z. Shao, X. Jin, W. Jiang, M. Chen, and M. Chiang, "Intra-data-center traffic engineering with ensemble routing," in *Proceedings of the 32nd IEEE International Conference on Computer Communications*, 2013, pp. 2148–2156.

20. J. W. Jiang, T. Lan, S. Ha, M. Chen, and M. Chiang, "Joint VM placement and routing for data center traffic engineering," in *Proceedings of the 31st IEEE International Conference on Computer Communications*, 2012, pp. 2876–2880.

21. S. Agarwal, M. Kodialam, and T. Lakshman, "Traffic engineering in software defined networks," in *Proceedings of the 32nd IEEE International Conference on Computer Communications*, 2013, pp. 2211–2219.

22. W. Lloyd, M. J. Freedman, M. Kaminsky, and D. G. Andersen, "Don't settle for eventual: Scalable causal consistency for wide-area storage with COPS," in *Proceedings of the 23rd ACM Symposium on Operating Systems Principles*, 2011, pp. 401–416.

23. S. Jain, A. Kumar, S. Mandal, J. Ong, L. Poutievski, A. Singh, S. Venkata et al., "B4: Experience with a globally-deployed software defined WAN," in *Proceedings of the ACM Special Interest Group on Data Communication*, 2013, pp. 3–14.

24. N. McKeown, T. Anderson, H. Balakrishnan, G. Parulkar, L. Peterson, J. Rexford, S. Shenker, and J. Turner, "OpenFlow: Enabling innovation in campus networks," *ACM SIGCOMM Computer Communication Review*, 38(2): 69–74, 2008.

25. D. Gross, J. F. Shortle, J. M. Thompson, and C. M. Harris, *Fundamentals of Queueing Theory*, 4th edition. New York: John Wiley & Sons, 2008.

26. C. Jung, H. Kim, and T. Lee, "A branch and bound algorithm for cyclic scheduling of timed petri nets," *IEEE Transactions on Automation Science and Engineering*, 12(1): 309–323, 2014.

27. H. Zhang, L. Cui, X. Zhang, and Y. Luo, "Data-driven robust approximate optimal tracking control for unknown general nonlinear systems using adaptive dynamic programming method," *IEEE Transactions on Neural Networks and Learning Systems*, 22(12): 2226–2236, 2011.

28. M. Alekhnovich, A. Borodin, J. Buresh-Oppenheim, R. Impagliazzo, Magen, and T. Pitassi, "Toward a model for backtracking and dynamic programming," *Computational Complexity*, 20(4): 679–740, 2011.

29. X. Zhao, "Simulated annealing algorithm with adaptive neighborhood," *Applied Soft Computing*, 11(2): 1827–1836, 2011.

30. R. C. Eberhart and Y. Shi, "Particle swarm optimization: Developments, applications and resources," in *Proceedings of the IEEE Congress on Evolutionary Computation*, 2001, pp. 81–86.

31. M. Bansal, K. Kianfar, Y. Ding, and E. Moreno-Centeno, "Hybridization of bound-and-decompose and mixed integer feasibility checking to measure redundancy in structured linear systems," *IEEE Transactions on Automation Science and Engineering*, 10(4): 1151–1157, 2013.

32. A. Neumaier and O. Shcherbina, "Safe bounds in linear and mixed-integer linear programming," *Mathematical Programming*, 99(2): 283–296, 2004.

33. J. Cao, K. Li, and I. Stojmenovic, "Optimal power allocation and load distribution for multiple heterogeneous multicore server processors across clouds and data centers," *IEEE Transactions on Computers*, 63(1): 45–58, 2014.

34. L. Leslie, Y.C. Lee, P. Lu, and A. Zomaya, "Exploiting performance and cost diversity in the cloud," in *Proceedings of the 6th IEEE International Conference on Cloud Computing*, 2013, pp. 107–114.
35. D. Xu, X. Liu, and Z. Niu, "Joint resource provisioning for internet datacenters with diverse and dynamic traffic," *IEEE Transactions on Cloud Computing*, 5(1): 71–84, 2015.
36. D. Xu, X. Liu, and A. Vasilakos, "Traffic-aware resource provisioning for distributed clouds," *IEEE Cloud Computing*, 2(1): 30–39, 2015.
37. Y. Xu, Z. Musgrave, B. Noble, and M. Bailey, "Workload-aware provisioning in public clouds," *IEEE Internet Computing*, 18(4): 15–21, 2014.
38. L. Wu, S. K. Garg, and R. Buyya, "SLA-based admission control for a Software-as-a-Service provider in cloud computing environments," *Journal of Computer and System Sciences*, 78(5): 1280–1299, 2012.
39. M. Arlitt and T. Jin, "A workload characterization study of the 1998 World Cup web site," *IEEE Network*, 14(3): 30–37, 2000.
40. R. Chiang and H. Huang, "TRACON: Interference-aware scheduling for data-intensive applications in virtualized environments," *IEEE Transactions on Parallel and Distributed Systems*, 25(5) 1349–1358, 2011.
41. X. Lin, Y. Wang, Q. Xie, and M. Pedram, "Task scheduling with dynamic voltage and frequency scaling for energy minimization in the mobile cloud computing environment," *IEEE Transactions on Services Computing*, 8(2): 175–186, 2015.
42. W. Tan, Y. Sun, L.X. Li, G. Lu, and T. Wang, "A trust service-oriented scheduling model for workflow applications in cloud computing," *IEEE Systems Journal*, 8(3): 868–878, 2014.
43. F. Zhang, J. Cao, K. Hwang, K. Li, and S. Khan, "Adaptive workflow scheduling on cloud computing platforms with iterative ordinal optimization," *IEEE Transactions on Cloud Computing*, 3(2): 156–168, 2015.
44. Q. Zhang, M. Zhani, Y. Yang, R. Boutaba, and B. Wong, "PRISM: Fine-grained resource-aware scheduling for MapReduce," *IEEE Transactions on Cloud Computing*, 3(2): 182–194, 2015.
45. A. Amokrane, R. Langar, M. Zhani, R. Boutaba, and G. Pujolle, "Greenslater: On satisfying green SLAs in distributed clouds," *IEEE Transactions on Network and Service Management*, 12(3): 363–376, 2015.
46. Y. Guo, Y. Gong, Y. Fang, P. Khargonekar, and X. Geng, "Energy and network aware workload management for sustainable data centers with thermal storage," *IEEE Transactions on Parallel and Distributed Systems*, 25(8): 2030–2042, 2014.
47. M. Lin, A. Wierman, L. Andrew, and E. Thereska, "Dynamic right-sizing for power-proportional data centers," *IEEE/ACM Transactions Networking*, 21(5): 1378–1391, 2013.
48. J. Bi, H. Yuan, M. Tie, and W. Tan, "SLA-based optimisation of virtualised resource for multi-tier web applications in cloud data centres," *Enterprise Information Systems*, 9(7): 743–767, 2015.
49. J. Cao, K. Hwang, K. Li, and A. Zomaya, "Optimal multiserver configuration for profit maximization in cloud computing," *IEEE Transactions on Parallel and Distributed Systems*, 24(6): 1087–1096, 2013.
50. S. Maguluri and R. Srikant, "Scheduling jobs with unknown duration in clouds," *IEEE/ACM Transactions on Networking*, 22(6): 1938–1951, 2014.
51. S. Ren and M. van der Schaar, "Dynamic scheduling and pricing in wireless cloud computing," *IEEE Transactions on Mobile Computing*, 13(10): 2283–2292, 2014.
52. H. Yuan, J. Bi, W. Tan, and B. Li, "CAWSAC: Cost-aware workload scheduling and admission control for distributed cloud data centers," *IEEE Transactions on Automation Science and Engineering*, 13(2): 976–985, 2015.
53. H. Narman, M. Hossain, and M. Atiquzzaman, "DDSS: Dynamic dedicated servers scheduling for multi priority level classes in cloud computing," in *Proceedings of the 2014 IEEE International Conference on Communications*, 2014, pp. 3082–3087.
54. L. Rao, X. Liu, L. Xie, and W. Liu, "Coordinated energy cost management of distributed internet data centers in smart grid," *IEEE Transaction on Smart Grid*, 3(1): 50–58, 2012.

55. G. Darzanos, I. Koutsopoulos, and G. Stamoulis, "A model for evaluating the economics of cloud federation," in *Proceedings of the 4th IEEE International Conference on Cloud Networking*, 2015, 291–296.

56. Y. Shi, X. Jiang, and K. Ye, "An energy-efficient scheme for cloud resource provisioning based on CloudSim," in *Proceedings of the 2011 IEEE International Conference on Cluster Computing*, 2011, pp. 595–599.

57. G. Pagani and M. Aiello, "Generating realistic dynamic prices and services for the smart grid," *IEEE Systems Journal*, 9(1): 191–198, 2015.

58. C. Ge, Z. Sun, N. Wang, K. Xu, and J. Wu, "Energy management in cross-domain content delivery networks: A theoretical perspective," *IEEE Transactions on Network and Service Management*, 11(3): 264–277, 2014.

59. M. Hosseinzadeh and F. Salmasi, "Robust optimal power management system for a hybrid AC/DC micro-grid," *IEEE Transactions on Sustainable Energy*, 6(3): 675–687, 2015.

60. M. Subathra, S. Selvan, T. Victoire, A. Christinal, and U. Amato, "A hybrid with cross-entropy method and sequential quadratic programming to solve economic load dispatch problem," *IEEE Systems Journal*, 9(3): 1031–1044, 2015.

61. A. Stachurski, "Oblique projections, broyden restricted class and limited-memory quasi-newton methods," *Optimization*, 63(1): 129–144, 2014.

62. Y. Zhang, B. Mu, and H. Zheng, "Link between and comparison and combination of Zhang neural network and quasi-newton BFGS method for time-varying quadratic minimization," *IEEE Transactions on Cybernetics*, 43(2): 490–503, 2013.

63. J. Torres-Jimenez, I. Izquierdo-Marquez, A. Garcia-Robledo, A. Gonzalez-Gomez, J. Bernal, and R.N. Kacker, "A dual representation simulated annealing algorithm for the bandwidth minimization problem on graphs," *Information Sciences*, 303: 33–49, 2015.

64. Q. Yuan and G. Yin, "Analyzing convergence and rates of convergence of particle swarm optimization algorithms using stochastic approximation methods," *IEEE Transactions on Automatic Control*, 60(7): 1760–1773, 2015.

65. Q. Zhang, M. F. Zhani, R. Boutaba, and J. L. Hellerstein, "Dynamic heterogeneity-aware resource provisioning in the cloud," *IEEE Transactions on Cloud Computing*, 2(1): 14–28, 2014.

66. I. Nafea, M. Younas, R. Holton, and I. Awan, "A priority-based admission control scheme for commercial web servers," *International Journal of Parallel Programming*, 42(5): 776–797, 2014.

67. M. Masa and E. Parravicini, "Impact of request routing algorithms on the delivery performance of content delivery networks," in *Proceedings of the IEEE International Conference Performance, Computing and Communications*, 2003, pp. 5–12.

68. D. Liao, K. Li, G. Sun, V. Anand, Y. Gong, and Z. Tan, "Energy and performance management in large data centers: A queuing theory perspective," in *Proceedings of the 2015 International Conference on Computing, Networking and Communications*, 2015, pp. 287–291.

69. New York Independent System Operator (NYISO), http://www.nyiso.com/public/index.jsp.

70. National wind technology center-wind datasets, http://www.nrel.gov/midc/nwtc_m2/.

71. NREL solar radiation research laboratory-solar datasets, http://www.nrel.gov/midc/srrl_bms/

72. C. Stewart and K. Shen, "Some joules are more precious than others: Managing renewable energy in the datacenter," in *Proceedings of the Workshop on Power-Aware Computing and Systems*, 2009, pp. 15–19.

73. P. Xiong, Y. Chi, S. Zhu, H.J. Moon, C. Pu, and H. Hacgumus, "SmartSLA: Cost-sensitive management of virtualized resources for CPU-bound database services," *IEEE Transactions on Parallel and Distributed Systems*, 26(5): 1441–1451, 2015.

14

Ultra-Low-Voltage Implementation of Neural Networks

Farooq Ahmad Khanday, Nasir Ali Kant, and Mohammad Rafiq Dar

CONTENTS

14.1 Introduction

Artificial Neural Network (ANN), being one of the fascinating fields, has become the focus of study from quite a long time. The concept of neural network is originated from neuroscience. It can be easily said that the field of ANNs has ascended from diverse fronts, stretching from the captivation of humankind with understanding and emulating the human brain, to the broader issues of copying human abilities, like speech, to the practical, scientific, commercial, and engineering disciplines of pattern recognition, modeling and prediction.

The simplest definition of ANN is provided by the inventor of one of the first neuro-computers, Dr. Robert Hecht-Nielsen [1]. He defines a neural network as:

> *A computing system made up of a number of simple, highly interconnected processing elements,*
> *which process information by their dynamic state response to external inputs.*

The definition of ANNs can be put forward in a number of ways. At one extreme, the answer could be that these are simply a class of mathematical algorithms, since a network can be regarded essentially as a graphic notation for a large class of algorithms. Such algorithms produce solutions to a number of specific problems. At the other end, the reply may be that these are synthetic networks that emulate the biological neural networks found in living organisms. In light of today's limited knowledge of biological neural networks and organisms, the more plausible answer seems to be closer to the algorithmic one. Thus, we can say ANNs are the result of academic investigation that uses mathematical formulations to model nervous system operation. These models formed by the interconnection of artificial neurons, which mimic the behavior of biological neurons. The models in the form of electronic circuits or simulators comprise hardware (analog or digital) or software (digital) components, which compute mathematical models of biological neurons and biological synapses.

Our knowledge about actual brain functions is still limited, and no model has been successful in duplicating exactly the performance of the human brain. Despite ANNs have undoubtedly been biologically inspired, and it has been argued that neural networks mirror to a certain extent the behavior of networks of neurons in the brain, but the close correspondence between the biological neural systems and ANN is still weak. One of the biggest challenges faced regarding the mimicking of the natural brain behavior is to portray the behavior exactly because despite the advancements achieved in almost every aspect of life we are yet unable to completely justify the human brain. One difficulty is to understand the basic physiological mechanisms underlying the generation of complex spatiotemporal spiking patterns observed during the normal functioning of the brain and the second difficulty is to understand how the produced spiking patterns provide substrate with the information to transport and encode. Then, again, the complexity associated is very high even the brains of the primitive animals are believed to be enormously complex structures. The human brain contains about 10^{11} neurons, each capable of storing more than a single bit of data. Thus, gigantic incongruities exist between both the architectures and capabilities of natural and artificial neural networks. However, the brain has been and still is only a metaphor for a wide variety of neural network configurations that have been developed [2].

14.2 Biological Neuron

The networks formed by biological neurons, which are connected to carry out the functionalities typical of the nervous system in biological life forms, are referred as biological neural networks. However, not all multicellular life forms on the earth have a nervous system. For example, sponges are very old life forms comprising colonies of cells, which do not have a nervous system to allow electrical communication between the various parts of the body [2]. In particular, the complexity of the nervous system evolved during the various historical eras.

In the evolutionary scale, with regard to the nervous system, after the sponges it is possible to find the "Radiata" branch. In these life forms, it is possible to identify a "top" and a "bottom," but not "left" and "right" sides; jellyfish are an example of this class. These life forms have a simple nerve net that allows reactions to external stimuli [3].

Life forms having the most complex neural system belong to the "Bilateria" branch. These are the life forms that are (approximately) symmetric with respect to a longitudinal axis, and for which it is possible to define a "left" and a "right" side in addition to a "top" and a "bottom." This branch includes the human race, which has the most complex nervous system known in nature [4,5].

More generally, in the "Bilateria" branch, the "Vertebrate" subphylum occurs. In individuals of this subphylum, the nervous system can be divided into two interconnected halves [6]: the peripheral nervous system and the central nervous system. However, these two parts differ only from an anatomical point of view as depicted in Figure 14.1.

Considering the human nervous system, the heart of the intelligence associated with living being's lies with the brain and the heart of the brain lies with the neurons. Besides neurons, brain contains the glia cell, which is only for the energy supply and the structural stabilization of brain tissue. It is the neuron that forms the elementary processing unit in the central nervous system. Human brain consists of approximately 10^{11} neurons. The neurons communicate through a connection network of axons and synapses having a density of approximately 10^4 synapses per neuron. The hypothesis regarding the modeling of the natural nervous system is that neurons communicate with each other by means of electrical impulses [7]. The neurons operate in a chemical environment that is even more important in terms of actual brain behavior. Thus, the brain can be considered a densely connected electrical switching network conditioned largely by the biochemical processes. The vast neural network has an elaborate structure with very complex interconnections. The input to the network is provided through sensory receptors. Receptors deliver stimuli both from within the body, as well as from sense organs when the stimuli originate in the external world. The stimuli are in the form of electrical impulses that convey the information into the network of neurons. Because of information processing in the central nervous systems, the effectors are controlled and give human responses in the form of diverse actions. Thus, we have a three-stage system, consisting of receptors, neural network, and effectors, in control of the organism and its actions.

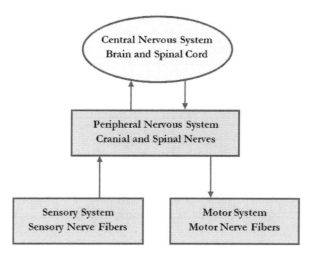

FIGURE 14.1
Simplified structure of the nervous system in vertebrate animals.

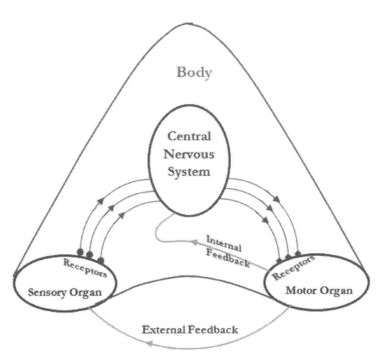

FIGURE 14.2
Information flow in nervous system.

A lucid, although rather approximate idea, about the information links in the nervous system is shown in Figure 14.2. As we can see from the figure, the information is processed, evaluated, and compared with the stored information in the central nervous system. The necessary commands are generated there and transmitted to the motor organs. Notice that motor organs are monitored in the central nervous system by feedback links that verify their action. The implementation of commands is controlled with the help of both internal and external feedbacks. As can be seen, the overall nervous system structure has many of the characteristics of a closed-loop control system.

As mentioned above, the neurons communicate with each other by means of electrical impulses. This was recognized by Galvani in 1791 and is regarded as the first major contribution in neural networks. Later came the groundbreaking contribution of Santiago Ramón y Cajal, wherein he showed that the neural system is made of an assembly of the well-defined cells, which he named neurons.

Five decades later, in 1952, the mechanisms involved in the creation and the propagation of neuronal electric signals were explained by Hodgkin and Huxley [8]. They introduced the model for neuron wherein ionic mechanism and electrical current on membrane surface of the neuron were taken into account. The communication of the neurons among themselves takes place through tiny processes, the synapses, and the synaptic transmission was thoroughly studied, in particular, by Katz. The pictorial view of a single neuron is given in Figure 14.3 and the synaptic connection between two neurons is shown in Figure 14.4.

To describe the dynamical behavior of a neuron, here is a brief look at the process that takes place inside the neuron. A neuron is a living cell immersed in an interstitial fluid. Due to the uneven distribution of the electrolytes inside and outside of the cell membrane, there is a difference in the voltage level from the inside and the outside of the neuron cell.

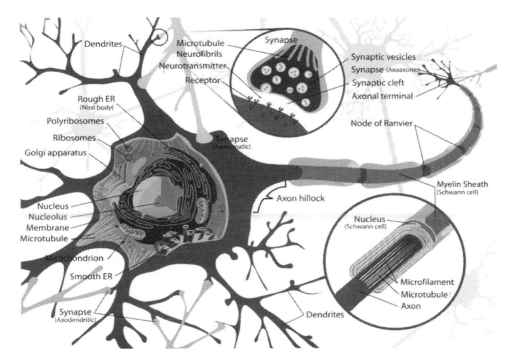

FIGURE 14.3
Pictorial view of the neuron. (From LadyofHats, Wikimedia Commons.)

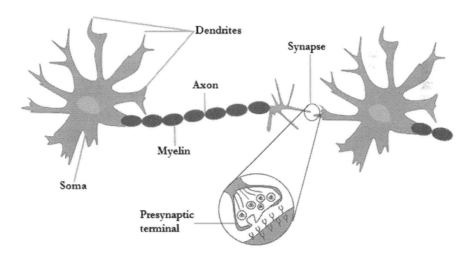

FIGURE 14.4
Pictorial view of the synaptic connection between neurons.

This unevenness in the ion distribution is a consequence of the different permeability factors for each of the ions in case of the cell membrane. Inside the cell, there are mainly K^+ ions and large organic A^- ions. While outside the cell, there are mainly Cl^- and Na^+ ions. In addition, it is important to note that for A^- ions, the cell membrane is always impermeable, and for that reason they always remain trapped inside the cell.

During the resting state, the cell membrane is permeable only to K^+ and Cl^- ions, which results in the diffusion of the K^+ and Cl^- ions outward and inward of the cell, respectively.

An electrical field gets generated (negative inside with respect to the outside), as a consequence of this diffusion. The generated field opposes the diffusion of ions down their concentration gradient. This causes the cell to attain the equilibrium state in which the chemical force, which makes the ions diffuse, equals the force caused by the electrical field against the ions.

At this equilibrium state, typically, a voltage difference of 60–80 mV is present between the internal and external walls of the cell membrane. This potential difference is referred to as the membrane potential. Because the membrane of the neuron in its resting state is impermeable to Na$^+$, for this reason the resting potential of the neuron is approximately -75 mV (inside with respect to outside). However, the Na$^+$ permeability factor of the membrane can be made nonzero, and this is what happens when the neuron cell is producing its *action potential*. If the permeabilities of the membrane change, then the voltage drop, V_m, between inside and outside of the cell changes according to the following Goldman equation:

$$V_m = 58\,\mathrm{mV}\log\frac{P_K[K^+]_{out} + P_{Na}[Na^+]_{out} + P_{Cl}[Cl^-]_{in}}{P_K[K^+]_{in} + P_{Na}[Na^+]_{in} + P_{Cl}[Cl^-]_{out}} \tag{14.1}$$

where, P_K, P_{Na}, and P_{Cl} are the relative membrane permeabilities for K$^+$, Na$^+$, and Cl$^-$, respectively. $[K^+]_{out}$, $[Na^+]_{out}$, $[Cl^-]_{out}$, $[K^+]_{in}$, $[Na^+]_{in}$, and $[Cl^-]_{in}$ are the concentration of the ions K$^+$, Na$^+$, and Cl$^-$ outside and inside the cell membrane, respectively.

During every action potential, there is a flow of Na$^+$ inward to the cell, which increases the membrane voltage, followed by the K$^+$ outward flow to reestablish the resting state potential. Naturally, there is a mechanism that after the action potential is going to pump Na$^+$ outside and K$^+$ inside, so that the resting concentrations of these ions are recovered. This is performed independently by the so-called Na$^+$/K$^+$ pumps, which are very complex organic molecules that literally pump Na$^+$ out and K$^+$ in against their concentration gradients, by means of a sequence of chemical metabolic reactions that consume energy.

The real biological neuron exhibits variety of dynamics, and the understanding of these dynamical behaviors is of paramount significance to the study of function of the brain in information processing and possible engineering applications. The dynamical behavior of the neurons includes quiescent, spiking, bursting, and chaotic behaviors.

- *Quiescent behavior* is achieved when the input signal applied to the neuron is below a threshold value, the neuron does not generate an output response.
- *Spiking behavior* is the case when there is a regular series generated at the output of the neuron.
- *Bursting behavior* is said to be exhibited if the output resembles the ensembles of spikes separated by a certain period.
- *Chaotic behavior* is said to be exhibited if the output signal of the neuron is produced in the chaos mode. Among the various dynamics of neural networks, chaotic behavior is of much interest and has been regarded as an important phenomenon that plays a vital role in memory storage and retrieval mechanism for the storage, retrieval and creation of information in neural networks and has, therefore, received a considerable attention in the recent past [9–15].

14.3 Artificial Neural Networks (ANNs)

The roots and inspiration for ANNs are drawn from biological nervous system. Such biological system or wetware consists of a multitude of simple processing elements that are connected in a massively parallel architecture. Before going further into the ANNs, here are few of the important questions first:

- Why study and develop Artificial Neural Networks at all?
- What task or tasks could they be used to perform?
- What benefit can they offer beyond a traditional von Neumann architecture machine?

By answering the latter two questions, hopefully a more complete reason for the study of ANNs will become apparent. Within a traditional computer, a failure in a processing section is catastrophic in terms of system performance; this is not necessarily the case with a neural network. Benefits of ANN are their potential robustness and only a gradual degradation in performance, but not the stoppage in network working, if an area of the network meets a fault. Furthermore, certainly for rapid exact algorithmic or mathematical operations, a traditional computer is excellent but this is not the case for noisy, inexact information processing. In addition, the interesting and powerful thing about the neural networks is their ability to adapt and learn from the data presented to them. Neural networks without learning are rather uninteresting. If the weights of a network were fixed from the beginning, neural networks could be implemented using any programming language in conventional computers. But an important point about neural networks is that they are not static i.e., the strengths of inter connections vary with time, new ones are formed and old ones may decay away. Due to the large quantity of parallelism, there is redundancy built in the system and a level of fault tolerance is available. Rather than being explicitly programmed, a neural networks evolves to perform an action by learning and adaption. Thus, given that the network changes through damage or the network has to increase its functionality, it is able to adapt to the new situation. Therefore, in contrast to a simple computer, the ANN design need not to be re-programmed once it is trained. The ANN has recently been applied in process control, identification, diagnostics, character recognition, sensory prediction, robot vision, and forecasting [8,16–20] etc.

Owing to their importance, a tremendous amount of work is going on ANNs. In the open literature, several models/designs have been reported to mimic the behavior of the natural neurons and have been widely used in many fields. The modeling of neurons comes under the discipline of computational neurobiology or computational neuroscience. The field has a long history, at least we can say it goes back to 1952 when the ground breaking kinetic model of Hodgkin and Huxley [8], for the generation of the nerve action potential was presented. The Hodgkin–Huxley model dealt with events at the molecular and ionic levels on a unit area of axon membrane. After this, various other models also have examined neural behavior on a more macroscopic level, preserving neural components such as synapses, dendrites, soma, axon, etc. These models include Morris–Lecar model [16], Hindmarsh–Rose model [17], Chay model [18], Izhikevich model [19], and FitzHugh–Nagumo model [20] where the focus has been on achieving the exact behavior of the biological neuron.

After having the mathematical model for the biological neurons, their implementation can be achieved either by software or by hardware approach. It is worth to mention here that the key features of neural networks are asynchronous parallel processing, continuous-time dynamics, and global interaction of network elements [21]. However, the parallel nature of the neural system contrasts with the sequential nature of computer systems, resulting in slow and complex software simulation design. Unfortunately, the portion of work found in literature, demonstrating the ANNs, is dominated by employing the software approach. The direct hardware implementation of these neural networks holds out the promise of faster emulation both because it is inherently faster than software implementation and the operation is much more parallel.

Hardware implementation of neural networks can be achieved in different ways. The implementations of neural networks can be analog, digital, or hybrid depending upon the way signals is processed. A brief introduction of these is given in the next sections.

14.3.1 Analog ANNs

In the analog implementation of neural networks, a coding method is used in which signals are represented by currents or voltages. This allows us to think of these systems as operating with real numbers during the neural network simulation. The basic operation of an ANN processing element can be summarized as

$$out = F\left[\sum(\Pi)\right] \qquad (14.2)$$

Therefore, three operations of multiplication, summation and activation function are to be performed within analog hardware. Graf and Jackal [22] and Foo et al. [23] provide a general introduction into analog implementations, while Mead [24] provides a greater depth and more specialized viewpoint for using analog circuits.

14.3.2 Digital Artificial Neural Networks

In a digital implementation of an ANN, it is obviously necessary to perform the same operations as with an analog approach. A number of approaches can be taken: one way is to form all the components of a neuron separately using digital technology, second way is to generate digital architectures and processors tailored toward ANN implementation and application, i.e., to design neuro-computer devices/accelerator boards and the third way is to make use of existing high performance parallel computers and devices to construct purpose built machines, for example using transputers or parallel DSP devices [25,26]. Atlas and Suziki have introduced digital Neural Network (NN) systems in [27].

Yet another approach using digital circuits is to use pulse-coded computation as exemplified by Murray et al. [28], Tomlinson et al. [29], and Leaver [30].

14.3.3 Hybrid Artificial Neural Networks

Analog and digital techniques for the hardware implementation of ANNs could be combined to provide a hybrid solution. This could lead to the best of both disciplines being combined. In a hybrid system, actual computation could be performed using analog processing technique as it often provides the smaller, faster circuits while as weight storage and update can be performed digitally. Inter-element communication could be a mixture of digital and analog. Analog communication links could be used internally within an individual neural chip. Alternatively, pseudo analog systems could be realized using digital signals by means of pulse encoding.

From the last few years, there has been a considerable increase in the hardware implementation of neural networks as they provide simple and more exact solutions and maintain many of the important features of neural networks. The implementation has also unveiled many important applications of neural networks such as dynamically associative memories etc. Considering, the hardware implementation of neural networks, it is important to consider flexibility and power consumption in order to obtain a wide range of applications. Therefore, while designing ANN, the focus is to use the circuits that consume very little power per connection allowing for a high number of connections per neuron. However, the literature related to the hardware implementation of the neural networks is mostly dominated by their realization using the off-the-shelf components, which include operational amplifiers and passive components. A comprehensive review of the electronic neuron models constructed with discrete electronic components may be found in [31] and [32]. In addition, few of the papers related to the hardware realization of ANN consist of [33–35]. Besides, these realizations are fine in verifying the behavior of the neurons but are of no use when their integration on an Integrated Circuits (IC) is considered. In addition, with neuron models, the formation of assembly of these neurons is essential to form a network, then only we can train them and put them to use for their application in various fields. However, this eventually increase the required number of components and hence increase the power consumption. Therefore, it is very important to keep a check on the power and the supply voltage of the designed circuits especially in the case of neural networks owing to their inherent complexity and computational requirements.

Keeping the abovementioned facts in view, the low-voltage electronic implementation of the ANNs is an important field of research at present. There are various techniques in the literature with which the design with reduced power requirements and reduced voltage supply can be achieved, including Rail-to-Rail design approach, Bulk-driven, Multistage, Adaptive Biasing design approaches, and companding.

Besides the power and voltage constrains, unless the design is integrated on an IC along with the other contemporary designs, its applications cannot be exploited to full extent. In addition, along with integrating the neurons models on the IC using contemporary technology, their behavior must be controlled externally to provide design with the feature of training. In order to achieve these features, companding technique is a very good design approach. The companding technique is a current-mode technique in which the input current is first converted into a compressed voltage by an appropriate compressor, the compressed voltage is then processed by the companding core and finally the compressed voltage from the core is converted into output current using an appropriate expander. Broadly classified, there are two main types of companding namely, instantaneous companding and syllabic companding. Out of the two techniques, instantaneous companding has been studied in detail and various sub classes of instantaneous companding have been reported in the literature. In this chapter, we have employed sinh-domain companding techniques. Sinh-domain design offers lot of advantages. These features include the class-AB nature that allows the handling of bidirectional currents significantly larger than the DC bias currents and thereby leading to more power saving [36,37]. The common features offered by the sinh-domain technique are: (a) electronic tunability of the characteristics parameters, (b) resistorless designs, (c) employment of only grounded components, and (d) low-power consumption. Besides, sinh-domain implementation using sub-threshold MOSFETs also offers the advantage of low-voltage operation.

14.4 Sinh-Domain (SD) Companding

Sinh-Domain (SD) companding is an important technique for realizing analog circuits with inherent class-AB nature. In contrast to Log Domain (LD), where a pseudo class-AB operation is realized by establishing two identical class-AB signal paths and employing a current splitter at the input, the required current splitting is simultaneously realized with the compression of the linear input current and its conversion into a non-linear voltage. The produced intermediate output currents are then subtracted in order to derive the final output. Besides, the aforementioned feature, SD technique also offers the capability for electronic adjustment of their frequency characteristics because the realized time constants are controlled by a dc current. Because of the companding nature, SD circuits also allow the capability of operation under a low-voltage environment. SD technique compared to their corresponding counterparts (Log-Domain and Squre-Root-Domain), offer more power efficient realizations but the price to be paid may be an increased circuit complexity [38–56].

The basic building blocks of the SD technique are Sinh and Sinh^{-1} operators, lossy and lossless integrators, and algebraic summation/subtraction blocks, which are introduced in the following section.

14.4.1 Sinh-Cosh (SC) Transconductor Cells

The weak inversion MOSFET (WIMOSFET)-based Sinh and Cosh cell form the basic block for SD circuit design. The transistorized implementation of the cell can be achieved using the translinear principle [57]. For the simplification of the circuit complexities, we can make use of a multiple output transconductor cell where we get all the functions from the same circuit as depicted in Figure 14.5. In a similar way with slight change in the topology

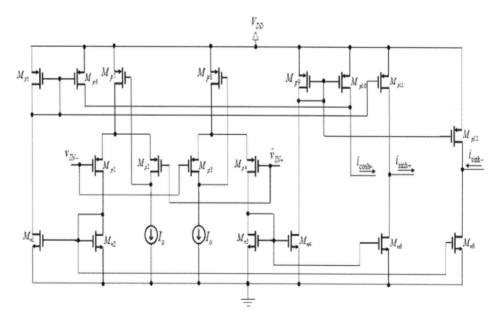

FIGURE 14.5
Multiple output non-linear transconductor cell.

of circuit of Figure 14.5, we can achieve the circuit that will give multiple Sinh outputs, multiple Cosh outputs and so on.

For Figure 14.5, utilizing the translinear principle and performing a routine algebraic analysis, it can be easily obtained that the output currents for sinh and cosh are given by the following equations:

$$i_{\text{sinh}} = 2I_0 Sinh\left(\frac{\hat{v}_{in+} - \hat{v}_{in-}}{U_T}\right) \tag{14.3}$$

$$i_{\text{cosh}} = 2I_0 Cosh\left(\frac{\hat{v}_{in+} - \hat{v}_{in-}}{U_T}\right) \tag{14.4}$$

where, $U_T = nV_T$, n is the sub-threshold slope factor of a WIMOSFET and V_T is the well-known thermal voltage.

14.4.2 SD Complementary Operators

Using Equation 14.3 and inspecting the topology in Figure 14.6a, it is readily obtained that the voltage ($\hat{v}_{IN}$) is given by Equation 14.5; that is, a linear input current is converted into a compressed voltage. In addition, from the configuration in Figure 14.6b and the employment of Equation 14.3, it is derived that the expression in Equation 14.6 is realized. In other words, the topology in Figure 14.6b performs an expansion of a compressed voltage and simultaneously a conversion of it into a linear current. Consequently, the topologies in Figure 14.6 perform two complementary operations as those described by SINH^{-1} and SINH operators introduced in Equations 14.5 and 14.6, respectively.

$$\hat{v}_{IN} = SINH^{-1}(i_{in}) = U_T Sinh^{-1}\left(\frac{i_{in}}{2I_0}\right) \tag{14.5}$$

$$i_{out} = SINH(\hat{v}_{OUT}) = 2I_0 Sinh\left(\frac{\hat{v}_{OUT}}{U_T}\right) \tag{14.6}$$

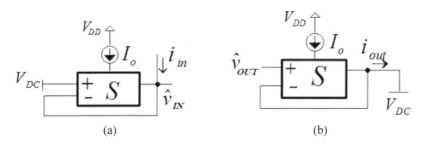

(a)   (b)

FIGURE 14.6
Realization of the SD operators: (a) SINH^{-1} and (b) SINH.

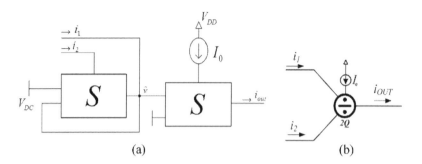

FIGURE 14.7
(a) Two-quadrant multiplier/divider and (b) Employed symbol.

14.4.3 Low-Voltage Two-Quadrant SD Divider

Another important block is a two-quadrant multiplier/divider block. Following a similar concept to that introduced in [38], the derivation of this class-AB divider will be performed. The topology and the corresponding symbol are depicted in Figure 14.7, where the label 2Q depicts the two-quadrant operation capability of the cell. This originates from the fact that both currents I_0 and i_2 are dc bias currents and, consequently, they must be strictly positive.

The configuration of the cell S_2 establishes that its output current could be written as,

$$i_{out} = I_o \cdot \frac{i_1}{i_2} \tag{14.7}$$

14.4.4 SD Summation/Subtraction Block

The realization of SD algebraic summation block with a weighted input is that given in Figure 14.8. Applying the KCL at the output node, it is derived that

$$2I_0 Sinh\left(\frac{\hat{v}_{OUT} - V_{DC}}{U_T}\right) = 2I_0 Sinh\left(\frac{\hat{v}_{IN1} - V_{DC}}{U_T}\right) - a \cdot 2I_0 Sinh\left(\frac{\hat{v}_{IN2} - V_{DC}}{U_T}\right) \tag{14.8}$$

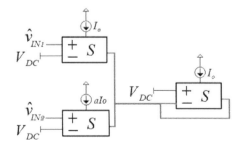

FIGURE 14.8
SD algebraic summation/subtraction block with weighted input.

Using Equation 14.5, Equation 14.8 can be written as

$$SINH(\hat{v}_{OUT}) = SINH(\hat{v}_{IN1}) - a \cdot SINH(\hat{v}_{IN2})$$ (14.9)

14.4.5 SD Integrators

A typical configuration of SD two-input lossless integrator, constructed from blocks mentioned is demonstrated in Figure 14.9. its governing equation is given in Equation 14.10 as

$$\hat{\tau} \cdot \frac{d}{dt} SINH(\hat{v}_{OUT}) = SINH(\hat{v}_{IP}) - SINH(\hat{v}_{IN})$$ (14.10)

where, $\hat{\tau} = CU_T/2I_o$, is the time-constant. As the value of $\hat{\tau}$ is dependent on I_o, which can be changed externally, the SD circuits possess the inherent property of tunability.

In Laplace domain, the input-output relationships of the SD two-input lossless integrator can be given by Equation 14.11.

$$\frac{i_{out}}{i_{ip} - i_{in}} = \frac{g_m/\hat{C}}{S}$$ (14.11)

where, $g_m = \frac{2I_0}{U_T}$, is the trans-conductance and $(g_m/\hat{C})$ is the reciprocal of the integrator's time-constant.

Now that the basic idea about the SD companding is presented, in the next section, the SD realization of the perceptron and its application for the achievement of various digital functions is presented. Later the SD implementation of the inertial neuron is also presented. The performance of the designed topologies in the chapter have been evaluated through simulation results, using MOS transistor models provided by the 130 nm TSMC CMOS process and HSPICE software. The power supply voltages and DC current sources have been chosen as $V_{DD} = 0.5V$, $V_{DC} = 0.1V$ and $I_0 = 50pA$. The DC current sources have been realized by PMOS transistors with aspect ratio 55 µm/1.5 µm. Also, to mention here that the aspect ratios of the transistors of the used transconductance cell of Figure 14.5 are given in Table 14.1.

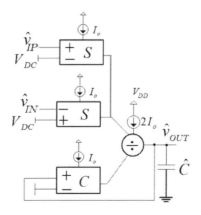

FIGURE 14.9
SD Two-input SD lossless integrator.

TABLE 14.1

Aspect Ratio of MOS Transistors
in Figure 14.5

Transistor	Aspect Ratio
Mp1–Mp4	35 μm/0.55 μm
Mp5–Mp12	58 μm/0.6 μm
Mn1–Mn6	21 μm/1 μm

14.5 Perceptron

The simplest neuron model given by McCulloch and Pitts [58], which is believed to endure resemblances to the human neuron, consists of a weighted sum of its inputs followed by a non-linear activation function. The block diagram of the McCulloch and Pitts model is given in Figure 14.10, and its mathematical representation is given by Equation 14.12

$$OUT = AF\left(\sum_i w_{i,j} IN_i + b_i\right) \tag{14.12}$$

Number of neurons can be connected in a network and it was Rosenblatt [59] who studied the capabilities of groups of neurons in a single layer acted upon by the same input vectors. The structure was termed the "Perceptron," and its learning rule to attain the appropriate weights for classification problems was proposed by Rosenblatt [59]. Minsky and Papert pointed out that many of the real-world problems cannot be solved using the single-layer perceptron while as two-layer perceptron structure [60] is very useful in these cases. The block diagram of the two-layer perceptron is given in Figure 14.11, which is governed by the mathematical expression given in Equation 14.13.

$$OUT = AF\left[\sum_{j=1}^{2} W_{j,3}.AF\left(\sum_{i=1}^{2} W_{i,j} IN_i + b_j\right) + b_3\right] \tag{14.13}$$

Perceptron is a powerful tool and Funahashi [61] showed that the two-layered perceptron is a universal function approximator and can approximate an arbitrary continuous mapping almost perfectly if there is no limit to the number of hidden nodes. The concept of Multi-Layer Perceptron (MLP) was developed from two-layer perceptron to enhance the working of the neural network. The problem faced was that perceptron learning rule proposed by Rosenblatt [59] could not be generalized to find weights for MLP structure.

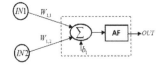

FIGURE 14.10
Single neuron cell.

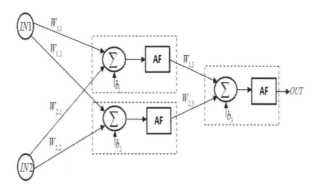

FIGURE 14.11
A two-layer Perceptron.

A learning rule for an MLP structure was later developed that defines a notion of back-propagation of error derivatives through the network [62]. This rule enabled the training of a large class of models with different architectures or connection structures.

The SD design of the single perceptron can be achieved simply by employing the building blocks introduced earlier and the activation function blocks introduced in the next section.

14.6 Activation Functions

Activation Functions (AFs) form an important block of ANNs. AFs have a wide range of applications in Sigma-pi and Hopfield neural networks in addition to being employed in multi-layer perceptron neural network, chaotic and inertial neural networks. Among the AFs, the sigmoid and hyperbolic tangent functions are most often used in the design of neural networks [21]. Another function found in the literature, which was introduced by Shukai Duan and Xiaofeng Liao [63], is named as Liao's AF. Liao's AF is a non-monotonously increasing AF, which is well suited for the design of neural networks showing chaotic behaviours.

In open literature, we can find several different approaches for the hardware implementation of the AFs, including piecewise linear approximation, piecewise non-linear approximation, and lookup tables [64–66]. According to the published manuscripts, only few attempts have been made in the open literature to obtain direct realizations of AFs [68–71]. Most of these implementations are either using high voltage circuits or floating gate MOSFETs, which is a costly technique in itself. Shukai Duan and Xiaofeng Liao [63] achieved the Liao's function by using op-amp, which again is a high voltage circuit design.

Therefore, the low-voltage SD realization of various analog AFs, namely Tanh, bipolar sigmoidal, unipolar sigmoidal, and the Liao's function are presented hereunder.

14.6.1 Tanh Function

The Tanh function produce a curve with an "S" shape. Mathematically, Equation 14.14 defines the Tanh function

$$\text{Tanh} = \left(\frac{e^x - e^{-x}}{e^x + e^{-x}} \right) \tag{14.14}$$

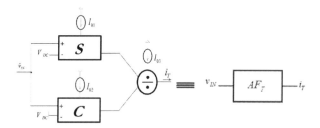

FIGURE 14.12
SD realization of Tanh function.

The proposed idea of its SD implementation is expressed in Equation 14.15.

$$\text{Tanh} = K_1 \left(\frac{\sinh(x)}{\cosh(x)} \right) \tag{14.15}$$

The SD realizations of Tanh AF is given in Figures 14.12. After routine algebraic manipulations, the equation for Tanh function can be obtained as given by Equation 14.16.

$$i_T = K_1 \left(\tanh \left(\frac{\hat{v}_{IN} - v_{DC}}{nV_T} \right) \right) \tag{14.16}$$

where K_1, is given by

$$K_1 = I_{031} \left(\frac{I_{011}}{I_{021}} \right) \tag{14.17}$$

For the achievement of the Tanh function, the parameters I_{01}, I_{02} and I_{03} were equal to 200 pA. The simulation results of the function are given in Figure 14.13. The result in Figure 14.13a shows the DC response while as the result in Figure 14.13b gives the Monte-Carlo analysis results of AF circuits to the effect of process variations and mismatching calculated for N = 70 runs assuming 5% deviation (with Gaussian distribution). Three important performance parameters i.e., upper saturation, lower saturation levels and switching threshold voltage (voltage corresponding to mean of upper and lower saturation currents), have been considered for the study. The simulated values of the upper saturation level, lower saturation level and switching threshold voltage standard deviations for Tanh AF were 1.20 pA, 1.73 pA and 269 μV, respectively.

14.6.2 Sigmoidal Function

The sigmoidal function also produces a curve with an "S" shape. The sigmoidal function can further be classified as unipolar and bipolar depend upon whether the output response is positive only or can have both positive and negative values. Mathematically, Equations 14.18 and 14.19, respectively, can define the functions as

$$\text{Unipolar Sigmoidal} = \left(\frac{1}{1 + e^{-x}} \right) \tag{14.18}$$

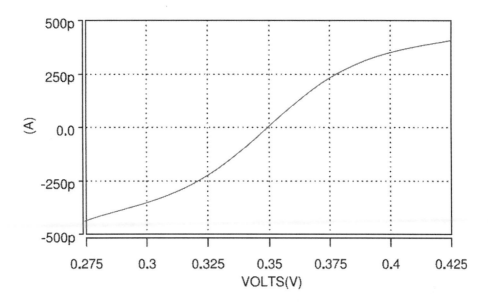

(A) : VOLTS(V) I_TANH

(a)

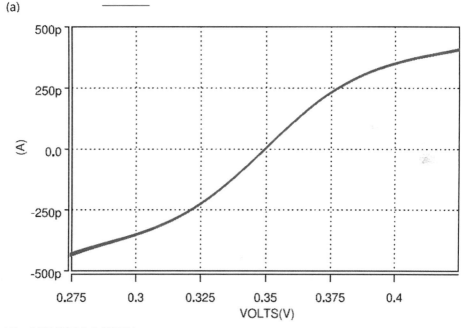

(A) : VOLTS(V) I_TANH

(b)

FIGURE 14.13

Tanh function (a) DC sweep response, and (b) Monte-Carlo analysis results of circuit to the effect of process variations and mismatching.

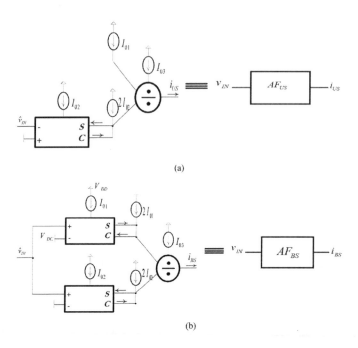

(a)

(b)

FIGURE 14.14
SD realization of activation functions (a) Unipolar Sigmoidal, and (b) Bipolar Sigmoidal.

$$Bipolar\ Sigmoidal = \left(\frac{1-e^{-x}}{1+e^{-x}}\right) \tag{14.19}$$

The SD implementations of the expressions are expressed in Equations 14.20 and 14.21.

$$Unipolar\ Sigmoidal = K_2\left(\frac{1}{1+\cosh(x)-\sinh(x)}\right) \tag{14.20}$$

$$Bipolar\ Sigmoidal = K_3\left(\frac{1-\cosh(x)+\sinh(x)}{1+\cosh(x)-\sinh(x)}\right) \tag{14.21}$$

The SD realizations of Unipolar Sigmoidal and Bipolar Sigmoidal AFs are given in Figures 14.14a and b, respectively. After routine algebraic manipulations, the equations for Unipolar Sigmoidal and Bipolar Sigmoidal activation functions can be obtained as given by Equations 14.22 and 14.23, respectively.

$$i_{US} = K_2\left(\frac{1}{1+e^{-\left(\frac{\hat{v}_{IN}+-v_{DC}}{nV_T}\right)}}\right) \tag{14.22}$$

$$i_{BS} = K_3\left(\frac{1-e^{-\left(\frac{\hat{v}_{IN}-v_{DC}}{nV_T}\right)}}{1+e^{-\left(\frac{\hat{v}_{IN}-v_{DC}}{nV_T}\right)}}\right) \tag{14.23}$$

where K_2 and K_3, respectively, and given by Equations 14.24 and 14.25, as

$$K_2 = I_{03}\left(\frac{I_{01}}{I_{02}}\right) \tag{14.24}$$

$$K_3 = I_{03}\left(\frac{I_{01}}{I_{02}}\right) \tag{14.25}$$

For the achievement of the Unipolar Sigmoidal function, the parameters I_{01}, I_{02} and I_{03} were taken to be 200 pA, 100 pA, and 200 pA, respectively, and for the Bipolar Sigmoidal the parameters I_{01}, I_{02}, and I_{03} were taken to be 200 pA each. The simulation results of sigmoidal functions are drawn in Figure 14.15a–d. The simulated standard deviations values of the upper saturation level, lower saturation level and switching threshold voltage for Unipolar Sigmoidal AF were 3.53 pA, 2.56 pA, and 9.1 mV, respectively, and that for the Bipolar Sigmoidal AF were 7.81 pA, 2.62 pA, and 780.23 µV, respectively.

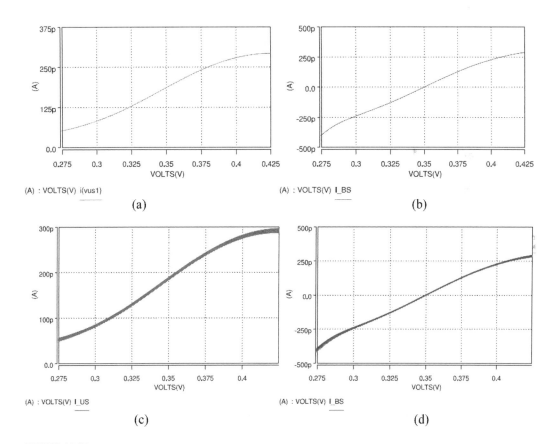

FIGURE 14.15
(a) DC sweep response of Unipolar sigmoidal, (b) DC sweep response of Bipolar sigmoidal, (c) Monte-Carlo analysis results of Unipolar sigmoidal AF circuits to the effect of process variations and mismatching, and (d) Monte-Carlo analysis results of Bipolar Sigmoidal AF circuits to the effect of process variations and mismatching.

14.6.3 Liao's Function

In their paper, Shukai Duan and Xiaofeng Liao [63] gave a non-monotonously increasing activation function described by the equation as given in Equation 14.26.

$$f(x) = \sum_{i=1}^{n} \alpha_i \left[\tanh(x + k_i) - \tanh(x - k_i) \right] \tag{14.26}$$

where n, α_i and k_i are constants determining the shape, positive, and negative amplitudes of the AF. The block diagram representation of Equation 14.26 is shown in Figure 14.16. From Figure 14.16, it is clear that in order to implement the Liao activation function, Tanh and scaled summation blocks are required.

The SD realization of Liao's function is given in Figure 14.17.

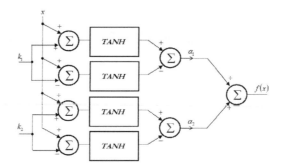

FIGURE 14.16
Block Diagram implementation of Liao's function.

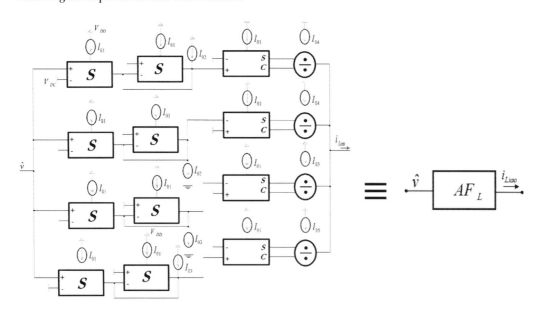

FIGURE 14.17
SD Implementation for Liao's function.

In order to demonstrate the performance of Liao's AF of Figure 14.17, the simulations were carried out in HSPICE using a TSMC 130nm CMOS process file. The results were obtained for the biasing values given by $V_{DD} = 0.5V$, $V_{DC} = 0.18V$, and $I_0 = 10pA$. The values of n, α_1, α_2, k_1, k_2 are kept to be equal to 2, 2, −1.5, 1, and 1.33, respectively. The corresponding values for the parameters for the SD design of the Liao's function I_{01}, I_{02}, I_{03}, I_{04}, and I_{05} were taken to be 200 pA, 200 pA, 266 pA, 400 pA, and 300 pA, respectively.

The obtained MATLAB Simulink and PSPICE simulation results are shown in Figure 14.18. From the derived results, it is clear that the simulation results are in close agreement with the MATLAB Simulink results. To check the matching of MATLAB Simulink and HSPICE simulator results, R-sq data was calculated and the value for this fitting test came out to be equal to 0.89, which shows that the design gives a very good fit to the result obtained from MATLAB Simulink. The small variation is due to the small inexactness in the function of the SD blocks. The results for the effect of process variations and mismatching are calculated using Monte-Carlo analysis for N = 100 runs assuming 5% deviation (with Gaussian distribution) and are shown in Figure 14.19. The standard deviation values of the maxima and minima were 0.373 pA and 1.2 pA, respectively. The mean values of the maxima and minima were 13.7 pA and −6.37pA, and for desired confidence level of 95%. The values of confidence interval and the range for the true mean for maxima are ±0.07 pA and 13.63 pA to 13.77 pA, respectively, while as in case of minima, the values are ±0.24 pA and −6.61 pA to −6.13 pA, respectively.

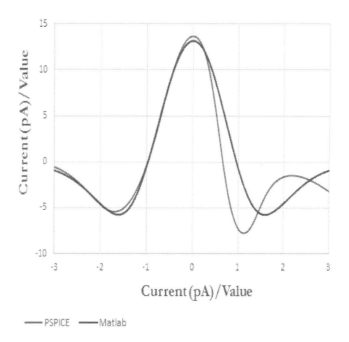

FIGURE 14.18
DC sweep response of Liao's function.

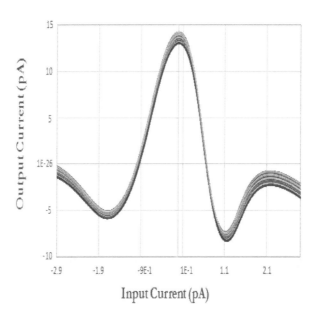

FIGURE 14.19
Monte-Carlo result of Liao's function.

14.7 Effects of Mismatching between the DC Current Sources

To study the effect of mismatch between the dc current sources in the presented building blocks, we assume that the values of the dc current source is given by $I_o(1+\varepsilon)$ instead of I_o. Employing this value, the expression of the output currents for hyperbolic sine, hyperbolic cosine, and inverted hyperbolic sine are given by Equations 14.27 to 14.29, respectively.

$$i = 2I_o(1+\varepsilon)\sinh\left(\frac{\hat{v}_{IN+} - \hat{v}_{IN-}}{nV_T}\right) \tag{14.27}$$

$$i = 2I_o(1+\varepsilon)\cosh\left(\frac{\hat{v}_{IN+} - \hat{v}_{IN-}}{nV_T}\right) \tag{14.28}$$

$$i = -2I_o(1+\varepsilon)\sinh\left(\frac{\hat{v}_{IN+} - \hat{v}_{IN-}}{nV_T}\right) \tag{14.29}$$

This intern means that the non-linear transconductance cells (hyperbolic sine, hyperbolic cosine, and inverted hyperbolic sine) can be represented as their ideal circuits biased at a dc current $(1+\varepsilon)I_o$.

In case of the two-quadrant divider following the routine algebraic manipulations and neglecting the second-order error terms, the output current can be written by Equation 14.30.

$$i_{OUT} = I_o(1+\varepsilon_D)\frac{i_1}{i_2} \tag{14.30}$$

where error factor $\varepsilon_D = (\varepsilon_2 - \varepsilon_1)/(1 + \varepsilon_1)$ and ε_1 and ε_2 are the error factors corresponding to the two sinh cells of the divider. Similar to the non-linear transconductance cell, the two-quadrant divider can be represented by an ideal two-quadrant divider biased at a DC current $(1 + \varepsilon_D)I_o$.

The output of the Tanh block will be given by Equation 14.31.

$$I_{out} = AI_0 (1 + \varepsilon_D) \frac{(1 + \varepsilon_1)}{(1 + \varepsilon_2)} \left(\tanh\left(\frac{\hat{v}_{IN} - V_{DC}}{nV_T} \right) \right) \tag{14.31}$$

where ε_1, ε_2, and ε_D are the error factors corresponding to the Sinh, Cosh, and the Divider cells in the Tanh circuitry.

The outputs for the unipolar and Bipolar Sigmoidal functions can be given by Equations 14.32 and 14.33, respectively.

$$I_{out} = AI_0 (1 + \varepsilon_D) \left(1 / (1 + (\varepsilon_2 - \varepsilon_1)) \right) \tag{14.32}$$

$$I_{out} = AI_0 (1 + \varepsilon_D) \left(\left. (1 - (\varepsilon_2 - \varepsilon_1)) \middle/ (1 + (\varepsilon_2 - \varepsilon_1)) \right. \right) \tag{14.33}$$

where ε_1, ε_2, and ε_D are the error factors corresponding to the Sinh, Cosh, and the Divider cells.

Similarly, after some algebraic manipulations, the output of the Liao's function can be given as in Equation 14.34.

$$I_{out} = \alpha_1 I_0 (1 + \varepsilon_{D1}) \left(\tanh\left(\frac{\hat{v}_x + V_{k1}}{nV_T} \right) \right) - \alpha_1 I_0 (1 + \varepsilon_{D2}) \left(\tanh\left(\frac{\hat{v}_x - V_{k1}}{nV_T} \right) \right)$$

$$- \alpha_2 I_0 (1 + \varepsilon_{D3}) \left(\tanh\left(\frac{\hat{v}_x + V_{k2}}{nV_T} \right) \right) + \alpha_2 I_0 (1 + \varepsilon_{D4}) \left(\tanh\left(\frac{\hat{v}_x - V_{k2}}{nV_T} \right) \right) \tag{14.34}$$

where, $\varepsilon_{D1}, \varepsilon_{D2}, \varepsilon_{D3}$, and ε_{D4} are the error factors associated with the divider topologies used to achieve the desired function. Thus, the mismatch in the current sources change scale factors of the Liao's function. It is worth to mention here that the error factors in the blocks could be easily compensated through an adjustment of the bias current I_o.

14.8 Perceptron Applications

Now as we are already aware of the fact that neural networks can be used to solve any mathematical/engineering problem, the application of the perceptron, realized in SD companding, for the implementation of various logic functions is presented in this section. These logic functions include various logic gates, multiplexer, encoder, and so on. Among these designs, the logic gates are designed using four different AFs (discussed above). It is worth to mention here that the Boolean logic functions can be implemented by state-of-the-art digital technology, which often is simpler than the analog neuron based implementation. However, the main drawback of the conventional digital Very Large Scale Integration (VLSI) designs is that there is a specific circuit for each logic gate. Therefore, from implementation point of view, there is no reconfigurabilty of the functions. Reconfiguration of the logic gate, however, is an attractive subject for different applications. The reconfigurability property can be obtained with the help of the neuron based implementation. In this case, all the logic functions (AND, OR, NOT, NAND, NOR) can be obtained from a single circuit (single-layer perceptron) just by changing the weights and biases of the neuron. The results for this are obtained and presented in Figures 14.20 through 14.24. For this purpose, the training of the perceptron was first done in MATLAB environment to obtain the weights for the neurons. The training was carried out using the backpropagation algorithm. The corresponding weights for attaining these results calculated through MATLAB program are adjusted properly for the SD design to overcome the non-idealities. The values of these weights are given in Tables 14.2 and 14.3.

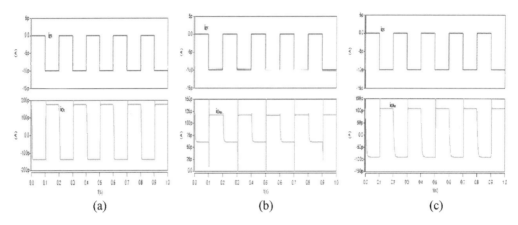

FIGURE 14.20
Simulation results of single-layer perceptron trained for NOT function with (a) Tanh function, (b) Unipolar sigmoidal function, and (c) Bipolar Sigmoidal function.

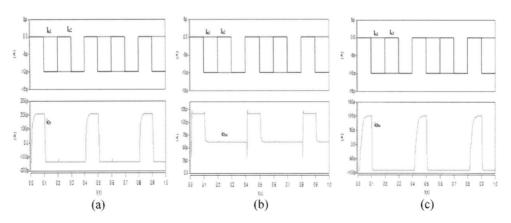

FIGURE 14.21
Simulation results of single-layer perceptron trained for AND function with (a) Tanh function, (b) Unipolar sigmoidal function, and (c) Bipolar Sigmoidal function.

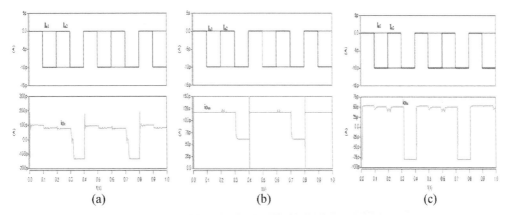

FIGURE 14.22
Simulation results of single-layer perceptron trained for OR function with (a) Tanh function, (b) Unipolar sigmoidal function, and (c) Bipolar Sigmoidal function.

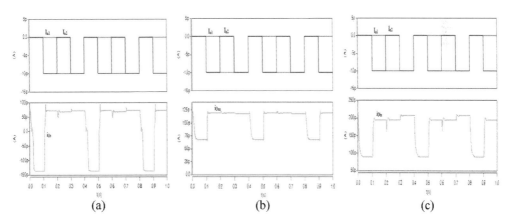

FIGURE 14.23
Simulation results of single-layer perceptron trained for NAND function with (a) Tanh function, (b) Unipolar sigmoidal function, and (c) Bipolar Sigmoidal function.

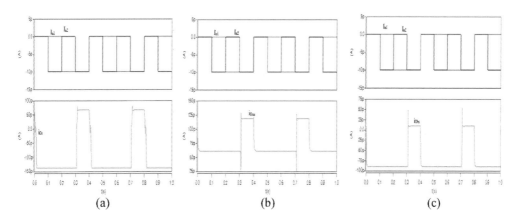

FIGURE 14.24
Simulation results of single-layer perceptron trained for NOR function with (a) Tanh function, (b) Unipolar sigmoidal function, and (c) Bipolar Sigmoidal function.

TABLE 14.2

Weights and Bias for Perceptron with Tanh, Unipolar Sigmoidal Function, and Bipolar Sigmoidal Function to Obtain NOT Logics

Weights & Bias	Tanh Function	Unipolar Sigmoidal	Bipolar Sigmoidal
$W_{1,1}$	−17.7pA	−38.9pA	−18.7pAdc
B	45.6pA	83.9pA	35.9pA

TABLE 14.3

Weights and Bias for Perceptron with Tanh, Unipolar Sigmoidal Function, and Bipolar Sigmoidal Function to Obtain AND, OR, NAND, and NOR Logics

	Weight and Bias	Logic Function			
		AND	OR	NAND	NOR
Tanh	$W_{1,1}$	6.98pA	9.25pA	−7.99pA	−9.25pA
	$W_{2,1}$	6.98pA	11.25pA	−7.48pA	−11.25pA
	B	−21.87pA	−111.07pA	38.87pA	111.07pA
Unipolar Sigmoidal	$W_{1,1}$	5.83pA	10.8pA	−5.8pA	−10.8pA
	$W_{2,1}$	5.98pA	12.88pA	−5.88pA	−12.88pA
	h	−17.97pA	−110.87pA	18.87pA	110.87pA
Bi-polar Sigmoidal	$W_{1,1}$	10.88pA	6.8pA	−6.98pA	−11.8pA
	$W_{2,1}$	10.88pA	6.88pA	−6.98pA	−12.88pA
	b	−15.98pA	−92.87pA	15.87pA	110.87pA

For the case of the Liao's function the weights and bias for achieving various functions is given in Tables 14.4 and 14.5 and the corresponding results are shown in Figures 14.25 and 14.26.

In addition to the implementation of the logic gates, the implementation of the XOR, which comes under the brand of non-separable problem, and hence it was believed that this problem cannot be solved with the help of the single-layer perceptron. However,

TABLE 14.4

Weights and Bias for Perceptron with Liao's Function to Obtain NOT and Buffer Logics

Weights & Bias	NOT Logic	Buffer Logic
$W_{1,1}$	−10 pA	30 pA
B	40 pA	−92 pA

TABLE 14.5

Weights and Bias for Perceptron with Liao's Function to Obtain AND, OR, NAND, and NOR Logics

Weight and Bias	Logic Function			
	AND	OR	NAND	NOR
$W_{1,1}$	8.1 pA	10 pA	−18.8 pA	−7.9 pA
$W_{2,1}$	8.1 pA	10 pA	−18.8 pA	−7.9 pA
b	−21.9 pA	−50 pA	42.8 Pa	100 pA

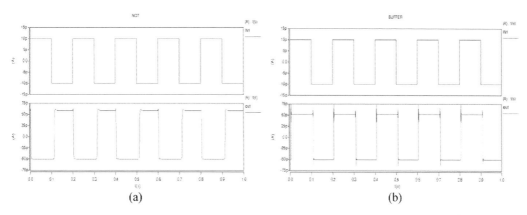

(a)

(b)

FIGURE 14.25

Simulation result of single neuron trained with Liao's function (a): NOT (b) Buffer Logic.

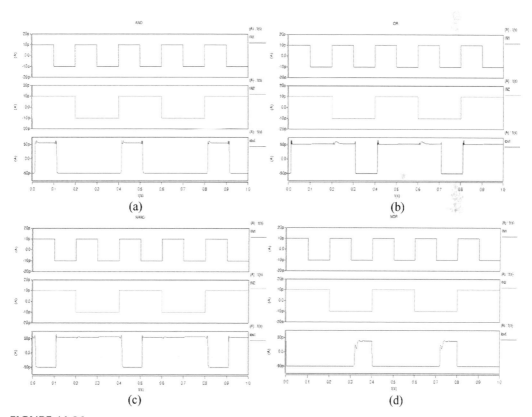

(a)

(b)

(c)

(d)

FIGURE 14.26

Simulation result of single neuron trained for with Liao's AF: (a) AND, (b) OR, (c) NAND, and (d) NOR operations.

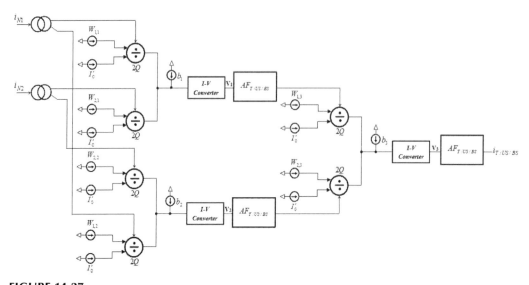

FIGURE 14.27

SD realization of two-layer perceptron.

it is shown here that this problem can also be solved with a single-layer perceptron by using the Liao's AF. For rest of the above discussed functions, namely Tanh, unipolar sigmoidal, and bipolar sigmoidal, it is impossible to achieve XOR with a single-layer perceptron and, thus, for achieving the XOR logic with these functions, two-layer perceptron of Figure 14.27 is utilized.

The weights and bias values for this are given in Table 14.6. The simulation results are given in Figure 14.28. In order to demonstrate the achievement of XOR problems with a single-layer perceptron with the Liao's AF, the results of single-layer perceptron trained for XOR logic are given in Figure 14.29. The values of weights "$W_{1,1}$," "$W_{2,1}$," and bias "b" for single-layer perceptron for non-linear separable problem are -40.8 pA, -40.8 pA, and 92.8 pA. The calculation of the power dissipation in case of various circuits is given in Table 14.7.

TABLE 14.6

Weights and Bias for Two-Layer Perceptron to Obtain Non-Linear Separable XOR Function with Tanh, Unipolar Sigmoidal function, and Bipolar Sigmoidal Function

Weights and Bias	Tanh Function	Unipolar Sigmoidal	Bipolar Sigmoidal
$W_{1,1}$	-30.83 pA	-28.99 pA	-30.83 pA
$W_{2,1}$	11.98 pA	11.64 pA	11.98 pA
b_1	-7.97 pA	-7.97 pA	-7.97 pA
$W_{1,2}$	28.21 pA	27.11 pA	28.21 pA
$W_{2,2}$	-42.96 pA	-40.44 pA	-42.96 pA
b_2	-9.7 pA	-9.7 pA	-9.7 pA
$W_{1,3}$	-9.99 pA	-5.82 pA	-3.57 pA
$W_{2,3}$	-8.98 pA	-5.58 pA	-3.56 pA
b3	101.89 pA	191.89 pA	284.89 pA

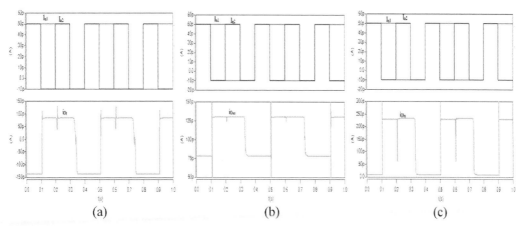

FIGURE 14.28

Simulation results of trained two-layer perceptron for XOR function with (a) Tanh function, (b) Unipolar sigmoidal function, and (c) Bipolar Sigmoidal function.

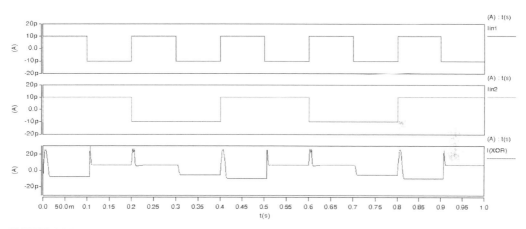

FIGURE 14.29

Simulation result of single-layer perceptron trained for non-linearly separable XOR operation with Liao's AF.

TABLE 14.7

Total Static Power Consumption of Various Circuits for: (a) NOT (b) AND, OR, NAND, and NOR

	NOT	AND	OR	NAND	NOR	XOR
Tanh function	2.82 nW	5.24 nW	17.18 nW	2.82 nW	2.85 nW	7.42 nW
Unipolar Sigmoidal	1.34 nW	4.36 nW	13.90 nW	1.36 nW	2.55 nW	8.46 nW
Bipolar Sigmoidal	3.58 nW	2.62 nW	16.81 nW	4.10 nW	3.64 nW	8.25 nW

Then to confirm the functionality of the designed SD perceptron for higher number of layers, the 3-bit grey to binary converter shown in Figure 14.30 is considered and the simulation results for this circuit are shown in Figure 14.31.

The next problem that has been considered is Multiplexer. Multiplexer, frequently termed as MUX, is an important logic building block that finds its application in almost all the digital integrated circuits. MUXs find their application in the field of telecommunications and are extremely important as they are very helpful in reducing network complexity by minimizing the number of communications links needed between two points and, in turn, reducing the cost. Along with all the computing systems, multiplexers have evolved with time. Each new generation has additional intelligence, and additional intelligence brings more benefits. Few of the accrued benefits are [23]

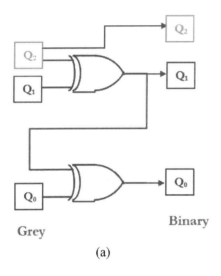

Gray			Binary		
Q_2	Q_1	Q_0	Q_2	Q_1	Q_0
0	0	0	0	0	0
0	0	1	0	0	1
0	1	0	0	1	1
0	1	1	0	1	0
1	0	0	1	1	1
1	0	1	1	1	0
1	1	0	1	0	0

(a)   (b)

FIGURE 14.30
(a) Logic diagram of 3-bit grey to binary converter and (b) Truth table of 3-bit grey to binary converter.

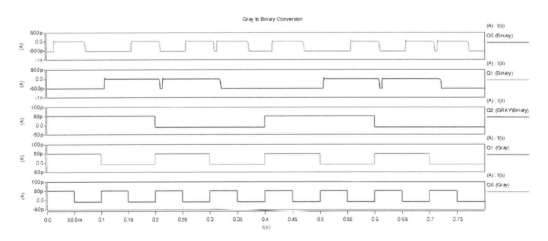

FIGURE 14.31
Simulation results of 3-bit grey to binary converter.

- Data compression: The capability to do data compression that enables us to encode certain characters with smaller number of bits than normally required. The freed capacity can be utilized for movement of other information.

- Error detection and correction: Data integrity and accuracy are maintained by error detection and correction between the two points.

- Managing transmission resources: The capability to manage transmission resources on a dynamic basis, by introducing the additional features such as priority levels.

Device cost versus line cost is an important issue in communication for transferring signals from one point to the other. We can provide extremely high levels of service by ensuring everybody always has a continuous and live communications link. However, this has a drawback of extremely high cost. In the other option, we can opt to offset the costs associated with providing large numbers of lines by using devices, such as multiplexers, that help in making more intelligent use of a smaller group of lines. The more intelligent the multiplexer, the more perceptively and vigorously it can work on our behalf to dynamically make use of the available transmission resources.

Keeping the above in view, the implementation of 2:1 MUX using SD perceptron is shown in Figure 14.32.

The simulation results for the perceptron designed to perform the operation of the 2:1 MUX is given in Figure 14.33. The total current source power dissipation of the 2:1 multiplexer network was 213.13 pW while as total voltage source power dissipation was 10.2736 nW, which establishes the fact that the low-power design of the 2:1 MUX has been achieved.

Furthermore, as already mentioned, the designs can be extended to multi-layer perceptron where complex logic functions can be obtained from a single circuit. In order to verify this, a three-layer perceptron given in Figure 14.34 has been considered. The AF used,

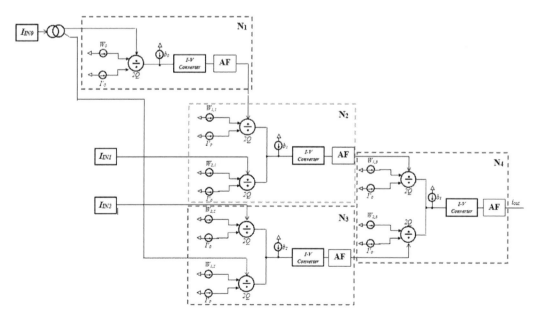

FIGURE 14.32
SD realization of ANN based 2:1 MUX.

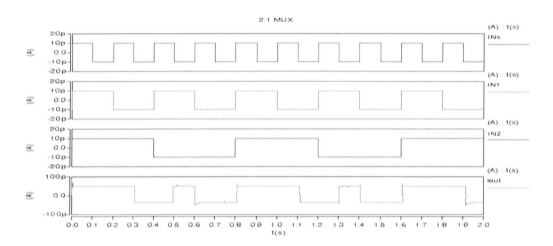

FIGURE 14.33
Simulation results for SD NN based 2:1 MUX.

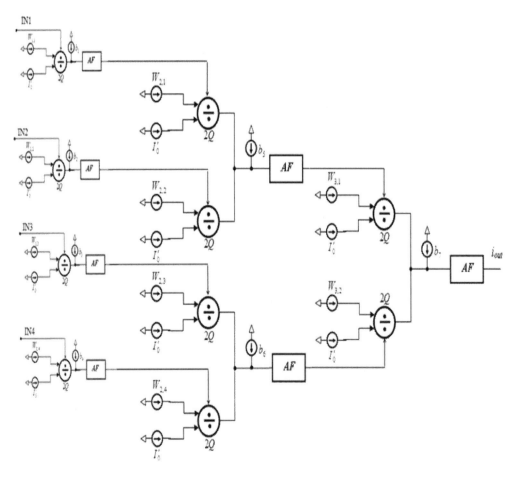

FIGURE 14.34
SD realization of three-layer perceptron.

in this case, is the Liao's function. The perceptron has been trained for three logic functions given in Table 14.8. The corresponding obtained responses for the three logic functions are given in Figure 14.35a–c. These designs can be obtained for other functions too; however, the values of the weights will be altered.

TABLE 14.8

Weights and Bias Associated with Neurons for Obtaining Different Logical Functions from Three-Layer Perceptron

	Weights									
Function	$W_{1,1}$ (pA)	$W_{1,2}$ (pA)	$W_{1,3}$ (pA)	$W_{1,4}$ (pA)	$W_{2,1}$ (pA)	$W_{2,2}$ (pA)	$W_{2,3}$ (pA)	$W_{2,4}$ (pA)	$W_{3,1}$ (pA)	$W_{3,2}$ (pA)
$F_1 = I_{N1}.I_{N2} + I_{N3}.I_{N4}$	30	30	30	30	8.1	8.1	8.1	8.1	10	10
$F_2 = (I_{N1} + I_{N2}).$ $(I_{N3} + I_{N4})$	30	30	30	30	10	10	8.1	8.1	8.1	8.1
$F_3 = I_{N1}.I_{N2} + \bar{I}_{N1}.I_{N4}$	30	30	−10	30	8.1	8.1	8.1	8.1	10	10

	Bias						
Function	b1 (pA)	b2 (pA)	b3 (pA)	b4 (pA)	b5 (pA)	b6 (pA)	b7 (pA)
$F_1 = I_{N1}.I_{N2} + I_{N3}.I_{N4}$	−92	−92	−92	−92	−21.9	−21.9	−50
$F_2 = (I_{N1} + I_{N2}).$ $(I_{N3} + I_{N4})$	−92	−92	−92	−92	−50	−50	−21.9
$F_3 = I_{N1}.I_{N2} + \bar{I}_{N1}.I_{N4}$	−92	−92	40	−92	−21.9	−21.9	−50

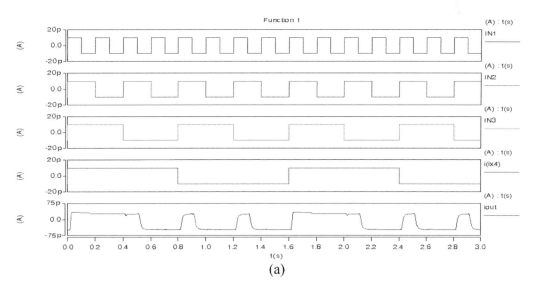

(a)

FIGURE 14.35

Simulation results of three-layer perceptron trained for function (a) Function F1, *(Continued)*

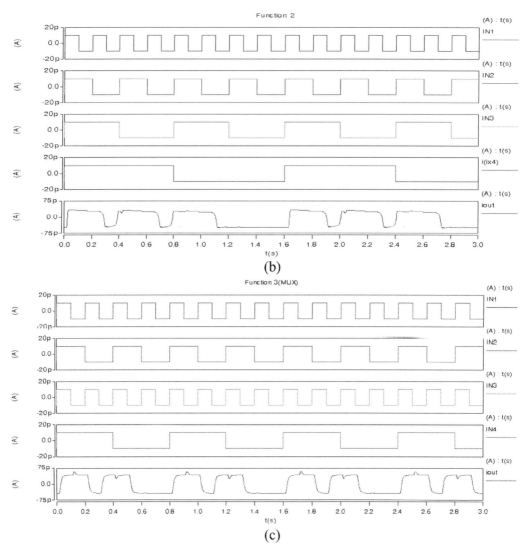

FIGURE 14.35 (Continued)
Simulation results of three-layer perceptron trained for function (b) Function F2, and, (c) Function F3 of Table 14.3.

14.9 Inertial Neuron

The perceptron forms the basic and the simplest model for the neural network but as we have already discussed that actual neurons exhibits various behaviours and one of the behaviours of the natural neuron is chaotic behaviour. Chaos is an important phenomenon and plays a vital role in memory storage and retrieval. Therefore, we find number of ANNs proposed in literature, which show chaotic dynamical behavior [72–74]. One such neuron model is the inertial neuron [75]. The association of the inertial term with the neuron finds a lot of background in the literature. In ANNs, the inertia has been modeled by an inductance term. The theory behind including the inertial term in the neuron equation was put forth by Kotch [76] where he stated that under certain conditions neurons exhibit a quasi-active membrane, which can be modeled by an inductance that allows the membrane to behave like a bandpass filter enabling electrical tuning, or temporal differentiation, or spatio-temporal filtering. The inductance has also been associated with the membrane of a hair cell in the semicircular canals of some animals [77,78], axon of the squid [79], and so on. Subsequently, it was shown that including inertia in single/coupled neuron(s) also shows chaos in their dynamical behavior [80,81]. The inertial (inductive) term added to the simple neuron results in the ringing transient response, spontaneous oscillation, intertwined basin boundaries, and even chaotic response to an external drive. It was also shown that two-neuron system exhibits chaos with one or two inertial terms [81]. In addition, Tain et al. [82,83] added inertia to neural equations as a way of chaotically searching for memories in neural networks.

The differential equation governing the inertial neuron can be written in the form

$$\ddot{x} + 2\xi\dot{x} + bx = cf(x) + U \tag{14.35}$$

where, "ξ" is the damping factor and "b" is a constant, while "c" represents the overall gain of the neuron and determines the strength of the nonlinearity and "U" is the excitation. The above expression describes the motion of a damped, two-well, nonlinear mechanical oscillator in case the value of "c" is greater than unity. The system produces a variety of nonlinear, chaotic, and intermittent phenomena when driven by a periodic signal.

In order to obtain the SD design for the inertial neuron of Equation 14.35, we need integrator and activation function block. Furthermore, for the design of the said neuron, the Liao's activation function has been utilized. The SD representation of the inertial neuron is given in Figure 14.36.

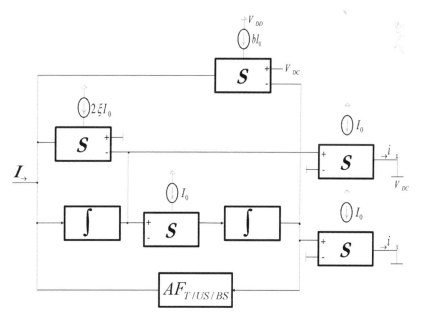

FIGURE 14.36
SD Implementation of inertial neuron.

For the demonstration of the performance of the inertial neuron topology of Figure 14.36, three different cases are considered. For case 1, the values of the parameters ξ, b, c, and $\hat{C}$ are taken to be -0.15, -0.5, 1.5, and 600 pF, while as the external excitation applied is the pulse of duration 0.01 ms. The derived transient and phase responses using MATLAB Simulink and HSPICE are shown in Figures 14.37 and 14.38a,b, respectively. Note that in all the responses X_1 mean $\dot{x}$ and X_2 mean x. For case 2, the values of the parameters ξ, b, c, and $\hat{C}$ are taken to be -0.15, -0.4, 5, and 600 pF, while the external excitation applied is the sinusoidal source of 10 pA amplitude and 1 Hz frequency. The derived transient and phase responses using MATLAB Simulink and HSPICE are shown in Figures 14.39a,b and 14.40a,b, respectively. For case 3, the values of the parameters ξ, b, c, and $\hat{C}$ are taken to be -0.175, -0.35, 6, and 600 pF and the external excitation applied is same as in case 2. The derived transient and phase responses using MATLAB Simulink and HSPICE are shown in Figures 14.41a,b and 14.42a,b, respectively. It is

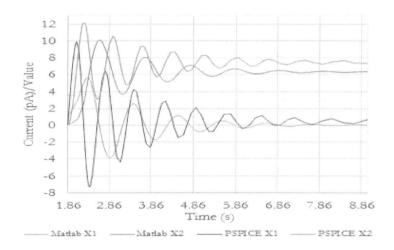

FIGURE 14.37
Transient responses using MATLAB Simulink and PSPICE for Case 1.

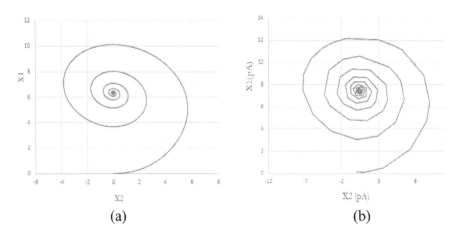

FIGURE 14.38
Phase responses obtained in case 1 using: (a) MATLAB Simulink, and (b) PSPICE.

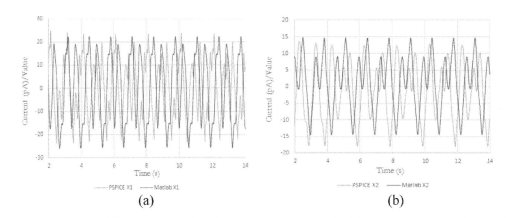

(a)

(b)

FIGURE 14.39

Transient responses obtained in case 2 using MATLAB Simulink and PSPICE for: (a) X1, and (b) X2.

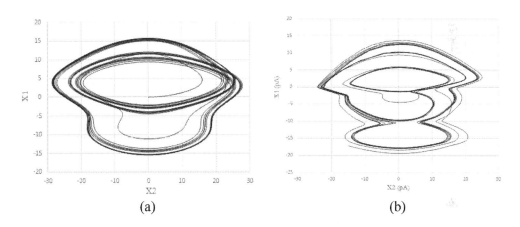

(a)

(b)

FIGURE 14.40

Phase responses obtained in case 2 using (a) MATLAB Simulink, and (b) PSPICE.

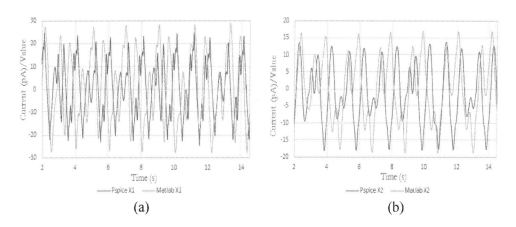

(a)

(b)

FIGURE 14.41

Transient responses obtained in case 3 using MATLAB Simulink and PSPICE for: (a) X1, and (b) X2.

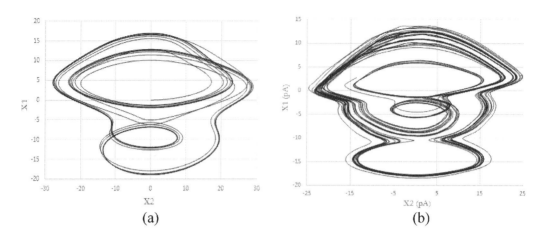

FIGURE 14.42
Phase responses obtained in case 3 using (a) MATLAB Simulink and (b) PSPICE.

worth to mention here that multiplying factor (c) required for the Liao's AF has been achieved by modifying α_1 to $c\alpha_1$ and α_2 to $c\alpha_2$ in the function. Again, from the achieved results, it is clear that the simulation results are in close agreement with the MATLAB Simulink results. The deviations in the MATLAB and HSPICE results are due to the fact that the two valleys in the designed Liao function are not having the same amplitude. As a result, instead of a single bulge, two bulges are seen in the HSPICE results of Figures 14.40b and 14.42b.

It is pertinent to mention here that the largest Lyapunov exponent (LLE) values for different parameter variation was found greater than one in all cases ranging from 1.123 to 1.214, for the considered neuron model, which affirms that the system will show chaotic behavior with parameter variations as well. The total power dissipation of the Liao's AF and inertial neuron was calculated and were equal to 14 nW and 58 nW, respectively.

14.10 Conclusion

The chapter presents the ultra-low-voltage implementation of the perceptron and the inertial neuron. The said designs of the neurons are achieved using sinh-domain companding, which also provides the feature of electronic tunability of the designs, important for hardware neural network design (easy adjustment of the weights). More importantly, the designs are compatible with modern-day VLSI technologies and, thus, implementable in IC form. In addition, to show the application areas of the designed perceptron, small problems (various digital logic functions) are considered and presented in the chapter. For the students or researchers who want to pursue the field of hardware implementation of neural networks, the chapter can be of paramount importance.

References

1. M. Caudill. Neural Network Primer: Part I. AI Expert, 1989.
2. D. H. Sanes, T. A. Reh, and W. A. Harris, *Development of the Nervous System*. Amsterdam, the Netherlands: Academic press, Chapter 1, 2000.
3. A. Garm, M. O'Connor, L. Parkefelt, and D. E. Nilsson, Visually guided obstacle avoidance in the box jellyfish tripedalia cystophora and chiropsella bronzie. *Journal of Experimental Biology* 210 (20), 3616–3623, 2007.
4. R. W. Williams and K. Herrup, The control of neuron number. *Annual Review of Neuroscience* 11 (1), 423–453, 1988.
5. B. Pakkenberg, D. Pelvig, L. Marner, M. J. Bundgaard, H. J. J. Gundersen, J. R., Nyengaard, and L. Regeur, Aging and the human neocortex. *Experimental Gerontology* 38 (1–2), 95–99, 2003.
6. E. R. Kandel, J. H. Schwartz, and T. M. Jessell, *Principles of Neural Science*, 4th ed. New York: McGraw-Hill Medical, Chapter 17, 2000.
7. M. A. Arbib, *Brains, Machines and Mathematics*, 2nd ed. New York: Springer Verlag, 1987.
8. A. Hodgkin and A. Huxley, A quantitative description of membrane current and its application to conduction and excitation in nerve. *Journal of Physiology*, 117, 500–544, 1952.
9. A. Babloyantz and C. Lourenco, Brain chaos and computation. *International Journal of Neural Systems* 7 (4), 461–471, 1996.
10. H. Bersini, The frustrated and compositional nature of chaos in small Hopfield networks. *Neural Networks* 11 (6), 1017–1025, 1998.
11. H. Bersini and P. Sener, The connections between the frustrated chaos and the intermittency chaos in small Hopfield networks. *Neural Networks* 15 (10), 1197–1204, 2002.
12. A. Das, P. Das, and A. B. Roy, Chaos in a three-dimensional general model of neural network. *International Journal of Bifurcation and Chaos* 12 (10), 22271–22281, 2000.
13. G. Dror and M. Tsodyks, Chaos in neural networks with dynamic synapses. *Neurocomputing* 32–33, 365–370, 2000.
14. M. R. Guevara, L. Glass et al., Chaos in neurobiology. *IEEE Transactions on Systems, Man, and Cybernetics* 13, 790–798, 1983.
15. X.-S. Yang and Q. Li, Horseshoe chaos in cellular neural networks. *International Journal of Bifurcation and Chaos* 16 (1), 2006 to appear.
16. C. Morris and H. Lecar, Voltage oscillations in the barnacle giant muscle fiber. *Biophysical Journal* 35, 193–213, 1981.
17. J. L. Hindmarsh and R. M. Rose, A model of neural bursting using three couple first order differential equations. *Proceedings of the Royal Society London B, Biological Sciences* 221 (1222), 87–102, 1984.
18. T. R. Chay, Chaos in a three-variable model of an excitable cell. *Physica D: Nonlinear Phenomena* 16, 233–242, 1985.
19. E. M. Izhikevich, Simple model of spiking neurons. *IEEE Transactions on Neural Networks*, 14 (6), 2003.
20. R. Fitzhugh, E. Izhikevich, FitzHugh-Nagumo model. *Scholarpedia* 1 (9), 1349, 2006.
21. S. Haykin, *Neural Networks: A Comprehensive Foundation*, 2nd ed. Upper Saddle River, NJ: Prentice Hall, 1999.
22. H. P. Grafand and L. D. Jackal, Analog electronic neural network circuits. *IEEE Circuits and Devices Magazine* 5 (4), 44–49, 1989.
23. S. Y. Foo, L. R. Anderson, and Y. Takefufi, Analog components for the VLSI of neural Networks. *IEEE Circuits and Devices Magazine* 6 (4) 18–26, 1990.
24. C. Mead, *Analog VLSI and Neural Systems*. Reading, MA: Addison-Wesley, 1989.
25. C. Chao-Ming, L. Chih-Min, C. Ching-Tsan, and D. S. Yeung, Hardware implementation of CMAC neural network using FPGA approach. *IEEE International Conference on Machine Learning and Cybernetics*, Hong Kong, China, 4, 2005–2011, 2007.

26. J. Seul and K. Sung, Hardware implementation of a real-time neural network controller with a DSP and an FPGA for nonlinear systems. *IEEE Transactions on Industrial Electronics*, 54, 265–271, 2007.

27. L. E. Atlas and Y. Suziki, Digital systems for artificial neural networks. *IEEE Circuits and Devices Magazine*, 5 (6), 20–24, 1989.

28. A. F. Murray and L. Tarassenko, *Analogue Neural VLSI: A Pulse Stream Approach*. London, UK: Chapman & Hall, 1994.

29. M. S. Tomlinson, D. J. Walker and M. A. Sivilotti, A digital neural network architecture for VLSI. In *IEEE International Joint Conference on Neural Networks*, San Diego, CA, 2, 545–550, 1990.

30. R. A. Leaver, Stochastic arrays and learning networks, PhD thesis, University of Durham, Durham, UK, 1988.

31. R. F. Reiss, H. J. Hamilton et al. Neural theory and modelling. In *Proceedings of the 1962 Ojai Symposium*, p. 427, Stanford University Press, Palo Alto, CA, 1964.

32. J. A. Malmivuo, Bioelectric function of a neuron and its description with electronic models. Helsinki University of Technology, Department of Electrical Engineering, Espoo, Finland, p. 195, 1973.

33. Z. Hrubos, T. Gotthans, and J. Petrzela, Circuit realization of the Inertial Neuron. *IEEE, 21st International Conference Radioelektronika* 2011. doi:10.1109/RADIOELEK.2011.5936435.

34. S. Duan and X. Liao, An electronic implementation for Liao's chaotic delayed neuron model with non-monotonous activation function. *Physics Letters A*, 369, 37–43, 2007.

35. T. Wang, X. He, and T. Huang, Complex dynamical behavior of neural networks in circuit implementation. *Neurocomputing* 2016. doi:10.1016/j.neucom.2016.01.030i.

36. F. A. Khanday et al., Shin-Domain linear transformation filters. *International Journal of Electronics*, 101, 241–254, 2014.

37. F. A. Khanday, E. Pilavaki, and C. Psychalinos, Ultra low-voltage, ultra low-power Sinh domain wavelet filter for ECG analysis. *ASP Journal of Low Power Electronics*, 9, 1–7, 2013.

38. P. Poort et al., A 1-V class AB translinear integrator for filters applications. *Analog Integrated Circuits and Signal Processing*, 21, 79–90, 1999.

39. W. Serdijn, et al., Design of high dynamic range fully integrable translinear filters. *Analog Integrated Circuits and Signal Processing*, 19, 223–239, 1999.

40. A. Katsiamis, K. K. Glaros, and E. Drakakis. Insights and advances on the design of CMOS Sinh companding filters. *IEEE Transactions on Circuits and Systems-I*, 55, 2539–2550, 2008.

41. C. Sawigun and W. Serdijn, Ultra-low power, class-AB, CMOS four quadrant current multiplier. *Electronics Letters*, 45, 483–484, 2009.

42. F. A. Khanday, N. A. Kant and M. R. Dar, Low-voltage realization of neural networks using non-monotonic activation function for digital applications. *Recent Advances in Electrical & Electronic Engineering*, 11 (1), 2018. doi:10.2174/2352096511666180312144420.

43. I. N. Beigh, F. A. Khanday, and C. Psychalinos, 0.5V log-domain realization of tinnitus detection system. *Indian Journal of Pure and Applied Physics*, 55, 595–603, 2017.

44. I. N. Beigh, F. A. Khanday, and C. Psychalinos, Log-domain implementation of QRS detection system using the Pan-Tompkins algorithm with fractional-order differentiator for improved noise rejection. *ASP Journal of Low Power Electronics*, 12 (4), 1–9, 2016.

45. F. A. Khanday and C. Psychalinos, Realization of square-root domain integrators with large time-constant. *Indian Journal of Pure and Applied Physics*, 53, 321–326, 2016.

46. K. Roumelioti, C. Psychalinos, F. A. Khanday, and N. A. Shah, 1.2V Sinh-Domain allpass filter. *International Journal of Circuit Theory and Applications*, 43 (01), 22–35, 2015. doi:10.1002/cta.1922.

47. G. Tsirimokou, C. Psychalinos, F. A. Khanday, and N. A. Shah, 0.5V Sinh-Domain differentiator. *International Journal of Electronics Letters*, 3(01), 34–44, 2015. doi:10.1080/00207217.2014.901425.

48. G. D. Skotis, F. A. Khanday, and C. Psychalinos, Sinh-Domain complex integrators. *International Journal of Electronics*, 102 (7), 1073–1090, 2015. doi:10.1080/00207217.2014.963891.

49. F. A. Khanday, C. Psychalinos, and N. A. Shah, Universal filters of arbitrary order and type employing square-root-domain technique. *International Journal of Electronics*, 101, 894–918, 2014. doi:10.1080/00207217.2013.805357.

50. F. Kafe, F. A. Khanday, and C. Psychalinos, A 50 mHz Sinh-Domain high-pass filter for realizing an ECG signal acquisition system, *Circuits, Systems and Signal Processing*, 33, 3673–3696, 2014. doi:10.1007/s00034-014-9826-1.

51. N. A. Kant, F. A. Khanday, C. Psychalinos, and N. A. Shah, 0.5V Sinh-Domain design of activation functions and neural networks, *ASP Journal of Low Power Electronics*, 10, 1–13, 2014.

52. N. A. Shah and F. A. Khanday, Realisation of low-voltage square-root-domain all-pass filters, *Maejo International Journal of Science and Technology*, 7 (3), 422–432, 2013.

53. F. A. Khanday, C. Psychalinos, and N. A. Shah, Design of low-voltage Sinh-domain n-th order multifunction FLF filter topology for EEG signal recognition. *Journal of Active and Passive Electronic Devices*, 8 (4), 295–308, 2013.

54. M. Panagopoulou, C. Psychalinos, F. Khanday, and N. A. Shah, Sinh-Domain multiphase sinusoidal oscillator. *Microelectronics Journal*, 44, 834–839, 2013. doi:10.1016/j.mejo.2013.06.017.

55. F. A. Khanday, C. Psychalinos, and N. A. Shah, Square-rootdomain realization of single-cell architecture of complex TDCNN, *Circuits, Systems and Signal Processing*, 32, 959–978, 2013. doi:10.1007/s00034-012-9503-1.

56. F. A. Khanday and N. A. Shah, A low-voltage and low-power Sinh-Domain universal biquadratic filter for low-frequency applications. *Turkish Journal of Electrical Engineering and Computer Sciences*, 21, 2205–2217, 2013. doi:10.3906/elk-1203-128.

57. F. A. Khanday, Realization of integrable low-voltage companding filters for portable system applications, PhD thesis, University of Kashmir, Srinagar, India, 2013.

58. W. S. McCulloch and W. A. Pitts, Logical calculus of the idea immanent in nervous activity. *Bulletin of Mathematical Biophysics*, 5, 115–133, 1943.

59. F. Rosenblatt, The perceptron: A probabilistic model for information storage and organization in the brain. *Psychological Review*, 65, 386–408, 1958.

60. M. Minsky and S. Papert, *Perceptron: An Introduction to Computational Geometry*, 1st ed. Cambridge, MA: MIT Press, 1969.

61. K. Funahashi, On the approximate realization of continuous mapping by neural networks. *Neural Networks*, 2, 183–192, 1989.

62. G. E. Hinton, D. E. Rumelhart, and R. J. Williams, Learning representations by back-propagating errors. *Nature*, 323, 533, 1986.

63. S. K. Duan and X. F. Liao, An electronic implementation for Liao's chaotic delayed neuron model with non-monotonous activation function. *Physics Letters A*, 369, 37–43, 2007.

64. K. Basterretxea, J. M. Tarela and I. Del Campo, Approximation of sigmoid function and the derivative for hardware implementation of artificial neurons. *IEE Proceedings—Circuits, Devices and Systems*, 151, 18–24, 2004.

65. H. K. Kwan, Simple sigmoid-like activation function suitable for digital hardware implementation. *Electron Letters*, 28, 1379–1380, 1992.

66. F. Piazza, A. Uncini, and M. Zenobi, Neural networks with digital LUT activation functions. In *Proceedings of International Joint Conference on Neural Networks (IJCNN)*, Japan, 1401–1404, October 1993.

67. G. Khodabandehloo, M. Mirhassani, and M. Ahmadi, Analog implementation of a novel resistive-type sigmoidal neuron. *IEEE Transactions on Very Large Scale Integration (VLSI) Systems*, 20, 750–754, 2012.

68. S. Tabarce, V. G. Tavares, and P. G. de Oliveira, Programmable analogue VLSI implementation for asymmetric sigmoid neural activation function and its derivative. *Electronics Letters*, 41, 863–864, 2005.

69. N. Nedjah, R. M. Da Silva, Analog hardware implementations of artificial neural networks. *Journal of Circuits, Systems and Computers*, 20, 349–373, 2011.

70. A. I. Khuder and S. H. Husain, Hardware realization of artificial neural networks using analogue devices. *Al-Rafidain Engineering*, 21, 77–90, 2013.

71. F. Keles and L. Yildirim, Low voltage low power neuron circuit design based on subthreshold FGMOS transistors and XOR implementation. In *Proceedings of the XIth International Workshop on Symbolic and Numerical Methods, Modeling and Applications to Circuit Design (SM2ACD)*, Tunisia, 1–5, 2010.
72. K. Aihara, T. Takabe, and M. Toyoda, Chaotic neural networks. *Physics Letters A*, 144, 333–340, 1990.
73. P. Arena, S. Baglio et al., Hyperchaos from cellular neural networks. *Electron Letters*, 31, 250–251, 1995.
74. D. W. Wheeler and W. C. Schieve, Stability and chaos in an inertial two-neuron system. *Physica D: Nonlinear Phenomena*, 105, 267–284, 1997.
75. N. A. Kant, M. R. Dar, and F. A. Khanday, An ultra-low-voltage electronic implementation of inertial neuron model with non-monotonous Liao's activation function. *Network: Computation in Neural Systems*, 26, 116–135, 2015.
76. C. Koch, Cable theory in neurons with active, linearized membranes. *Biological Cybernetics*, 50, 15–33, 1984.
77. J. F. Ashmore and D. Attwell, Models for electrical tuning in hair cells. *Proceedings of Royal Society London B, Biological Sciences*, 226, 325–334, 1985.
78. D. E. Angelaki and M. J. Correia, Models of membrane resonance in pigeon semicircular canal type II hair cells. *Biological Cybernetics*, 65, 1–10, 1991.
79. A. Mauro, F. Conti, F. Dodge et al., Subthreshold behavior and phenomenological impedance of the squid giant axon. *The Journal of General Physiology*, 55, 497–523, 1970.
80. K. L. Babcock and R. M. Westervelt, Stability and dynamics of simple electronic neural networks with added inertia. *Physica D: Nonlinear Phenomena*, 23, 464–469, 1986.
81. K. L. Babcock and R. M. Westervelt, Dynamics of simple electronic neural networks. *Physica D: Nonlinear Phenomena*, 28, 305–316, 1987.
82. J. Tani and M. Fujita, Coupling of memory search and mental rotation by a non-equilibrium dynamics neural network. *IEEE Transection on Fundamental Electronics Communication and Computer Sciences*, 75, 578–585, 1992.
83. J. Tani, Model-based learning for mobile robot navigation from the dynamical systems perspective. *IEEE Transactions on Systems, Man, and Cybernetics, Part B: Cybernetics*, 26, 421–436, 1996.

15

Multi-Pattern Matching Based Dynamic Malware Detection in Smart Phones

V. S. Devi, S. Roopak, Tony Thomas, and Md. Meraj Uddin

CONTENTS

15.1 Introduction

Smart phones are being widely used due to the availability of many new services, such as games, Internet, and location-based services, apart from conventional services, such as voice calls and SMS. Among smart phones, Android smart phones are popular due to its open source operating system (OS) and its application programming interface (API). The increasing popularity of Android devices attracted malware authors, and there is a big rise in Android malware apps in every year. Commercial anti-malware companies have released different anti-malwares, and they claim to provide complete protection for smart phones against the malwares. The two main methods used for malware detection are static analysis and dynamic analysis. Commercial virus scanners use static analysis methods to analyze the presence of a malware in an app. However, these methods are vulnerable to various transformation attacks, which are popular nowadays. Another method is the dynamic analysis that uses system calls generated by an application for malware detection. It is commonly used for detecting malicious applications at the repository level and is able to detect transformed malwares. However, they are vulnerable to system call evasion attacks. In this work, we first evaluate the accuracy of two existing dynamic analysis techniques to detect malwares, which perform system call evasion, and show that these methods are not capable of detecting such malware apps at the repository level. Finally, we propose a new pattern matching based dynamic analysis method for improving the current state of malware detection in smart phones. The method uses a multi-pattern matching technique for analyzing system calls. From the implementation results, we found that the proposed method offers a high level of accuracy in detecting malwares that use system call evasion to avoid detection.

15.1.1 Smart Phone as Low Power Device

Mobile devices use batteries to derive the energy required for their operations. Mobile phone batteries are usually limited in size due to the constraints in the size and weight of the device. Despite the higher cost, recent mobile phones use lithium-ion rechargeable batteries as their power source. Both battery capacity and battery volume are strictly linked to each other. For having high capacity batteries, mobile phones should have a large physical size. The physical size of mobile phone batteries varies from model to model; however, the variation is not significantly large. Almost all mobile phones are roughly of the same size. Manufacturers make the mobile phones as light as possible, because smaller mobile phones sell better in the market. Due to these reasons, the battery capacity of most mobile phones (whether a smart phone or conventional mobile phone) are fixed within an average. A mobile phone consumes most of the energy when it is on standby. Normal standby power consumption is 18–20 mW. Battery capacity has been stabilized at around 1000 mAh. Battery thickness is 3–5 mm for products developed after 2006, because of the reduction in device thickness. During charging, the power consumption of a normal mobile phone charger is rated around 3–7 W. This means that even if a phone takes 2 hours for charging, its electricity consumption is only 0.006 to 0.014 units or kWh.

15.1.2 Popularity of Smart Phones

Popularity of smart phone has increased over the last few years due to various reasons. One among the important reasons is that the interface of smart phone is attractive, simple and easy to use. As a result, new users are able to easily learn the use of different functions quickly. Smart phones allow users to stay connected with their personal and business life. Internet access is available to users all the time, and so they can conveniently send emails, access Facebook, access bank accounts, pay bills, and so on. These factors save the time of the users. Nowadays, smart phone systems are affordable to everyone. Smartphone companies always focus on improving the capabilities and functions of the smart phone systems, while at the same time lowering prices. The most important reason for smart phone popularity is the widespread use of third-party applications. Users are allowed to develop and run applications on their devices. These applications provide more functionality and make smart phones so desirable and useful. Due to these functions, services, and popularity of smart phones, users store confidential information and make confidential operations and transactions on their devices. The popularity of smart phone systems and their ability to store confidential information are the two main factors that attracted cyber criminals and malware developers. It has become very complex and difficult to detect sophisticated smart phone malwares nowadays. Since malwares use advanced techniques to attack smart phones, it is crucial to develop effective solutions to protect the smart phones against malware attacks.

15.1.3 Android Smart Phones

Android smart phones are the most commonly available smart phones in the market. It uses Google's Android as its operating system. Android is an open source initiative. It is mainly programmed in C, C++, and JAVA. It is open source and can be used by any company that wishes to manufacture a handheld device with certain restrictions and regulatory measures to control the quality of the product. Google Inc. is involved with a group of manufacturers called the Open Handset Alliance which includes device manufacturers like HTC, Samsung, LG, Motorola, etc. This chapter is focused on a detailed study of malware detection in smart phones with Android operating system. Android operating system has created a benchmark for smart phone operating systems. It offers advanced computing capabilities to a user such as touch screen interface, Internet access, and an operating system capable of running downloaded applications (apps). The numbers of applications are increasing rapidly. The Android OS gives us access to apps that allow us to take information from the web, check location on the map, play music and videos, take photos using device's camera etc. In 2008 the first Android smart phone was released. Cupcake was the first official version of the Android. After that Google released several versions of Android each after certain time period. New versions are associated with bug fixes, new features and functionality, and improved performance.

15.1.4 Android Architecture

Android is being developed under Android Open Source Project (AOSP), maintained by Google and promoted by the Open Handset Alliance (OHA). Android apps are written in Java, and the native code and shared libraries are developed in C or C++. Typical Android architecture is illustrated in Figure 15.1. The bottom layer is the Linux kernel which provides a level of abstraction between the hardware and the upper layers of the Android software stack. The kernel provides preemptive multitasking, low-level core system services such as memory, process and power management and provides device drivers for hardware such as the device display, Wi-Fi, and audio [20]. The Dalvik Virtual Machine (DVM) [11] depends on the Linux kernel for low-level functionality. It is designed to allow multiple instances to run efficiently using the low resource constraints of a mobile device. The core libraries of Android contain DVM specific libraries, java interoperability libraries, Android libraries and C or C++ libraries. The application

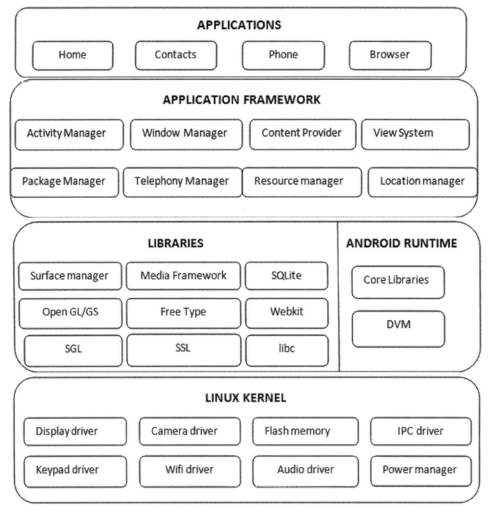

FIGURE 15.1
Android architecture.

framework contains a set of services that creates an environment for Android applications to run. Its key services include activity manager, location manager, telephony manager, etc. Applications are located at the top of the Android architecture. This includes both the native applications and third-party applications, which are installed by the user after purchasing the device.

Android applications which are written in java language are translated to Dalvik byte code that runs under the DVM. Once the operating system boots, a process called zygote process initializes the DVM by loading the core libraries. Finally, the application framework provides a uniform and concise view of the java libraries to the app developer. Android protects the sensitive functionality such as telephony, GPS, network, power-management, radio, and media as system services using its permission-based model.

15.1.4.1 App Structure

Android app is packaged into the .apk zip archive, which consists of several files and folders [11]. The AndroidManifest.xml stores metadata such as package name, permissions required, and definitions of one or more components, such as Activities, Services, Broadcast Receivers or Content Providers, minimum and maximum version support, and libraries to be linked. The res folder stores icons, images, constants, menus, and animations that are compiled into a binary code. The assets folder contains non-compiled resources. Executable file classes.dex stores the Dalvik byte code to be executed on the DVM. Meta-Inf stores the signature of the app developer certificate to verify the identity of a third-party developer. Since Android apps are developed in Java, the compiled Java code generates a number of class files and intermediate byte code of the classes. To speed up the execution by the DVM, the class files are merged into a single Dalvik executable called the .dex file.

15.1.4.2 App Components

An Android app is composed of one or more components [13] such as Activity, Service, Broadcast Receiver, and Content Provider. Activity is the user interface component of an app. Any number of activities can be declared within the manifest depending on the developer requirements. The Service component performs background tasks without any user interface, such as playing audio or downloading data from the network. The Broadcast Receiver component listens to the Android system generated events. Other apps can broadcast their own events, which can be handled by other apps using the Service component. The Content provider component, also known as the data store, provides a consistent interface for data access between different apps.

15.1.4.3 Android Permissions

In order to restrict an app from accessing the sensitive functionalities of a device, such as telephony, network, contacts, SD card, and location, Android provides a permission-based security model in its application framework [13]. Android permissions are classified into four protection-levels. Normal permissions are granted by default during the installation of the app. Some permissions are considered high-risk permissions since it has the capability to access the private data and important device sensors, such as the camera. Signature permissions are the permissions that are granted only if the requesting app is signed with the same developer certificate. They are granted automatically at the time of installation.

Signature or system permissions are granted only if the requesting app is signed with the same certificate as the Android system image. They are also granted automatically during the installation time.

15.1.4.4 *Android Vulnerabilities*

In the case of an Android smart phone, a vulnerable application (app) in it can perform various security attacks. The common security vulnerabilities include privacy escalation, capability leaks, permission re-delegation, content leaks and pollution, and component hijacking. Android applications run over permission-based framework, and so each application's access is controlled according to the permission that it gets during the installation phase. Much of the recent research shows that the Android permission framework is weak by its design. By launching a privilege escalation attack [4], a malicious application can obtain privileged permissions. An application can gain permission to perform a privileged task, which is not authorized to it. In capability leak, an app gains access to permissions without requesting it. This kind of attack includes exploiting publicly-accessible interfaces or services and acquiring permission from another application. Permission re-delegation vulnerability occurs when an application with certain permissions performs a privileged task for an application that does not have permissions. Content leaks include vulnerabilities that are rooted in an unprotected Android component (i.e., content provider) inside vulnerable apps. Such apps can disclose various types of private in-app data or can manipulate certain security-sensitive in-app settings or configurations that may further cause system-wide side effects (e.g., blocking all incoming phone calls or SMS messages). An app that has component hijacking vulnerability is able to access sensitive information or tamper with sensitive data in a critical data repository on behalf of the malicious app.

15.1.5 Malware in Android Smart Phones

Smart phones contain lots of personal data, such as financial account details, images, documents, and much more. Hence, nowadays, smart phones have become an important target for malware developers. A malware/virus is a piece of software that attaches to an executable file, and, when such a file starts running, the viral code gets executed. Malware is able to perform various functionalities such as overwriting programs on the system, destroying or deleting data, or erasing everything in the memory. They spread when the file is copied or moved from one phone to another using the network, file sharing, or e-mail attachments. There are different types of viruses. Among them, worms and trojans are more common. A worm is a stand-alone malicious program. It is a self-replicating program that uses a communication channel to send its copy to other systems. A worm can delete files, encrypt files, send junk e-mail, and consume network bandwidth. The first smart phone malware was a worm called Cabir, which propagated through Bluetooth. A trojan horse is a program that presents itself as a legitimate program, while it actually contains malicious code segments. A trojan always stays hidden on the infected system and performs many types of attacks, such as generating pop-up windows, deleting files, stealing data, or activating and spreading other malware, like viruses or bots. Trojan can open back doors for providing access to the system for malicious activities. It spreads through e-mail message attachments, copying files from storage media, such as USB keys, or during file downloads from the Internet.

In recent years, there has been a huge growth of malicious programs that use super-user rights by employing advertising apps embedded with trojans. For example, certain trojans are capable of hiding within the on-screen display of a legitimate banking app with a fake copy that is being created for phishing purposes. When a user sees a familiar interface, he enters the bank account details and gets hacked simultaneously. OpFake [2] is one of the most notable examples of such an app. It can imitate the interface of more than 100 legitimate banking and financial apps. Sometimes trojans can come along with the official apps from a bank. The SmsThief [7] malware, detected in 2015, was embedded in a legitimate banking app. This trojan steals victim's messages and sends them to the attackers together with other information like the device model and personal data. SMS trojans which subscribes people to unnecessary services are also commonly found. These programs send paid text messages from an infected device or subscribe the victims to paid services. Nowadays, distribution of malwares usually happens through Google Play and advertising services.

15.2 String Matching Mechanisms

The main aim of a string matching algorithm is to find all occurrences of a string pattern **S** in text **T**. String matching algorithms are used in software applications like virus scanners (anti-virus) or intrusion detection systems for improving data security over the Internet. Other fields such as music technology, computational linguistics, artificial intelligence, and artificial vision have been using string matching algorithms for problem solving. There are two types of string matching mechanisms—single pattern matching and multi-pattern matching. In a single pattern matching, the algorithm finds all occurrences of a pattern in a given input text. Multi-pattern matching algorithms can search multiple patterns in a text at the same time. If more than one pattern needs to be matched against the given text simultaneously, we can use multiple pattern matching. The pattern matching algorithms are widely used in network security environments. In network security, a pattern can be a string indicating a network intrusion attack, virus signature, packet flow information, and so forth.

15.2.1 Multi-Pattern Matching

The multi-pattern matching technique finds all occurrences of the patterns from a finite set **X** in a given text **T** of length **n**. They are used commonly in searching large pattern sets. Because of its high performance, they are more useful than the single pattern matching algorithms. Multi-pattern matching technique has many applications, such as data filtering (data mining) to find selected patterns in anti-virus scanning, intrusion detection, content scanning, and filtering. It is used in security applications to detect certain suspicious keywords. Aho–Corasick [1] and Commentz–Walter [3] are the most significant algorithms for multiple pattern matching.

15.3 Malware Detection Techniques in Smart Phones

Research related to malware detection in contexts other than Android is being carried out day by day. Our work is also inspired by some of these approaches, but we primarily focus on the related work covering Android malware detection. Ever since Android became popular, numerous research works have been carried out for detecting malicious Android applications. An overview of the current malware is provided in the studies of Felt et al. [12] and Zhou et al. [19]. Several concepts and techniques have been proposed to detect malwares.

15.3.1 Common Detection Mechanisms

The two popular methods of malware detection in smart phones are static analysis and dynamic analysis. Typically, the virus scanners use the static analysis method, which relies on a database of descriptions, or signatures, that characterize known malware instances. Whenever an unknown malware sample is found, usually it is necessary to update the signature database accordingly so that the new malware piece can be detected by the antivirus software. Another method of malware detection is called dynamic analysis. It uses system calls for detecting malicious applications on a repository level. By using dynamic analysis, it is possible to detect most of the malwares that have undergone transformations.

15.3.1.1 Static Analysis

In the static analysis technique, the analysis is performed based on structural characteristics of a malware. Suspicious binary code is analyzed without executing it. The first step is to disassemble the code. The disassembled code is parsed to collect information, which is pushed into the stack, stored in processor registers or memory, and so on. In basic static analysis, a malware is examined without viewing its actual code or instructions. Such analysis employs different tools and techniques to easily determine whether a sample is malicious or not. It can provide information about the functionality of the malwares. Malware normalization, prior to static analysis, enhances the detection of obfuscated malwares. Several methods have been proposed that statically inspect applications and disassemble their code. Static techniques are typically based on source code or binary analysis that search for malicious patterns. For example, static approaches include analyzing permission requests for application installation control flow, signature-based detection, and static taint-analysis as shown in Figure 15.2.

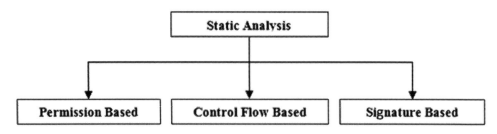

FIGURE 15.2
Classification of static analysis.

The main technique used during static analysis is signature-based detection. A malware signature is a byte sequence that uniquely identifies a specific malware. Signature-based detection is the most common method that anti-virus software uses to identify a malware. When anti-virus software scans a sample for viruses, it checks the contents of the sample against a dictionary of virus signatures. If it finds a virus signature in the sample, the anti-virus software takes some action to remove the virus. Such a signature-based detection approach requires frequent updates in its virus signature dictionary since new viruses are being created each day. SAVE (Static Analyzer for Vicious Executable) [15] is a strong signature-based malware detection tool capable of detecting obfuscated malwares. Signature-based anti-virus software typically examines samples when the computer's operating system creates, opens, closes, or e-mails them and immediately detects a known virus upon its receipt. Signature-based detection examines a malware sample to find its key aspects and creates a static fingerprint of known malware.

15.3.1.2 Dynamic Analysis

This technique detects a malware at runtime. Dynamic analysis is mainly applied for online detection of malware, such as scanning and analyzing large collections of Android applications. A detection system, called Bouncer, is currently operated by Google. Such dynamic analysis systems are suitable for filtering malicious applications from Android markets. Due to the openness of the Android platform, applications can also be installed from other sources, such as web pages, which require detection mechanisms operating on the smart phone. ParanoidAndroid [7] is one of the few detection systems that employ dynamic analysis and can spot malicious activity on the smart phone. Dynamic analysis techniques run applications in a sandbox environment or on real devices for gathering information about the application's behavior. Dynamic taint analysis and behavior-based detection are examples of dynamic approaches, as shown in Figure 15.3.

Android applications are analyzed using dynamic analysis by capturing their behavior based on the execution pattern of the system calls. The difficulty of manually creating and updating the system call patterns of Android malware has motivated the application of machine learning. Several methods have been proposed that analyze applications automatically using machine learning methods.

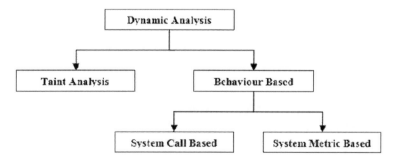

FIGURE 15.3
Classification of dynamic analysis.

15.3.1.3 Limitations

All the existing approaches used in static analysis lack the ability to analyze code that is obfuscated or loaded dynamically at runtime. Static analysis uses signature-based detection. A major limitation of signature-based detection is its inability to flag malicious samples whose signatures have not yet been developed. With this in mind, modern attackers frequently mutate their creations to retain malicious functionality by changing the sample's signature. Although the signature-based approach is effective, a malware developer can create polymorphic or metamorphic viruses to overcome this.

Dynamic analysis overcomes the abovementioned drawbacks of static analysis. Although dynamic analysis is better when compared to the static analysis in many aspects, dynamic analysis also has some drawbacks. Dynamic analysis mainly uses system call traces to detect malwares. An application produces system calls based on an instruction in its source code. It is difficult to fetch all system calls corresponding to a source code due to the low code coverage in dynamic analysis. Dynamic analysis uses machine learning classifiers to detect malwares, and hence they require more resources and are less efficient. Another drawback of existing dynamic analysis system is that recent malwares attempt to detect the emulator and the detection system, and hence they are prone to analysis evasion. Zhang et al. [18] points out that traditional system call analysis is not appropriate for characterizing the behaviors of Android apps, as it misses high-level android specific semantics.

15.4 Energy Efficient Algorithms

The main problem with a smart phone is that it runs on limited battery for a long time, and its smart features consume more battery to work [14]. The increasing popularity of smart phones is accompanied by the cost of higher energy consumption, which makes power supply a crucial issue in developing advanced functionalities. Nowadays, the limited battery lifetime of smart phones is incapable of meeting the demands of sustaining the full functionality on smart phones without frequent recharge. Most of the energy consumed by a system is converted into heat, which results in reduced reliability of hardware components. Hence, energy has become a design constraint for computing devices due to these reasons. Therefore, to increase the reliability of smart phones by extending power supply, the mobile industry is focusing on power management. During the past years, there has been much research interest in developing algorithmic techniques for saving energy. For a given problem, the goal is to design energy-efficient algorithms that reduce energy consumption, while providing reliable service. An important factor to focus on during such a design is that these algorithms must always achieve good performance.

15.4.1 Landauer's Principle

The energy complexity of algorithms (in addition to time and space complexity) can be studied systematically using Landauer's principle [10] in physics. Landauer's principle gives a lower bound on the amount of energy a system must dissipate if it destroys information. Most of the computing systems we use today follow the principle of irreversibility. They found that an irreversible computation places a lower limit (Landauer's limit) on energy consumption. That is, for each irreversible operation we do, we have to dissipate a minimum amount of energy. CPU power efficiency (number of computations per kilowatt hour of energy) is doubling every two years. It will hit Launder's limit in next 15–60 years [5].

15.4.2 Irreversible vs Reversible Computations

Irreversibility can be understood better by using an example shown in Figure 15.4a. Consider a machine that can add two numbers. Suppose we give it two inputs say, 2 and 2, it will produce an output 4. But, if we try to invert this machine; that is, to run it backwards by giving the result 4 as input and ask to find the actual inputs, the machine won't be able to reproduce 2 and 2. This is because there exists an uncertainty regarding the possible inputs. The possible inputs may be 1 and 3, 2 and 2, 0 and 4, and so on. Here, addition is an irreversible computation. However, we can make this machine a reversible one by adding an additional functionality to it; that is, change the mode of operation by including a sub-traction operation along with the addition operation as shown in Figure 15.4b. Suppose, if we give 5 and 3 as input it will give two outputs: 8(sum) and 2(difference). Now, it's not difficult for the machine to perform a reversible operation to find the exact inputs.

The only way to cross the Landauer limit is to do reversible computations. In revers-ible computations, the inputs can be reconstructed from their outputs, using adiabatic cir-cuits [5]. A comparison of irreversible and reversible computation with Landauer limit is shown in Figure 15.5 [15].

Over the last 40 years, the amount of energy per bit computation has been exponen-tially decreased. Even though the limit placed by Landauer's principle is very small, we are just a few orders of magnitude away from it. Since we use irreversible computa-tions, the projection shown in Figure 15.5 has to bend off without hitting the Landauer

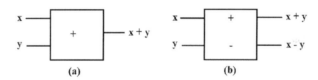

FIGURE 15.4
Example for irreversible and reversible computation.

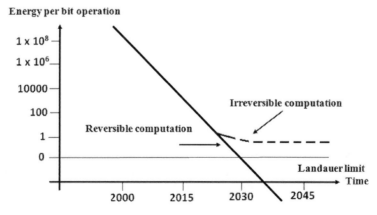

FIGURE 15.5
Comparison with Landauer limit.

limit. But reversible computations can lower the energy consumption per operation without a limit, making it possible to go beyond Landauer limit.

15.5 Multi-Pattern Based Malware Detection Mechanism

In Section 15.3, we reviewed two forms of smart phone malware detection techniques— static and dynamic analysis. In the discussion, we analyzed the current trends in smart phone malware detection techniques and identified their advantages and limitations. The main drawback of them is that the efficiency and accuracy of existing Android malware detection techniques are very low in detecting modern transformed malwares. Dynamic analysis usually makes use of system calls for malware detection. It generally uses the techniques of machine learning. Many of such detection systems, based on system call traces, still suffer from some important limitations affecting their performance and adaptability. Even though system call monitoring is out of reach of many malicious applications, some of them are designed for evading system calls so that they can escape from the monitoring process. Our approach is capable of detecting malwares that use system call evasion techniques (Section 15.5.1.2) for hiding their presence from dynamic analysis. Our work mainly focuses on improving the efficiency and accuracy of dynamic Android malware detection by proposing a new methodology based on pattern matching. It can be used for detecting malwares at the repository level itself. By guarding the users at the repository level, a malicious application is detected before it is made publicly available for installation. This approach protects user's data, privacy, and more.

15.5.1 Preliminaries

15.5.1.1 System Calls

System calls can also be considered as direct entry points into the kernel. Using system calls, programs request services from the kernel. Linux system calls are the interface between user applications and the kernel services. They allow programs to perform tasks that would not normally be permitted. It allows user-level processes to request some services from the operating system. The operating system will enter in the kernel mode and perform desired service on behalf of the user-level process. System calls provide various functions to user applications. It includes file operations, such as open, read, and write; network-based operations, such as connect, send, and receive; or process operations, which include creating a new process or killing a process. Information regarding an application's behavior can be found from system call traces. Therefore, it is widely used in the field of anomaly detection. The list of common Linux system calls is given in Table 15.1.

15.5.1.2 System Call Evasion

Many of the dynamic analysis-based detection systems, which uses system call traces, suffer from some important limitations affecting their performance and adaptability. One such thing is the system call evasion. Using system call evasion, a malware can change its system call sequence and become undetectable to dynamic analysis. System call sequence can be evaded easily by inserting no-ops or nullified system calls. The method of inserting such system calls is illustrated in Figure 15.6.

TABLE 15.1

System Calls in Linux

Purpose	Name
Process Control	fork(), exit(), wait()
File Manipulation	open(), read(), write(), close()
Device Manipulation	ioctl(), read(), write()
Information Maintenance	getpid(), alarm(), sleep()
Communication	pie(), shmget(), mmap()
Protection	chmod(), umask(), chown()

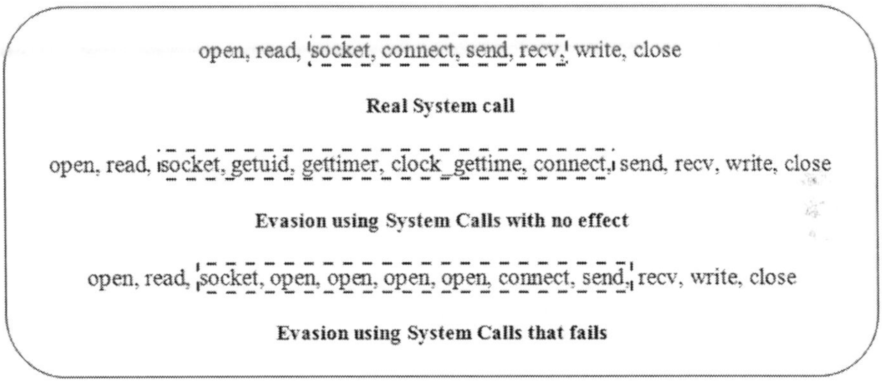

open, read, ¦socket, connect, send, recv,¦ write, close

Real System call

open, read, ¦socket, getuid, gettimer, clock_gettime, connect,¦ send, recv, write, close

Evasion using System Calls with no effect

open, read, ¦socket, open, open, open, open, connect, send,¦ recv, write, close

Evasion using System Calls that fails

FIGURE 15.6
System call evasion.

The first sequence is a harmful real system call generated by a malicious app. This sequence can be easily altered by inserting system calls having no effect, like system calls returning information about process or file status or returning time or clock status. Similarly, system calls that can fail can also be used (e.g., opening non-existing files).

15.5.1.3 System Call Filtering

The system call traces generated by an application is referred to as low-level events. These traces represent the application-kernel interactions, such as information maintenance, memory management, inter-processes communications, and hardware interrupts. Filtering is the process of refining system call traces according to a set of well-defined rules and producing clearer system call traces. It can also be defined as the process of removing nonsignificant system calls in order to characterize the main behavior of the application. Once the size of the traces is reduced, the anomaly detection techniques would consume lesser computational resources and will improve their detection accuracy.

15.5.1.3.1 Filtering Phase

From the system call categories mentioned in Table 15.1, we can infer which system calls are irrelevant to describe the main behavior of Android applications. We are categorizing those system calls into four classes: inter-process communication, memory management, information maintenance system calls, and unsuccessful system calls.

Each Android application runs in an isolated virtual machine. Thus, an application cannot manipulate the data of another application directly. However, in many cases, there is a need to exchange data between applications. Android contains a kernel module called Binder for inter-process communications. All Binder transactions happen through the ioctl system call. Thus, all the standard Linux inter-process communication (IPC) mechanisms can be ignored.

The brk memory management system call grabs a large chunk of memory and then splits it to smaller sizes according to the need. The brk system call is able to change the size of the heap. The mmap system call is used when very large memory space is allocated. Two applications with the same behavior may have very different memory management system call sequences. Therefore, we can ignore these calls because they do not characterize the application's main behavior.

Many system calls do not add any valuable information to describe the application behaviors. They exist simply for the purpose of transferring information between an application and the operating system. For example, system calls returning the current time and date (e.g., time, gettimeofday), system calls returning information about the system, such as available or free system resources (e.g., getcpu, getrlimit), system calls returning file status (e.g., state, fstatfs, getxattr, listxattr), system calls returning process information, such as the user id and group id of the process (e.g., getuid, getgid), and system calls synchronizing the process (e.g., wait, epoll_wait). Therefore, these system calls can be ignored.

Unsuccessful system calls do not have any impact on the application behaviors. For example, if an application fails to open a file in the first two attempts and succeeds in the third one, the open system call is used three times. In such a case, the first two open calls should be ignored.

15.5.1.3.2 Abstraction Phase

There are many system calls having the similar and overlapping functionalities but different names. For example, the readv (int fd, const struct iovec, int iovcnt) system call has almost the same function as read (int fd, void buf, size_t count). The only difference lies in the fact that the former call fills multiple buffers instead of only one. Hence, the abstraction process considers system calls of the similar and overlapping functionalities as equivalent. This process not only simplifies the traces and reduces the required resources to analyze them, but also protects against attacks where an attacker replaces system call with equivalent ones.

15.5.1.4 File Descriptors

In UNIX and related computer operating systems, a file descriptor is an abstract indicator used to access a file or other resource, such as a pipe or network socket. It is illustrated in Figure 15.7.

A file descriptor is a non-negative integer. Negative values are reserved to indicate an error condition. A process always keeps track of its file descriptors. When a new process is started, three file descriptors are created by default. They are called the standard file descriptors. They are further given the numbers 0, 1, 2, respectively. Everything in a Unix system (including the hardware devices, such as monitor and keyboard) are considered as files. 0 is the index corresponding to the keyboard file, and 1 and 2 are indices corresponding to the monitor file. Every running process has a file descriptor table that contains pointers to all open input and output streams. In the first three cells of the table, three entries are created when a process starts. Entry 0 points to standard input, entry 1 points to standard output, and entry 2 points to standard error. Whenever a file or other stream is opened, in the first available empty slot of the table a new entry is created.

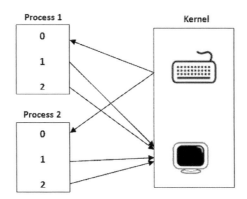

FIGURE 15.7
File descriptor.

15.5.1.5 Polling Process

Polling is a mechanism of continuous monitoring by a controlling device or process to check the current state of other devices, processes, queues, and so on in some defined sequence. Pooling is used to know whether the devices need attention; such as whether they are still connected, want to communicate, and contains tasks to be executed. An alternative to polling is the use of interrupts, which are signals generated by devices or processes to indicate that they need attention and want to communicate. It monitors multiple file descriptors to see if input or output is possible on any of them.

15.5.1.6 Performance Metrics

In this work, we statistically measured the performance of existing malware detection methods and the newly proposed method by testing it with malicious and benign data samples. The statistical measures include true positive rate, false positive rate, true negative rate, false negative rate, and accuracy. Let G be the number of goodware samples, M be the number of malware samples, and N be the total number of samples collected. Therefore, $N = G + M$.

True Positive (**TP**) refers to the number of tested samples correctly detected as malware. The **TP Rate** is calculated as:

$$TP\ Rate = TP/M.$$

False Positive (**FP**) refers to the number of tested samples falsely detected as malware. The FP Rate is calculated as:

$$FP\ Rate = FP/G$$

True Negative (**TN**) refers to the number of tested samples correctly detected as goodware. The **TN Rate** is calculated as:

$$TN\ Rate = TN/G.$$

False Negative (**FN**) refers to the number of tested samples falsely detected as goodware. The **FN Rate** is calculated as;

$$FN\ Rate = FN/M.$$

Finally, the Accuracy (**A**) of the detection method is calculated as;

$$A = (TP+TN)/N.$$

15.5.2 Proposed Mechanism

The proposed methodology involves two steps: creating a string database and comparing a new string pattern with the string database using any of the multi-pattern matching algorithms. We traced system calls of 50 malware samples belonging to different malware families, taken from the Drebin dataset [20] using the *strace* command. The extracted system calls contained both evaded as well as original system calls. The extracted system calls were filtered (described in Section 15.1.3) in order to avoid memory management system calls, error system calls, and information maintenance system calls from our system call sequence. From those system call sequences, we extracted patterns to create a string database. The patterns that we used for matching includes all the basic system calls that occur in between two busy-wait system calls in Linux (epoll-wait). The epoll-wait system call denotes the polling process (Section 15.5.1.5). The resulting sequence is converted into a string pattern and is stored in a text file to create the string database. This is illustrated in Figure 15.8.

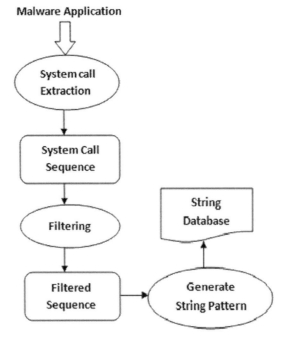

FIGURE 15.8
Creating string database.

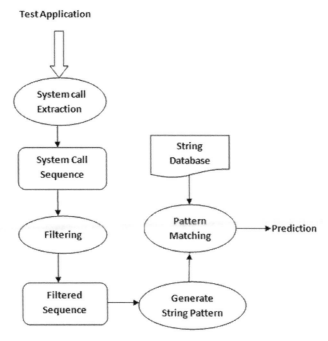

FIGURE 15.9
Prediction using multi-pattern matching algorithm.

When a test application arrives, this procedure is repeated to create its corresponding system call string pattern. Further, we apply the chosen multi-pattern string matching algorithm to find out whether any of the string sequence in the database matches with that of the system call pattern of the test application. When a match is found, we can conclude that the test application is a malware. This is illustrated in Figure 15.9. The performance of this approach relies on the performance of the chosen multi-pattern matching algorithm. Consider an input text **T** (system call trace of test application) of length **n** and a set **S** of **m** keywords (patterns in string database). The problem is to find all occurrences of these keywords in the input text **T**. The solution to this problem is to scan the input text **T** for each keyword, which requires a total scanning time of **nM**, where **M** is the sum of keyword lengths. This solution is efficient with large input texts and large keyword set.

This approach rectifies the resource constraint problem of machine learning as it requires only a few lines of string matching source code to be executed. The problem with low code coverage is also handled in this approach as it compares most of the system call patterns in a test application.

15.6 Experimental Results and Analysis

During experimental analysis, two existing works on machine learning-based dynamic malware detection were evaluated, and their efficiencies were compared with our proposed methodology. Two different classifiers were implemented and tested using various data that represent the behavior of the applications.

TABLE 15.2

Experimental Results

Parameter	ANN [6]	SVM [17]	Proposed Method
True Positive	80%	80%	96%
False Positive	16%	20%	8%
True Negative	84%	80%	92%
False Negative	20%	20%	4%
Accuracy	82%	80%	94%

We have acquired an initial dataset of 100 samples comprising of benign as well as malicious from the Drebin dataset [20]. To determine malicious and benign applications, we inspected the output of five common anti-virus scanners (Avast, BitDefender, Zonar, McAfee, and CM Security). We flag all applications as malicious that are detected by at least two of the scanners. This procedure ensures that our data has been equally divided into benign and malicious samples. In this work, we statistically measured the performance of existing malware detection methods and the proposed method by testing it with malicious and benign data samples.

First we implemented and evaluated an Android malware detection based on system calls by Dimjasevic et al. [6]. This paper suggests some techniques for detecting malicious Android applications on a repository level. The techniques perform automatic classification based on tracking system calls. We implemented those techniques and performed evaluation on a set of around 100 samples, which includes both benign and malicious applications. We used Artificial Neural Network (ANN) as the classifier.

Next, we implemented and evaluated a behaviour-based malware detection by Wei et al. [17]. In this, a malware detection system was proposed that uses a behavior based detection approach to detect a malware. Here, system calls are captured to examine the runtime behavior of an application and used the machine learning approach called Support Vector Machine (SVM) to learn the dynamic behavior of its execution. The results of the experiments are shown in Table 15.2.

We implemented our methodology and performed an evaluation on a suite of around 100 applications. The accuracy rate of ANN-based detection method is 82%, and that of SVM-based analysis is 80%, whereas the proposed pattern matching method gave an accuracy of 94%. From the evaluation results, we found that our proposed pattern matching method for dynamic analysis has high performance and accuracy, when compared to the commonly used machine learning techniques.

15.7 Energy Reduction Techniques

The only method to handle power efficiency is to convert the proposed methodology and algorithm into a reversible program. In order to handle energy efficiency, the multi-pattern matching algorithm, which we select, should first be converted into a reversible algorithm (described in Section 15.4). In this section, we discuss some general techniques that can be used in constructing reversible algorithms. The main techniques among them are discussed in the sub-sections below.

15.7.1 Reversible Primitives

The basic primitives of programming such as jumps, branches, conditionals, for loops, and function calls should be made reversible for making an algorithm reversible. According to Demaine et al. [5], jumps (goto statements), conditional statements (if statement), for loops, and function calls can be implemented reversibly with constant-factor increases in time and space.

15.7.2 Memory Management and Garbage Collection

Allocating and reclaiming free memory are critical tasks to the function of any modern computer. Some basic forms of memory allocation and garbage collection can be done reversibly with only constant factor overhead in time and space. Allocation is performed by removing the pointer from the free list and handing it to the program requesting the memory. To deallocate a block of memory, the pointer to that block is put on the end of the free list. We assume that all the deallocated memory blocks have been returned to their zeroed-out state.

15.7.3 Reversible Subroutine

Function calls can be implemented reversibly. Further, some reversible subroutines can be efficiently used. If we have a fully reversible subroutine, whose only effect on the program is through its return value, one need to only store the inputs and outputs to this subroutine later to unroll it with only a constant-factor overhead in time.

15.7.4 Data Structure Rebuilding

When attempting to implement data structures, which support insert and delete operations reversibly, we run into a new challenge. Often the insertion or deletion operation will create some amount of garbage data, which is necessary to reverse it in the future. We also need the result of the operation to remain in place. Hence, we cannot immediately reverse the operation. According to Demaine et al. [5], doubly-linked lists can be implemented reversibly with constant-factor overheads in time and space.

15.7.5 Implementation

For expressing reversibility, we require a new programming language. Tyagi et al. [16] introduced a new programming language for expressing reversibility. It is called Energy-Efficient Language (Eel) and it can be used for algorithm design and implementation. Eel is more suitable for partially reversible computation model, where programs can have both reversible and irreversible operations. In partially reversible computation model, irreversible operations cost energy for every bit of information created or destroyed. Eel introduces many powerful control logic operators including conditionals and loops. We suggest implementing our pattern matching based mechanism using a reversible language like Eel and execute the program in a reversible machine in order to properly handle the energy efficiency.

15.8 Conclusions

In this chapter, we reviewed the existing techniques in dynamic malware detection. For that, we implemented two main approaches in machine learning and tested it with the collected malware samples. While using ANN, the accuracy was found to be 82%. With the SVM classifier, the accuracy was only 80%. Finally, we tested the samples with our proposed mechanism and found that the accuracy rate is comparatively higher (94%) than these existing machine learning approaches. From this, we can conclude that there are numerous ways for malwares to get hidden from existing defense mechanisms, and we need better mechanisms to protect our smart phones from them. The proposed mechanism demands a reversible multi-pattern matching algorithm to handle energy efficiency. More studies have to be carried out to make use of this methodology at the user level (as anti-virus), rather than in the repository level. In the future, a mechanism for detecting all possible system call evasions, including system call interchange and system call substitution, needs to be found. Moreover, there can be more unknown vulnerabilities/limitations in the current detection mechanisms, which we have not considered in this work. They have to be further studied for improving the current status of Android security.

References

1. Aho A. V., and M. J. Corasick, "Efficient string matching: An aid to bibliographic search," *Proceedings of ACM Communications*, pp. 333–340, 1975.
2. Canfora, G., A. De Lorenzo, E. Medvet, F. Mercaldo, and C. A. Visaggio, "Effectiveness of opcode ngrams for detection of multifamily android malware," *International Conference on Availability, Reliability and Security (ARES)*, pp. 333–340. IEEE, 2015.
3. Commentz-Walter, B., "A string matching algorithm fast on the average," *Automata, Languages and Programming*, pp. 118–132, 1979.
4. Davi L., A. Dmitrienko, A.-R. Sadeghi, and M. Winandy. "Privilege escalation attacks on android," *Proceedings of the International Conference on Information Security*, pp. 346–360. Springer, Berlin, Germany, 2010.
5. Demaine E., J. Lynch, G. J. Mirano, and N. Tyagi, "Energy-efficient algorithms," *Proceedings of the ACM Conference on Innovations in Theoretical Computer Science*, pp. 321–332, 2016.
6. Dimjasevic M., S. Atzeni, I. Ugrina, and Z. Rakamaric, "Evaluation of android malware detection based on system calls," *Proceedings of ACM International Workshop on Security and Privacy Analytics*, pp. 1–8, 2016.
7. Emm D., M. Garnaeva, A. Ivanov, D. Makrushin, and R. Unuchek, "IT threat evolution in Q2 2015," Moscow, 125212, Russian Federation: Kaspersky Lab HQ, 2015.
8. Felt A. P., M. Finifter, E. Chin, S. Hanna, and D. Wagner," A survey of mobile malware in the wild," *Proceedings of ACM Workshop on Security and Privacy in Smartphones and Mobile Devices*, pp. 3–14, 2011.
9. Jin R., Q. Wei, P. Yang, and Q. Wang, "Normalization towards instruction substitution metamorphism based on standard instruction set," *Conference on Computational Intelligence and Security Workshops*, pp. 795–798, IEEE, 2007.
10. Landauer R. W., "Irreversibility and heat generation in the computing process," *IBM Journal of Research and Development*, pp. 261–269, 1961.
11. Lee J. K., and J. Y. Lee, "Android programming techniques for improving performance," *IEEE Conference on Awareness Science and Technology*, pp. 386–389, 2011.

12. Portokalidis G., P. Homburg, K. Anagnostakis, and H. Bos, "Paranoid Android: Versatile protection for smartphones," In *Proceedings of the 26th Annual Computer Security Applications Conference*, pp. 347–356, ACM, 2010.

13. Rogers R., J. Lombardo, Z. Mednieks, and B. Meike, "Android application development: Programming with the Google SDK," O'Reilly Media, Newton, MA, 2009.

14. Stephen O., and L. Onwuka, "Mobile terminals energy: A survey of battery technologies and energy management techniques," *International Journal of Engineering and Technology*, pp. 282–286, 2013.

15. Sung A. H., J. Xu, P. Chavez, and S. Mukkamala, "Static analyzer of vicious executables (save)," *IEEE conference on Computer Security Applications*, pp. 326–334, 2004.

16. Tyagi N., J. Lynch, and E. Demaine, "Toward an energy efficient language and compiler for (partially) reversible algorithms," *Proceedings of the International Conference on Reversible Computing*, pp. 121–136, 2016.

17. Wei Y., H. Zhang, L. Ge, and R. Hardy, "On behavior-based detection of malware on android platform," *IEEE Global Communications Conference*, pp. 814–819, 2013.

18. Zhang Y., M. Yang, B. Xu, Z. Yang, G. Gu, P. Ning, S. Wang, and B. Zang, "Vetting undesirable behaviors in android apps with permission use analysis," *Proceedings of the ACM SIGSAC Conference on Computer and Communications Security*, pp. 611–622, 2013.

19. Zhou W., Y. Zhou, X. Jiang, and P. Ning, "Detecting repackaged smart phone applications in third-party android marketplaces," *ACM conference on Data and Application Security and Privacy*, pp. 317–326, 2012.

20. Drebin: The Drebin Dataset. [Online]. Available: https://www.sec.cs.tu-bs.de/~danarp/drebin/. Accessed: June 25, 2017.

Index

Note: Page numbers in *italic* and **bold** refer to figures and tables respectively.